LEARNING GUIDE

Kathleen Prezbindowski
College of Mount St. Joseph

PRINCIPLES OF
ANATOMY AND PHYSIOLOGY

NINTH EDITION

Gerard J. Tortora
Bergen Community College

Sandra Reynolds Grabowski
Purdue University

John Wiley & Sons, Inc.
New York • Chichester • Weinheim • Brisbane • Singapore • Toronto

To Amy and Joost, Laurie and Christian, and Maureen and Alexis

To order books or for customer service call 1-800-CALL-WILEY (225-5945).

ISBN 0-471-37467-9

Printed in the United States of America

10 9 8 7 6 5 4 3 2

Printed and bound by Courier Kendallville, Inc.

C O N T E N T S

TO THE STUDENTS

This learning guide is designed to help you do exactly what its title indicates – learn. It will serve as a step-by-step aid to help you bridge the gap between goals (objectives) and accomplishments (learning). The 29 chapters of the *Learning Guide* parallel the 29 chapters of Tortora and Grabowski's *Principles of Anatomy and Physiology*, Ninth Edition.

Take a moment to turn to Chapter 1 and look at the major headings in this learning guide. Now consider each one.

Overview. As you begin your study of each chapter, read this brief introduction, which presents the topics of the chapter. The overview is a look at the "forest" (or the "big picture") before examination of the individual "trees" (or details).

Framework. The Framework arranges key concepts and terms in an organizational chart that demonstrates interrelationships and lists key terms. Refer back to the Framework frequently to keep sight of the interrelationships.

Topic Outline and Objectives. Objectives that are identical to those in the text organize the chapter into "bite-size" pieces that will help you identify the principal subtopics that you must understand. Preview objectives as you start the chapter, and refer back to them as you complete each one. Objectives are arranged according to major chapter divisions in the Topic Outline.

Wordbytes present word roots, prefixes, and suffixes that will help you master terminology of the chapter and facilitate understanding and remembering of key terms. Study these; then check your knowledge. (Cover meanings and write them in the lines provided.) Try to think of additional terms in which these important bits of words are used.

Checkpoints offer a variety of learning activities arranged according to the sequence of topics in the chapter. At the start of each new Checkpoint section, note the related pages in the Tortora and Grabowski text. Read these pages carefully and then complete this group of Checkpoints. Checkpoints are designed to help you to "handle" each new concept, almost as if you were working with clay. Through a variety of exercises and activities, you can achieve a depth of understanding that comes only through active participation, not passive reading alone. You can challenge and verify your knowledge of key concepts by completing the Checkpoints along your path of learning.

A diversity of learning activities is included in Checkpoints. You will focus on specific facts in the chapter by doing exercises that require you to provide definitions and comparisons, and you will fill in blanks in paragraphs with key terms. You will also color and label figures. Fine-point, color, felt-tip pens or pencils will be helpful so that you can easily fill in related color-coded ovals. Your understanding will also be tested with matching exercises, multiple-choice questions, and arrange-in-correct-sequence activities.

The black triangle (▲) beside selected Checkpoint activities within each chapter refers you to Appendix 1, where you will find a cross-reference guide and directions for using A.D.A.M.® Interactive Anatomy (AIA) software. If you have access to this software, try one of the AIA

Checkpoints for an in-depth view and in-depth learning of anatomy. Read more about the AIA Checkpoints in the preface and on page LG 677.

Three special types of Checkpoints are **For Extra Review**, **Clinical Correlations**, and **The Big Picture**. For Extra Review provides additional help for difficult topics or refers you by page number to specific activities earlier in the *Learning Guide* so that you may better integrate related information. Applications for enrichment and interest in areas particularly relevant to nursing and other health science students are found in selected Clinical Correlation exercises. The Big Picture: Looking Ahead (or Looking Back) refers you to relevant sections of the Guide or text to help you integrate concepts within the entire text. When you come upon a Big Picture Checkpoint, take a few moments to turn to those sections of the book that may reinforce content you studied earlier and also reduce your anxiety about upcoming content.

When you have difficulty with an exercise, refer to the related text pages (listed with each section of the *Learning Guide*) before proceeding further. You will find specific references to many text figures and tables.

Answers to Selected Checkpoints. As you glance through the activities, you will notice that a number of questions are marked with a check mark (✓). Answers to these Checkpoints are given at the end of each chapter, in order to provide you with some immediate feedback about your progress. Incorrect answers alert you to review related objectives in the *Learning Guide* and corresponding text pages. Each *Learning Guide* (LG) figure included in Answers to Selected Checkpoints is identified with an "A", such as Figure LG 13.1A. Activities without answers are also purposely included in this book. The intention is to encourage you to verify some answers independently by consulting your textbook and also to stimulate discussion with students or instructors.

More Critical Thinking provides several suggestions for essays to improve critical thinking and application skills.

Mastery Test. This 25-question self-test provides an opportunity for a final review of the chapter as a whole. Its format will assist you in preparing for standardized tests or course exams that are objective in nature, because the first 20 questions are multiple-choice, true-false, and arrangement questions. The final five questions of each test help you to evaluate your learning with fill-in and short-answer questions. Answers for each mastery test are placed at the end of each chapter.

I wish you success and enjoyment in learning concepts relevant to your own anatomy and physiology and in applying this information in clinical settings.

Kathleen Schmidt Prezbindowski

A B O U T   T H E   A U T H O R S

Kathleen Schmidt Prezbindowski, Ph.D., M.S.N., R.N., is a professor of biology at the College of Mount St. Joseph, where she was named Distinguished Teacher of the Year. As a visiting professor at the University of Cincinnati, she has taught anatomy and physiology, pathophysiology, and biology of aging for 30 years. Her Ph.D. is from Purdue University. Within the past decade, Dr. Prezbindowski earned a B.S.N. and an M.S.N. in both Gerontological Nursing and Psychiatric/Mental Health Nursing. Through a grant from the U.S. Administration on Aging, she produced the first video series on the human body ever designed especially for older persons, as well as a training manual for health promotion and aging. Kathleen is a delegate to the Health Promotion Institute of the National Council on the Aging. She is the author of two other learning guides.

Gerard J. Tortora is a professor of biology and teaches human anatomy and physiology and microbiology at Bergen Community College in Paramus, New Jersey. He received his B.S. in biology from Fairleigh Dickinson University in 1962 and his M.A. in biology from Montclair State College in 1965. The author of a number of best-selling anatomy and physiology, anatomy, and microbiology textbooks, laboratory manuals, and a new photographic atlas of the human body, Jerry belongs to numerous biology organizations: the American Association for the Advancement of Science (AAAS), The Human Anatomy and Physiology Society (HAPS), the American Association of Microbiology (ASM), and the Metropolitan Association of College and University Biologists (MACUB).

FRAMEWORK 1
An Introductory Tour of the Human Body

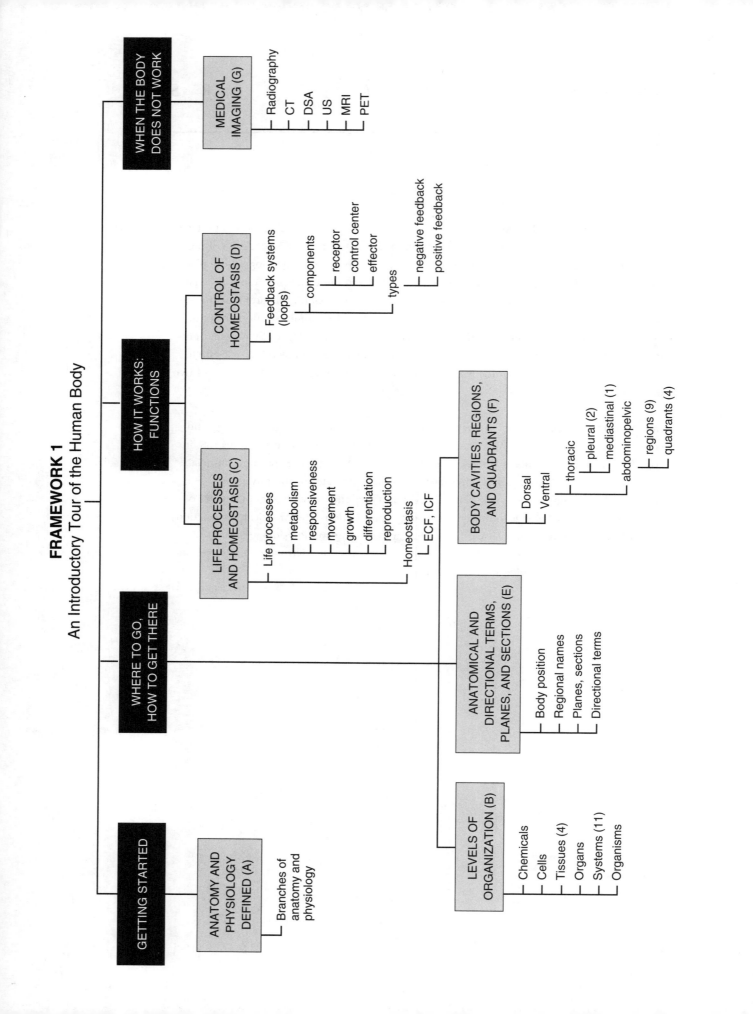

C H A P T E R **1**

An Introduction to the Human Body

You are about to embark on a tour, one that you will enjoy for at least several months and hopefully for many years. Your destination: the far-reaching corners and fascinating treasures of the human body. Imagine the human body as an enormous, vibrant world, abundant in architectural wonders and bustling with activity. You are the traveler, the visitor, the learner on this tour.

Just as the world is composed of individual buildings, cities, countries, and continents, so the human body is organized into cells, tissues, organs, and systems. In fact, a close-up look at a cell (comparable to a detailed study of a building) will reveal structural intricacies—the array of chemicals that comprise cells. Any tour requires a sense of direction and knowledge of the language: directional and anatomical terms that guide the traveler through the body. A tour is sure to enlighten the visitor about the culture in each region—in the human body, the functions of each body part or system. By the time you leave each area (body part or system), you will have gained a sense of how that area contributes to overall stability and well-being (homeostasis). You will also learn about how feedback systems (loops) contribute to homeostasis.

In planning a trip, an itinerary serves as a useful guide. A kind of itinerary is found below in the objectives (same as in your text) listed according to a topic outline that can help you organize your study. It can also give you a well-deserved feeling of accomplishment as you complete each objective. Then, at the end of the trip/chapter, look back at this itinerary/outline to put all aspects of the chapter in context. Also look at the Framework for Chapter 1, study it carefully, and refer back to it as often as you wish. It contains the key terms of the chapter and serves as your map. Bon voyage!

TOPIC OUTLINE AND OBJECTIVES

A. Anatomy and physiology defined

 1. Define anatomy and physiology and several subdisciplines of these sciences.

B. Levels of organization

 2. Describe the levels of structural organization that make up the human body.
 3. Define the 11 systems of the human body, the organs that comprise each, and their general functions.

C. Life processes and homeostasis

 4. Define the important life processes of the human body.
 5. Define homeostasis and explain its relation to interstitial fluid.

D. Control of homeostasis

 6. Describe the components of a feedback system.
 7. Contrast the operation of negative and positive feedback systems.
 8. Explain why homeostatic imbalances cause disorders.

E. Anatomical and directional terms, planes, and sections

9. Describe the orientation of the body in the anatomical position.
10. Relate the common names and corresponding anatomical descriptive terms used to refer to various regions of the human body.
11. Describe the anatomical planes and sections and directional terms used to describe the human body.

F. Body cavities, regions, and quadrants

12. Describe the principal body cavities, the organs they contain, and their associated linings.
13. Name and describe the nine abdominopelvic regions and the four abdominopelvic quadrants.

G. Medical imaging

14. Describe the principles and importance of medical imaging procedures in the evaluation of organ functions and diagnosis of disease.

WORDBYTES

Now become familiar with the language of this chapter by studying each wordbyte, its meaning, and an example of its use within a term. After you study the entire list, self-check your understanding by writing the meaning of each wordbyte on the line. As you continue through the *Learning Guide*, identify and write in other examples of terms that contain the same wordbyte.

Wordbyte	Self-check	Meaning	Example(s)
ana-	_____	apart	*ana*tomy
ante-	_____	before	*ante*rior
dis-	_____	part	*dis*section
homeo-	_____	sameness	*homeo*stasis
inter-	_____	between	*inter*cellular
intra-	_____	within	*intra*cellular
-logy	_____	study of	physio*logy*
-meter	_____	measure	milli*meter*
pariet-	_____	wall	*pariet*al
physio-	_____	nature (or function) of	*physio*logy
post-	_____	after	*post*erior
sagitta-	_____	arrow	*sagitta*l
-section	_____	act of cutting	dis*section*
-stasis	_____	standing still	homeo*stasis*
-tomy	_____	process of cutting	ana*tomy*
viscero-	_____	body organs	*viscera*l

CHECKPOINTS

A. Anatomy and physiology defined (pages, 1–2)

✓ **Al.** Anatomy is the study of _____ and physiology is the study

of _____.

✓ **A2.** Identity the subdivisions of anatomy and physiology by selecting the term that best
fits the definitions provided below.

C.	Cytology	H.	Histology
DA.	Developmental anatomy	I.	Immunology
Embr.	Embryology	PA.	Pathological anatomy
Endo.	Endocrinology	PP.	Pathophysiology
EP.	Exercise physiology	RP.	Renal physiology
GA.	Gross anatomy		

_____ a. Study of cells

_____ b. Study of tissues

_____ c. Study of structures that can be examined
without the aid of a microscope, such as
study of an entire cadaver

_____ d. Study of development during the first two
months of life in utero

_____ e. Study of function of kidneys

_____ f. Study of functions of hormones

_____ g. Study of functional changes associated
with disease

_____ h. Study of the defense mechanisms
of the body

A3. An intimate relationship exists between structure and function: one determines the
other. Explain how the structure of each of the following determines the functions
for which it can be used.

a. Spoon/fork

b. Hand/foot

c. Incisors (front teeth)/molars

▲　*For A.D.A.M.®-users.* Wherever you see the A.D.A.M.® Interactive Anatomy (AIA) logo
(shown here in the margin), turn to the appendix (beginning on page 677) and do the
AIA activities for that section of the chapter. Do AIA Chapter 1, Checkpoint A4 now.

✓▲ **A4.** *A clinical correlation.* Identify the procedures most likely to be used by a nurse in assessments of patients. Select from these answers:

Auscultation	Palpation	Percussion

a. Listening via stethoscope to hear "heart murmurs" caused by

damaged heart valves: _____
b. Pressing of fingers on breast tissue to check for the presence of hard,

immobile masses or for softer, mobile masses: _____
c. Examining for possible liver enlargement by pressing fingers just

below the right side of the rib cage: _____
d. While tapping finger over chest wall, listening for changes in sounds

that reveal fluid-filled lungs: _____

B. Levels of organization (page 2–4)

✓ **B1.** Arrange the terms listed in the left box from highest to lowest in level of organization by writing terms on lines provided at left below. One is done for you. Then match the examples in the right box with the related level of organization by writing those terms on lines at the right.

Cell	Organism
Chemical	System
Organ	Tissue

DNA	Smooth muscle tissue
Reproductive system	Uterus
Smooth muscle cell	21-year-old woman

Levels of organization

a. _____ (highest)

b. **System**_____

c. _____

d. _____

e. _____

f. _____ (lowest)

Examples

a. _____ (highest)

b. _____

c. _____

d. _____

e. _____

f. _____ (lowest)

✓ **B2.** List the four basic tissue types.

_____   _____

_____   _____

B3. Complete Table LG 1.1 on the next page describing systems of the body. Name two or more organs in each system. Then list one or more functions of each system.

Table LG 1.1 Systems of the human body

System	Organs	Functions
a.	Skin, hair, nails	
b. Skeletal		
c. Muscular		
d.		Regulates body by nerve impulses
e.	Glands that produce hormones	
f.	Blood, heart, blood vessels	
g. Lymphatic		
h.		Supplies oxygen, removes carbon dioxide, regulates acid-base balance
i.		Breaks down food and eliminates solid wastes
j.	Kidney, ureters, urinary bladder, urethra	
k. Reproductive		

✓▲ **B4.** *The Big Picture: Looking Ahead.* Look ahead to upcoming chapters as you identify organs that function as parts of more than one system. Text page numbers in parentheses may guide you to answers

a. The pancreas is part of the _____ system because the pancreas secretes enzymes that digest foods (page 840). Because it produces hormones such as insulin and glucagon, the pancreas.

   is also part of the _____ system (page 592). List two other organs that are parts of these same two systems:

   _____ and _____ (pages 567 and 819)

b. Because the pituitary releases many hormones, such as human growth hormone (hGH) and antidiuretic hormone (ADH), it is part

   of the _____ system (page 573). Yet the connections of the pituitary

   with the brain make it a component of the _____ system (page 460).

c. Because ovaries produce female sex cells (ova) and also hormones (estrogens and

   progesterone), ovaries are components of both the female _____ system

   (page 989) and the _____ system (page 595).

d. The large bone of your thigh (the femur) provides strong support for your body and it also forms blood cells, so the femur is part of two systems:

   _____ (page 231) and _____ (page 614).

e. The urethra of the male serves as a passageway for both urine (page 947) and semen,

   (page 985) so it is part of both the _____ system and the

   _____ system.

C. Life processes and homeostasis. (pages 4–7)

C1. Six characteristics distinguish you from nonliving things. List these characteristics below and give a brief definition of each. One is done for you.

a. _____

b. _____

c. _____

d. **Growth: increase in size and complexity** _____

e. _____

f. _____

✓ **C2.** Refer to the list of life processes in the box. Demonstrate your understanding of these processes by selecting the answer that best fits each of the activities in your own body, described below.

D.	**Differentiation**	**Move.**	**Movement**
G.	**Growth**	**Repro.**	**Reproduction**
Meta.	**Metabolism**	**Resp.**	**Responsiveness**

_____ a. Your hunger at 8:00 A.M. prompts you to head towards breakfast.

_____ b. During breakfast you chew your toast with jelly and an egg; your stomach and intestine then contract to help break apart the food and propel it through the digestive tract.

_____ c. Your body utilizes the starch, sugars, and proteins in your breakfast foods to provide building blocks for more muscle protein and to provide energy to your eyes and brain for studying anatomy and physiology.

_____ d. After you work out four days a week for a month, you note that your arm and thigh muscles (biceps and quadriceps) are larger (two answers).

_____ e. As you work out, "stem cells" in your bone marrow are stimulated to undergo changes to become mature red blood cells (RBC) so you experience a healthy increase in your RBC count.

✓ **C3.** *Critical Thinking* about homeostasis.
  a. Using your list of wordbytes, write a literal definition of homeostasis.

  b. Explain why a broader definition of homeostasis is actually more accurate.

  c. Besides control of blood glucose, what other aspects of your body need to be kept relatively constant, within a narrow range?

Figure LG 1.1   Internal environment of the body: types of fluid.
Complete as directed. Areas labeled A to G refer to Checkpoint C5.

✓ **C4.** Describe the types of fluid in the body and their relationships to homeostasis by completing these statements and Figure LG 1.1.

a. Fluid inside of cells is known as _____ fluid (ICF). Color areas containing this type of fluid yellow. Color yellow the corresponding color code oval.

b. Fluid in spaces between cells is called _____. It surrounds cells and is one form of *(intracellular? extracellular?)* fluid. Color the spaces containing this fluid (as well as the related color code oval) light green.

c. Another form of extracellular fluid is that located in _____ vessels and

_____ vessels. Color these areas (and the color code oval) dark green.

d. The body's "internal environment" (that is, surrounding cells) is _____ -cellular fluid (ECF) (all green areas in your figure). The condition of maintaining

ECF in relative constancy is known as _____.

✓ **C5.** Refer again to Figure LG 1.1. Show the pattern of circulation of body fluids by drawing arrows connecting letters in the figure in alphabetical order (A → B → C, etc.). Now fill in the blanks below describing this lettered pathway:

A (arteries and arterioles)   → B (blood capillaries)   →

C (_____) → D (_____) →

E (_____) → F (_____) →

G (_____).

D. Control of homeostasis (pages 7–10)

D1. Think of a recent situation in which your mind and body have experienced stress. Identify the two body systems that are most often responsible for initiating changes that bring you back into homeostatic balance. (*Hint:* Refer to Checkpoint D5.)

✓ **D2.** Are you seated right now? For safety, hold on to the back of a chair or a desk and then stand up quickly. As you do, conceptualize what happens to your blood as you change position: As a response to gravity, blood tends to fall to the lower part of your body. If nothing happened to offset that action, your blood pressure (BP) would drop and you would likely faint because of insufficient blood supply to the brain. This usually does not happen because homeostatic mechanisms act to quickly elevate blood pressure back to normal.

Match the answers in the box with factors and functions that helped maintain your blood pressure when you stood up. (*Hint*: Figure 1.3 [page 8 your text] describes a situation that is just the opposite of this one.)

CCen.	Control center	O.	Output
CCond.	Controlled condition	Recep.	Receptors
E.	Effector	Respo.	Response
I.	Input	S.	Stimulus

_____ a. Your blood pressure (BP): a factor that must be maintained homeostatically

_____ b. Standing up; blood flowed by gravity to lower parts of your body

_____ c. Pressure-sensitive nerve cells in large arteries in your neck and chest

_____ d. Nerve impulses bearing the message "BP is too low" (since less blood remained in the upper part of your body as you were standing)

_____ e. Your brain

_____ f. Nerve impulses from your brain to your heart, bearing the message, "Beat faster!"

_____ g. Your heart (which beats faster)

_____ h. Elevated BP

✓ **D3.** *A clinical correlation.* In certain populations, such as malnourished or frail elderly, the mechanisms necessary for maintaining blood pressure occur more slowly than normal. As a result, blood pressure may drop as the person suddenly changes position (from lying to sitting or sitting to standing). This condition, known as postural hypotension, may lead to faintness and falls. What might a person subject to postural hypotension do to minimize the chance of falling?

✓ **D4.** Consider this example of a feedback system loop. Ten workers on an assembly line are producing handmade shoes. As shoes come off the assembly line, they pile up around the last worker, Worker Ten. When this happens, Worker Ten calls out, "We have hundreds of shoes piled up here." This information (input) is heard by (fed back to) Worker One, who determines when more shoes should be begun. Worker One is therefore the ultimate controller of output. This controller could respond in either of two ways. In a *negative feedback system*, Worker One says, "We have an excess of unsold shoes. Let us slow down or stop production until the excess (stress) is relieved—until these shoes are sold." What might Worker One's response be if this were a *positive feedback system?*

✓ **D5.** Distinguish the two types of feedback in this exercise.
   a. When blood pressure is abnormally low, the following mechanism should occur to raise blood pressure. Nerve impulses cause many tiny blood vessels that flow into stomach, intestine, and skin to become narrowed so that blood must stay in major arteries—raising blood pressure. This is an example of a *(negative? positive?)* feedback mechanism since the lowered blood pressure *(was raised? decreased even more?)*.
   b. The increase of uterine contraction during labor as a response to increase of the hormone oxytocin is an example of a *(negative? positive?)* feedback mechanism because the uterine contractions are *(decreased? enhanced?)*.
   c. Most feedback mechanisms in the human body are *(negative? positive?)*. Why do you think this might be advantageous and directed toward maintenance of homeostasis?

✓ **D6.** Circle the terms below that are classified as signs rather than as symptoms.

   chest pain   fever   headache   diarrhea   swollen ankles   skin rash   toothache

D7. *A clinical challenge.* Contrast the two terms in each set.
   a. *Local disease/systemic disease*

   b. *Epidemiology/pharmacology*

   c. *Medical history/diagnosis*

E. Anatomical and directional terms, planes, and sections (pages 10–13)

  E1. How would you describe anatomical position? Assume that position yourself.

  Contrast these body positions: *prone* and *supine*.

✓▲**E2.** Complete the table relating common terms to anatomical terms. (See Figure 1.5, page 11 in your text, to check your answers.) For extra practice use common and anatomical terms to identify each region of your own body and that of a study partner.

Common Term	Anatomical Term
a.	Axillary
b. Fingers	
c. Arm	
d.	Popliteal
e.	Cephalic
f. Mouth	
g.	Inguinal
h. Chest	
i.	Cervical
j.	Antebrachial
k. Buttock	
l.	Calcaneal

✓▲ **E3.** Match each of the following planes with the phrase telling how the body would be divided by such a plane.

| F. | Frontal (coronal) | P. | Parasagittal |
| M. | Midsagittal (median) | T. | Transverse |

_____ a. Into superior and inferior portions   _____ c. Into anterior and posterior portions

_____ b. Into equal right and left portions   _____ d. Into unequal right and left portions

✓▲ **E4.** Using your own body, a skeleton, and Exhibit 1.1 and Figure 1.9 determine relationships among body parts. Write correct directional terms to complete these statements.

a. The liver is _____ to the diaphragm.

b. Fingers (phalanges) are located _____ to wrist bones (carpals).
c. The skin on the dorsal surface of your body can also be said to be located on

your _____ surface.

d. The great (big) toe is _____ to the little toe.

e. The little toe is _____ to the great toe.

f. The skin on your leg is _____ to muscle tissue in your leg.

g. Muscles of your arm are _____ to skin on your arm.
h. When you float face down in a pool, you are lying on your

_____ surface.

i. The lungs and heart are located _____ to the abdominal organs.
j. Because the stomach and the spleen are both located on the left side of the

abdomen, they could be described as _____ -lateral.

▲ **E5.** To do the AIA activities for this section of the chapter, go to Chapter 1 AIA Checkpoints B1b–E4g in the Appendix now.

F. Body cavities, regions, and quadrants (pages 13, 16–19)

✓▲ **F1.** After you have studied Figures 1.8, 1.10, and 1.11 in your text, complete this exercise about body cavities. Circle the correct answer in each statement.
a. The *(dorsal? ventral?)* cavity consists of the cranial cavity and the vertebral canal.
b. The viscera, including such structures as the liver, lungs, and stomach, are all located in the *(dorsal? ventral?)* cavity.
c. Of the two body cavities, the *(dorsal? ventral?)* appears to be better protected by bone.
d. Pleural, mediastinal, and pericardial are terms that refer to regions of the *(thorax? abdominopelvis?)*.

   e. The *(heart? lungs? esophagus and trachea?)* are located in the pleural cavities.
   f. Which organs are located in the mediastinum? *(Esophagus? Major airways such as the trachea? Heart and its pericardial covering? Stomach? Thyroid? Thymus?)*.
   g. The stomach, pancreas, small intestine, and most of the large intestine are located in the *(abdomen? pelvis?)*.
   h. The urinary bladder, rectum, and internal reproductive organs are located in the *(abdominal? pelvic?)* cavity.
   i. Which is larger: the *(abdominal? pelvic?)* region of the abdominopelvic cavity.

✓ **F2.** Complete Figure LG 1.2 according to the directions below.
   a. Color each of the organs listed on the figure. Select different colors for each organ, and be sure to use the same color for the related color code oval ◯.
   b. Next, label each organ on the figure.
   c. *For extra review.* Draw lines dividing the abdomen into the nine regions, and note which organs are in each region.

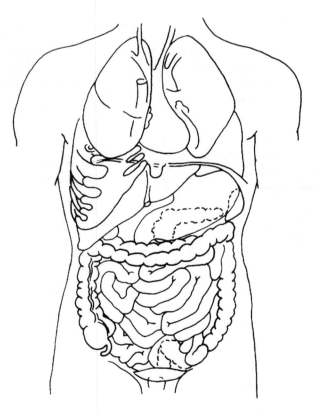

◯ Appendix	◯ Heart	◯ Lungs	◯ Spleen
◯ Diaphragm	◯ Large intestine	◯ Pancreas	◯ Stomach
◯ Gallbladder	◯ Liver	◯ Small intestine	

Figure LG 1.2 Regions of the ventral body cavity. Complete as directed in Checkpoint F2.

✓▲ **F3.** Using the names of the nine abdominal regions and quadrants, complete these statements. Refer to the figure you just completed and to Figures 1.11 and 1.12 in the text.

a. From superior to inferior, the three abdominal regions on the right side are

_____ , _____ and _____ .

b. The stomach is located primarily in the two regions named

_____ and _____ .

c. The navel (umbilicus) is located in the _____ region.

d. The region immediately superior to the urinary bladder is named the

_____ region.

e. The liver and gallbladder are located in the _____ quadrant.

G. Medical imaging (pages 19–21)

✓ **G1.** Match the radiographic anatomy techniques listed in the box with the correct descriptions.

CR.	**Conventional radiograph**		**MRI.**	**Magnetic resonance imaging**
CT.	**Computed tomography**		**PET.**	**Positron emission tomography**
DSA.	**Digital subtraction angiography**		**US.**	**Ultrasound**

_____ a. This method provides a cross-sectional picture of an area of the body by means of an x-ray source moving in an arc. Results are processed by a computer and displayed on a video monitor.

_____ b. This technique produces a two-dimensional image (radiograph) via a single barrage of x-rays. Images of organs overlap, making diagnosis difficult.

_____ c. A method used to study blood vessels by comparing a region before and after injection of a contrast substance (dye).

_____ d. Utilizes injected radioisotopes that assess function as well as structure.

_____ e. Two techniques that are noninvasive and do not utilize ionizing radiation (two answers).

ANSWERS TO SELECTED CHECKPOINTS: CHAPTER 1

A1. Structure, function.
A2. (a) C. (b) H. (c) GA. (d) Embr (also a part of DA).
(e) RP. (f) Endo. (g) PP. (h) I.
A4. (a) Auscultation. (b-c) Palpation. (d) Percussion.
B1. (a) Organism; 21-year-old woman; (b) (System);
reproductive system. (c) Organ; uterus. (d) Tissue;
smooth muscle tissue. (e) Cell; smooth muscle cell.
(f) Chemical; DNA.
B2. Epithelial tissue, connective tissue, muscle tissue,
nervous tissue.
B4. (a) Digestive; endocrine; stomach, small intestine.
(b) Endocrine; nervous. (c) Reproductive,
endocrine. (d) Skeletal, cardiovascular. (e) Urinary,
reproductive
C2. (a) Resp. (b) Move. (c) Meta. (d) G, Repro. (e) D.
C3. (a) Literal definition: standing still or staying the
same. (b) Broader definition: maintaining a
dynamic (ever-changing) state—yet over a narrow
range. (c) Examples: blood levels of other chemicals such as potassium or calcium ions, water, fats;
body temperature; heart rate and blood pressure.
C4. (a) Intracellular. (b) Interstitial (or intercellular or
tissue) fluid, extracellular. (c) Blood, lymph. (d)
Extra-; homeostasis.
C5. C, interstitial (intercellular) fluid; D, intracellular
fluid; E, interstitial (intercellular) fluid again; F,
blood or lymph capillaries; G, venules and veins or
lymph vessels
D2. (a) CCond. (b) S. (c) Recep. (d) I. (e) CCen. (f) O.
(g) E. (h) Respo.
D3. Change position gradually, for example, by sitting
on the side of the bed before getting out of bed.
Support body weight on a nightstand or chair for a
moment after standing. Wear support hose. Drink
adequate fluid.
D4. "Hurray! We have produced hundreds of shoes; let
us go for thousands! Do not worry that the shoes
are not selling. Step up production even more."
D5. (a) Negative; was raised. (b) Positive, enhanced.
(c) Negative; they permit continual fine-tuning so
changes do not become excessive in one direction.

D6. Fever, diarrhea, swollen ankles, skin rash.
E2. (a) Armpit. (b) Phalangeal or digital. (c) Brachial.
(d) Hollow area at back of knee. (e) Head. (f) Oral.
(g) Groin. (h) Thoracic. (i) Neck. (j) Forearm.
(k) Gluteal. (l) Heel.
E3. (a) T. (b) M. (c) F. (d) P.
E4. (a) Inferior (caudal). (b) Distal. (c) Posterior or
dorsal (d) Medial. (e) Lateral. (f) Superficial.
(g) Deep. (h) Anterior (ventral). (i) Superior
(cephalic or cranial) (j) Ipsi-.
F1. (a) Dorsal. (b) Ventral. (c) Dorsal. (d) Thorax.
(e) Lungs. (f) Esophagus, major airways such as
the trachea, heart and its pericardial covering,
thymus. (g) Abdomen. (h) Pelvic. (i) Abdominal.
F2. (a-c) See Figure LG 1.2A.

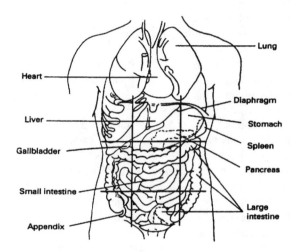

Figure LG 1.2A Regions of the ventral body cavity.

F3. (a) Right hypochondriac, right lumbar, right iliac
(inguinal). (b) Left hypochondriac, epigastric.
(c) Umbilical. (d) Hypogastric (pubic). (e) Right
upper.
G1. (a) CT. (b) CR. (c) DSA. (d) PET. (e) MRI, US.

MORE CRITICAL THINKING: CHAPTER 1

1. Think of a person you know who has a chronic (ongoing) illness. Explain how changes
in anatomy (abnormal body structure) in that person are associated with changes in
physiology (inadequate or altered body function).
2. Cancer is a disease of abnormal cell production. Explain how alterations at the cellular
level affect higher levels of organization, such as specific organs, systems, and the entire
organism.
3. Suppose you decide to run a quarter mile at your top speed. Tell how your body would
respond to this stress created by exercise. List according to system the changes your body
would probably make in order to maintain homeostasis.

MASTERY TEST: CHAPTER 1

Questions 1—11: Circle the letter preceding the one best answer to each question.

1. In a negative feedback system, when blood pressure decreases slightly, the body will respond by causing a number of changes which tend to:
   A. Lower blood pressure
   B. Raise blood pressure

2. The following structures are all located in the ventral cavity *except*:
   A. Spinal cord   D. Gallbladder
   B. Urinary bladder   E. Esophagus
   C. Heart

3. Which pair of common/anatomical terms is mismatched?
   A. Eye/ocular   D. Neck/cervical
   B. Skull/cranial   E. Buttock/gluteal
   C. Armpit/brachial

4. Which science studies the microscopic structure of tissues?
   A. Endocrinology   D. Pharmacology
   B. Histology   E. Epidemiology
   C. Cytology

5. Which is most inferiorly located?
   A. Abdomen   D. Diaphragm
   B. Pelvic cavity   E. Pleural cavity
   C. Mediastinum

6. The spleen, tonsils, and thymus are all organs in which system?
   A. Nervous   D. Digestive
   B. Lymphatic   E. Endocrine
   C. Cardiovascular

7. Which of the following structures is located totally outside of the upper right quadrant of the abdomen?
   A. Liver   D. Spleen
   B. Gallbladder   E. Pancreas
   C. Transverse colon

8. Choose the FALSE statement. (It may help you to mark each statement T for true or F for false as you read it.)
   A. Stress disturbs homeostasis.
   B. Homeostasis is a condition in which the body's internal environment remains relatively constant.
   C. The body's internal environment is best described as extracellular fluid.
   D. Extracellular fluid consists of plasma and intracellular fluid.

9. The system responsible for providing support, protection, and leverage; for storage of minerals; and for production of most blood cells is:
   A. Urinary   D. Reproductive
   B. Integumentary   E. Skeletal
   C. Muscular

10. Which is most proximally located?
    A. Ankle   C. Knee
    B. Hip   D. Toe

11. Which of the following describes pathological anatomy?
    A. Study of a specific region of the body such as the chest or arm
    B. Study of markings of the surface of the body
    C. Study of the body according to specific systems such as skeletal or respiratory
    D. Study of structural changes associated with disease

Questions 12—20: Circle T (true) or F (false). If the statement is false, change the underlined word or phrase so that the statement is correct.

T F   12. In order to assume the anatomical position, you should <u>lie down with arms at your sides and palms turned backwards</u>.

T F   13. The appendix is usually located in the <u>left iliac</u> region of the abdomen.

T F   14. <u>Anatomy</u> is the study of how structures function.

T F   15. In a negative feedback system, the body will respond to an <u>increased blood glucose level by increasing blood glucose to an even higher level</u>.

T F   16. A <u>tissue</u> is a group of cells and their intercellular substances that are similar in both origin and function.

T F   17. The right kidney is located <u>mostly in the right iliac region of the abdomen</u>.

T F   18. The study of the microscopic structure of cells is <u>cytology</u>.

T F   19. A parasagittal section is <u>parallel to and lateral to</u> a midsagittal section.

T F   20. The term homeostasis means that the body exists <u>in a static state</u>.

Questions 21—25: Fill-ins. Write the word that best fits the description.

_____ 21. Region of thorax located between the lungs. Contains heart, thymus, esophagus.

_____ 22. Name the fluid located within cells.

_____ 23. Life process in which unspecialized cells, as in the early embryo, change into specialized cells such as nerve or bone.

_____ 24. A painless diagnostic technique in which sound waves are generated and detected by a transducer, resulting in a sonogram.

_____ 25. Name one or more extracellular fluids.

ANSWERS TO MASTERY TEST: CHAPTER 1

Multiple Choice

1. B
2. A
3. C
4. B
5. B
6. B
7. D
8. D
9. E
10. B
11. D

True—False

12. F. Stand erect facing observer, arms at sides, palms forward
13. F. Right iliac
14. F. Physiology
15. F. Decreasing blood glucose to a lower level
16. T
17. F. At the intersection of four regions of the abdomen: right hypochondriac, right lumbar, epigastric, and umbilical
18. T
19. T
20. F. In a state of dynamic equilibrium

Fill-ins

21. Mediastinum
22. Intracellular
23. Differentiation
24. Ultrasound (US)
25. Plasma, interstitial fluid, lymph, serous fluid, mucus, fluid within eyes or ears, cerebrospinal fluid (CSF)

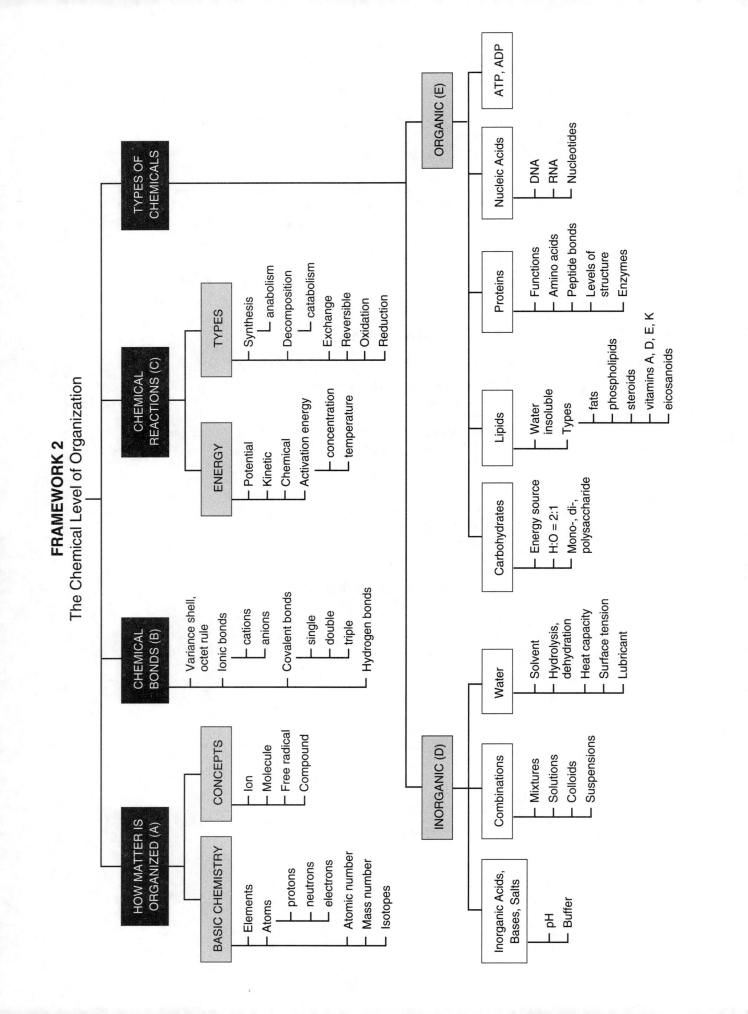

FRAMEWORK 2
The Chemical Level of Organization

C H A P T E R **2**

The Chemical Level of Organization

Chemicals comprise the ultrastructure of the human body. They are the submicroscopic particles of which cells, tissues, organs, and systems are constructed. Chemicals are the minute entities that participate in all of the reactions that underlie bodily functions. Every complex compound that molds a living organism is composed of relatively simple atoms held together by chemical bonds. Two classes of chemicals contribute to structure and function: inorganic compounds, primarily water, but also inorganic acids, bases, and salts, and organic compounds including carbohydrates, lipids, proteins, nucleic acids, and the energy-storing compound ATP.

As you begin your study of the chemistry of life, carefully examine the Chapter 2 Framework and note the key terms associated with each section. Also refer to the Chapter 2 Topic Outline and check off objectives as you meet each one.

TOPIC OUTLINE AND OBJECTIVES

A. How matter is organized

1. Identify the principal chemical elements of the human body.
2. Describe the structure of atoms, ions, molecules, free radicals, and compounds.

B. Chemical bonds

3. Describe how valence electrons form chemical bonds.
4. Distinguish between ionic, covalent, and hydrogen bonds.

C. Chemical reactions

5. Define a chemical reaction.
6. Describe the various forms of energy.
7. Compare the basic differences between exergonic and endergonic reactions.
8. Describe the role of activation energy and enzymes.
9. Distinguish between synthesis, decomposition, exchange, reversible, and oxidation-reduction reactions.

D. Inorganic compounds and solutions

10. Describe the properties of inorganic acids, bases, salts, and water.
11. Distinguish among solutions, colloids, and suspensions.
12. Define pH.
13. Explain the role of buffer systems in homeostasis.

E. Organic compounds

14. Describe the functional groups of organic molecules.
15. Identify the building blocks and functions of carbohydrates, lipids, proteins, enzymes, deoxyribonucleic acid (DNA), ribonucleic acid (RNA), and adenosine triphosphate (ATP).

WORDBYTES

Now become familiar with the language of this chapter by studying each wordbyte, its meaning, and an example of its use within a term. After you study the entire list, self-check your understanding by writing the meaning of each wordbyte on the line provided. As you continue through the *Learning Guide*, identify and write in additional terms that contain the same wordbyte.

Wordbyte	Self-check	Meaning	Example(s)
-ase	_____	enzyme	lip*ase*
de-	_____	remove	*de*hydrogenase
di-	_____	two	*di*saccharide
hex-	_____	six	*hex*ose
mono-	_____	one	*mono*saccharide
pent-	_____	five	*pent*ose
poly-	_____	many	*poly*peptide
prot-	_____	first	*prot*ein
-saccharide	_____	sugar	mono*saccharide*
tri-	_____	three	*tri*phosphate

CHECKPOINTS

A. How matter is organized? (pages 26–30)

A1. Why is it important for you to understand the "language of chemistry" in order to communicate about life processes?

✓ **A2.** Complete this exercise about matter and chemical elements.
   a.   All matter is made up of building units called chemical _____.
   b.   Of the total 109 different chemical elements found in nature, four elements make up

   96% of the mass of the human body. These elements are _____ (O),

   _____ (C), hydrogen (_____), and

   _____ (_____).
   c.   Circle the nine chemical symbols of elements listed here that together make up 3.9% of human body mass:
   Ca Cl Co Cu Fe I K Mg Mn Na P S Si Sn Zn
   d.   Define this term: *trace elements*.

Now write on these lines the symbols of six trace elements from the list of elements

above (in d): _____ _____ _____ _____ _____ _____

✓ **A3.** Refer to Table 2.1 (page 27) in your text. Match the elements listed in the box with descriptions of their functions in the body. The first one has been done for you.

C.	Carbon	K.	Potassium
Ca.	Calcium	Na.	Sodium
Cl.	Chlorine	N.	Nitrogen
Fe.	Iron	O.	Oxygen
F.	Fluorine	P.	Phosphorus
I.	Iodine		

____**C**____ a.   Found in every organic molecule

_____ b.   Component of all protein molecules, as well as DNA and RNA

_____ c.   Vital to normal thyroid gland function

_____ d.   Essential component of hemoglobin, which gives red color to blood

_____ e.   Composes part of every water molecule; functions in cellular respiration

_____ f.   Contributes hardness to bone and teeth; required for blood clotting and for muscle contraction

_____ g.   Component of ATP, DNA, and RNA; also component of bone and teeth

_____ h.   Anion (negatively charged ion) in table salt (NaCl)

_____ i.   Cation (positively charged ion) in table salt (NaCl); helps control water balance in the body; important in nerve and muscle function

_____ j.   A trace element that contributes to tooth structure

_____ k.   Most abundant positively charged ion (cation) in intracellular fluid (inside of cells)

A4. Contrast *element* with *atom*.

✓ **A5.** Complete this exercise about atomic structure.

a.   An atom consists of two main parts: _____ and _____.

b.   Within the nucleus are positively charged particles called _____ and

uncharged (neutral) particles called _____.

c.   Electrons possess *(positive? negative?)* charges. The number of electrons forming a charged cloud around the nucleus is *(greater than? equal to? smaller than?)* the number of protons in the nucleus of the atom. Therefore each atom is said to be electrically

(charged? neutral?) with a total charge of _____.

d.   Electrons are grouped in regions known as electron _____. The shell

closest to the nucleus can hold a maximum of _____ electrons; the next two shells

outward may hold up to _____ and _____ electrons, respectively.

e. The largest element in the human body is *(iron? iodine?)*, which has _____ protons

and _____ electrons. Fill in the number of electrons in its five shells from innermost

to outermost: _____, _____, _____, _____, and _____.

✓ **A6.** Refer to Figure 2.2 in your text and complete this exercise about atomic structure.

a. The *atomic number* of potassium (K) is _____. Locate this number below the

symbol K on the figure. This means that a potassium atom has _____ protons and

_____ electrons.

b. Identify the *atomic mass* associated with potassium in the figure. This num-

ber is _____, and it represents the average of the mass numbers of
the potassium atom. This indicates the total number of protons and neutrons in the
potassium nucleus.

✓ **A7.** Complete this exercise about isotopes.

a. If you subtract the *atomic number* from the *mass number,* you will identify the

number of neutrons in an atom. Most potassium atoms have _____ neutrons.
Different isotopes of potassium all have 19 protons, but have varying numbers of
(electrons? neutrons?).

b. The isotope 14C differs from 12C in that 14C has _____ neutrons.

c. Radioisotopes have a nuclear structure that is unstable and can decay, emitting

_____. Each radioactive isotope has its own distinctive

_____, which is the time required for the radioactive isotope to
emit half of its original amount of radiation.

d. List one helpful and one harmful effect of radioactive isotopes.

✓ **A8.** Complete this exercise about several different categories of chemicals. Select all
answers from the box that fit descriptions below. All answers except two will be used.

Ca^{++}	Cl$^-$	C$_6$H$_{12}$O$_6$ (glucose)	H
H$_2$O	Na$^+$	O$_2$	O$_2^-$ (superoxide)

_____ a. Molecule   _____ d. Ion that has gained one
                                              electron
_____ b. Compound
                                _____ e. Ion that has lost one
_____ c. Free radical   electron

A9. A *clinical correlation.* Discuss roles of free radicals in disease and aging by completing this Checkpoint.
a.   State how free radicals may form.

b.   List six or more disorders or other changes associated with free radicals.

c.   Name a category of chemicals that can slow effects of free radicals.

B. Chemical bonds (pages 30–34)

✓ **B1.** The type of chemical reacting (or bonding) that occurs between atoms depends on the number of electrons possessed by the bonding atoms. Refer to Figure 2.2 in your text and to the atoms listed in the box below; then do this Checkpoint about electrons and bonding.

C	Cl	H	He	Na	Ne

a.   Which of these atoms contain only one electron shell? _____ and _____. Write the

maximum number of electrons in this shell: _____.

b.   The other atoms in the box all have *(1? 2? 3?)* electron shells. As you refer to Figure 2.2, write the number of electrons present in the outermost electron shell of these

atoms: Na: _____ Cl _____ C: _____. The maximum number of electrons that can be

present in the outer shell of each of these atoms is _____. (For help, refer to Check-point A5d.)

c.   The outer shell of any electron is called its _____ shell. Atoms can achieve stability by filling this outer shell to the maximum; this process is normally achieved by bonding. The likelihood that an atom will form chemical bonds is directly related to the number of extra or deficient electrons in the valence shell of that atom. (Visu-alize that this valence shell is what atom A presents to nearby atom B, which may be a possible "bonding partner.")

d.   Find sodium (Na) in Figure 2.2. Sodium has one *(extra? missing?)* electron in its valence shell. It is therefore more likely to become an electron *(donor? acceptor?)*. When sodium gives up the electron (which is a negative entity), sodium becomes more *(positive? negative?)*. In other words, sodium forms Na^+, which is a(n) *(anion? cation?)*.

e.   Chlorine has _____ electrons in its outer shell, so it has one *(extra? missing?)* electron compared to a full valence shell. As sodium chloride is formed, chlorine *(accepts? gives up?)* an electron and so becomes the *(anion? cation?)* Cl^-.

f.   Ions are defined as atoms that have *(gained or lost? shared?)* one or more electrons into or from their valence shells. A cation, such as K^+, Na^+, or H^+, is an atom that has *(gained? lost?)* an electron (a negative entity) by ionic bonding. A(n) *(anion? cation?)*, such as Cl^-, has gained an electron from another atom. Ions in solution, such as table

salt (Na^+Cl^-) in water, are called _____ because _____.

g.   The fact that most atoms interact in ways that permit them to achieve stable outer shells of eight electrons is known as the *(triplet? septet? octet?)* rule.

h.   Some atoms exist in nature with the valence shell already filled; these atoms *(do? do not?)* not attempt to bond, and so are said to be _____. Two examples of inert atoms are *(H? He?)* and *(Ne with the atomic number of 10? Na with the atomic number of 11?)*

✓ **B2.** Match the types of bonds listed in the box with the descriptions below.

C. Covalent	**H. Hydrogen**	**I. Ionic**

_____ a.   Atoms lose electrons if they have just one or two electrons in their valence shells; they gain electrons if they need just one or two electrons to complete the valence shell, as in NaCl, (see Activity B1 above).

_____ b.   This bond is a bridgelike, weak link between a hydrogen atom and another atom such as oxygen or nitrogen. The hydrogen atom has a partial positive charge (+) that attracts the partial negative charge (–) of the electronegative (atom such as oxygen or nitrogen).

_____ c.   A pair of electrons is *shared* in this type of bond, for example, in a C–H bond; the two electrons (one from C and one from H) spend most of their time in the region between the nuclei (of C and of H).

_____ d.   These bonds are only about 5% as strong as covalent bonds and so are easily formed and broken. They are vital for holding large molecules like protein or DNA in proper configurations.

_____ e.   Double bonds, such as O = O (O_2), or triple bonds, such as N = N (N_2).

_____ f.   This type of bond is most easily formed by carbon (C) because its outer orbit is half-filled.

_____ g.   These bonds may be polar, such as H_2O (H — OH), or nonpolar, such as O = O (O_2).

A. _____   B. _____   C. _____   D. _____
   (   )   (   )   (7)   (   )

Figure LG 2.1 Atomic structure of four common atoms in the human body. Complete as directed in Checkpoint B3.

✓ **B3.** Refer to Figure LG 2.1 and complete this exercise about covalent bonding.
  a.   Write atomic numbers in the parentheses ( ) under each atom. (*Hint:* Count the number of electrons or protons = atomic number.) One is done for you.
  b.   Write the name or symbol for each atom on the line under that atom. (*Hint:* You may identify these by atomic number in Figure 2.2 in your text.)
  c.   Based on the number of electrons that each of these atoms is missing from a complete outer electron shell, draw the number of covalent bonds that each of the atoms can be expected to form. Note that this is the valence, or combining capacity, of these atoms. One is done for you.

H   O   N   C

✓ **B4.** *Critical thinking* about covalently bonded molecules. Complete this exercise.
  a.   Which of the following molecules involves a triple (covalent) bond? (*Hint:* Refer to Figure 2.5 of your text.)

A. Oxygen molecule (O_2)	**B. Nitrogen molecule (N_2)**

  b.   In the margin (or on separate paper) draw the atomic structure of the covalently bonded molecule, ammonia (NH_3).

C. Chemical reactions (pages 34–37)

✓ **C1.** Complete this Checkpoint about chemical reactions and energy.
  a.   When two reactants (such as hydrogen and oxygen) react, they form a product

  (such as _____). The fact that the same number of atoms are in the reactants and the product is a statement of the law of conservation of *(mass? energy?)*.

  b.   The term _____ refers to all of the chemical reactions in the body.
  c.   Chemical reactions involve changes in energy, which is defined as the capacity to do

  _____. Now identify three types of energy. The energy within chemical bonds

  is known as _____ energy. Inactive or stored energy is known as *(kinetic?*

  potential?) energy, whereas the energy of motion is _____ energy.
  d.   Chemical reactions in which more energy is released from chemical bonds than is used to break apart those bonds is called an *(endergonic? exergonic?)* reaction. Most reactions involved in breakdown of chemicals in food nutrients are *(endergonic? exergonic?)* reactions. Much of this released energy is stored in the chemical

  adenosine triphosphate (or)_____; this energy is stored for later use in

  fueling _____ -gonic reactions, such as muscle building.

C2. Define *activation energy*

  Now write a short paragraph describing three factors facilitating collisions between atoms, increasing the likelihood that chemical reactions will take place. Which one of these three factors actually lowers the activation energy needed for a reaction to occur?

✓ **C3.** Identify the kind of chemical reaction described in each statement below.

> **D.** **Decomposition reaction**   **R.** **Reversible reaction**
>
> **E.** **Exchange reaction**   **S.** **Synthesis reaction**

_____ a. The end product can revert to the original combining molecules.

_____ b. Two or more reactants combine to form an end product; for example, many glucose molecules bond to form glycogen.

_____ c. Such a reaction is partly synthesis and partly decomposition.

_____ d. All chemical reactions involve making or breaking of bonds. This type of reaction involves only breaking of bonds.

_____ e. This type of reaction is catabolic, as in digestion of foods, such as starch digestion to glucose.

✓ **C4.** We recognize the importance of oxygen for our survival. Oxygen is ultimately required for processes called biological _(oxidation? reduction?)_ reactions that provide energy for all types of body activities. Describe such reactions in this exercise.

a. Most biological oxidation reactions involve _(gain? loss?)_ of hydrogen atoms from chemicals in foods we have eaten. For example, high energy sugars we eat, with their many C–H (carbon-to-hydrogen) bonds, will lose H's and ultimately form C–O (carbon-to oxygen) bonds, as in the low energy waste we exhale, carbon dioxide (CO_2).

b. Look at the figures of the 3-carbon sugars, lactic acid and pyruvic acid, on page 37 of your text. Count the atoms in each chemical and write them here. (The exercise is already started for you.)

	Carbons (C)	Hydrogens (H)	Oxygens (O)
Lactic acid	_____	6	_____
Pyruvic acid	3	_____	_____

c. Which atom has more hydrogens? _(Lactic? Pyruvic?)_ acid. This acid loses _____ hydrogen atoms as it is converted to pyruvic acid. In this way lactic acid is _(oxidized? reduced?)_ to form pyruvic acid.

✓ **C5.** _The Big Picture: Looking Ahead._ You can expand on what you have just learned about oxidation if you turn to Figure 25.2 of your text. There you will see a number of examples of biological oxidation in metabolism (or utilization) of the foods.

a. Think of the structure of each hydrogen atom: it consists of _(one? two?)_ proton(s) with _(one? two?)_ electrons(s). Suppose a hydrogen atom suddenly loses its electron, which takes off to whirl around another hydrogen's proton. What happens to that first

hydrogen? It becomes: _____.
A. H^+ (a hydrogen ion, which is simply a proton with no orbiting electrons)
B. H^- (a hydride ion, which is a proton with two orbiting electrons)

b. It is possible then for a chemical, such as a sugar, to lose two hydrogens in the ionized forms, one as H^+ and one as H^-. In fact, this is exactly what happens in biological oxidation: removal not simply of two hydrogen atoms (2H), but removal

of a pair of hydrogens in the _____ forms, that is, ($H^+ + H^-$).

c. Whenever one chemical gives up ($H^+ + H^-$), another chemical must _gain_ those two ionized hydrogens; this type of reaction (the opposite of oxidation) is called

_____. In fact, oxidation reactions must always be

_____ with reduction reactions.

d. In summary, oxidation reactions involve *(gain? loss?)* of hydrogens (or their equiva-

lents) with release of energy; oxidation reactions are _____-gonic, so they result

in a(n) _____-crease of ATP in the body. Reduction reactions involve *(gain? loss?)*

of hydrogens (or equivalents) and are _____-gonic.

D. Inorganic compounds and solutions (pages 38–42)

✓ **D1.** Write I next to phrases that describe inorganic compounds and write O next to
      phrases that describe organic compounds.

_____ a. Contain carbon and hydrogen; held together almost entirely by covalent
              bonds

_____ b. Tend to be very large molecules that serve as good building blocks for
              body structures

_____ c. The class that includes carbohydrates, proteins, fats, and nucleic acids

_____ d. The class that includes water, the most abundant compound in the body,
              and carbon dioxide (CO_2)

✓ **D2.** Fill in the blanks in this exercise about acids, bases, and salts.

a. NaOH (sodium hydroxide) is an example of a(n) _____, that is, a
    chemical that breaks apart into hydroxide ions (OH^-) (or proton acceptors) and one
    or more cations.

b. H_2SO_4 (hydrogen sulfate) is an example of a(n) _____, that is, a

    chemical that dissociates into hydrogen ions (H^+) (or proton _____) and one or
    more anions.

c. Salts, such as NaCl (sodium chloride), are chemicals that dissolve in water to form

    cations and anions, neither of which is _____ or _____.

✓ **D3.** Mixtures are combinations of *(chemically bonded? physically blended?)* elements or

compounds. Name three common liquid mixtures :_____,

_____, and _____. Now take a closer look at
mixtures in this Checkpoint.

a. An oil-based salad dressing is a *(solution? suspension?)* that is formed as the dressing is
    shaken. Upon sitting for a few seconds, the lighter weight oil rises to the top as the
    herbs settle out. Blood is another example of a *(solution? suspension?)* because blood
    cells are likely to settle out of blood plasma if the blood is centrifuged.

b. Coffee, tea, and tears are examples of *(solutions? suspensions?)*; water serves as the
    (solute? solvent?) in which ions and other *(solutes? solvents?)* (such as crystals of coffee
    or tea) are dissolved. Colloids are forms of solutions with very *(large? small?)* particles
    as the solutes. Plasma proteins in blood are examples of colloids.

✓ **D4.** Concentrations of solutions may be expressed in several ways. Describe them in this
      Checkpoint. *(Hint:* Refer to atomic masses for Na and Cl in Figure 2.2, in your text.)

a. A mole per liter (1 mol/liter) means that a number of *(milligrams? grams?)* equal to
    the molecular weight of the chemical should be dissolved in enough solvent to total

    one _____ of solution. Because the atomic masses of Na and Cl are _____ and _____, then *(12.45? 58.44? 1000?)* g of NaCl would be needed to prepare a liter of a "one mole per liter" (1 mol/liter) solution of NaCl.

b.   A one millimole per liter solution (1 mmol/liter) has *(1000 times the? 10 times the? the same? one-thousandth the?)* concentration of a one mole per liter (1 mol/liter) solution.

To prepare a liter of a 1 mmol/liter solution of NaCl, how much NaCl is needed? _____.

✓ **D5.** Do this exercise on roles of water in the body.

a.   Water *(is? is not?)* a common solvent in the human body. This characteristic is due to two factors. One is related to the *(straight? bent?)* shape of water atoms, which provides positive regions (o^+) present on the *(oxygen? hydrogen?)* atoms and negative regions (o^-) present on the *(oxygen? hydrogen?)* atom. The other factor is based upon the *(polar covalent? nonpolar covalent? ionic?)* bonds of water. As a result, water will interact best with chemicals that are hydro-*(philic? phobic?)*, since those solutes also contain polar covalent or ionic bonds.

b.   When added to water, salt is readily dissolved as the Na^+ ions are attracted to the *(positively charged hydrogens? negatively charged oxygens?)* and Cl^- ions are attracted to the *(positively charged hydrogens? negatively charged oxygens?)* of water.

c.   Besides NaCl, name several other chemicals in the human body that are dissolved in the excellent solvent, water.

d.   In organs of the digestive system, water breaks apart large food molecules via the

breakdown process called _____. In the opposite type of reaction, the

hydrogen ion (H^+) and the hydroxide (_____) components of water are removed from two small molecules, which can then bond to form a larger chemical; this

process is known as _____.

e.   Write two or more examples of how water functions as a lubricant in the human body.

✓ **D6.** Do this exercise about how two important properties of water are related to the large number of hydrogen bonds in water.

a.   Water has a *(high? low?)* heat capacity, meaning that water can absorb much heat yet experience a *(large? small?)* increase in temperature. The reason for this phenomenon is that as water absorbs heat, some of the energy is used to break _____ bonds so less energy is left over to increase water temperature. Since the composition of the human body is normally more than 50% water, this large amount of water *(enhances? lessens?)* the impact of high environmental temperatures. Also, evaporation of water from skin permits the body to release heat

by the process of _____.

b.   Each water molecule can link to *(2? 3? 4?)* neighboring water molecules by

_____ bonds *(Hint:* Refer to Figure 2.7 in your text). Consequently, water

molecules exhibit the property of _____, defined as the tendency for similar particles to stay together. Such cohesion creates a *(high? low?)* surface tension. Each time you breathe, such high surface tension of water molecules lining air sacs (alveoli) of lungs, _____-creases the work required to breathe. As you will discover in

Chapter 23, surface tension-lowering agents (or surfactants) help to _____-crease the work required for breathing.

D7. Draw a diagram showing the pH scale. Label the scale from 0 to 14. Indicate by arrows increasing acidity (H^+ concentration) and increasing alkalinity (OH^- concentration).

✓ **D8.** Choose the correct answers regarding pH.
  a. Which pH is most acid?
      A. 4   B. 7   C. 10
  b. Which pH has the highest concentration of OH^- ions?
      A. 4   B. 7   C. 10
  c. Which solution has pH closest to neutral?
      A. Gastric juice (digestive juices of the stomach)
      B. Blood
      C. Lemon juice
      D. Milk of magnesia
  d. A solution with pH 8 has 10 times *(more? fewer?)* H^+ ions than a solution with pH 7.
  e. A solution with pH 5 has *(10? 20? 100?)* times *(more? fewer?)* H^+ ions than a solution with pH 7.

D9. Write a sentence or two explaining each of the following:
  a. How pH is related to homeostasis

  b. How buffers help to maintain homeostasis

E. Organic compounds (pages 42–55)

E1. More than 50% of the human body is composed of the *(organic? inorganic?)*

compound named _____. Organic compounds all contain carbon (C) and hydrogen (H). State two reasons why carbon is an ideal element to serve as the primary structural component for living systems.

✓ **E2** Carbohydrates carry out important functions in your body.
  a. State the principal function of carbohydrates.

b.   State two secondary roles of carbohydrates.

✓ **E3.** Complete these statements about carbohydrates.

a.   The ratio of hydrogen to oxygen in all carbohydrates is _____:_____. The general

formula for carbohydrates is _____.

b.   The chemical formula for a *hexose* is _____. Two common hexoses

are _____ and _____.

c.   One common pentose sugar found in DNA is _____. Its name
and its formula ($C_5H_{10}O_4$) indicate that it lacks one oxygen atom from the typical

pentose formula, which is _____.

d.   When two glucose molecules ($C_6H_{12}O_6$) combine, a disaccharide is formed. Its

formula is _____, indicating that in a synthesis reaction such as
this, a molecule of water is *(added? removed?)*. Refer to Figure LG 2.2. Identify which
reaction *(1? 2?)* on that figure demonstrates synthesis.

e.   Continued dehydration synthesis leads to enormous carbohydrates called

_____. One such carbohydrate is _____.

f.   When a disaccharide such as sucrose is broken, water is introduced to split the bond
linking the two monosaccharides. Such a decomposition reaction is termed

_____ which literally means "splitting using water." (See Figure
2.16 in your text.) On Figure LG 2.2, hydrolysis is shown by reaction *(1? 2?)*.

Figure LG 2.2 Dehydration synthesis and hydrolysis reactions.
Refer to Checkpoint E3.

✓ **E4.** Contrast carbohydrates with lipids in this exercise. Write C next to any statement
true of carbohydrates and write L next to any statement true of lipids.

_____ a.   These compounds are insoluble in water (hydrophobic). (*Hint:* Water alone is
ineffective for washing such compounds out of clothes.)

_____ b. These compounds are organic.

_____ c. These substances have a hydrogen to oxygen ratio of about 2:1.

_____ d. Very few oxygen atoms (compared to the numbers of carbon and hydrogen atoms) are contained in these compounds. (Look carefully at Figure 2.17 in your text.)

✓ **E5.** Circle the answer that correctly completes each statement.
  a. A triglyceride consists of:
      A. Three glucoses bonded together
      B. Three fatty acids bonded to a glycerol
  b. A typical fatty acid contains (see Figure 2.17 in your text):
      A. About 3 carbons
      B. About 16 carbons
  c. A polyunsaturated fat contains _____ than a saturated fat.
      A. More hydrogen atoms
      B. Fewer hydrogen atoms
  d. Saturated fats are more likely to be derived from:
      A. Plants, for example, corn oil or peanut oil
      B. Animals, as in beef or cheese
  e. Oils that are classified as saturated fats include:
      A. Tropical oils such as coconut oil and palm oil
      B. Vegetable oils such as safflower oil, sunflower oil, and corn oil
  f. A diet designed to reduce the risk for atherosclerosis is a diet low in:
      A. Saturated fats
      B. Polyunsaturated fats

✓ **E6.** Match the types of lipids listed in the box with their functions given below.

B.	**Bile salts**	**E.**	**Estrogens**
Car.	**Carotene**	**Pho.**	**Phospholipids**
Cho.	**Cholesterol**	**Pro.**	**Prostaglandins**

_____ a. Present in egg yolk and carrots; leads to formation of vitamin A, a chemical necessary for good vision

_____ b. The major lipid in cell membranes

_____ c. Present in all cells, but infamous for its relationship to "hardening of the arteries"

_____ d. Substances that break up (emulsify) fats before their digestion

_____ e. Steroid sex hormones produced in large quantities by females

_____ f. Important regulatory compounds with many functions, including modifying responses to hormones and regulating body temperature

✓ **E7.** Complete this exercise about fat-soluble vitamins.
  a. Name four fat-soluble vitamins (consult Table 2.7 in your text, for help):

      _____ _____ _____ _____

  b. Because these four vitamins are absorbed into the body along with fats, any problem with fat absorption is likely to lead to:

  c. The fat-soluble vitamin that helps with calcium absorption and so is necessary for

      normal bone growth is vitamin _____.

d. Vitamin _____ may be administered to a patient before surgery in order to prevent excessive bleeding because this vitamin helps in the formation of clotting factors.

e. Normal vision is associated with an adequate amount of vitamin _____.

f. Vitamin _____ has a variety of possible functions including the promotion of wound healing and prevention of scarring.

✓ **E8.** Match the types of proteins listed in the box with their functions given below.

Cat.	**Catalytic**	**R.**	**Regulatory**
Con.	**Contractile**	**S.**	**Structural**
I.	**Immunological**	**T.**	**Transport**

_____ a. Hemoglobin in red blood cells

_____ b. Proteins in muscle tissue, such as actin and myosin

_____ c. Keratin in skin and hair; collagen in bone

_____ d. Hormones such as insulin

_____ e. Defensive chemicals such as antibodies and interleukins

_____ f. Enzymes such as lipase, which digests lipids

✓ **E9.** Contrast proteins with the other organic compounds you have studied so far by completing this exercise.

a. Carbohydrates, lipids, and proteins all contain carbon, hydrogen, and oxygen. A

fourth element, _____, makes up a substantial portion of proteins.

b. Just as large carbohydrates are composed of repeating units (the _____),

proteins are composed of building blocks called _____.

c. As in the synthesis of carbohydrates or fats, when two amino acids bond together, a water molecule must be *(added? removed?)*. This is another example of *(hydrolysis? dehydration?)* synthesis.

d. The product of such a reaction is called a _____-peptide. (See Figure 2.21 in your

text.) When many amino acids are linked together in this way, a _____-peptide

results. One or more polypeptide chains form a _____.

e. Refer to Figure 2.20 of your text. Which part of the molecular structure varies among the 20 different kinds of amino acids to result in their unique identities? *(Amino group? Side chain = R group? Carboxyl group?)*

E10. There are _____ different kinds of amino acids forming the human body, much like a 20-letter alphabet that forms protein "words." It is the specific sequence of amino acids that determines the nature of the protein formed. In order to see the significance of this fact, do the following activity. Arrange the five letters listed below in several ways so that you form different five-letter words. Use all five letters in each word.

e i l m s

___ ___ ___ ___ ___   ___ ___ ___ ___ ___

___ ___ ___ ___ ___   ___ ___ ___ ___ ___

Note that although all of your words contain the same letters, the sequence of letters differs in each case. The resulting words have different meanings, or functions. Such is the case with proteins, too, except that protein "words" may be thousands of "letters" (amino acids) long, involving all or most of the 20-amino acid "alphabet." Proteins must be synthesized with accuracy. Think of the drastic consequences of omission or misplacement of even a single amino acid. (Think how different are *limes, miles, smile, and slime!*)

✓ **E11.** Match the description below with the four levels of protein organization in the box.

P. **Primary structure**	**S.** **Secondary structure**
Q. **Quaternary structure**	**T.** **Tertiary structure**

_____ a. Specific sequence of amino acids

_____ b. Twisted and folded arrangement (as in spirals or pleated sheets) due to hydrogen bonding

_____ c. Irregular three-dimensional shape; disulfide bonds play a role in this level of structure

_____ d. Relative arrangement of two or more polypeptide chains

✓ **E12.** Describe changes in protein structure by completing this activity.
  a. Replacement of one amino acid (valine) with another (glutamate) in the protein portion of hemoglobin results in a condition known as _____.

  b. Cooking an egg white results in destruction (or _____) of the protein albumin. Extreme increases in body temperature, for example, increases caused by _____, can denature body proteins such as those in skin or blood plasma.

✓ **E13.** Do this exercise about roles of enzymes related to chemical reactions.
  a. Explain why it is necessary for chemical reactions to take place rapidly within the human body.

  b. Increase in body temperature will affect rates of chemical reactions. Discuss pros and cons of increasing body temperature.

  c. Explain how enzymes increase chemical reaction rates without requiring a raised body temperature.

  d. Because enzymes speed up chemical reactions (yet are not altered themselves), they are known as _____.

e. Chemically, all enzymes consist mostly of *(carbohydrates? lipids? proteins?)*. Each

   enzyme reacts with a specific molecule called a _____.

f. The substrate interacts with a specific region on the enzyme called the

   _____, producing an intermediate known as an

   _____ – _____ complex. Very quickly the
(enzyme? substrate?) is transformed in some way (for example, is broken down or
transferred to another substrate), while the *(enzyme? substrate?)* is recycled for

   further use. A _____ number of 1 to 600,000 molecules of product
per second may be expected for each enzyme molecule.

g. Whole enzymes are sometimes known as *(apo? holo?)*-enzymes. These consist of a
protein called the *(apo? holo?)*-enzyme, as well as a nonprotein cofactor.

h. Cofactors may be of two types. Some are metal ions, such as _____.

   So, for example, a protein (apoenzyme) plus calcium ion (Ca^{2+}) may serve as a

   holoenzyme. The other type of cofactor is known as a(n) _____-enzyme. Most
coenzymes are derived from *(vitamins? minerals?)*.

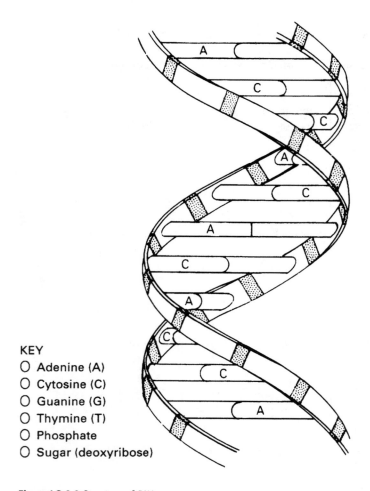

KEY
- ◯ Adenine (A)
- ◯ Cytosine (C)
- ◯ Guanine (G)
- ◯ Thymine (T)
- ◯ Phosphate
- ◯ Sugar (deoxyribose)

Figure LG 2.3 Structure of DNA.
Complete as directed in Checkpoint E14.

✓ **E14.** Refer to Figure LG 2.3 and fill in the blanks in this exercise about DNA.

a. DNA has a structure much like a ladder, but one that is arranged in a spiral; this

arrangement is called a double _____. The sides of the ladder are formed of *(bases? alternating sugars and phosphates?)*. Color these according to the color code ovals on the figure.

b. "Rungs" of this DNA ladder are formed of _____. There are *(2? 4? 8? 20?)* different bases in DNA; their sequence "writes" the genetic code. Note the dire consequences of omission or substitution in this "four-letter" alphabet.

c. Bases in DNA are paired _____ (A) to _____

(_____) and _____ (C) to _____ (_____). Label complementary nucleotides on the right side of the DNA in the figure. Then color all nucleotides in the figure according to the color code ovals.

d. A DNA _____ is one monomer of DNA, and it consists of three

parts: _____ (a sugar), a base, and a _____.
Circle one nucleotide in the figure.

e. Describe the structure and function of a gene.

E15. How does a strand of RNA differ from DNA?

✓ **E16.** Describe the structure and significance of ATP by completing these statements.

a. ATP stands for adenosine triphosphate. Adenosine consists of a base that is a

component of DNA, that is, _____, along with the five-carbon sugar

named _____.

b. The "TP" of ATP stands for _____. The final *(one? two? three?)* phos-phates are bonded to the molecule by high energy bonds.

c. When the terminal phosphate is broken, a great deal of energy is released as ATP is

split into _____ + _____.

d. ATP, the body's primary energy-storing molecule, is constantly reformed by the reverse of this reaction as energy is made available from foods you eat. Write this reversible reaction.

e. The anaerobic phase of cellular respiration yields _____ molecules of ATP, whereas

the aerobic phase yields _____ molecules of ATP.

ANSWERS TO SELECTED CHECKPOINTS: CHAPTER 2

A2. (a) Elements. (b) Oxygen, carbon, H, nitrogen (N). (c) Ca, Cl, Fe, I, K, Mg, Na, P, S. (d) Thirteen elements found in very small amounts, together comprising only 0.1% of the total body mass; Co, Cu, Mn, Si, Sn, Zn.

A3. (b) N. (c) I. (d) Fe. (e) O. (f) Ca. (g) P. (h) Cl. (i) Na. (j) F. (k) K.

A5. (a) Nucleus, electrons. (b) Protons (p^+), neutrons (n^0). (c) Negative; equal to neutral; zero. (d) Shells; 2; 8, 18. (e) Iodine, 53, 53; 2, 8, 18, 18, 7.

A6. (a) 19; 19, 19. (b) 39.10 (rounded off to 39).

A7. (a) 20 (39–19); neutrons. *Note:* Varied masses of isotopes account for the uneven mass number, such as 39.10 for potassium. (b) 8. (c) Radiation half-life. (d) Helpful: radioisotopes are used for medical imaging (as for PET and DSA, discussed in Chapter 1 of your text) and to treat cancer (such as cobalt treatments); harmful: radioisotopes can alter cells and lead to cancer

A8. (a) $C_6H_{12}O_6$ (glucose), H_2O, O_2; O_2-(superoxide). (b) $C_6H_{12}O_6$ (glucose), H_2O. (c) O_2- (superoxide). (d) Cl^-. (e) Na^+.

B1. (a) H and He; 2. (b) 2 or 3; Na: 1, Cl: 7, C: 4; 8. (c) Valence. (d) Extra; donor; positive; cation. (e) 7, missing; accepts, anion. (f) Gained or lost; lost; anion; electrolytes (because) they can conduct an electrical charge. (g) Octet. (h) Do not, inert; He, Ne with atomic number of 10 (since Ne has 2 electrons in the inner shell and 8 electrons in the outer or valence shell).

B2. (a) I. (b) H. (c) C. (d) H. (e-g) C.

B3. (a, b: See Figure LG 2.1A) (c)

A. Hydrogen (1) B. Oxygen (8) C. Nitrogen (7) D. Carbon (6)

Figure LG 2.1A Atomic structure of four common atoms in the human body.

B4. (a) B. (b) See Figure at right: Atomic structure of ammonia (NH_3)

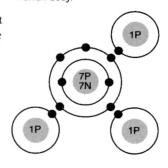

C1. (a) Water; mass. (b) Metabolism. (c) Work; chemical; potential, kinetic. (d) Exergonic; exergonic; ATP; ender-.

C3. (a) R. (b) S. (c) E. (d) D. (e) D.

C4. Oxidation. (a) Loss. (b) Lactic: 3 C, 6 H, 3 O; pyruvic: 3 C, 4 H, 3 O. (c) Lactic; 2; oxidized.

C5. (a) One, one; A. (b) Ionized. (c) Reduction; coupled. (d) Loss, exer-, in; gain, ender.

D1. (a–c) O. (d) I.

D2. (a) Base. (b) Acid, donors. (c) H^+ or OH^-.

D3. Physically blended; solutions, colloids, and suspensions. (a) Suspension; suspension. (b) Solutions, solvent, solutes; large.

D4. (a) Grams, liter; 22.99, 35.45, 58.44. (b) One thousandth the; 58.44 mg = 0.05844 g (or 0.058 g).

D5. (a) Is; bent, hydrogen, oxygen; polar covalent; philic. (b) Negatively charged oxygen, positively charged hydrogens. (c) Some of the O_2 and CO_2 gas molecules in blood, wastes such as urea, and small sugars such as glucose. (d) Hydrolysis; OH, synthesis. (e) Water in mucus permits foods to slide along the digestive tract and wastes to be flushed out in the urinary tract; water in serous fluid and synovial fluids decreases friction.

D6. (a) High, small; hydrogen; lessens; sweating or perspiration. (b) 4, hydrogen; cohesion; high; in; de.

D8. (a) A. (b) C. (c) B. (d) Fewer. (e) 100, more.

E2. (a) Provide a readily available source of energy. (b) Converted to other substances; serve as food reserves.

E3. (a) 2:1; $(CH_2O)n$. (b) $C_6H_{12}O_6$ glucose, fructose, galactose. (c) Deoxyribose; $C_5H_{10}O_5$. (d) $C_{12}H_{22}O_{11}$, removed; 1. (e) Polysaccharides; starch, glycogen, cellulose. (f) Hydrolysis; 2.

E4. (a) L. (b) C, L. (c) C. (d) L.

E5. (a) B. (b) B. (c) B. (d) B. (e) A. (f) A.

E6. (a) Car. (b) Pho. (c) Cho. (d) B. (e) E. (f) Pro.

E7. (a) A, D, E, K. (b) Symptoms of deficiencies of fat-soluble vitamins. (c) D. (d) K. (e) A. (f) E.

E8. (a) T. (b) Con. (c) S. (d) R. (e) I. (f) Cat.

E9. (a) Nitrogen. (b) Monosaccharides, amino acids. (c) Removed; dehydration. (d) Di; poly; protein. (e) Side chain = R group.

E11 (a) P. (b) S. (c) T. (d) Q.

E12. (a) Sickle cell anemia. (b) Denaturation; burns or fever (hyperthermia).

E13.(a) Thousands of different types of chemical reactions are necessary daily to maintain homeostasis. (b) Chemical reactions occur more rapidly at higher temperature (one reason for increased catabolism and weight loss when you have a fever). However, very high temperatures will destroy body tissues. (c) Enzymes orient molecules so that they are more likely to react, lower activation energy needed for reactions, and increase frequency of

collisions. (d) Catalysts. (e) Proteins; substrate.
(f) Active site, enzyme-substrate; substrate, enzyme;
turnover. (g) Holo; apo. (h) Magnesium (Mg_2^+),
zinc (Zn^+), or calcium (Ca_2^+); co; vitamins.

E14.(a) Helix; alternating sugars and phosphates; see
Figure LG 2.2A. (b) Bases; 4. (c) Adenine (A),
thymine (T), cytosine (C), guanine (G); see Figure
LG 2.2A. (d) Nucleotide, deoxyribose, phosphate;
see Figure LG 2.2A. (e) Segment of DNA that
determines one trait or regulates one activity.

E16.(a) Adenine, ribose. (b) Triphosphate; two. (c) ADP
+ phosphate (P). (d) ATP $\rightleftharpoons$ ADP + P +
energy. (e) 2, 36–38.

Figure LG 2.3A Structure of
DNA. P, phosphate; S, sugar
(deoxyribose); A, adenine; T,
thymine; G, guanine; C, cytosine.

MORE CRITICAL THINKING: CHAPTER 2

1. Discuss the roles of the following in human anabolism and catabolism: (a) exergonic and
   endergonic reactions; (b) synthesis, decomposition, exchange, and reversible reactions.
2. How does an enzyme enhance the rate of chemical reactions in the human body? Exactly
   how does this "biological catalyst" do its job? Is the enzyme permanently altered during
   the chemical reaction?
3. Contrast the roles of carbohydrates and proteins in the human body.

MASTERY TEST: CHAPTER 2

*Questions 1–12: Circle the letter preceding the one best
answer to each question.*

1. All of the following statements are true *except*:
   A. A reaction in which two amino acids join to
      form a dipeptide is called a dehydration
      synthesis reaction.
   B. A reaction in which a disaccharide is digested
      to form two monosaccharides is known as a
      hydrolysis reaction.
   C. About 65 to 75% of living matter consists of
      organic matter.
   D. Strong acids ionize more easily than weak
      acids do.

2. Choose the one *true* statement.
   A. A pH of 7.5 is more acidic than a pH of 6.5.
   B. Anabolism consists of a variety of
      decomposition reactions.
   C. An atom such as chlorine (Cl), with seven elec-
      trons in its outer orbit, is likely to be an
      electron donor (rather than electron receptor)
      in ionic bond formation.
   D. Polyunsaturated fats are more likely to reduce
      cholesterol level than saturated fats are.

3. In the formation of the ionically bonded salt NaCl,
   Na^+ has:
   A. Gained an electron from Cl^-
   B. Lost an electron to Cl^-
   C. Shared an electron with Cl
   D. Formed an isotope of Na^+

4. Over 99% of living cells consist of just six elements.
   Choose the element that is NOT one of these six.
   A. Calcium   E. Nitrogen
   B. Hydrogen   F. Oxygen
   C. Carbon   G. Phosphorus
   D. Iodine

5. Which of the following describes the structure of a
   nucleotide?
   A. Base-base   D. Dipeptide
   B. Phosphate-sugar-base   E. Adenine-ribose
   C. Enzyme

6. Which of the following groups of chemicals
   includes only polysaccharides?
   A. Glycogen, starch
   B. Glycogen, glucose, galactose
   C. Glucose, fructose
   D. RNA, DNA
   E. Sucrose, polypeptide

7. $C_6H_{12}O_6$ is most likely the chemical formula for:
   A. Amino acid   D. Polysaccharide
   B. Fatty acid   E. Ribose
   C. Hexose

8. All of the following answers consist of correctly
   paired terms and descriptions related to enzymes
   except:
   A. Active site—a place on an enzyme that fits the
      substrate
   B. Substrate—molecule(s) upon which the
      enzyme acts
   C. -ase—ending of most enzyme names
   D. Holoenzyme—another name for a cofactor

9. Which of the following substances are used mainly for structure and regulatory functions and are not normally used as energy sources?
   A. Lipids   C. Carbohydrates
   B. Proteins

10. Which is a component of DNA but not of RNA?
    A. Adenine   D. Ribose
    B. Phosphate   E. Thymine
    C. Guanine

11. All of the following answers consist of pairs of proteins and their correct functions except:
    A. Contractile—actin and myosin
    B. Immunological—collagen of connective tissue
    C. Catalytic—enzymes
    D. Transport—hemoglobin

12. All of the following answers consist of pairs of elements and their correct chemical symbols except:
    A. Nitrogen (N)   D. Calcium (Ca)
    B. Sodium (Na)   E. Potassium (P)
    C. Carbon (C)

Questions 13–20: Circle T (true) or F (false). If the statement is false, change the underlined word or phrase so that the statement is correct.

T F   13. Oxygen, water, NaCl, and glucose are inorganic compounds.
T F   14. There are about four different kinds of amino acids found in human proteins.
T F   15. The number of protons always equals the number of neutrons in an atom.
T F   16. K+ and Cl⁻ are both cations.
T F   17. Oxygen can form three bonds since it requires three electrons to fill its outer electron shell.
T F   18. Carbohydrates constitute about 18 to 25% of the total body weight.
T F   19. ATP contains more energy than ADP.
T F   20. Prostaglandins, ATP, steroids, and fats are all classified as lipids.

Questions 21–25: Fill-ins. Identify each organic compound below. Write the names of the compounds on the lines provided.

_____ 21.

_____ 22.

_____ 23.

_____ 24.

_____ 25.

```
        H        H  H  H  H  H  H  H  H  H  H  H  H  H  H  H
        |        |  |  |  |  |  |  |  |  |  |  |  |  |  |  |  |
    H－C－O－C－C－C－C－C－C－C－C－C－C－C－C－C－C－C－C－H
        |        ‖  |  |  |  |  |  |  |  |  |  |  |  |  |  |  |
        |        O  H  H  H  H  H  H  H  H  H  H  H  H  H  H  H

                 H  H  H  H  H  H  H  H  H  H  H  H  H  H  H
                 |  |  |  |  |  |  |  |  |  |  |  |  |  |  |  |
    H－C－O－C－C－C－C－C－C－C－C－C－C－C－C－C－C－C－C－H
        |        ‖  |  |  |  |  |  |  |  |  |  |  |  |  |  |  |
        |        O  H  H  H  H  H  H  H  H  H  H  H  H  H  H  H

                 H  H  H  H  H  H  H  H  H  H  H  H  H  H  H
                 |  |  |  |  |  |  |  |  |  |  |  |  |  |  |  |
    H－C－O－C－C－C－C－C－C－C＝C－C－C－C＝C－C－C－C－C－H
        |        ‖  |  |  |  |  |  |        |  |        |  |  |  |
        H        O  H  H  H  H  H        H  H        H  H  H  H
```

ANSWERS TO MASTERY TEST: CHAPTER 2

Multiple Choice
1. C
2. D
3. B
4. D
5. B
6. A
7. C
8. D
9. B
10. E
11. B
12. E

True–False
13. F. Water and NaCl (Oxygen is not a compound; glucose is organic.)
14. F. 20
15. F. Electrons
16. F. K^+ is a cation.
17. F. Two bonds because it requires two electrons
18. F. 2–3%
19. T.
20. F. Prostaglandins, steroids, and fats

Fill-ins
21. Amino acid
22. Monosaccharide or hexose or glucose
23. Polysaccharide or starch or glycogen
24. Adenosine triphosphate (ATP)
25. Fat or lipid or triglyceride (unsaturated)

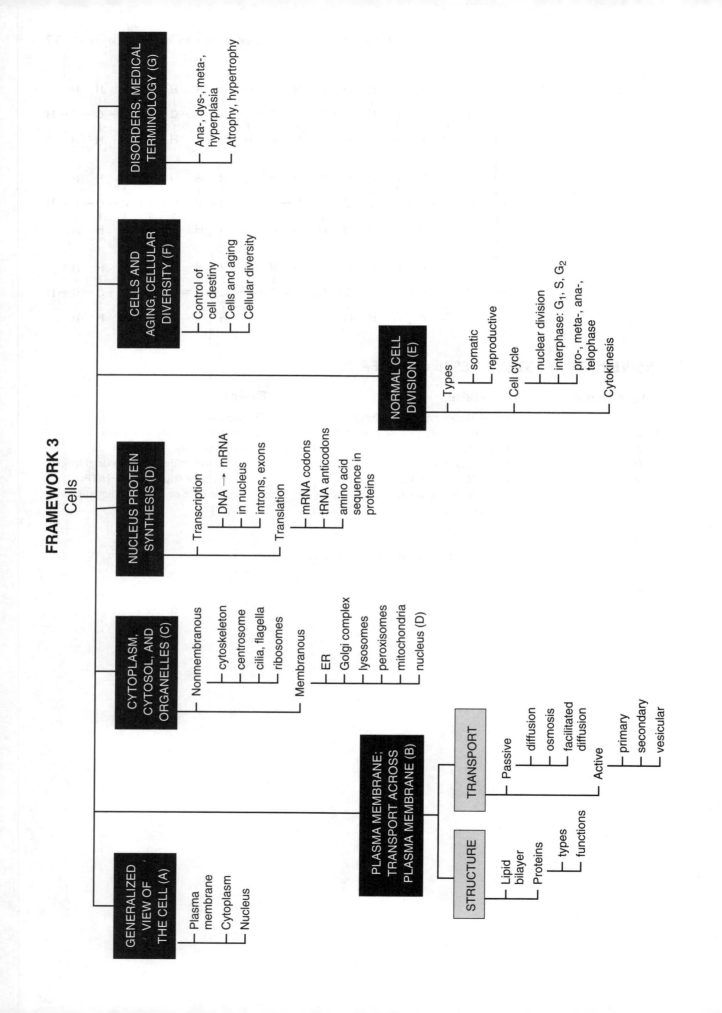

C H A P T E R **3**

The Cellular Level of Organization

Cells are the basic structural units of the human body, much as buildings comprise the cities of the world. Every cell possesses certain characteristics similar to those in all other cells of the body. For example, membranes surround and compartmentalize cells, just as walls support virtually all buildings. Yet individual cell types exhibit variations that make them uniquely designed to meet specific body needs. In an architectural tour, the structural components of a cathedral or mansion, lighthouse or fort, can be recognized as specific for activities of each type of building. So, too, the number and type of organelles differ in muscle or nerve, cartilage or bone cells. Cells carry out significant functions such as movement of substances into, out of, and throughout the cell; synthesis of the chemicals that cells need (like proteins, lipids, and carbohydrates); and cell division for growth, maintenance, repair, and formation of offspring. Just as buildings change with time and destructive forces, cells exhibit aging changes. In fact, sometimes their existence is haphazard and harmful, as in cancer.

As you begin your study of cells, carefully examine the Chapter 3 Topic Outline and Objectives and note relationships among concepts and key terms in the Framework.

TOPIC OUTLINE AND OBJECTIVES

A. A generalized view of the cell

1. Name and describe the three principal parts of a cell.

B. The plasma membrane; transport across the plasma membrane

2. Describe the structure and functions of the plasma membrane.
3. Describe the processes that transport substances across the plasma membrane.

C. Cytoplasm, cytosol, and organelles

4. Describe the structure and function of cytoplasm, cytosol, and organelles.

D. Nucleus; protein synthesis

5. Describe the structure and function of the nucleus.
6. Describe the sequence of events involved in protein synthesis.

E. Normal cell division

7. Discuss the stages, events, and significance of somatic cell division.
8. Discuss the signals that induce cell division, and describe apoptosis.

F. Cells and aging, cellular diversity

9. Explain the relationship of aging to cellular processes.

G. Disorders, medical terminology

WORDBYTES

Now become familiar with the language of this chapter by studying each wordbyte, its meaning, and an example of its use within a term. After you study the entire list, self-check your understanding by writing the meaning of each wordbyte on the line. As you continue through the *Learning Guide*, identify (and fill in) additional terms that contain the same wordbyte.

Wordbyte	Self-check	Meaning	Example(s)
a-	_____	without	*a*trophy
auto-	_____	self	*auto*phagy
cyto-	_____	cell	*cyto*logist
-elle	_____	small	organ*elle*
homo-	_____	same	*homo*logous
hydro-	_____	water	*hydro*static
hyper-	_____	above	*hyper*tonic
hypo-	_____	below	*hypo*tonic
iso-	_____	equal	*iso*tonic
lyso-	_____	dissolving	*lyso*some
meta-	_____	beyond	*meta*stasis
neo-	_____	new	*neo*plasm
-oma	_____	tumor	carcin*oma*
-phagy	_____	eating	auto*phago*some
-philic	_____	loving	hydro*philic*
-phobic	_____	fearing	hydro*phobic*
-plasia, -plasm	_____	growth	dys*plasia*
reticulum	_____	network	endoplasmic *reticulum*
-some	_____	body	lyso*some*
-stasis, -static	_____	stand, stay	meta*stasis*
-tonic	_____	pressure	hyper*tonic*
-trophy	_____	nourish	hyper*trophy*

CHECKPOINTS

A. A generalized view of the cell (pages 60–61)

A1. Write principal functions next to these parts of a cell:

a. Plasma membrane (two functions)

b. Cytosol

c. Organelles

d. Nucleus

A2. Contrast terms in each pair:

a. *Cytology* with *cell physiology.*

b. *Cytoplasm* with *cytosol.*

B. The plasma membrane: transport across the plasma membrane (pages 61–75)

✓ **B1.** Refer to Figure LG 3.1 to complete the following exercise about the fluid mosaic model of the plasma membrane. First color each part of the membrane listed below (a-e) with a different color; then use the same colors to fill in the color code ovals next to the letters below. Next label these parts on the figure. Now use the lines provided next to the name of each membrane part to write one or more functions for each part.

○ a. Phospholipid polar "head" _____

○ b. Phospholipid nonpolar "tail" _____

○ c. Integral protein _____

○ d. Glycoprotein _____

○ e. Cholesterol _____

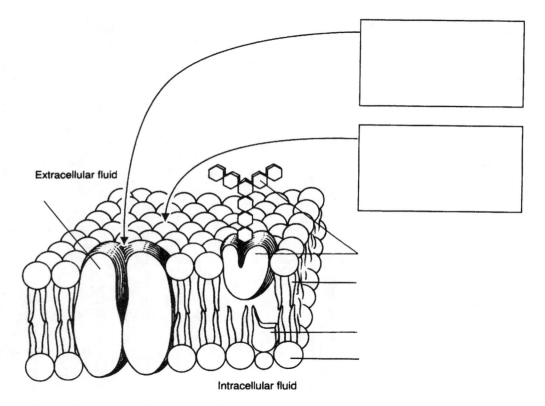

Figure LG 3.1 Plasma membrane, according to the fluid mosaic model. Color as directed in Checkpoint B1. Fill in answers in boxes according to Checkpoint B4.

✓ **B2.** *For extra review.* Complete this exercise about the plasma membrane.

   a. Plasma membranes are made up of about _____% proteins and _____% lipids by weight. About 20% of the lipids are *(cholesterol? glycolipids? phospholipids?)*.

   b. About 75% of the lipids are *(cholesterol? glycolipids? phospholipids?)*. Because phospho-

      lipids have polar and nonpolar parts, they are said to be _____. The polar part of a phospholipid molecule faces toward the *(interior? surface?)* of the membrane. The polar parts are hydro-*(philic? phobic?)*, meaning water-*(loving? fearing?)*.

   c. Glycolipids appear only in the layer that faces *(extracellular? intracellular?)* fluid. This

      location is logical because these lipids may be important in _____.

d. Which type of membrane molecule is larger? *(Lipid? Protein?)*. In fact, related to this size difference, about one protein molecule is present for each *(2? 5? 50?)* lipid molecules. In different types of membranes, there is more variety in *(lipids? proteins?)*.

Some integral proteins serve as receptors for chemicals such as _____

or _____. Other membrane proteins known as _____ form catalysts for chemical reactions. The fact that your body can accept only certain types of blood\transfusions is related to proteins called cell-identity

_____. And finally, proteins called _____ help cells adhere to each other.

B3. *Critical thinking.* Discuss membrane fluidity in this exercise.
a. What structural features of cellular membranes account for fluidity?

b. What advantages are provided by membrane fluidity?

c. What factor accounts for the fact that the two halves of the bilayer seldom "flip-flop"?

d. What advantages are provided by the lack of "flip-flopping" of the two bilayers?

e. How does excessive cholesterol in the body lead to "hardening of the arteries"?

✓ **B4.** Complete this activity about selective permeability of membranes.
a. Plasma membranes are *(quite? not very?)* permeable to nonpolar chemicals such as O_2 and CO_2 because these molecules dissolve readily in the *(protein? phospholipid?)* portion of the membrane.
b. Charged chemicals (such as Na^+ K^+, or Ca^{2+}) and most polar chemicals pass through the *(protein? phospholipid?)* portions of the membrane. Two such proteins are

_____ and _____.
c. Refer again to Figure LG 3.1. Write the names of the following chemicals in the correct boxes on the figure to differentiate chemicals that pass through the phospholipid bilayer from those that pass through protein channels:
(Hint: One of these types of chemicals will not normally pass across either part of healthy plasma membranes. Consult p. 69 of the text for information about transport of Ca^{++}, Cl^-, vitamins A and D, and ammonia [NH_3].)

Ca++   Cl-   K+   Na+ H_2O protein

O_2 CO_2 vitamins A and D   ammonia (NH3)

B5. Complete this Checkpoint about how differences (or gradients) are maintained across plasma membranes.

a. Refer to Figure LG 3.2. Write inside the plasma membrane the symbols for the major chemicals concentrated in intracellular fluid (ICF). Write outside the plasma membrane the symbols for major ions or electrolytes concentrated in extracellular fluid (ECF). Use the following symbols. (*Hint:* proteins are made inside of cells and cannot easily pass through the plasma membrane. It may help to refer to Figure 27.6, page 962 in your text.)

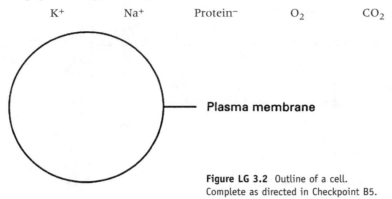

Figure LG 3.2 Outline of a cell.
Complete as directed in Checkpoint B5.

b. Chemicals tend to move from where they are more concentrated to where they are less concentrated, in other words, *(up? down?)* the concentration gradient.

c. The plasma membrane normally exhibits a difference between the number of positively and negatively charged ions on either side of the membrane. This is known

as a(n) *(chemical? electrical?)* gradient or a membrane _____.
Which side of the membrane normally has a more negative charge *(Inside? Outside?)*

d. Combined, these two gradients across plasma membranes are known as the

_____ gradient.

B6. *Critical thinking.* Explain why substances must move across your plasma membranes in order for you to maintain homeostasis and survive. (What types of chemicals must be allowed into cells or kept out of cells?)

✓ **B7.** Contrast different categories of transport by circling the correct answers next to each description below.

a. A transporter protein is not used. *(Mediated? Nonmediated?)*

b. A chemical is moved "down" its concentration gradient, a process that requires only the kinetic (or movement-related) energy of the chemicals. *(Active? Passive?)*

c. A chemical is moved "uphill," against its concentration gradient, a process that requires cellular energy. *(Active? Passive?)*

d. Transport utilizing a transporter of a single substance. *(Uniporter–mediated? Coupled transporter–mediated?)*

e. Transport utilizing a transporter of two substances in opposite directions. *(Antiporter-mediated? Symporter-mediated?)*

✓ **B8.** *Clinical correlation.* Circle which person in each pair of hospitalized clients is likely to have a *lower* rate of diffusion across plasma membranes, based upon different factors that influence diffusion rate.

   a. Factor: surface area for diffusion of oxygen from lung alveoli into blood vessels there.
      A. Mr. O'Loughlin: has had two-thirds of a lung removed due to lung cancer.
      B. Ms. Yesberger: about the same size as Mr. O'Loughlin, but has both lungs intact.

   b. Factor: thickness of membrane in lungs for diffusion of oxygen from lung alveoli into blood vessels there.
      A. Mr. Claeys: has lungs free of infection.
      B. Ms. Clemen: due to pneumonia, respiratory membranes of lungs are thickened by fluid.

   c. Factor: concentration gradient of oxygen from blood into brain cells.
      A. Mrs. Thielens has a normal level of oxygen (Po_2 = 96 mm Hg) in her blood.
      B. Ms. Taylor has a low level of oxygen (Po_2 = 66 mm Hg) in her blood since she has a serious case of the flu.

   d. Factor: body temperature affecting rate of diffusion of oxygen and nutrients from blood into muscle cells of the right thigh.
      A. Six-year-old Amy: 98.6°F
      B. Six-year-old Paula: 100.6°F

   e. *For extra review.* Of the four factors listed above (a–d), three are *directly* related to

   diffusion rate. Which one is *inversely* related to diffusion rate? _____

✓ **B9.** Select from the following list of terms to identify passive transport processes described below.

FD. Facilitated diffusion	O. Osmosis	D. Diffusion

_____ a. Net movement of any substance (such as cocoa powder in hot milk) from region of higher concentration to region of lower concentration; membrane not required

_____ b. Same as (a) except movement across a semipermeable membrane with help of a carrier; ATP not required

_____ c. Net movement of water from region of high water concentration (such as 2% NaCl) to region of low water concentration (such as 10% NaCl) across semipermeable membrane; important in maintenance of normal cell size and shape

✓ **B10.** Complete the following exercise about osmosis in blood.

   a. Human red blood cells (RBCs) contain intracellular fluid that is osmotically similar to _____ NaCl, which is known as normal saline.
      A. 2.0%   B. 0.90%   C. 0% (pure water)

   b. A solution that is hypertonic to RBCs contains *(more? fewer?)* solute particles and *(more? fewer?)* water molecules than blood.

   c. Which of these solutions is hypertonic to RBCs?
      A. 2.0% NaCl   B. 0.90% NaCl

   d. If RBCs are surrounded by hypertonic solution, water will tend to move *(into? out of?)* them, so they will *(crenate? hemolyze?)*.

   e. A solution that is _____ -tonic to RBCs will maintain the shape and

   size of the RBC. An example of such a solution is _____

   f. Which solution will cause RBCs to hemolyze?
      A. 2.0% NaCl   B. 0.90% NaCl   C. Pure water

   g. Which solution has the highest osmotic pressure?
      A. 2.0% NaCl   B. 0.90% NaCl   C. Pure water

✓ **B11.** Check your understanding of facilitated transport in this Checkpoint.
   a. List two categories of chemicals that undergo facilitated transport.

   b. This process *(does? does not?)* involve a transporter protein. This chemical *(moves across the membrane? undergoes a change in structure?)*, which allows the transported solute to reach the opposite side of the membrane.

   c. Name the transporter protein for glucose: _____. Name two other sugars

       transported by facilitated transport: _____ and _____.
   d. The number of transporter molecules in a plasma membrane places an upper limit, or

       "transport _____," on the rate at which this type of transport can

       occur. The hormone named _____ promotes the insertion of many copies of a glucose transporter protein into plasma membranes of some cells, thereby permitting transport of glucose into those cells.
   e. Once transported to the inside of the cell, the glucose is attached to a

       _____ group with the help of an enzyme called a _____. As a result, the intracellular concentration of "plain" glucose stays *(low? high?)*, a factor that favors additional transport of glucose into cells.

✓ **B12.** Describe active transport in this exercise.
   a. Active transport may be either primary or _____. The sodium pump,

       also known as the" _____ / _____ pump," is the most abundant form of *(primary? secondary?)* active transport. This pump requires energy derived from splitting ATP

       with an enzyme known as _____-ase. A typical cell may use up to _____% of its ATP to fuel primary active transport. When cells are energy-deprived, this pump is

       likely to stop working. As a result, cells _____.
   b. The sodium pump requires a transporter protein that changes shape when a

       _____ group is added to it or removed from it. The effect of the pump is to concentrate sodium (Na^+) *(inside? outside?)* of the cell and to concentrate potassium (K^+) *(inside? outside?)* of the cell. (Refer to Figure 27.6).
   c. For every three Na^+ ions pumped out of the cell, *(two? three? six?)* K^+ ions are returned to the cell. This factor partially accounts for the negative electrical charge *(inside? outside?)* of the cell. (See Checkpoint B5c, page LG 46.) Concentration of these

       ions in ICF and ECF also contributes to _____ pressure of these fluids.
   d. Secondary active transport derives its energy *(directly from ATP? indirectly from ATP by utilizing concentration gradients [such as that for Na] developed by primary active transport?)*. List four or more chemicals transported by secondary active transport.

   e. Digitalis (or digoxin or Lanoxin) is a medication used for a failing heart. This chemical works by *(increasing? decreasing?)* the action of the sodium pump, which leads to a *(larger? smaller?)* concentration gradient of Na^+. As a result, Na^+/Ca^{2+} antiporters *(speed up? slow down?)*. Since more Ca^{2+} remains in heart cells, contractions of the heart are *(strengthened? weakened?)*.

B13. Endocytosis is a form of vesicular transport in which specific chemicals are transported *(into? out of?)* cells. Outline the main steps of receptor-mediated endocytosis, exemplified by HIV infection. Be sure to include these terms: *ligand, receptor (such as a CD4 receptor), vesicles and endosomes, lysosome, and HIV virus.*

B14. Contrast these types of active transport: *phagocytosis* and *pinocytosis.*

C. Cytoplasm, cytosol, and organelles (pages 75–85)

✓ **C1.** Cytosol is the *(fluid? solid?)* part of cytoplasm. Cytosol consists of about _____ to

_____% water. Name six or more types of dissolved or suspended components of cytosol.

C2. Check your understanding of organelles in this Checkpoint.
a. Define *organelles.*

b. Of what advantage is compartmentalization offered by organelles?

c. Circle the four types of organelles in this list that are membranous.

      centrosomes   cilia   cytoskeleton   endoplasmic reticulum

      flagella   Golgi complex   lysosomes   mitochondria

✓ **C3.** Discuss the *cytoskeleton* in this activity.
a. State two categories of functions of the cytoskeleton.

b. Describe three types of movements provided by the cytoskeleton.

c. Arrange these components of the cytoskeleton according to thickness, from thickest to thinnest.

    ____ ____ ____   If. Intermediate filaments   Mf. Microfilaments   Mt. Microtubules

✓ **C4.** Match the nonmembranous organelles in the box with the best descriptions of their functions. Use each answer once.

Cen. Centrosomes	**If.** intermediate filaments	**R. Ribosomes**
Cil. Cilia	**Mf.** Microfilaments	
F. Flagella	**Mt.** Microtubules	

_____ a. Site of assembly of microtubules and organizing center for the mitotic spindle; consists of a pair of cylinders (with "9 + 0 array") arranged perpendicular to each other.

_____ b. Long, hairlike structures found only on sperm; like cilia, each is anchored to a basal body.

_____ c. Short, hairlike structures that move particles over cell surfaces, such as cells lining the uterine (Fallopian) tubes.

_____ d. Sites of protein synthesis; found free in the cytoplasm or bound to membranes (such as endoplasmic reticulum).

_____ e. Composed of the protein tubulin; their movements are powered by motor proteins known as kinesins and dyneins.

_____ f. Composed of as many as six kinds of proteins; are quite strong so help cells withstand mechanical stress.

_____ g. Form microvilli; assist in cell movements such as migration of skin cells during healing or of embryonic tissue during development, as well as invasion of tissues by white blood cells.

✓ **C5.** Match the membranous organelles in the box with the best descriptions of their functions. Use each answer once.

G. Golgi complex	**P.** Peroxisomes
L. Lysosomes	**RER.** Rough endoplasmic reticulum
M. Mitochondria	**SER.** Smooth endoplasmic reticulum
N. Nucleus	

_____ a. Site of direction of cellular activities by means of DNA in chromosomes located here.

_____ b. Network of tubules that is extension of rough ER; site of synthesis of steroids and phospholipids.

_____ c. "Rough" appearance is due to presence of ribosomes on these flattened sacs; sites of protein and glycoprotein synthesis.

_____ d. Stacks of cisternae; involved in synthesis of glycolipids, glycoproteins, and lipoproteins.

_____ e. Vesicles that contain powerful enzymes that can digest bacteria, organelles, or entire cells by processes of phagocytosis, autophagy, or autolysis.

_____ f. Structurally similar to lysosomes; contain enzymes that help metabolize amino acids, fatty acids, and toxic chemicals such as alcohol.

_____ g. Cristae-containing structures, called "powerhouses of the cell" because ATP production takes place here.

✓ **C6.** *For extra review.* Choose the organelles that best fit descriptions below. Select answers from lists in Checkpoints C4 and C5. Those answers may be used once, more than once, or not at all.

_____ a. Abundant in white blood cells, which use these organelles to help in phagocytosis of bacteria

_____ b. Present in large numbers in muscle and liver cells, which require much energy

_____ c. Located on surface of cells of the respiratory tract; help to move mucus

_____ d. Absence of a single enzyme from these organelles causes Tay-Sachs, an inherited disease affecting nerve cells

_____ e. An acid pH must be present inside these organelles for enzymes to function effectively

_____ f   Sites of cellular respiration requiring oxygen (O_2); contain DNA located within 37 genes used for self-replication

_____ g. Known as sarcoplasmic reticulum in muscle cells; detoxification centers in liver cells

_____ h. Described as having a "nine and two" structure based on arrangement of nine pairs of microtubules around a core of two single microtubules (two answers).

_____ i. Might be described as a "clean-up service" since enzymes released from these organelles help to remove debris within and around injured cells (two answers)

_____ j. Sites of phospholipid synthesis (two answers)

✓ **C7.** Describe the sequence of events in synthesis, secretion, and discharge of protein from

the cell by listing these five steps in connect order: _____, _____, _____, _____, _____.

A. Proteins accumulate and are modified, sorted, and packaged in cisternae of Golgi complex.

B. Proteins pass within a transport vesicle to *cis* cisternae of Golgi complex.

C. Secretory granules move toward cell surface, where they are discharged.

D. Proteins are synthesized at ribosomes on rough ER.

E. Vesicles containing protein pinch off of Golgi complex *trans* cisternae to form secretory vesicles.

O Rough endoplasmic reticulum
O Chromatin
O Nucleolus
O Plasma (cell) membrane
O Mitochondrion
O Cytosol

O Centriole
O Microtubule
O Lysosome
O Nuclear membrane
O Smooth endoplasmic reticulum
O Golgi complex

Figure LG 3.3 Generalized animal cell. Complete as directed in Checkpoint C8.

C8. Color and then label all parts of Figure LG 3.3 listed with color code ovals.

D. Nucleus; protein synthesis (pages 85–91)

✓ **D1.** Describe parts of the nucleus by completing this exercise.

a.   The nuclear membrane has pores that are *(larger? smaller?)* than channels in the plasma membrane. Of what significance is this?

b.   Ribosomes are assembled at sites known as _____ , where protein, DNA, and RNA are stored.

c.   Chromosomes consist of beadlike subunits known as _____ which

are made of double-stranded _____ wrapped around core proteins

called _____. Nucleosome "beads" are held together, necklace-fashion, by *(histones? linker DNA?)*. Nucleosomes are then held into chromatin fibers

and folded into _____.

d.   Before cell division, DNA duplicates and forms compact, coiled conformations known as *(chromatin? chromatids?)* that may be more easily moved around the cell during the division process.

✓ **D2.** State the significance of protein synthesis in the cell.

✓ **D3.** Refer to Figure LG 3.4 and describe the process of protein synthesis in this exercise.

a.   Write the names of the two major steps of protein synthesis on the two lines in Figure LG 3.4.

b.   Transcription takes place in the *(nucleus? cytoplasm?)* with the help of the enzyme

called _____. How does this enzyme know precisely where on DNA to start the transcription process? The "start" region consists of a specific

nucleotide sequence known as a _____. At this site,

_____ attaches to the DNA. Similarly, the "stop" section on the

gene is a nucleotide sequence called a _____, at which point the enzyme detaches from the DNA.

c.   In the process of transcription, a portion of one side of a double-stranded DNA serves

as a mold or _____ and is called the *(sense? antisense?)* strand.

Nucleotides of _____ line up in a complementary fashion next to DNA nucleotides (much as occurs between DNA nucleotides in replication of DNA prior to mitosis). Select different colors and color the components of DNA nucleotides (P, D. and bases T, A, C. and G).

d.   Complete the strand of mRNA by writing letters of complementary bases. Then color the parts of mRNA (P. R, and bases U, A, C, and G).

e.   The second step of protein synthesis is known as _____ because it involves translation of one "language" (the nucleotide base code of

_____ RNA) into another "language" (the correct sequence of the

20 _____ to form a specific protein).

D5. List six or more therapeutic substances that have been produced by bacteria containing recombinant DNA.

E. Normal cell division (pages 91–95)

E1. Once you have reached adult size, do your cells continue to divide? *(Yes? No?)* Of what significance is this fact?

✓ **E2.** Describe aspects of cell division by completing this exercise.
  a. Division of any cell in the body consists of two processes: division of the

  _____ and division of the _____.
  b. In the formation of mature sperm and egg cells, nuclear division is known as

  _____. In the formation of all other body cells, that is, *somatic* cells,

  nuclear division is called _____.
  c. Cytoplasmic division in both somatic and reproductive division is known as

  _____. In *(meiosis? mitosis?)* the two daughter cells have the same
  hereditary material and genetic potential as the parent cell.

✓ **E3.** Describe interphase in this activity.
  a. Interphase occurs *(during cell division? between cell divisions?)*.
  b. Chromosomes are replicated during phase *(G_1? S? G_2?)* of interphase. In fact "S"

  stands for _____ of new DNA. During this time, the DNA helix
  partially uncoils and the double DNA strand separates at points where

  _____ connect so that new complementary nucleotides can attach
  here. (See Figure LG 2.3.) In this manner the DNA of a single chromosome is

  doubled to form two identical _____.
  c. Two periods of growth occur during interphase, preparing the cell for cell division.
  The *(G_1? G_2?)* phase immediately precedes the S phase, whereas the *(G_1? G_2?)*
  phase follows the S portion of interphase. Which of these phases is usually longest?
  (G_1? S? G_2?)
  d. Most nerve cells *(are? are not?)* capable of mitosis after birth; their development is

  permanently arrested in the _____ phase, and are then said to be in the _____
  phase.

✓ **E4.** Carefully study the phases of the cell cycle shown in Figure 3.33b, page 93 in your text. Then check your understanding of the process by identifying major events in each phase.

A. **Anaphase**	**P.** **Prophase**
I. **Interphase**	**T.** **Telophase**
M. **Metaphase**	

_____ a. This phase immediately follows inter-phase.

_____ b. Chromatin condenses into distinct chromosomes (chromatids) each held together by a centromere complexed with a kinetochore.

_____ c. Nucleoli and nuclear envelope break up; the two centrosomes move to opposite poles of the cell, and formation of the mitotic spindle begins.

_____ d. Chromatids line up with their centro-meres along the equatorial plate.

_____ e. Centromeres split; chromatids (now called daughter chromosomes) are dragged by microtubules to opposite poles of the cell. Cytokinesis begins.

_____ f. Events of this phase are essentially a reversal of prophase; cytokinesis is completed.

E5. Contrast roles of three types of microtubules in mitosis and identify which type shortens as daughter chromosomes are drawn toward opposite poles of the cell.

E6. Write the full names and a brief description of the roles of the following three chemicals that regulate cell division.

a. MPF

b. cdc2 proteins

c. Cyclin

E7. Contrast these two terms: *necrosis* and *apoptosis*.

E8. Tumor suppressor genes normally *(stimulate? inhibit?)* cell division. Loss or alteration of such a gene leads to *(cancer? atrophy?)*. The most common genetic change leading to cancer is a change in the tumor suppressor gene called *(cdc? cyclin? p53?)*: this gene is located on chromosome number _____.

F. Cells and aging, cellular diversity (pages 95–96)

F1. Describe changes that normally occur in each of the following during the aging process.

a. Number of cells in the body. *Note:* which three types of cells are permanently arrested in the G_0 phase?

b. Glucose cross-links

F2. In a sentence or two, summarize the main points of each of these theories of aging.

a. An "aging gene," affecting rate of mitosis

b. Free radicals

c. Immune system

F3. Describe the diversity of human cells in this Checkpoint.

a. The human body is composed of about _____ different types of cells with a total of about _____ cells. These range in size from a red blood cell (_____ mu) to an ovum (_____ mu).

b. List several different shapes of cells. (More about this in Chapter 4.)

c. State the advantages afforded to cells by their own specific shapes:

sperm with a long "tail" (flagellum) _____

intestinal cells with microvilli _____

thin, flat cells forming walls of the tiniest blood vessels _____

G. Disorders, medical terminology (pages 97–98)

✓ **G1.** Complete this exercise about cancer.
  a. A term that means "tumor or abnormal growth" is _____. A cancer-
      ous growth is a *(benign? malignant?)* neoplasm, whereas a noncancerous tumor is a

      (benign? malignant?) neoplasm. The term _____ refers to the study
      of cancer.
  b. Tumors grow quickly by increasing their rate of cell division, a process known as

      _____. Cancer cells also compete with normal body cells by robbing

      those cells of _____. Tumors also favor their own growth by synthe-

      sizing chemicals known as TAFs. TAFs refer to _____

      _____ _____, indicating that these chemicals stimu-

      late growth of new _____ around tumor cells.
  c. Malignant tumors are *(more? less?)* likely than benign tumors to spread and possibly

      cause death. A term that means "spread" of cancer cells is _____.
      By what routes do cancers reach distant parts of the body?

  d. What may cause pain associated with cancer?

✓ **G2.** Match terms in the box with the definitions below.

C.	**Carcinoma**	**O.**	**Osteogenic sarcoma**
L.	**Lymphoma**	**S.**	**Sarcoma**

_____ a. General term for malignant tumor arising   _____ c. General term for malignant tumor
             from any connective tissue   of epithelial tissue

_____ b. Malignant tumor derived from bone   _____ d. A malignancy of lymph tissue
             (a connective tissue)

G3. Relate the following terms to cancer:

a. Carcinogens

b. Proto-oncogenes

c. Oncogenes

d. Growth factors

e. Mutation

G4. *Clinical thinking.* Are all cancer cells alike within a single tumor? *(Yes? No?)* What significance is this in planning cancer treatment?

✓ **G5.** Match the terms in the box describing alterations in cells or tissues with the descriptions below.

A.	**Atrophy**	**HT.**	**Hypertrophy**
D.	**Dysplasia**	**M.**	**Metaplasia**
HP.	**Hyperplasia**		

_____ a. Increase in size of tissue or organ by increase in size (not number) of cells, such as growth of your biceps muscle with exercise

_____ b. Increase in size of tissue or organ due to increase in number of cells, such as a callus on your hand or growth of breast tissue during pregnancy

_____ c. Change of one cell type to another normal cell type, such as change from single row of tall (columnar) cells lining airways to multilayers of cells as response to constant irritation of smoking

_____ d. Abnormal change in cells in a tissue due to irritation or inflammation. May revert to normal if irritant removed, or may progress to neoplasia

_____ e. Decrease in size of cells with decrease in size of tissue or organ

ANSWERS TO SELECTED CHECKPOINTS: CHAPTER 3

B1. See Figure LG 3.1A. (a) Water-soluble and facing the membrane surface, so compatible with watery environment surrounding the membrane. (b) Forms a water-insoluble barrier between cell contents and extracellular fluid; nonpolar molecules such as lipids, oxygen, and carbon dioxide pass readily between "tails." (c) Channels for passage of water or ions; transporters; receptors; linkers (helping to bind cells to one another. (d) Form sugary coat (glycocalyx) that is important in cell recognition, immunity, protection, adhesion of cells; this coating also helps make red blood cells slippery so they can easily pass through tiny blood vessels. (e) Strengthens the membrane.

B2. (a) 50, 50; cholesterol. (b) Phospholipids; amphipathic; surface; philic, loving. (C) Extracellular; cell adhesion, recognition, and/or communication. (d) Protein; 50; proteins; hormones (such as insulin) or nutrients (such as glucose); enzymes; markers; linkers.

B4. (a) Quite, phospholipid. (b) Protein; channels with pores and transporters. (c) See Figure LG 3.1A.

B5. (a) See Figure LG 3.2A. (b) Down. (c) Electrical, potential; inside. (d) Electrochemical.

B7. (a) Nonmediated. (b) Passive. (c) Active. (d) Uniporter-mediated. (e) Antiporter-mediated.

B8. (a) A. (b) B. (c) B. (d) A. (e) b: membrane thickness.

B9. (a) D. (b) FD. (c) O.

B10. (a) B. (b) More, fewer. (c) A. (d) Out of crenate. (e) Iso; 0.90% NaCl or 5% glucose. (f) C. (g) A.

B11. (a) Chemicals too polar or charged to go through the lipid bilayer, or too large to pass through membrane channels. (b) Does; undergoes a change in structure. (c) GluT; fructose and galactose. (d) Maximum; insulin. (e) Phosphate, kinase; low.

B12. (a) Secondary; Na$^+$/K$^+$, primary; ATP; 40; experience Na$^+$/K$^+$ pump failure and can die. (b) Phosphate; outside, inside. (c) Two, inside; osmotic. (d) Indirectly from ATP by utilizing concentration gradients [such as that for Na$^+$] developed by primary active transport; Ca^{2+}, H$^+$, glucose, and amino acids from foods. (e) Decreasing, smaller; slow down; strengthened.

C1. Fluid; 75, 90; ions, glucose, amino acids, fatty acids, proteins, lipids, ATP, waste products, and temporary storage molecules such as lipid droplets and glycogen granules.

C3. (a) Provides a scaffold or structural framework that organizes the cell and gives it shape; permits movement. (b) Movement of the entire cell, movement of chromosomes during cell division, and movement of organelles and some chemicals within the cell. (c) Mt If Mf.

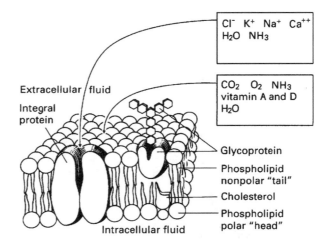

Figure LG 3.1A Plasma membrane

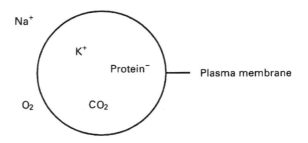

Figure LG 3.2A Major ions in intracellular fluid (ICF) and extracellular fluid (ECF).

C4. (a) Cen. (b) F. (c) Cil. (d) R. (e) Mt. (f) If. (g) Mf.
C5. (a) N. (b) SER. (c) RER. (d) G. (e) L. (f) P. (g) Mit.
C6. (a) L. (b) Mit. (c) Cil. (d) L. (e) L. (f) Mit. (g) SER. (h) Cil, F. (i) L, P. (j) RER, SER.
C7. D B A E C.
C8.

Figure LG 3.3A Generalized animal cell.

D1. (a) Larger; permit passage of large molecules such as RNA and proteins. (b) Nucleoli. (c) Nucleosomes. DNA, histones; linker DNA; loops. (d) Chromatids.

D2. Proteins serve a number of significant functions. Refer to Chapter 2, page 32 of the *Learning Guide*, Checkpoint E8.

D3. (a) See Figure LG 3.4A. (b) Nucleus, RNA polymerase; promotor; RNA polymerase; terminator. (c) Template, sense; messenger RNA (mRNA) (see Figure LG 3.4A). (d) See Figure LG 3.4A. (e) Translation, m (messenger), amino acids. (f) Cytoplasm, ribosome; small, the start codon; A U G (g) t (transfer), anticodon; U C A, see Figure LG 3.4A. (h) t (transfer); peptide, large. (i) It is released and may be recycled to transfer another amino acid. (j) Stop; it splits into its large and small subunits. (k) Polyribosome.

D4. (a) Exons, introns; both exons and introns. (b) Introns, before; small nuclear ribonucleoproteins or snRNPs (c) A group of nucleotides on a DNA molecule that codes for synthesis of a specific sequence of bases in mRNA (by transcription) and ultimately, a specific sequence of amino acids (by transcription). (d) 300, 3000, 100, 1000.

E2. (a) Nucleus, cytoplasm. (b) Meiosis; mitosis. (c) Cytokinesis; mitosis.

E3. (a) Between cell divisions. (b) S; synthesis; nitrogen bases; chromatids. (c) G_1; G_2; G_1. (d) Are not, G_1 G_0.

E4. (a) P. (b) P. (c) P. (d) M. (e) A. (f) T.

E8. Inhibit; cancer; p53; 17.

G1. (a) Neoplasm; malignant, benign; oncology. (b) Hyperplasia; nutrients and space; tumor angiogenesis factors, blood vessels (*angio* = vessel). (c) More; metastasis; blood or lymph, or by detaching from an organ, such as liver, and "seed-

Figure LG 3.4A Protein synthesis.

ing" into a cavity (as abdominal cavity) to reach other organs which border on that cavity. (d) Pressure on nerve, obstruction of passageway, loss of function of a vital organ.

G2. (a) S. (b) 0. (c) C. (d) L.

G5. (a) HT. (b) HP. (c) M. (d) D. (e) A.

MORE CRITICAL THINKING: CHAPTER 3

1. Discuss clinical aspects of tonicity—for example, how would a client be affected by the accidental administration of 5% NaCl solution (which is hypertonic to human body fluids) instead of an 0.9% NaCl solution (which is isotonic to the human body)?

2. Inadequate dietary protein can rob the body of nutrients for critical functions. If plasma membranes lack adequate protein, what membrane functions are affected?

3. Explain how cells are compartmentalized and what advantages are provided to cells by compartmentalization.

4. Tell the "life history" of a digestive enzyme made within a pancreas cell. Remember that most of any enzyme consists of protein. Start with the "birth" (synthesis) of this protein and outline its "travels" from one part of the cell to another, including its exit from the cell via exocytosis.

5. In Chapter 2, you studied DNA and ATP. Describe the roles of each of these chemicals in cellular functions.

6. Since virtually all human cells contain the same type of DNA in their 46 chromosomes, explain what accounts for the difference between a muscle and a bone cell.

MASTERY TEST: CHAPTER 3

Questions 1-9: Circle the letter preceding the one best answer to each question.

1. The organelle that carries out the process of autophagy in which old organelles are digested so that their components can be recycled is:
   A. Peroxisome   D. Lysosome
   B. Mitochondrion   E. Endoplasmic reticulum
   C. Centrosome

2. Which statement about proteins is *false*?
   A. Proteins are synthesized on ribosomes.
   B. Proteins are so large that they normally will not pass out of blood that undergoes filtration in kidneys.
   C. Proteins are so large that they can be expected to create osmotic pressure important in maintaining fluid volume in blood.
   D. Proteins are small enough to pass through pores of typical plasma (cell) membranes.

3. Choose the *false* statement about prophase of mitosis.
   A. It occurs just after metaphase.
   B. The nuclear membrane breaks down then.
   C. The nucleolus disappears then.
   D. The mitotic spindle forms then and pushes centrosomes to the poles of the cell.

4. Choose the *false* statement about the Golgi complex.
   A. It consists of stacked membranous sacs known as cisternae.
   B. It is more extensive in cells that secrete proteins into ECF.
   C. It is involved in lipoprotein secretion.
   D. It pulls chromosomes towards poles of the cell during mitosis.

5. Choose the *false* statement about genes.
   A. Genes contain DNA.
   B. Genes contain information that controls heredity.
   C. Genes are located only in the nucleus of the cell.
   D. Genes are transcribed by messenger RNA during the first step of protein synthesis.

6. Choose the *false* statement about protein synthesis.
   A. Translation occurs in the cytoplasm.
   B. Messenger RNA picks up and transports an amino acid during protein synthesis.
   C. Messenger RNA travels to a ribosome in the cytoplasm.
   D. Transfer RNA is attracted to mRNA due to their complementary bases.

7. All of the following structures are part of the nucleus *except*:
   A. Nucleolus   C. Chromosomes
   B. Chromatin   D. Centrosome

8. The Na^+/K^+ pump is an example of:
   A. Active transport   C. Diffusion
   B. Osmosis   D. Phagocytosis

9. Bone cancer is an example of what type of cancer?
   A. Sarcoma   C. Carcinoma
   B. Lymphoma   D. Leukemia

Questions 10-20: Circle T (true) or F (false). If the statement is false, change the underlined word or phrase so that the statement is correct.

T F   10. Plasma membranes consist of a double layer of <u>carbohydrate molecules</u> with proteins embedded in the bilayer.

T F   11. Free ribosomes in the cytosol are concerned primarily with synthesis of <u>proteins for insertion in the plasma membrane or for export from the cell</u>.

T F   12. Glucose in foods you eat is absorbed into cells lining the intestine by the process of <u>facilitated transfusion</u>.

T F   13. <u>Cristae</u> are folds of membrane in mitochondria.

T F   14. Sperm move by lashing their long tails named <u>cilia</u>.

T F   15. Hypertrophy refers to increase in size of a tissue related to increase in the <u>number</u> of cells.

T F   16. <u>Diffusion, osmosis, and facilitated diffusion</u> are all passive transport processes.

T F   17. A 5% glucose solution is <u>hypotonic</u> to a 10% glucose solution.

T F   18. <u>The free radical theory of aging</u> suggests that electrically charged, unstable, and highly reactive free radicals cause oxidative damage to chemical components of aging cells.

T F   19. The mitotic phase that follows metaphase is <u>anaphase</u>.

T F   20. <u>Cytokinesis</u> is another name for mitosis.

Questions 21-25: Fill-ins. Complete each sentence with the word which best fits.

_____ 21. Cells of cardiac muscle, skeletal muscle, and nerves do not divide throughout a lifetime since they are arrested permanently in the _____ phase.

_____ 22. Active processes involved in movement of substances across cell membranes are those that use energy from the splitting of _____.

_____ 23. White blood cells engulf large solid particles by the process of _____.

_____ 24. _____ DNA consists of DNA combined from several sources, such as from human cells and from bacteria.

_____ 25. The sequence of bases of mRNA which would be complementary to DNA bases in the sequence A-T-T-C-A-C would be _____.

ANSWERS TO MASTERY TEST: CHAPTER 3

Multiple Choice

1. D
2. D
3. A
4. D
5. C
6. B
7. D
8. A
9. A

True-False

10. F. Phospholipid molecules
11. F. Proteins for use inside of the cell
12. F. Facilitated diffusion (or facilitated transport).
13. T
14. F. Flagella
15. F. Size
16. T
17. T
18. T
19. T
20. F. Nuclear division (Cytokinesis is cytoplasmic division.)

Fill-ins

21. G_0 or G_1
22. ATP
23. Phagocytosis
24. Recombinant
25. U-A-A-G-U-G

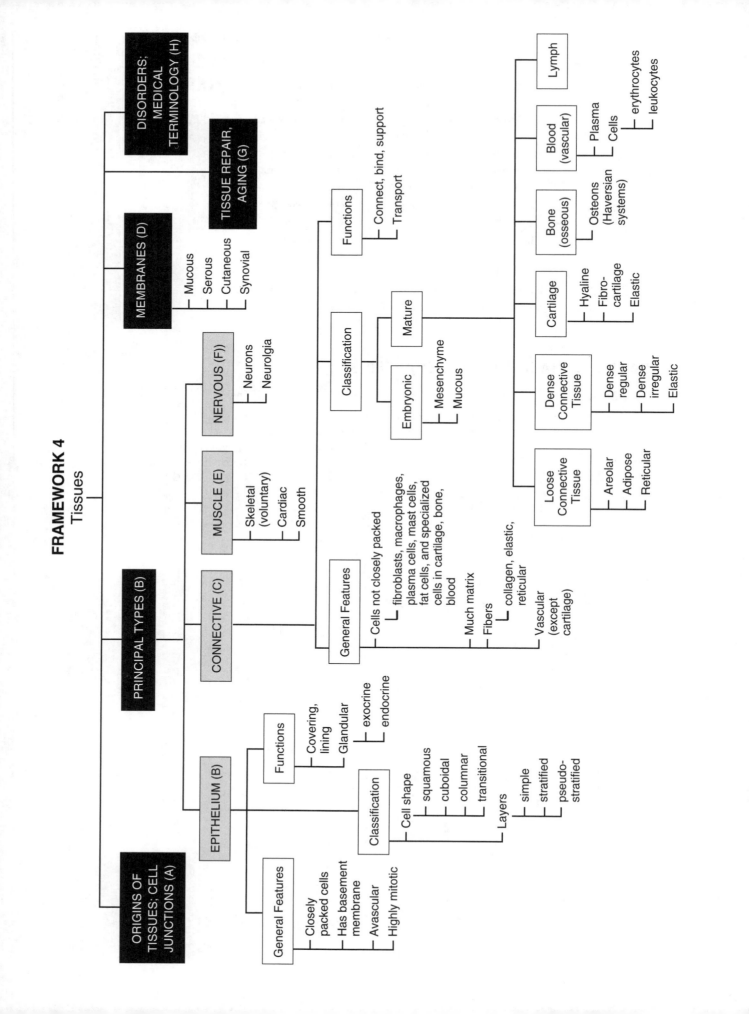

FRAMEWORK 4
Tissues

Tissues

ORIGINS OF TISSUES; CELL JUNCTIONS (A)

PRINCIPAL TYPES (B)

MEMBRANES (D)
- Mucous
- Serous
- Cutaneous
- Synovial

DISORDERS; MEDICAL TERMINOLOGY (H)

TISSUE REPAIR, AGING (G)

EPITHELIUM (B)

General Features
- Closely packed cells
- Has basement membrane
- Avascular
- Highly mitotic

Functions
- Covering, lining
- Glandular
  - exocrine
  - endocrine

Classification
- Cell shape
  - squamous
  - cuboidal
  - columnar
  - transitional
- Layers
  - simple
  - stratified
  - pseudo-stratified

CONNECTIVE (C)

General Features
- Cells not closely packed
  - fibroblasts, macrophages, plasma cells, mast cells, fat cells, and specialized cells in cartilage, bone, blood
- Much matrix
- Fibers
  - collagen, elastic, reticular
- Vascular (except cartilage)

Functions
- Connect, bind, support
- Transport

Classification
- Embryonic
  - Mesenchyme
  - Mucous
- Mature
  - Loose Connective Tissue
    - Areolar
    - Adipose
    - Reticular
  - Dense Connective Tissue
    - Dense regular
    - Dense irregular
    - Elastic
  - Cartilage
    - Hyaline
    - Fibro-cartilage
    - Elastic
  - Bone (osseous)
    - Osteons (Haversian systems)
  - Blood (vascular)
    - Plasma
    - Cells
      - erythrocytes
      - leukocytes
  - Lymph

MUSCLE (E)
- Skeletal (voluntary)
- Cardiac
- Smooth

NERVOUS (F)
- Neurons
- Neurolgia

C H A P T E R **4**

The Tissue Level of Organization

Tissues consist of cells and the intercellular (between cells) materials secreted by these cells. All of the organs and systems of the body consist of tissues that may be categorized into four classes or types: epithelium, connective, muscle, and nerve. Each tissue exhibits structural characteristics (anatomy) that determine the function (physiology) of that tissue. For example, epithelial tissues consist of tightly packed cells that form the secretory glands and also provide protective barriers, as in the membranes that cover the surface of the body and line passageways or cavities. Tissues are constantly stressed by daily wear and tear and sometimes by trauma or infection. Tissue repair is therefore an ongoing human maintenance project.

As you begin your study of tissues, carefully examine the Chapter 4 Framework and note relationships among concepts and key terms there. Also refer to the Topic Outline and Objectives; you may want to check off each objective as you complete it.

TOPIC OUTLINE AND OBJECTIVES

A. Types and origins of tissues; cell junctions

1. Describe the characteristics of the four basic types of tissue that make up the human body.
2. Describe the structure and function of the five main kinds of cell junctions.

B. Epithelial tissue

3. Describe the general features of epithelial tissue and the structure, location, and function for the different types of epithelium.

C. Connective tissue

4. Identify the various components of connective tissue.
5. Describe the structure, function, and location of the various types of connective tissues.

D. Membranes

6. Define a membrane, and describe the location and function of mucous, serous, and synovial membranes.

E. Muscle tissue

7. Describe the general features of muscle tissue, and contrast the structure, location, and mode of control for the three types of muscle tissue.

F. Nervous tissue

8. Describe the structural features and function of nervous tissue.

G. Tissue repair: restoring homeostasis; aging and tissues

9. Describe the role of tissue repair in restoring homeostasis.

H. Disorders and medical terminology

WORDBYTES

Now become familiar with the language of this chapter by studying each wordbyte, its meaning, and an example of its use within a term. After you study the entire list, self-check your understanding by writing the meaning of each wordbyte on the line. As you continue through the *Learning Guide*, identify (and fill in) additional terms that contain the same wordbyte.

Wordbyte	Self-check	Meaning	Example(s)
a-	_____	without	*a*vascular
acin	_____	berry	*acin*ar gland
adip(o)-	_____	fat	*adip*ose
apex, apic-	_____	tip	*apic*al
areol-	_____	small space	*areol*ar
ciliaris	_____	resembling an eyelash	*cilia*
-crine	_____	to secrete	endo*crine*
endo-	_____	within	*endo*crine
exo-	_____	outside	*exo*crine
hist -	_____	tissue	*hist*ology
lupus	_____	wolf	systemic *lupus* erythematosus
macro-	_____	large	*macro*phage
multi-	_____	many	*multi*cellular
path-	_____	disease	*path*ology
pseudo-	_____	false	*pseudo*stratified
squam-	_____	thin plate	*squam*ous
strat-	_____	layer	*strat*ified
uni-	_____	one	*uni*cellular
vasc-	_____	blood vessel	a*vasc*ular
xeno-	_____	strange, foreign	*xeno*transplantation

CHECKPOINTS

A. Types and origins of tissues; cell junctions (pages 104–105)

A1. Define the following terms.

a. Tissue

b. Histology

c. Pathologist

d. Biopsy

A2. Complete the table about the four main classes of tissues.

Tissue	General Functions
a. Epithelial tissue	
b. Connective tissue	
c.	Movement
d.	Initiates and transmits nerve impulses that help coordinate body activities

✓ **A3.** *The Big Picture: Looking Ahead*. Refer to Exhibit 29.1 in you text. Then do the following matching exercise on the three embryonic germ layers that form all body organs.

Ecto. Ectoderm	Endo. Endoderm	Meso. Mesoderm

_____ a. Epithelial lining of the organs of digestion and of respiratory passageways

_____ b. Brain and nerves

_____ c. Muscles and bones

_____ d. Outer layers of skin

✓ **A4.** Complete this exercise about extracellular materials and connections between cells.

a. Name one type of cell that typically moves around within extracellular areas.

_____.

b. Most body cells stay in one place. They are held in place by points of contact between

adjacent cell membranes; such points are known as cell _____. For example, *(anchoring gap? tight?)* junctions are helpful in forming a tight seal between

cells, such as that required in the lining of the stomach or the _____.

c. *(Desmosome? Gap? Tight?)* junctions are common in areas subject to stretching or friction. Write one or more example of this type of junction.

d. A gap junction consists of proteins called _____ that form tunnels between cells. Passage of ions through these tunnels permits spread of

_____ within the nervous system or heart muscle.

e. Adherens junctions consist of plaque, which is a dense layer of *(lipid? protein?)* and from which integral membrane proteins extend to other cells to help resist separation of the tissue. In epithelial tissue, such junctions are extensive and are called

_____ belts.

B. Epithelial tissue (pages 105–118)

✓ **B1.** Describe structural features of epithelium in this Checkpoint.

a. Epithelium consists of *(closely packed cells? intercellular material with few cells?)* *(Many? Few?)* cell junctions are present, securing these cells *(tightly? loosely?)* to each other.

b. Blood vessels *(do? do not?)* typically penetrate epithelium. A term meaning "lack of

blood vessels" is _____. State the significance of this feature of epithelium.

c. Epithelium *(has? lacks?)* a nerve supply.

d. Epithelium *(does? does not?)* have the capacity to undergo mitosis.

✓ **B2.** Complete these sentences describing basement membrane.

a. The *(apical? basal?)* surfaces of epithelial cells attach by a _____ membrane to underlying connective tissue.

b. This membrane consists of *(cells? extracellular material?)*. The part of the membrane adjacent to epithelium is secreted by *(epithelium? connective tissue?)*. It consists of the protein named *(collagen? reticulin?)* along with proteoglycans and is known as the *(reticular lamina? basal lamina?)*.

c. The second underlying layer of the membrane is secreted by *(epithelium? connective*

tissue?) and is called the _____ lamina.

B3. Complete the table by describing five or more locations and functions of epithelial tissue. One is done for you.

Location	Functions
a. Eyes, nose, outer layer of skin	Sensory reception
b.	
c.	
d.	
e.	

✓ **B4.** Name the two main subtypes of epithelial tissue based on location and function.

_____ and _____.

✓ **B5.** Choose from the terms in the box once to complete the sentences below.

Columnar	Simple	Stratified
Cuboidal	Squamous	Transitional

a. Epithelial cells arranged in one layer are known as _____

epithelium. _____ epithelium is composed of many layers of cells.

b. _____ epithelium consists of cells so flat that diffusion readily occurs

across them. Tall, cylindrical or rectangular cells are called _____
epithelium. Intermediate between these two types of cells are the cells that form

_____ epithelium.

c. A type of epithelium that can stretch; lines urinary bladder. _____

✓ **B6.** Study diagrams of epithelial tissue types (A-F) in Figure LG 4.1 and then write the
following information on the lines provided on the figure:
a. The name of each tissue
b. One or more locations of each type of tissue
c. One or more functions of each type of tissue

Now color the following structures on each diagram:

◯ Nucleus   ◯ Intercellular material

◯ Cytoplasm   ◯ Basement membrane

A Name _____

   Location _____

   Function _____

B Name _____

   Location _____

   Function _____

C Name _____

   Location _____

   Function _____

D Name _____

   Location _____

   Function _____

E Name _____

   Location _____

   Function _____

F Name _____

   Location _____

   Function _____

G Name _____

   Location _____

   Function _____

H Name _____

   Location _____

   Function _____

Figure LG 4.1 Diagrams of selected tissues types. Complete as directed.

I Name _____

Location _____

Function _____

J Name _____

Location _____

Function _____

K Name _____

Location _____

Function _____

L Name _____

Location _____

Function _____

M Name _____

Location _____

Function _____

N Name _____

Location _____

Function _____

O Name _____

Location _____

Function _____

P Name _____

Location _____

Function _____

✓ **B7.** *The Big Picture: Looking Ahead.* Check your understanding of the most common types of epithelium listed in the box by writing the name of the type after the phrase that describes it. To confirm your answers, refer to text figures or tables of upcoming chapters.

Pseudostratified ciliated columnar	**Simple cuboidal**
Simple columnar, ciliated	**Simple squamous**
Simple columnar, nonciliated	**Stratified squamous**

   a. Lines the inner surface of the stomach (Figure 24.12) and intestine (Figure 24.22):

   _____.

   b. Forms the outer surface of skin (Figures 5.1, 5.3, 5.4, and 5.6): _____.
   c. Single layer of cube-shaped cells; found in some glands, such as hormone-producing follicles of the thyroid gland (Figure 18.13c), as well as ducts of glands:

   _____.

   d. Endothelium, as in capillaries (Figure 21.1), as well as mesothelium:

   _____.

   e. Lines air sacs of lungs where thin cells are required for diffusion of gases into blood

   (Figure 23.12): _____.
   f. Forms the delicate Bowman's capsule surrounding each filtering apparatus

   (glomerulus) (Figure 26.8): _____.

   g. Forms most of renal tubule walls (Table 26.1): _____.

✓ **B8.** *For extra review.* Complete this exercise on epithelium.
   a. Pseudostratified epithelium is *(one? many?)* layer(s) thick with *(all? only some?)* cells attached to the basement membrane and *(all? only some?)* cells reaching the apical surface.
   b. Stratified epithelium is named after the cells that form the *(superficial? deepest?)* layer of that tissue.
   c. Stratified squamous epithelium is likely to offer *(more? less?)* protection than simple squamous epithelium.

B9. Complete the table describing distinguishing characteristics of specialized epithelial cells. Also fill in sites where such modified cells would likely be located.

Modification	Description	Location
a. Cilia		
b. Microvilli		Cells lining digestive tract
c. Keratin production	Waterproofing protein	
d. Goblet cells		

B10. Define glands? Why are glands studied in this section on epithelium?

✓ **B11.** Write EXO before descriptions of *exocrine* glands, and ENDO before descriptions of *endocrine* glands. (Endocrine glands will be studied further in Chapter 18.)

_____ a. Their products are secreted into ducts that lead either directly or indirectly to the outside of the body.

_____ b. Their products are secreted into ECF and then the blood and so stay within the body; they are ductless glands.

_____ c. Examples are glands that secrete sweat, oil, mucus, and digestive enzymes.

_____ d. Examples are glands that secrete hormones.

✓ **B12.** Contrast classes of glands in this exercise. Circle the best answer in each case.
a. A gland with a duct that does branch is called a *(simple? compound?)* gland.
b. A gland consisting of one cell, such as a goblet cell, is a *(multi? uni?)*-cellular gland.
c. A gland with the secreting portions all shaped like flasks (or berries) is a(n) *(acinar? tubular? tubuloacinar?)* gland.
d. Most glands in the human body fit into the functional class of *(holo? mero?)*-crine gland. This means that *(only the secretion? the entire secretory cell, including the secretion?)* is released.

C. Connective tissue (pages 118–129)

✓ **C1.** Write *connective tissue* or *epithelial tissue* following the description below that correctly describes that tissue.
a. Consists of many cells with little intercellular substance (fibers and ground

substance): _____
b. Penetrated by blood vessels (vascular): _____
c. Does not cover body surfaces or line passageways and cavities but is more internally

located; binds, supports, protects: _____
✓ **C2.** List four or more categories of connective tissue matrix. (Note that *intercellular* means "between cells" or "extracellular.") What is the source of these materials?

✓ **C3.** Do this activity about connective tissue cells.
a. Cells in connective tissues derive from *(endo? meso? ecto?)*-derm cells in the

embryo. These cells are called _____ cells.
b. Most matrix is secreted by *(immature? mature?)* connective tissue cells. Which type of cells have greater capacity for mitosis? *(Mature? Immature?)*

c. Names of immature cells end in *(-blast? -cyte?)*, whereas names of mature cells end in *(-blast? -cyte?)*. To check your knowledge further, write correct names for the following cells: immature bone cell: _____; immature cartilage cell:

_____; mature bone cell: _____

✓ **C4.** Identify characteristics of each connective tissue cell type by writing the name of the correct cell type after its description. (Note that several of these cell types are shown in diagrams G, H, and I of Figure LG 4.1.)

Adipocyte	**Fibroblast**	**Macrophage**	**Mast cell**	**Plasma cell**	**WBC**

a. Derived from B lymphocyte; gives rise to antibodies; helpful in defense:

_____

b. Phagocytic cell that engulfs bacteria and cleans up debris; important during infection;

formed from monocyte: _____

c. Fat cell: _____
d. Forms collagenous and elastic fibers in injured tissue: _____
e. Abundant along walls of blood vessels; produces histamine, which dilates blood

vessels: _____
f. Neutrophils and eosinophils are examples: _____

✓ **C5.** The ground substance of connective tissue consists of *(large? small?)* molecules that are combinations of polysaccharides and *(lipids? proteins?)*. Identify components of ground substance by selecting the answers listed in the box that best fit the descriptions below.

AP.	**Adhesion proteins**	**HA.**	**Hyaluronic acid**
CS.	**Chondroitin sulfate**	**KS.**	**Keratan sulfate**
DS.	**Dermatan sulfate**		

_____ a. Chemicals that stabilize position of cells; includes proteins such as fibronectin

_____ b. Viscous, slippery substance that binds cells together and lubricates joints

_____ c. Chemical component of bone, cartilage, and the cornea of the eye

_____ d. Jellylike; provides support and adhesiveness in bone, skin, blood vessels, and cartilage

_____ e. Found in bone, skin, blood vessels, and heart valves

✓ **C6.** Select the type of fiber in the box that best fit descriptions below.

C. Collagen	E. Elastic	R. Reticular

_____ a. Formed of collagen and glycoproteins; form branching networks that provide stroma for soft organs such as spleen and networks around fat, nerve, and muscle cells

_____ b. Tough fibers in bundles that provide great strength, as needed in bone and tendons; formed of the most abundant protein in the body

_____ c. Can be greatly stretched without breaking, an important quality of connective tissues forming skin, blood vessels, and lungs

_____ d. Composed of elastin surrounded by fibrillin (a glycoprotein that is defective in Marfan syndrome)

✓ **C7.** As described earlier, the embryonic connective tissue from which all of the

connective tissues arise is called _____. *(Much? Little?)* of this type of tissue is found in the body after birth. Another embryonic tissue, known as mucous

connective tissue or _____ jelly, gives support to the umbilical cord. Tissues such as areolar connective tissue, cartilage, or bone are known as *(embryonic? mature?)* connective tissues.

C8. After you study each mature connective tissue type in the text, refer to diagrams of connective tissues on Figure LG 4.1, G-L. Write the following information on the lines provided on the figure.

✓ a. The name of each tissue
   b. One or more locations of the tissue
   c. One or more functions of the tissue

Now color the following structures on each diagram, using the same colors as you did for diagrams A-F:

○ Nucleus   ○ Matrix
○ Cytoplasm   ○ Fat globule (a cellular inclusion), diagram H

✓   From this activity, you can conclude that connective tissues appear to consist mostly of *(cells? intercellular material [matrix]?)*.

C9. Label fibers in Figure LG 4.1, diagrams G, I, and J.
(Note that fibers *are* present in bone but not visible in diagram K.)

✓▲ **C10.** Which two types of tissue form subcutaneous tissue (superficial fascia)?

_____, _____

✓ **C11.** Answer these questions about adipose tissue.
   a. Explain what accounts for the "signet ring" appearance (like a class ring with a stone) of adipocytes, as in Figure LG 4.1H.

b. List three or more functions of fat in the body.

c. Explain what gives the brown color to "brown fat," and state how this type of fat may be helpful to newborns.

✓▲ **C12.** *The Big Picture: Looking Ahead.* Refer to figures in your textbook and Table 4.3 (pages 122-127) to help you identify the dense connective tissues described below. Use the answers in the box. Use each answer once.

A.	**Aponeurosis**	**P.**	**Periosteum**
C.	**Capsule**	**SL.**	**Suspensory ligament**
EWA.	**Elastic walls of arteries**	**T.**	**Tendon**
F.	**Fascia**	**TVC.**	**True vocal cords**
L.	**Ligament**	**V.**	**Valves**

_____ a. Figure 6.3 shows the dense covering over bone which affords a pathway for blood to penetrate bone and also serves as a site of osteoblast cells that build new bone. _____

_____ b. Strong but flexible bands that attach bone to bone are shown in Figure 9.11. _____

_____ c. In Figure 9.14 a strong, fibrous band is shown attaching the quadriceps femoris muscle across the patella and eventually inserting it into bone (tibia). Similar fibrous bands attach forearm muscles into bones of the hand (Figure 11.17). _____

_____ d. Figure 11.10 shows broad, sheetlike attachments of external and internal oblique muscles. _____

_____ e. A broad sheet of fibrous tissue that provides support to the floor of the pelvic cavity (urogenital diaphragm) is shown in Figure 11.12. _____

_____ f. Figure 20.4 demonstrates the strong connective tissue leaflets that direct the flow of blood within the heart. _____

_____ g. Figure 21.1 shows the tissue that provides blood vessel's walls with the elasticity needed as the heart regularly pumps blood into these vessels. _____

_____ h. Figure 23.7 demonstrates the structures that require strength yet flexibility as their continual alteration of tension changes vocal quality. _____

_____ i. Connective tissue surrounds, supports, and protects each kidney (Figure 26.4). _____

_____ j. Figure 28.3 shows the tissue that suspends and supports the penis. _____

✓ **C13.** Now identify the category to which each of the connective tissues listed in Activity C12 belongs. On the lines following each description above, write DR for dense regular, DI for dense irregular, or E for elastic.

✓ **C14.** Do this activity about cartilage.
a. In general, cartilage can endure *(more? less?)* stress than other connective tissues studied so far.
b. The strength of cartilage is due to its *(collagen? elastic? reticular?)* fibers, and its resilience is related to the presence of *(chondroitin? dermatan? keratan?)* sulfate.

c. Cartilage heals *(more? less?)* rapidly than bone. Explain why this is so.

d. Growth of cartilage occurs by two mechanisms. In childhood and adolescent years, cartilage increases in size by *(appositional? interstitial?)* growth by enlargement of existing cells within the cartilage.

e. Post adolescence, cartilage increases in thickness by *(appositional? interstitial?)*

    growth. This process involves cells in the _____ that covers cartilage. Here cells known as *(chondroblasts? chondrocytes? fibroblasts?)* differentiate

    into _____, which lay down matrix and eventually become mature chondrocytes.

✓▲ **C15.** Match the types of cartilage with the descriptions given.

E. Elastic	**F. Fibrous**	**H. Hyaline**

_____ a. Found where strength and rigidity are needed, as in discs between vertebrae and in symphysis pubis

_____ b. White, glossy cartilage covering ends of bones (articular), covering ends of ribs (costal), and giving strength to nose, larynx, and trachea

_____ c. Provides strength and flexibility, as in external part of ear

_____ d. Forms menisci (cartilage pads) of the knees

_____ e. Forms most of the skeleton of the human embryo

✓ **C16.** Complete this exercise about bone tissue.

a. Bone tissue is also known as _____ tissue. Compact bone consists of concentric rings or *(lamellae? canaliculi?)* with bone cells, called *(chondrocytes?*

    osteocytes?), located in tiny spaces called _____ . Nutrients in blood

    reach osteocytes by vessels located in the _____ canal and then via

    minute, radiating canals called _____ that extend out to lacunae. The basic unit of structure and function of compact bone is known as an

    _____ or Haversian system.

b. Spongy bone *(is? is not?)* arranged in osteons. Spongy bone consists of plates of bone

    called _____ .

c. List three or more functions of bone tissue.

✓ **C17.** Fill in the blanks about blood tissue.

    Blood, or _____ tissue, consists of a fluid called _____

    containing three types of formed elements. Red blood cells, or _____ , transport oxygen and carbon dioxide; white cells, also known as

    _____ , provide defense for the body; thrombocytes assist in the

    function of blood _____ .

C18. Now label the following cell types found in diagrams J, K, or L of Figure LG 4.1: *chondrocytes, osteocytes, erythrocytes, leukocytes, thrombocytes.*

D. Membranes (pages 129–130)

D1. Complete the table about the four types of membranes in the body.

Type of Membrane	Location	Example of Specific Location	Function (s)
a.	Lines body cavity leading to exterior		
b. Serous			Allows organs to glide easily over each other
c.		Lines knee and hip joints	
d.	Covers body surface	Skin	

✓▲ **D2.** *The Big Picture: Looking Ahead.* Refer to figures in your textbook to help you identify the types of membranes described below. Use the answers in the box.

Cutaneous	Pericardial	Pleural
Mucous	Peritoneal	Synovial

a. A _____ membrane is another name for skin, shown in Figure 5.1.

b. Figure 9.3 shows a _____ membrane that secretes a lubricating fluid that reduces friction at joints. This type of membrane contains connective (rather than epithelial) tissue.

c. On Figure 20.2, the serous membrane that covers the heart is shown. This is the

_____ membrane. The same figure shows the _____ membrane covering the lungs and lining the chest.

d. Figure 24.2 contrasts the membrane lining the inside of digestive organs, known as a

_____ membrane, with the _____ membrane that covers the outside of such organs. Figure 26.22 shows similar membranes of the

urinary bladder: the pink _____ membrane with a rippled pattern

(rugae) inside the organ and the _____ membrane over the outside of the bladder.

✓▲ **D3.** Describe specific layers of serous and mucous membranes in this exercise.
  a. The layer of a serous membrane that adheres to an organ, such as the liver, is the *(parietal? visceral?)* layer, whereas the portion lining the body cavity wall is the *(parietal? visceral?)* layer.

b. The epithelial layer of mucous membranes varies, with *(simple columnar? pseudostratified columnar? stratified squamous?)* lining the mouth, _____ lining the small intestine, and _____ lining the large airways to the lungs.

c. The connective tissue of a mucous membrane is known as the lamina _____. Unlike the surface epithelial layer, the lamina propria *(is? is not?)* vascular.

D4. *A clinical correlation.* Serous membranes are normally relatively free of microorganisms, whereas mucous and cutaneous membranes have many microorganisms on them. Based on this information, state a rationale for the administration of antibiotics to patients undergoing stomach or intestinal surgery.

E. Muscle tissue (pages 130–132)

✓ **E1.** After you study each muscle tissue type in the text, refer to diagrams of muscle tissues in Figure LG 4.1, M-O. Write the following information on the lines provided on the figure.

✓   a. The name of each muscle tissue

    b. One or more locations of the tissue

    c. One or more functions of the tissue

    Now color the following structures on each diagram, using the same colors as you did for diagrams A-L.

    ◯ Nucleus   ◯ Cytoplasm   ◯ Intercellular material

✓   From this activity, you can conclude that muscle tissues appear to consist mostly of *(cells? intercellular material [matrix]?)*

✓ **E2.** Check your understanding of the three muscle types listed in the box by selecting types that best fit descriptions below.

C. Cardiac	Sk. Skeletal	Sm. Smooth

_____ a. Tissue forming most of the wall of the heart

_____ b. Attached to bones

_____ c. Spindle-shaped cells with ends tapering to points

_____ d. Contains intercalated discs

_____ e. Found in walls of intestine, urinary bladder, and blood vessels

_____ f. Voluntary muscle

F. Nervous tissue (page 132)

F1. Refer to diagram P in Figure LG 4.1. Write the following information on the lines provided on the figure.

✓   a. The name of the tissue

    b. One or more locations of the tissue

    c. One or more functions of the tissue

✓ **F2.** *(Axons? Dendrites?)* carry nerve impulses toward the cell body of a neuron, whereas

_____ transmit nerve impulses away from the cell body.

G. Tissue repair: restoring homeostasis; aging and tissues (pages 132–134)

✓ **G1.** Circle the type of tissue in each case that is more capable of repair:
   a. Cardiac muscle - Smooth muscle
   b. Bone - Cartilage
   c. Nervous tissue - Epithelium
   d. Blood - Tendon

✓ **G2.** Define a *stem cell*.

   List three locations where stem cells serve as sources of new cells throughout life.

✓ **G3.** Choose the correct term to complete each sentence. Write P for parenchymal or
        S for stromal.
   a. The part of an organ that consists of functioning cells, such as secreting epithelial

       cells lining the intestine, is composed of _____ tissue.
   b. The connective tissue cells that support the functional cells of the organ are called

       _____ cells.

   c. If only the _____ cells are involved in the repair process, the repair will be close
       to perfect.

   d. If _____ cells are involved in the repair, they will lay down fibrous tissue known
       as a fibrosis or a scar.

   G4. Define each of the following terms related to tissue repair.
   a. Regeneration

   b. Granulation tissue

   c. Adhesions

✓ **G5.** List three factors that affect the quality of tissue repair.

Describe two roles of vitamin C in tissue repair.

G6. Explain what changes in collagen and elastin result in aging of tissues.

H. Disorders and medical terminology (pages 134–135)

✓ **H1.** Check your understanding of two tissue disorders, Sjögren's syndrome and systemic lupus erythematosus (SLE), by filling in the table in this Checkpoint.

Characteristic of disease	Sjögren's syndrome	SLE (lupus)
a. Autoimmune *(Yes? No?)*		
b. Type of tissue most affected?		
c. Gender most affected *(F? M?)*		
d. Distinguishing signs/symptoms		

H2. *Critical thinking.* An autograft refers to replacement of one part of a person's body (such as a traumatized bone or burned skin) with another part of the same person's body (such as a piece of hip bone or healthy skin). Contrast autograft with *xenotransplantation* (or *xenograft*) (see text page 135) and predict the success rate of the procedures.

ANSWERS TO SELECTED CHECKPOINTS: CHAPTER 4

A3. (a) Endo. (b) Ecto. (c) Meso. (d) Ecto.

A4. (a) Phagocyte. (b) Junctions; tight, urinary bladder. (c) Desmosome; epidermis of skin, cardiac muscle, lining of the digestive tract. (d) Connexons; impulses. (e) Protein; adhesion.

B1. (a) Closely packed cells; many, tightly. (b) Do not; avascular; epithelial tissue is protective because it does not bleed when injured; however, the avascular epithelial cells must draw nutrients from adjacent tissues. (c) Has. (d) Does.

B2. (a) Basal, basement. (b) Extracellular material; epithelium; collagen, basal lamina. (c) Connective tissue, reticular.

B4. Lining, glandular.

B5. (a) Simple; stratified. (b) Squamous; columnar; cuboidal. (c) transitional.

B6. (A) Simple squamous epithelium. (B) Simple cuboidal epithelium. (C) Simple columnar epithelium (unciliated). (D) Simple columnar epithelium (ciliated). (E) Stratified squamous epithelium. (F) Pseudostratified columnar epithelium, ciliated.

B7. (a) Simple columnar, nonciliated. (b) Stratified squamous. (c) Simple cuboidal. (d, e) Simple squamous. (f) Simple squamous. (g) Simple cuboidal.

B8. (a) One, all, only some. (b) Superficial. (c) More.

B11. (a) EXO. (b) ENDO. (c) EXO. (d) ENDO.

B12. (a) Compound. (b) Uni. (c) Acinar. (d) Mero; only the secretion.

C1. (a) Epithelial tissue. (b) Connective tissue (except cartilage). (c) Connective tissue.

C2. Fluid, semifluid, gelatinous, fibrous, or calcified; usually secreted by connective tissue (or adjacent) cells.

C3. (a) Meso-; mesenchymal. (b) Immature; immature. (c) -blast, -cyte; osteoblast, chondroblast, osteocyte.

C4. (a) Plasma cell. (b) Macrophage. (c) Adipocyte. (d) Fibroblast. (e) Mast cell. (f) WBCs (white blood cells).

C5. Large, proteins. (a) AP. (b) HA. (c) KS. (d) CS. (e) DS.

C6. (a) R. (b) C. (c) E. (d) E.

C7. (a) Mesenchyme; little; Wharton's, mature.

C8. (G) Loose (areolar) connective tissue. (H) Adipose tissue. (I) Dense connective tissue. (J) Cartilage. (K) Osseous (bone) tissue. (L) Vascular (blood) tissue. Intercellular material (matrix).

C10. Loose (areolar) connective tissue; adipose tissue.

C11. (a) Most of the cell consists of a globule of stored triglycerides (fat); the nucleus and the rest of the cytoplasm are located in a narrow ring around the fat globule. (b) Fat insulates the body, supports and protects organs, and serves as a major source of energy. (c) This type of fat has an abundance of blood vessels as well as mitochondria (which appear brown due to the presence of copper-containing cytochromes). Brown fat may help to generate heat to maintain proper body temperature in newborns.

C12. 13. (a) P, DI. (b) L, DR. (c) T, DR. (d) A, DR. (e) F, DI. (f) V, DI. (g) EWA, E. (h) TVC, E. (i) C, DI. (j) SL, E.

C14. (a) More. (b) Collagen, chondroitin. (c) Less; cartilage is avascular, so chemicals needed for repair must reach cartilage by diffusion from the perichondrium. (d) Interstitial. (e) Appositional; perichondrium; fibroblasts, chondroblasts.

C15. (a) F. (b) H. (c) E. (d) F. (e) H.

C16. (a) Osseous; lamellae, osteocytes, lacunae; central (Haversian), canaliculi; osteon. (b) Is not; trabeculae. (c) Supports and protects, plays a role in movement, stores mineral salts, houses red and yellow marrow, including lipids.

C17. Vascular, plasma; erythrocytes, leukocytes, clotting.

D2. (a) Cutaneous. (b) Synovial. (c) Pericardial; pleural. (d) Mucous, peritoneal (or serous); mucous, peritoneal (or serous).

D3. (a) Visceral, parietal. (b) Stratified squamous, simple columnar, pseudostratified columnar. (c) Propria; is.

D4. Microorganisms from skin or from within the stomach or intestine may be introduced into the peritoneal cavity during surgery; the result may be an inflammation of the serous membrane (peritonitis).

E1. (M) Skeletal muscle. (N) Cardiac muscle. (O) Smooth muscle. Cells [known as muscle fibers].

E2. (a) C. (b) Sk. (c) Sm. (d) C. (e) Sm. (f) Sk.

F1. Nervous tissue (neurons).

F2. Dendrites, axons (*Hint:* Remember A for Away from cell body.)

G1. (a) Smooth muscle. (b) Bone. (c) Epithelium. (d) Blood.

G2. Immature, undifferentiated cell; skin, lining of the GI tract, and bone marrow.

G3. (a) P. (b) S. (c) P. (d) S.

G5. Nutrition, circulation, and (younger) age; affects quality of collagen and blood vessels.

H1. (a) Yes, Yes. (b) Exocrine glands; connective tissues. (c) F, F. (d) Sjögren's: dry mouth and dry eyes; also arthritis. SLE: butterfly rash, sensitivity to light, painful joints; possibly serious damage to organs such as kidneys.

MORE CRITICAL THINKING: CHAPTER 4

1. The outer part (epidermis) of skin is composed of epithelium. Describe what makes epithelium suitable for this location.
2. Which tissue is likely to heal better: bone or cartilage? Explain why.
3. Discuss the process of differentiation in development of tissues by explaining the roles of mesenchyme and stem cells.
4. Surgical repair on fetuses (in utero) leaves no surgical scar. Explain why this is so.
5. Contrast hypertrophy and hyperplasia. (*Hint:* See Medical Terminology in Chapter 2.) Predict which types of tissues can grow in adults by hypertrophy but not by hyperplasia. State your rationale.

MASTERY TEST: CHAPTER 4

Questions 1-6: Circle the letter preceding the one best answer to each question.

1. A surgeon performing abdominal surgery will pass through the skin, and then loose connective tissue (subcutaneous), then fascia covering muscle and

   muscle tissue itself to reach the _____ membrane lining the inside wall of the abdomen.
   A. Parietal pleura   D. Visceral pleura
   B. Parietal pericardium   E. Visceral pericardium
   C. Parietal peritoneum   F. Visceral peritoneum

2. The type of tissue that covers body surfaces, lines body cavities, and forms glands is:
   A. Nervous   C. Connective
   B. Muscular   D. Epithelial

3. Which of the four types of tissue is the most abundant and most widely distributed tissue in the body?
   A. Nervous tissue   C. Connective tissue
   B. Muscle tissue   D. Epithelial tissue

4. Which statement about connective tissue is *false*?
   A. Cells are very closely packed together.
   B. Most connective tissues have an abundant blood supply.
   C. Intercellular substance is present in large amounts.
   D. It is the most abundant tissue in the body.

5. Modified columnar cells that are unicellular glands secreting mucus are known as:
   A. Cilia   C. Goblet cells
   B. Microvilli   D. Branched tubular glands

6. Which tissues are avascular?
   A. Skeletal muscle and cardiac muscle
   B. Bone and dense connective tissue
   C. Cartilage and epithelium
   D. Areolar and adipose tissues.

Questions 7-12: Match tissue types with correct descriptions

Adipose	**Simple columnar epithelium**
Cartilage	**Simple squamous epithelium**
Dense connective tissue	**Stratified squamous epithelium**

7. Contains lacunae and chondrocytes: _____

8. Forms fascia, tendons, and deeper layers of dermis of skin: _____
9. Forms thick surface layer of skin on hands and feet, providing extra protection:

   _____

10. Single layer of flat, scalelike cells: _____

11. Lines stomach and intestine: _____

12. Fat tissue: _____

Questions 13-20: Circle T (true) or F (false). If the statement is false, change the underlined word or phrase so that the statement is correct.

T F   13. Stratified transitional epithelium lines the <u>urinary bladder</u>.
T F   14. About 75% of lipids in plasma membranes are <u>glycolipids</u>.
T F   15. Hormones are classified as <u>exocrine</u> secretions.
T F   16. Elastic connective tissue, because it provides both stretch and strength, is found in <u>walls of elastic arteries, in the vocal cords, and in some ligaments</u>.
T F   17. The surface attachment between epithelium and connective tissue is called <u>basement membrane</u>.
T F   18. <u>Simple</u> squamous epithelium is most likely to line the areas of the body that are subject to wear and tear.
T F   19. Marfan syndrome is a <u>muscle tissue</u> disorder that results in persons being short with short extremities.
T F   20. Appositional growth of cartilage <u>starts later in life than interstitial growth and continues throughout life</u>.

Questions 21-25: Fill-ins. Complete each sentence with the word that best fits.

_____ 21. The type of cell junction that forms a "zip-lock" is a _____ junction.

_____ 22. _____ is a dense, irregular connective tissue covering over bone.

_____ 23. The kind of tissue that lines alveoli (air sacs) of lungs is _____.

_____ 24. Another term for intercellular is _____.

_____ 25. All glands are formed of the tissue type named _____.

ANSWERS TO MASTERY TEST: CHAPTER 4

Multiple Choice

1. C   4. A
2. D   5. C
3. C   6. C

Matching

7. Cartilage
8. Dense connective tissue
9. Stratified squamous epithelium
10. Simple squamous epithelium
11. Simple columnar epithelium
12. Adipose

True-False

13. T
14. F. Phospholipids
15. F. Endocrine
16. T
17. T
18. F. Stratified
19. F. Connective tissue; tall with long extremities
20. T

Fill-ins

21. Tight
22. Periosteum
23. Simple squamous epithelium
24. Extracellular or interstitial
25. Epithelium

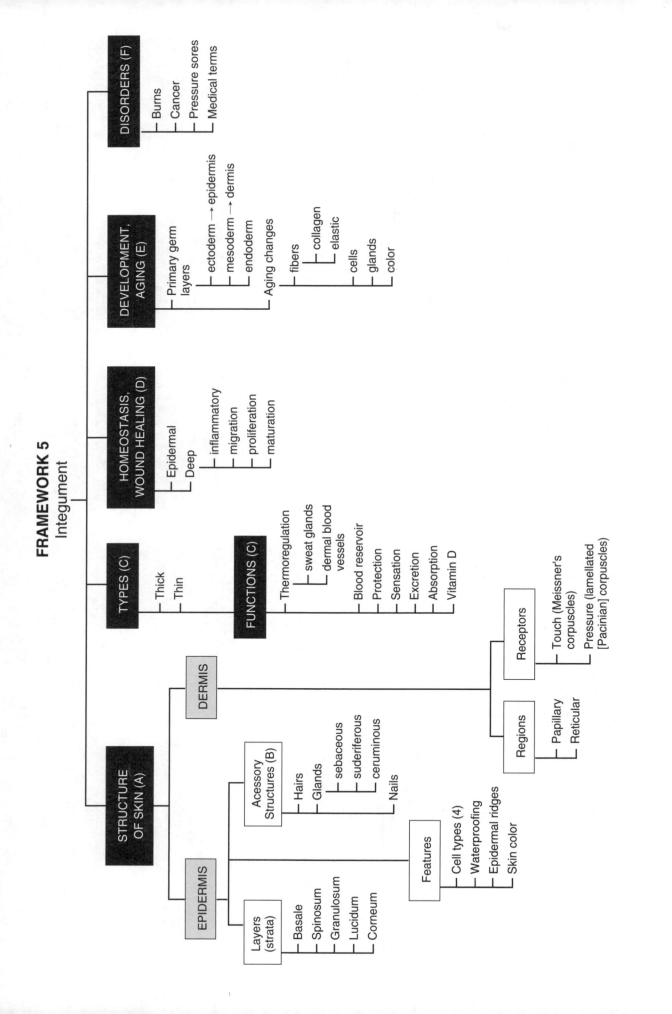

FRAMEWORK 5
Integument

The Integumentary System

In your human body tour you have now arrived at your first system, the integumentary system, which envelops the entire body. Composed of the skin, hair, nails, and glands, the integument serves as both a barrier and a link with the environment. The many waterproofed layers of cells, as well as the pigmentation of skin, afford protection to the body. Nails, glands, and even hairs offer additional fortification. Sense receptors and blood vessels in skin increase awareness of conditions around the body and facilitate appropriate responses. Even so, this microscopically thin surface cover may be traumatized, or invaded, or it may simply succumb to the passage of time. The integument then demands repair and healing for continued maintenance of homeostasis.

As you begin your study of the integumentary system, carefully examine the Chapter 5 Framework and note relationships among concepts and key terms there. Also refer to the Topic Outline and Objectives; you may want to check off each objective as you complete it.

TOPIC OUTLINE AND OBJECTIVES

A. Structure of the skin

1. Describe the various layers of the epidermis and dermis, and the cells that compose them.
2. Explain the basis for skin color.

B. Accessory structure of the skin

3. Compare the structure, distribution, and function of hair, skin glands, and nails.

C. Types of skin; functions of the skin

4. Compare structural and functional differences in thin and thick skin.
5. Describe how the skin contributes to thermoregulation, protection, sensation, excretion and absorption, and synthesis of vitamin D.

D. Maintaining homeostasis: skin wound healing

6. Explain how epidermal wounds and deep wounds heal.

E. Developmental anatomy and aging of the integumentary system

7. Describe the development of the epidermis, its accessory structures, and the dermis.
8. Describe the effects of aging on the integumentary system.

F. Disorders and medical terminology

WORDBYTES

Now become familiar with the language of this chapter by studying each wordbyte, its meaning, and an example of its use within a term. After you study the entire list, self-check your understanding by writing the meaning of each wordbyte on the line. As you continue through the *Learning Guide*, identify (and fill in) additional terms that contain the same wordbyte.

Wordbyte	Self-check	Meaning	Example(s)
albin-	_____	white	*albin*ism
basale	_____	base	stratum *basale*
cera-	_____	wax	*cer*umen
corneum	_____	horny	stratum *corneum*
cut-	_____	skin	*cut*aneous
derm-	_____	skin	*derm*atologist
epi-	_____	over	*epi*dermis
fer-	_____	carry	sudori*fer*ous
granulum	_____	little grain	stratum *granulosum*
lucidus	_____	clear	stratum *lucidum*
melan-	_____	black	*melan*oma
seb-	_____	fat	*seb*aceous
spinosum	_____	thornlike, prickly	stratum *spinosum*
sub-	_____	under	*sub*cutaneous
sudor-	_____	sweat	*sudor*iferous

CHECKPOINTS

A. Structure of the skin (pages 140–145)

A1. Name the structures included in the integumentary system.

A2. Skin may be one of the most underestimated organs in the body. What functions does your skin perform while it is "just lying there" covering your body? List some functions on the lines provided.

_____	_____
_____	_____

○　Arrector pili (hair) muscle
○　Blood vessel
○　Hair follicle
○　Sebaceous (oil) gland
○　Sudoriferous (sweat) gland
○　Touch (Meissner's) receptor
○　Pressure (lamellated or Pacinian) corpuscle
○　Nerve

Figure LG 5.1 Structure of the skin. Label as directed in Checkpoints A4, A7, and A11.

✓ **A3.** Skin typically covers about _____ square meters in the adult body. The thickness of

skin averages about _____ mm. In which part of the body is the thickest skin

located? _____

✓ **A4.** Answer these questions about the two portions of skin. Label them (bracketed
regions) on Figure LG 5.1.
　a. The outer layer is named the _____. It is composed of *(connective
tissue? epithelium?)*.
　b. The inner portion of skin, called the _____, is made of *(connective
tissue? epithelium?)*. The dermis is *(thicker? thinner?)* than the epidermis.

✓▲ **A5.** The tissue underlying skin is called *subcutaneous*, meaning _____.

This layer is also called _____. It consists of two types of tissue,

_____ and _____. What functions does subcutaneous tissue serve?

✓ **A6.** Epidermis contains four distinct cell types. Fill in the name of the cell type that fits each description.
  a. The most numerous cell type, this cell produces the protein keratin, which helps to

     waterproof skin: _____.
  b. This type of cell produces the pigments that give skin its color and protect against

     ultraviolet (UV) light: _____.
  c. These cells, derived from macrophages, function in immunity: _____.
  d. Located in the deepest layer of the epidermis, these cells may contribute to the

     sensation of touch: _____.

✓ **A7.** Label each of the five layers on the left side of Figure LG 5.1.

✓ **A8.** Match the names of the epidermal layers (strata) with the descriptions below.

B.	**Basale**	**L.**	**Lucidum**
C.	**Corneum**	**S.**	**Spinosum**
G.	**Granulosum**		

_____ a. Also known as the stratum germinativum because new cells arise from stem cells located here; attached to the basement membrane.

_____ b. Stratum immediately superficial to the stratum basale; "prickly spines" help attach cells to adjacent cells, providing strength and flexibility.

_____ c. Cells in this layer are undergoing apoptosis; they contain keratohyalin as well as lamellar granules—the latter serving as a waterproofing agent.

_____ d. A clear layer that is normally found only in thick layers of palms and soles.

_____ e. Uppermost layer, consisting of 25 to 30 rows of flat, dead cells filled with keratin.

A9. Explain what happens to skin to cause *psoriasis*, and specifically, *dandruff*.

✓ **A10.** Describe the dermis in this exercise.
  a. The outer one-fifth of the dermis is known as the *(papillary? reticular?)* region. It consists of *(loose? dense?)* connective tissue. Present in fingerlike projections known

     as dermal _____ are Meissner's corpuscles, sense receptors sensitive to *(pressure? touch?)*.
  b. The remainder of the dermis is known as the _____ region, composed of *(loose? dense?)* connective tissue. Skin is strengthened by *(elastic ? collagenous?)* fibers in the reticular layer. Skin is extensible and elastic due to

     _____ fibers.

c. Epidermal ridges on palms, fingers, soles, and toes form the basis for

_____ and _____. When are these ridges formed?

_____. Do these patterns change during life? _____ What
determines the patterns of these ridges?

✓ **A11.** Most sensory receptors, nerves, blood vessels, and glands are embedded in the *(epidermis? dermis?)*. Fill in all label lines on the right side of Figure LG 5.1. Then color those structures and their related color code ovals.

✓ **A12.** Explain what accounts for skin color by doing this exercise.
a. Dark skin is due primarily to *(a larger number of melanocytes? greater melanin production per melanocyte?)*. Melanocytes are in greatest abundance in the *(dermis? epidermis?)*.

Melanin is produced in organelles called _____.

b. Melanin is derived from the amino acid _____. The enzyme that

converts tyrosine to melanin is _____. This enzyme is activated by

_____ light. In the inherited condition known as

_____, melanocytes cannot synthesize tyrosinase, so skin, hair,
and eyes lack melanin.
c. What are freckles and liver spots?

d. Carotene contributes a _____ color to skin.
e. What accounts for the pinker color of skin during blushing and acts as a cooling mechanism during exercise?

This condition is known as *(cyanosis? erythema?)*.

f. Jaundiced skin has a more *(blue? red? yellow?)* hue, often due to _____
problems. Cyanotic skin appears more *(blue? red? yellow?)* due to lack of oxygen and excessive carbon dioxide in the blood vessels of the skin.

B. Accessory structures of the skin (pages 145–150)

✓ **B1.** List the three types of accessory structures of skin. _____

_____ _____. These structures all develop from the
embryonic *(dermis? epidermis?)*.

✓ **B2.** Complete this exercise about the structure of a hair and its follicle.

a. Arrange the parts of a hair from superficial to deep: _____ _____ _____
A. Shaft　　　　　B. Bulb　　　　　C. Root
b. Arrange the layers of a hair from outermost to innermost: _____ _____ _____
A. Medulla　　　　B. Cuticle　　　　C. Cortex

c. A hair is composed of *(cells? no cells, but only secretions of cells?)*. What accounts for the fact that hairs are waterproofed?

d. Surrounding the root of a hair is the hair _____. The follicle consists of two parts. The *(external? internal?)* root sheath is an extension of deeper layers of the *(epidermis? dermis?)*.

    The internal root sheath consists of cells derived from the _____.

e. The _____ is the part of a hair follicle where cells undergo mitosis, permitting growth of a new hair. What is the function of the papilla of the hair?

B3. Describe the relationship between the following pairs of terms related to hairs:
a. Round hair—oval hair

b. Sebaceous glands—root of hair

c. Arrector pili muscle—"goose bump"

d. Growth stage—resting stage (in the growth cycle of a hair)

B4. Answer these questions about hair growth.

a. Scalp hair normally grows for about _____ years, and then rests for _____ months.

b. Normal hair loss is about _____ hairs/day. List several factors that can increase the rate of hair loss.

B5. Describe the different chemicals in hairs that account for these colors:
a. Brown or black

b. Blonde or red

c. Graying

d. White

✓ **B6.** Answer these questions about sebaceous glands.
   a. In which parts of the body are these glands largest?

   b. In which body parts are sebaceous glands absent?

   c. What functions are attributed to these glands?

   d. What is a "blackhead"? Is the blackness of a blackhead due to accumulated dirt?

✓ **B7.** Contrast types of sudoriferous (sweat) glands in this activity. Use answers in the box.

Apo.	**Apocrine**	**Ecc.**	**Eccrine**

_____ a. Of apocrine and eccrine, the more
           common type of gland

_____ b. Responsible for "sweaty palms" because
           these glands are found in large numbers
           in palms

_____ c. Associated with foot odor because these
           glands are found in large numbers on
           soles

_____ d. Produce most of the sweat in armpits
           (axillae) and in pubic areas

_____ e. Of apocrine and eccrine glands, secrete
           more viscous (less watery) fluid

✓ **B8.** Write one major function and one minor function of sweat (perspiration).

✓ **B9.** *For extra review.* Write the answers (types of glands) listed in the box that best fit the descriptions below.

C.	**Ceruminous**	**Seb.**	**Sebaceous**	**Sud.**	**Sudoriferous**

_____ a. Sweat glands

_____ b. Inflamed in acne

_____ c. Ear wax

_____ d. Most numerous in skin of the palms and soles

_____ e. Usually most superficial in location

Figure LG 5.2 Structure of nails. Label according to Checkpoint B10.

B10. Refer to Figure LG 5.2. Now look at one of your own fingernails. Label the following parts of the nail on the figure: *free edge, nail body, nail root, lunula,* and *eponychium (cuticle)*. Check your accuracy by consulting Figure 5.5 page 149 in your text.

✓ **B11.** Answer these questions about nails and hair.
   a. Are nails formed of cells? *(Yes? No?)* Do nails contain keratin? *(Yes? No?)* What type of tissue forms nails? *(Dermis? Epidermis?)*
   b. How is the function of the *nail matrix* similar to that of the *matrix of a hair?*

   c. Why does the nail body appear pink, yet the lunula and free edge appear white?

C. Types of skin; functions of the skin (pages 150–152)

   C1. Contrast *thick skin* with *thin skin* by completing this table.

Characteristic	Thick skin	Thin skin
a. Locations	Palms, palmar surfaces of digits, and soles	
b. Structure of strata		
c. Epidermal ridges *(Yes? No?)*		
d. Sweat glands *(Many? Few?)*		
e. Sensory receptors *(Many? Few?)*		

✓ **C2.** Complete this exercise about temperature regulation.

a. When the body temperature is too hot, as during vigorous exercise, nerve messages from the brain inform sweat glands to *(in? de?)*-crease sweat production. In addition, blood vessels in the dermis are directed to *(dilate? constrict?)*. Skin blood vessels contain about _____ of the total body blood of a resting adult.

b. Temperature regulation is an example of a *(negative? positive?)* feedback system because the response (cooling) is *(the same as? opposite to?)* the stimulus (heating).

C3. List six or more ways in which skin protects the human body.

C4. List five or more types of sensations that arise in the skin.

C5. Check your understanding of the excretive and absorptive functions of skin in this Checkpoint.

a. How much water per day is likely to be lost from skin by evaporation?

_____ mL = _____ cup(s)

b. List several waste products excreted by skin.

c. *(Fat? Water?)*-soluble vitamins can be absorbed by skin. These include vitamins ____,

____, ____, and ___. Sunlight on skin helps to activate vitamin ____, which is needed

for absorption of the mineral _____ from foods.

d. Administration of medications by a patch (or paste) on skin is known as _____ drug administration. Name four medications that can be absorbed in this way.

D. Maintaining homeostasis: skin wound healing (page 152)

✓ **D1.** Describe the process of epidermal wound healing in this exercise.

a. State two examples of epidermal wounds.

b. Usually the deepest part of the wound is the *(central? peripheral?)* region.

c. In the process of repair, epidermal cells of the stratum *(corneum? basale?)* break contact from the basement membrane. These are cells at the *(center? periphery?)* of the wound.

d. These basal cells migrate toward the center of the wound, stopping when they meet other similar advancing cells. This cessation of migration is an example of the

phenomenon known as _____.

e. Both the migrated cells and the remaining epithelial cells at the periphery undergo

_____ to fill in the epithelium up to a normal (or close to normal) level.

Name a chemical that stimulates mitosis in the healing wound. _____

D2. The process of deep wound healing involves four phases. List the three or four major events that occur in each phase.

a. Inflammatory

b. Migratory

c. Proliferative

d. Maturation

✓ **D3.** For *extra review* of deep would healing, match the phases in the box with descriptions below.

I.	Inflammation		Mig.	Migration
Mat.	Maturation		P.	Proliferation

_____ a. Blood clot temporarily unites edges of wound; blood vessels dilate so neutrophils enter to clean up area.

_____ b. Clot forms a scab; epithelial cells migrate into scab; fibroblasts also migrate to start scar tissue; pink granulation tissue contains delicate new blood vessels.

_____ c. Epithelium and blood vessels grow; fibroblasts lay down many fibers.

_____ d. Scab sloughs off; epidermis grows to normal thickness; collagenous fibers give added strength to healing tissue; blood vessels are more normal.

D4. Contrast these two terms: *hypertrophied scar/keloid*

E. Developmental anatomy and aging of the integumentary system (pages 152–154)

✓ **E1.** *The Big Picture: Looking Ahead.* Do this exercise by referring to specific figures and exhibits in Chapter 29 in your text.

   a. Identify and list the three primary germ layers (Figure 29.5b) developed from a fertil-

      ized egg: _____-derm, _____-derm, and _____-derm.

   b. The epidermis of skin develops from the _____-derm (Table 29.1). All epidermal layers are formed by the *(second? fourth?)* month of the nine-month human gestation period (Table 29.2).

   c. Nails form from the *(dermis? epidermis?)* and reach the ends of fingertips by the *(fourth? ninth?)* month of gestation (Table 29.2).

   d. Delicate fetal hair called _____ covers the body during the middle trimester of fetal development. It is normally *(present at? shed before?)* birth (Table 29.2), so its presence on a newborn is an indicator of prematurity.

   e. The dermis and its associated blood vessels derive from the _____-derm (Table 29.1).

   f. What is the function of *vernix caseosa*?

✓ **E2.** Complete the table relating observable changes in aging of the integument to their causes.

Changes	Causes
a. Wrinkles; skin springs back less when gently pinched	
b.	Macrophages become less efficient; decrease in number of Langerhans cells
c. Dry, easily broken skin	
d.	Decrease in number and size of melanocytes

✓ **E3.** List and briefly describe two or more factors that can help keep your skin healthy throughout your lifetime.

F. Disorders and medical terminology (pages 155–156)

✓ **F1.** Circle the risk factor for cancer in each pair.

a. Skin type:
   A. Fair with red hair
   B. Tans well with dark hair

b. Sun exposure:
   A. Lives in Canada at sea level
   B. Lives in high-altitude mountainous area
      of South America

c. Age:
   A. 30 years old
   B. 75 years old

d. Immunological status:
   A. Taking immunosuppressive drugs for
      cancer or following organ transplant
   B. Normal immune system

✓ **F2.** Identify characteristics of the three different classes of burns by completing this exercise.

a. In a first-degree burn, only the superficial layers of the *(dermis? epidermis?)* are involved.

   The tissue appears _____ in color. Blisters *(do? do not?)* form. Give
   one example of a first-degree burn.

b. Which parts of the skin are injured in a second-degree burn?

   Blisters usually *(do? do not?)* form. Epidermal derivatives, such as hair follicles and glands,
   (are? are not ?) injured. Healing usually occurs in about three to four *(days? weeks?)*.

c. Third-degree burns are called *(partial? full?)*-thickness burns. Such skin appears *(red
   and blistered? white, brown, or black and dry?)*. Such burned areas are usually *(painful?
   numb?)* because nerve endings are destroyed. Regeneration is usually *(rapid? slow?)*.

F3. Contrast systemic effects with local effects of burns.

✓ **F4.** *A clinical challenge.* Answer these questions about the Lund-Browder method of burns
   assessment.

a. What does the Lund-Browder method estimate? *(Depth of burn? Amount of body surface
   area burned?)* This method is based upon differences in body *(size? proportions?)* of age
   groups.

b. If an adult and a one-year-old each experience burns over the entire anterior surface
   of both legs, who is more burned? *(Adult? One-year-old?)* If both are burned over the
   entire anterior surface of the head, who is more burned? *(Adult? One-year-old?)*

✓ **F5.** Match the name of the disorder with the description given.

A.	**Acne**	**M.**	**Malignant melanoma**
B.	**Basal cell carcinoma**	**N.**	**Nevus**
D.	**Decubitus ulcers**	**P.**	**Pruritus**
I.	**Impetigo**		

_____ a. Pressure sores

_____ b. Mole

_____ c. Staphylococcal or streptococcal infection
            that may become epidemic in nurseries

_____ d. Inflammation of sebaceous glands

especially in chin area; occurs under
hormonal influence

_____ e. Rapidly metastasizing form of cancer

_____ f. The most common form of skin cancer

_____ g. Itching

F6. Contrast the following pairs of terms.

a. *Topical/intradermal*

b. *Corn/wart*

ANSWERS TO SELECTED CHECKPOINTS: CHAPTER 5

A3. 2; 1-2; heels.

A4. See Figure LG 5.1A. (a) Epidermis; epithelium. (b) Dermis, connective tissue; thicker.

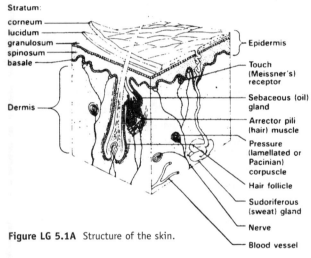

Figure LG 5.1A Structure of the skin.

A5. Under the skin; superficial fascia or hypodermis; areolar, adipose; anchors skin to underlying tissues and organs, insulates by fat here, contains blood vessels and pressure receptors.

A6. (a) Keratinocyte. (b) Melanocyte. (c) Langerhans cell. (d) Merkel cell.

A7. See Figure LG 5.1A above.

A8. (a) B. (b) S. (c) G. (d) L. (e) C.

A10. (a) Papillary; loose; papillae, touch. (b) Reticular, dense; collagenous; elastic. (c) Fingerprints and footprints; third and fourth fetal months; no; patterns of dermal papillae.

A11. Dermis. See Figure LG 5.1A.

A12. (a) Greater melanin production per melanocyte; epidermis; melanosomes. (b) Tyrosine; tyrosinase; ultraviolet (UV); albinism. (c) Accumulations of melanin. (d) Yellow. (e) Widening (dilation) of blood vessels; erythema. (f) Yellow, liver; blue.

B1. Hair, nails, and glands; epidermis.

B2. (a) A C B. (b) B CA. (c) Cells; keratin (especially in cuticle). (d) Follicle; external, epidermis; matrix. (e) Matrix (of bulb of hair follicle); contains blood vessels that nourish the hair.

B6. (a) In skin of the breast, face, neck, and upper chest. (b) Palms and soles. (c) Keep hair and skin from drying out, lower risk of dehydration and infection of skin. (d) Enlarged sebaceous gland with accumulated sebum; no, it is due to the presence of melanin and oxidation of the oil within sebum.

B7. (a-c) Ecc. (d-e) Apo.

B8. Major: regulation of body temperature; minor: eliminate wastes such as urea, uric acid, ammonia, lactic acid, and excess salts.

B9. (a) Sud. (b) Seb. (c) Cer. (d) Sud. (e) Seb.

B11. (a) Yes; yes; epidermis. (b) They bring about growth of nails and hairs, respectively. (c) The pink color is related to visibility of blood vessels deep to the nail body; the free edge has no tissue (so no blood vessels) deep to it, whereas the thickened stratum basale obscures blood vessels deep to the lunula.

C2. (a) In; dilate; 8–10(%). (c) Negative, opposite to.

D1. (a) Skinned knee, first- or second-degree burn. (b) Central. (c) Basale; periphery. (d) Contact inhibition. (e) Mitosis; epidermal growth factor (EGF).

D3. (a) I. (b) Mig. (c) P. (d) Mat.

E1. (a) Endo, meso, ecto. (b) Ecto; fourth (c) Epidermis; ninth. (d) Lanugo; shed before. (e) Meso. (f) This cheesy substance covers and protects fetal skin from amniotic fluid, and its slippery nature facilitates the birth process.

E2.

Changes	Causes
a. Wrinkles; skin springs back less when gently pinched	Elastic fibers thicken into clumps and fray
b. Increased susceptibility to skin infections and skin breakdown	Phagocytes become less efficient; decrease in number of Langerhans cells
c. Dry, easily broken skin	Decreased secretion of sebum by sebeceous glands; decreased sweat production
d. Gray or white hair; atypical skin pigmentation	Decrease in number and size of melanocytes

E3. Good nutrition, decreased stress, balance of rest and exercise, not smoking, protection from the sun.
F1. (a) A. (b) B. (c) B. (d) A.
F2. (a) Epidermis; redder; do not; typical sunburn.
(b) All of epidermis and possibly upper regions of

dermis; do; are not; weeks. (c) Full; white, brown, or black and dry; numb; slow.
F4. (a) Amount of body surface burned; proportions.
(b) Adult; one-year-old.
F5. (a) D. (b) N. (c) I. (d) A. (e) M. (f) B. (g) P.

MORE CRITICAL THINKING: CHAPTER 5

1. Recall (from Chapter 1) the definitions of an organ and a system. Is skin an organ or a system? State your rationale.
2. Skin may provide clues about the health of the body; in this case at least, you can tell a lot about a book by its cover. Discuss how a person's physical or emotional health state may be detected by inspection of skin.
3. Explain how skin helps you to regain and maintain homeostasis. State as many examples as you can think of.
4. Compare and contrast the structure of a hair with its follicle and a nail with its nail root.
5. Give scientific rationales for four readily observed aging changes of skin.

MASTERY TEST: CHAPTER 5

Questions 1-4: Circle the letter preceding the one best answer to each question.

1. "Goose bumps" occur as a result of:
A. Contraction of arrector pili muscles
B. Secretion of sebum
C. Contraction of elastic fibers in the bulb of the hair follicle
D. Contraction of papillae
2. Select the one *false* statement about the stratum basale.
A. It is the one layer of cells that can undergo cell division.
B. It consists of a single layer of squamous epithelial cells.
C. It is the stratum germinativum.
D. It is the deepest layer of the epidermis.

3. Select the one *false* statement.
A. Epidermis is composed of epithelium.
B. Dermis is composed of connective tissue.
C. Pressure-sensitive Pacinian corpuscles are normally more superficial in location than Meissner's touch receptors.
D. The amino acid tyrosine is necessary for production of the skin pigment melanin.
4. At about what age do epidermal ridges which cause fingerprints develop?
A. Months 3–4 in fetal development
B. Age 3–4 months
C. Age 3–4 years
D. Age 30–40 years

Questions 5–11: Arrange the answers in correct sequence.

_____ _____ _____ 5. From most serious to least serious type of burn:
A. First-degree
B. Second-degree
C. Third-degree
_____ _____ _____ 6. Deep wound healing involves four phases. List in order the phases following the inflammatory phase.
A. Migration
B. Maturation
C. Proliferation
_____ _____ _____ 7. From outside of hair to inside of hair:
A. Medulla
B. Cortex
C. Cuticle
_____ _____ _____ 8. From most superficial to deepest:
A. Dermis
B. Epidermis
C. Superficial fascia

__ __ __ 9. From most superficial to deepest:
  A. Stratum lucidum
  B. Stratum corneum
  C. Stratum germinativum

__ __ __ 10. From most to least abundant in epidermis:
  A. Melanocytes
  B. Merkel cells
  C. Keratinocytes

__ __ __ 11. From greatest to least in incidence (frequency of occurrence):
  A. Malignant melanoma
  B. Basal cell carcinoma
  C. Squamous cell carcinoma

Questions 12–20: Circle T (true) or F (false). If the statement is false, change the underlined word or phrase so that the statement is correct.

T F 12. The color of skin is due primarily to a pigment named <u>keratin</u>.
T F 13. The <u>outermost layers of epidermis</u> are composed of dead cells.
T F 14. Eccrine sweat glands are more numerous than apocrine sweat glands and are especially dense on <u>palms and soles</u>.
T F 15. The dermis consists of two regions; <u>the papillary region is more superficial, and the reticular region is deeper</u>.
T F 16. Temperature regulation is a <u>positive</u> feedback system.
T F 17. The <u>internal root sheath</u> is a downward continuation of the epidermis.
T F 18. <u>Both epidermis and dermis contain</u> blood vessels (<u>are vascular</u>).
T F 19. Hair, glands, and nails are all derived from the <u>dermis</u>.
T F 20. Hairs are <u>noncellular structures composed entirely of nonliving substances</u> secreted by follicle cells.

Questions 21–25: Fill-ins. Complete each sentence with the word or phrase that best fits.

__ 21. Skin contains a chemical that, under the influence of ultraviolet radiation, leads to formation of vitamin ___.
__ 22. The cells that are sloughed off as skin cells age and undergo keratinization are those of the stratum ___.
__ 23. Fingerprints are the result of a series of grooves called ___.
__ 24. The oily glandular secretion that keeps skin and hairs from drying is called ___.
__ 25. When the temperature of the body increases, nerve messages from brain to skin will decrease body temperature by ___.

ANSWERS TO MASTERY TEST: CHAPTER 5

Multiple Choice
1. A
2. B
3. C
4. A

Arrange
5. C B A
6. A C B
7. C B A
8. B A C
9. B A C
10. C A B
11. B C A

True-False
12. F. Melanin
13. T
14. T
15. T
16. F. Negative
17. F. External root sheath
18. F. Only the dermis contains, is vascular
19. F. Epidermis
20. F. Composed of different kinds of cells

Fill-ins
21. D
22. Corneum
23. Epidermal ridges or grooves
24. Sebum
25. Stimulating sweat glands to secrete and blood vessels to dilate

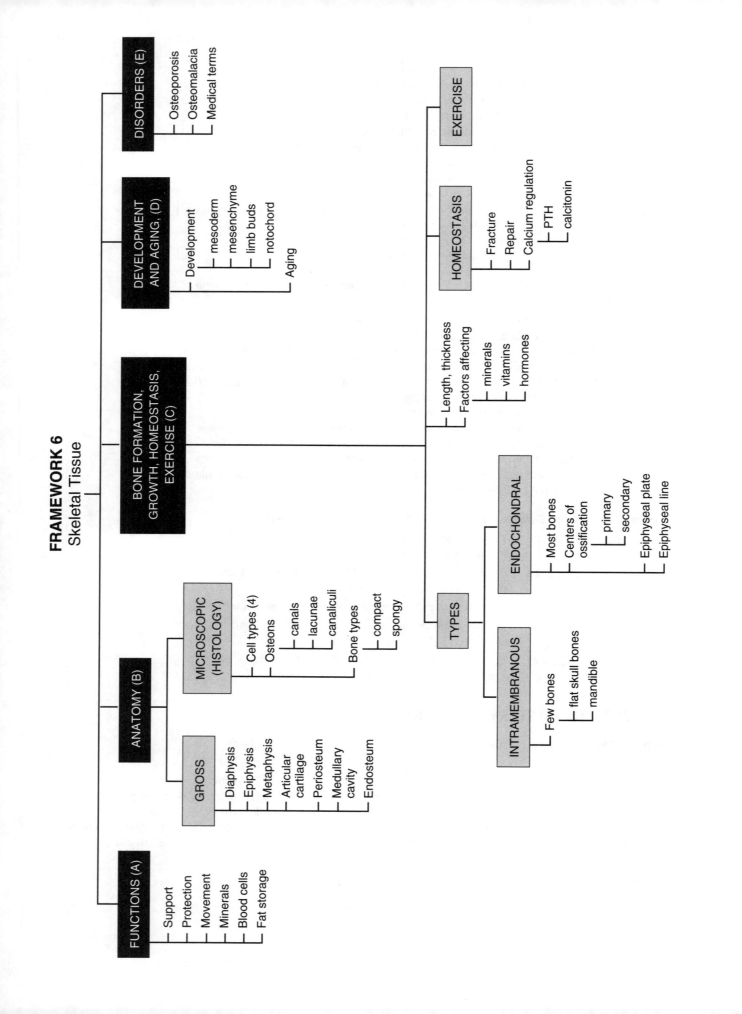

FRAMEWORK 6
Skeletal Tissue

FUNCTIONS (A)
- Support
- Protection
- Movement
- Minerals
- Blood cells
- Fat storage

ANATOMY (B)

GROSS
- Diaphysis
- Epiphysis
- Metaphysis
- Articular cartilage
- Periosteum
- Medullary cavity
- Endosteum

MICROSCOPIC (HISTOLOGY)
- Cell types (4)
- Osteons
  - canals
  - lacunae
  - canaliculi
- Bone types
  - compact
  - spongy

TYPES

INTRAMEMBRANOUS
- Few bones
  - flat skull bones
  - mandible

ENDOCHONDRAL
- Most bones
- Centers of ossification
  - primary
  - secondary
- Epiphyseal plate
- Epiphyseal line

BONE FORMATION, GROWTH, HOMEOSTASIS, EXERCISE (C)
- Length, thickness
- Factors affecting
  - minerals
  - vitamins
  - hormones

HOMEOSTASIS
- Fracture
- Repair
- Calcium regulation
  - PTH
  - calcitonin

EXERCISE

DEVELOPMENT AND AGING, (D)
- Development
  - mesoderm
  - mesenchyme
  - limb buds
  - notochord
- Aging

DISORDERS (E)
- Osteoporosis
- Osteomalacia
- Medical terms

The Skeletal System: Bone Tissue

The skeletal system provides the framework for the entire body, affording strength, support, and firm anchorage for the muscles that move the body. In this chapter, the first of Unit II, you will explore tissue that forms the skeleton. Bones provide rigid levers covered with connective tissue designed for joint formation and muscle attachment. Microscopically, bones present a variety of cell types intricately arranged in an osseous sea of calcified intercellular material. Bone growth or ossification occurs throughout life. Essential nutrients, hormones, and exercise regulate the growth and maintenance of the skeleton. Fractures and other disorders may result from a lack of such normal regulation.

As you begin your study of bone tissue, carefully examine the Chapter 6 Framework and note relationships among concepts and key terms there. Also refer to the Topic Outline and Objectives; you may want to check off each objective as you complete it.

TOPIC OUTLINE AND OBJECTIVES

A. Functions of the skeletal system

1. Discuss the functions of the skeletal system.

B. Structure of bone

2. Describe the components of a long bone.
3. Describe the histological features of compact and spongy bone.
4. Describe the blood and nerve supply of bone.

C. Bone formation and growth; bone and homeostasis; exercise and bone tissue

5. Describe the steps involved in intramembranous and endochondral ossification.
6. Describe how bone grows in length and thickness, and how hormones regulate bone growth.

7. Describe the processes involved in bone remodeling.
8. Describe the sequence of events in repair of a fracture.
9. Describe the role of calcium in bone homeostasis.
10. Describe how exercise and mechanical stress affect bone tissue.

D. Developmental anatomy and aging of the skeletal system

11. Describe the development of the skeletal system and the limbs.
12. Describe the effects of aging on bone tissue.

E. Disorders and medical terminology

WORDBYTES

Now become familiar with the language of this chapter by studying each wordbyte, its meaning, and an example of its use within a term. After you study the entire list, self-check your understanding by writing the meaning of each wordbyte on the line. As you continue through the *Learning Guide*, identify (and fill in) additional terms that contain the same wordbyte.

Wordbyte	Self-check	Meaning	Example(s)
-blast	_____	germ, to form	osteo*blast*
chondr-	_____	cartilage	peri*chondr*ium
-clast	_____	to break	osteo*clast*
dia-	_____	through	*dia*physis
endo-	_____	within	*endo*steum
epi-	_____	over	*epi*physis
meta-	_____	after, beyond	*meta*physis
os-, osteo-	_____	bone	*os*sification, *osteo*cyte
peri-	_____	around	*peri*osteum
-physis	_____	growing	dia*physis*

CHECKPOINTS

A. Functions of the skeletal system (page 160)

✓ **A1.** List six functions of the skeletal system.

_____   _____

_____   _____

_____   _____

✓ **A2.** What other systems of the body depend on a healthy skeletal system? Explain why in each case.

A3. Define *osteology*.

B. Structure of bone (pages 161–166)

B1. Name several types of tissue making up the skeletal system.

A single bone, such as one of your finger bones, is considered a(n) *(organ? system? tissue?)*.

○ Articular cartilage
○ Compact bone
○ Spongy bone
○ Periosteum
○ Endosteum

Figure LG 6.1 Diagram of a long bone that has been partially sectioned lengthwise. Color and label as directed in Checkpoints B2 and C5.

✓▲ **B2.** On Figure LG 6.1, label the *diaphysis, epiphysis, medullary (marrow) cavity*, and *nutrient foramen*. Then color the parts of a long bone using color code ovals on the figure.

✓ **B3.** Match the names of parts of a long bone listed in the box with the descriptions below.

A.	**Articular cartilage**	**M.**	**Metaphysis**
E.	**Endosteum**	**P.**	**Periosteum**

_____ a. Thin layer of hyaline cartilage at end of long bone

_____ b. Region of mature bone where diaphysis joins epiphysis

_____ c. Fibrous covering over bone into which tendons and ligaments attach; inner portion is site of osteogenic cells

_____ d. Layer of bone cells lining the marrow cavity

✓ **B4.** In the following exercise, bone cells are described in the sequence in which they form, and later destroy, bone. Write the name of the correct bone type after its description.

a. Derived from mesenchyme, these cells can undergo mitosis and differentiate into osteoblasts: _____ ↓

b. Form bone initially by secreting mineral salts and fibers; not mitotic: _____ ↓

c. Mature bone cells; maintain bone tissue; not mitotic. _____

d. Huge cells in the endosteum; powerful lysosomal enzymes permit these cells to resorb (degrade) bone in bone repair, remodeling, and aging. _____

✓ **B5.** Describe the components of bone by doing this exercise.
   a. Typical of all connective tissues, bone consists mainly of *(cells? intercellular material?)*.
   b. The intercellular substance of bone is unique among connective tissues. Protein fibers form about *(25? 50?)* % of the weight of bone, whereas mineral salts account for

      about _____% of the weight of bone.

   c. The two main salts present in bone are _____ and _____.
   d. The *(hardness? tensile strength?)* of bones is related to inorganic chemicals in bone,

      namely the mineral salts deposited around _____ fibers. Resistance to being stretched or torn apart, a property of bone known as *(hardness? tensile strength?)* is afforded by *(mineral salts? collagen fibers?)*.
   e. Bone *(is completely solid? contains some spaces?)*. Write two advantages of this structural feature of bone.

Figure LG 6.2 Osteons (Haversian systems) of compact bone.
Identify lettered structures and color as directed in Checkpoint B6.

✓ **B6.** Refer to Figure LG 6.2 and complete this exercise about bone structure.

   a. Compact bone is arranged in concentric circle patterns known as

      _____. Each individual concentric layer of bone is known as a

      _____ labeled with letter _____ in the figure.
   b. The osteon pattern of compact bone permits blood vessels and nerves to supply bone cells trapped in hard bone tissue. Blood vessels and nerves penetrate bone from the

      periosteum, labeled with letter _____ in the figure. These structures then pass through horizontal canals, labeled _____ and known as perforating (Volkmann's)

canals. These vessels and nerves finally pass into microscopic channels, labeled _____, in the center of each osteon. Color blood vessels red and blue in the figure.

c. Mature bone cells, known as _____, are located relatively far apart

in bone tissue. These are present in "little lakes," or _____, labeled

_____ in the figure. Color ten lacunae green.

d. Minute canals, known as canaliculi, are labeled with letter _____. What are the functions of these channels?

e. Compact bone is strengthened by the arrangement of osteons *(perpendicular to? along the same direction as?)* lines of stress. These lines of stress (and therefore the arrangement of osteons) *(remain constant? change?)* over a lifetime.

B7. Contrast these four types of lamellae: *concentric, interstitial, outer circumferential, inner circumferential.* Which type is found within osteons?

✓ **B8.** Do this exercise about spongy bone.
a. Spongy bone is arranged in *(osteons? trabeculae?)*, which may be defined as:

_____

b. Spongy bone makes up most of the *(diaphyses? epiphyses?)* of long bones. Spongy bone *(is not? is also)* located within bones that are short, flat, and irregular in shape.

c. *A clinical correlation.* Red blood cell formation (hemopoiesis) normally takes place in *(all? only certain areas of?)* spongy bone tissue. Name four or more bones in which this process takes place.

_____   _____

_____   _____

Of what clinical importance is this information?

B9. Describe a *bone scan* in this exercise.
a. Contrast the diagnostic interpretation of a "hot spot" versus a "cold spot."

b. What diagnostic advantages do bone scans offer over traditional x-rays?

B10. Discuss the blood and nerve supply to bones in this activity.

a. Contrast the blood supply *to* bones (arteries) with that *from* bones (veins).

b. Explain why cancer of the bone is particularly painful.

C. Bone formation and growth; bone and homeostasis; exercise and bone tissue (pages 166–176)

✓ **C1.** Complete this overview of bone formation.

a. Bone formation is also called _____.

b. All bones develop from preexisting *(epithelial? connective? muscle?)* tissue,

specifically from _____. This tissue then differentiates into either bone or cartilage, beginning at the sixth or seventh *(week? month?)* of development.

c. Two methods of bone formation occur in the human body. Name the bones that are formed by each of these methods of ossification:

endochondral _____

intramembranous _____

d. These two methods of bone formation described above *(do? do not?)* lead to different types of bone structurally.

✓ **C2.** Refer to Figure 7.4 in your text. Locate the parietal bone, a typical "flat bone" forming in the fetal skull. Describe the intramembranous formation of this bone by using each of the terms in the box once.

Compact	Mesenchyme	Osteogenic
Fontanel	Ossification	Protection
Growth	Osteoblast	Spongy
Marrow	Osteocyte	

a. Fibrous membranes surrounding the head of the developing baby contain embryonic

_____ cells. These cells cluster in areas known as centers of

_____, and these mark areas where bone tissue will form.

b. Differentiation of these cells leads to _____ cells and then into

_____ and _____ cells that surround themselves with organic and inorganic matrix forming spongy bone. (For help, look back at Checkpoint B4, page LG 105.)

c. Eventually, mesenchyme on the outside of the bone develops into a periosteal covering over the skull bone. This lays down some compact bone over spongy bone. The resulting typical flat bone is analogous to a flattened jelly sandwich: two firm (flat)

"slices of bread" comparable to _____ bone with a "jelly" of

_____ bone filled with red _____.

d. By birth, fetal skull bones are still not completed. Some fibrous membranes still

remain at points where skull bones meet, known as _____. Of what

advantage are these "soft spots"? They permit additional _____

of skull bones as well as safer passage (_____) of the fetal head during birth.

✓ **C3.** Endochondral ossification refers to bone formation from an initial model made of *(fibrous membranes? cartilage ?)*. Increase your understanding of this process by building your own outline in this exercise.

a. First identify the four major steps in endochondral ossification by filling in the five long lines (I–V) below with the correct terms selected from those in the box below.

Articular cartilage and epiphyseal plate	**Primary ossification center**
Development	**Secondary ossification center**
Growth	

I. _____ of the cartilage model   ____ ____

II. _____ of the cartilage model   ____ ____ ____

Ill. Development of the _____   ____ ____ ____

IV. Development of the _____   ____ ____ ____

V. Formation of the _____   ____

b. Now fill in details of the process by placing the following statements of events in the correct sequence. Write the letters on the short lines to the right of the correct phase (I–V) above. One is done for you.

A. The cartilage model grows in length as chondrocytes divide and secrete more cartilage matrix; it grows in thickness as new chondroblasts develop within the perichondrium.

B. Cartilage cells in the center of the diaphysis accumulate glycogen, enlarge, and burst, releasing chemicals that alter pH and trigger calcification of cartilage.

C. Cartilage cells die because they are deprived of nutrients; in this way spaces are formed within the cartilage model.

D. Blood vessels penetrate the perichondrium, stimulating perichondrial cells to form osteoblasts. A collar of bone forms and gradually thickens around the diaphysis. The membrane covering the developing bone is now called the periosteum.

E. A hyaline cartilage model of future bone is laid down by differentiation of mesenchyme into chondroblasts.

F. A perichondrium develops around the cartilage model.

G. Secondary ossification centers develop in epiphyses, forming spongy bone there about the time of birth.

H. Capillaries grow into spaces, and osteoblasts deposit bone matrix over disintegrating calcified cartilage. In this way spongy bone is forming within the diaphysis at the primary ossification center.

    I. Osteoclasts break down the newly formed spongy bone in the very center of the bone, thereby leaving the medullary (marrow) cavity.

    J. The original hyaline cartilage remains as articular cartilage at the ends of the new bone and also as the epiphyseal plate for as long as the bone grows.

C4. In growth of the cartilage model (Step A in Checkpoint C3), cartilage grows lengthwise by *(appositional? interstitial?)* growth; this means growth *(from within? On the cartilage surface?)*.
Cartilage grows in thickness by *(appositional? interstitial?)* growth, which means growth *(from within? on the cartilage surface?)*.

✓ **C5.** On Figure LG 6.1, label the *epiphyseal plate*. This plate of growing bone consists of four zones. First identify each region by placing the letter of the name of that zone next to the correct description below. Next place these same four letters in the boxes in the inset on Figure LG 6.1 to indicate the locations of these cartilage zones within the epiphyseal plate of the developing bone.

C.	**Calcified**	**P.**	**Proliferating**
H.	**Hypertrophic**	**R.**	**Resting**

_____ a. Zone of cartilage that is not involved in bone growth but anchors the epiphyseal plate (site of bone growth) to the bone of the epiphysis

_____ b. New cartilage cells form here by mitosis and are arranged like stacks of coins

_____ c. Cartilage cells mature and die here as matrix around them calcifies

_____ d. Osteoblasts and capillaries from the diaphysis invade this region to lay down bone upon the calcified cartilage remnants here

✓ **C6.** Complete this summary statement about bone growth at the epiphyseal plate. Cartilage cells multiply on the *(epiphysis? diaphysis?)* side of the epiphyseal plate, providing temporary new tissue. Cartilage cells then die and are replaced by bone cells on the *(epiphysis? diaphysis?)* side of the epiphyseal plate. In other words, as the bone grows in length, the *(diaphysis? epiphyses?)* grow(s) longer. If the epiphyseal plate of a growing bone is damaged, for example by fracture, the bone is likely to be *(shorter than normal? of normal length? longer than normal?)*.

✓▲ **C7.** Complete this activity about maturation of the skeleton.

  a. The epiphyseal *(line? plate?)* is the bony region in bones of adults that marks the original cartilaginous epiphyseal *(line? plate?)*.

  b. Ossification of most bones is completed by age *(18? 25?)*. If the epiphyseal plate of a growing bone is damaged, for example by fracture or by sex steroids (especially estrogens), the epiphyseal line is likely to form *(earlier? later?)* in development. This accounts for the *(shorter? longer?)* than normal bone length.

  c. Bones must grow in diameter to match growth in length. *(Osteoblasts? Osteoclasts?)* are cells that destroy bone to enlarge the medullary cavity, whereas *(osteoblasts? osteoclasts?)* are cells from the periosteum that increase the thickness of compact bone.

✓ **C8.** Fill in the names of hormones involved in bone growth in this exercise. Use the answers in the box.

hGH.	**Human growth hormone**	**INS.**	**Insulin**
IGFs.	**Insulinlike growth factor**	**SH.**	**Sex hormones**

_____ a. Hormone made by the pituitary. In excessive amounts, results in giantism; if deficient, leads to dwarfism.

_____ b. Produced by bone cells and also by the liver, these chemicals stimulate protein synthesis and growth of bones.

_____ c. Made by the pancreas; promotes normal bone growth.

_____ d. Produced by ovaries and testes, these hormones stimulate osteoblast activity and lead to growth spurt and changes in skeletal structure at puberty.

✓ **C9.** List 10 factors that are necessary for normal bone growth, remodeling, and repair of fractured bone.
  a. Six minerals: _____ _____ _____ _____ _____ _____

  b. Four vitamins: _____ _____ _____ _____

C10. Defend or dispute this statement: "Once a bone, such as your thighbone, is formed, the bone tissue is never replaced unless the bone is broken."

✓ **C11.** Discuss bone remodeling in this Checkpoint.
  a. Explain the roles of lysosomes and acids in osteoclastic activity.

  b. What fills in the tunnels formed by osteoclasts?

  c. Explain one mechanism by which estrogen helps to prevent bone breakdown.

C12. Contrast the terms related to types of fractures in each pair.
  a. _Simple/compound_

  b. _Pott's/Colles'_

  c. _Comminuted/greenstick_

✓ **C13.** Describe the four main steps in repair of a fracture by filling in terms from this list in the spaces below. One answer will be used twice.

Cell Type	Process or Stage	Time Period
Chondroblasts	Fracture hematoma	Hours
Fibroblasts	Inflammation and cell death	Months
Osteoblasts	Procallus	Weeks
Osteoclasts	Remodeling	
Phagocytes		

a. Blood flows from torn blood vessels, forming a clot known as a

   _____ within 6 to 8 _____ of the injury.

   Because of the trauma and lack of circulation, _____ occurs in

   the area. As a result, cells known as _____ and

   _____ enter the area and remove debris over a period of several

   _____.

b. Next, granulation tissue known as a _____ forms. Then cells

   called _____ produce collagen fibers that connect broken ends

   of the bone. Other cells, the _____, form a soft callus at a
   distance from the remaining healthy, vascular bone. This stage lasts about 3

   _____.

c. The hard callus stage occurs next as _____ cells form bone
   adjacent to the remaining healthy, vascular bone. This stage lasts 3 to 4

   _____.

   Callus is reshaped during the final (or _____) phase.

C14. *A clinical correlation.* What is the major difference between these two methods of setting a fracture: *closed reduction* and *open reduction?*

C15. *The Big Picture: Looking Ahead.* Refer to Figure 18.14 and other sections of your text, and answer these questions about regulation of calcium homeostasis.
a. Figures 1.2 and 1.3 in your text introduced the concept of negative feedback systems. Using that reference and Figure 6.11 in your text, complete Figure LG 6.3. Fill in the six boxes and the two lines using the list of answers provided in the figure.
b. Figure LG 6.3 shows that parathyroid gland cells respond to a lowered blood level of

   CA^{2+} by signaling increased production of cyclic _____. This INPUT then *("turns on"? "turns off"?)* the PTH cell gene in parathyroid glands. The increased blood level of PTH acts as OUTPUT to "tell" effectors to respond.

   Effectors include kidneys as well as _____ these organs *(raise? lower?)* the blood level of Ca^{2+} in several ways.

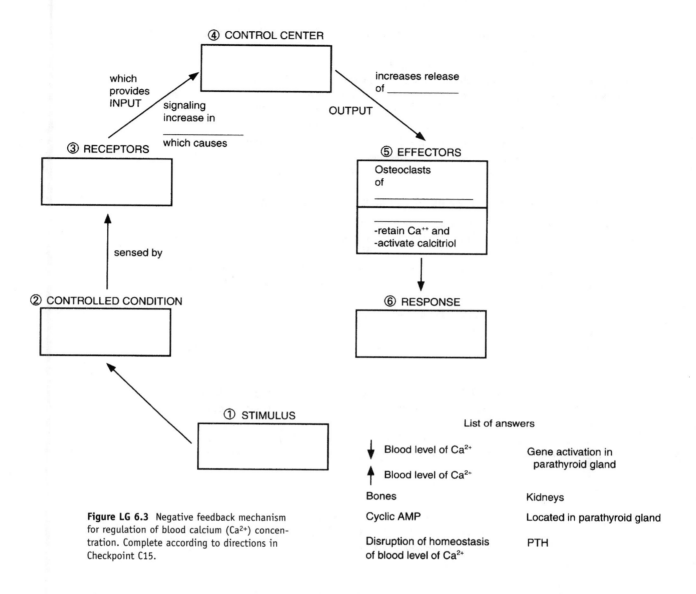

Figure LG 6.3 Negative feedback mechanism for regulation of blood calcium (Ca²⁺) concentration. Complete according to directions in Checkpoint C15.

List of answers

↓ Blood level of Ca²⁺

↑ Blood level of Ca²⁺

Bones

Cyclic AMP

Disruption of homeostasis of blood level of Ca²⁺

Gene activation in parathyroid gland

Kidneys

Located in parathyroid gland

PTH

c. PTH signals kidneys to *(eliminate? retain?)* calcium in blood so that less is lost in urine. PTH also activates kidneys to produce *(calcitriol? calcitonin?)*, which is vitamin

_____. This vitamin also raises blood levels of Ca²⁺ by increasing absorption of Ca²⁺ from foods in the intestine.

d. PTH activates osteo-*(blasts? clasts?)* of bone cells so that calcium is *(deposited in? resorbed from?)* bone.

e. Regulation of blood calcium (Ca²⁺) exemplifies a *(positive? negative?)* feedback cycle

because a decrease in blood Ca²⁺ ultimately causes PTH to _____-crease blood Ca²⁺.

f. The effect of PTH is *(the same as? opposite to?)* that of calcitonin (Figure 18.14, page 586). Calcitonin causes calcium to be *(deposited in? resorbed from?)* bone. In other words, calcitonin, which is released by the *(thyroid? parathyroid?)* gland, causes the

blood level of Ca²⁺ to _____-crease.

g. Chapter 27 (page 964) points out that *(hypo? hyper?)*-parathyroidism refers to a deficiency of PTH. This condition is likely to lead to *(high? low?)* blood calcium levels, known as *(hypo? hyper?)*-calcemia. Symptoms of this disorder include increased *(hardness? fragility?)* of bones because bones have *(lost? gained?)* calcium and of *(hyper? hypo?)*-active muscles because low blood levels of calcium increase activity of nerves.

    h. List two or more other functions of the human body affected by the blood concentration of calcium.

✓ **C16.** Circle the correct answers about the effects of exercise upon bones.

    a. Exercise *(strengthens? weakens?)* bones. A fractured bone that is not exercised is likely to become *(stronger? weaker?)* during the period that it is immobilized.

    b. The pull of muscles upon bones, as well as the tension on bones as they support body weight during exercise, causes *(increased? decreased?)* production of the protein collagen.

    c. The stress of exercise also stimulates *(osteoblasts? osteoclasts?)* and increases production of the hormone calcitonin, which inhibits *(osteoblasts? osteoclasts?)*.

D. Developmental anatomy and aging of the skeletal system (pages 176–178)

✓ **D1.** Complete this learning activity about development of the skeleton.

    a. Bones and cartilage are formed from *(ecto? meso? endo?)*-derm, which later different-

iates into the embryonic connective tissue called _____. Some of these cells become chondroblasts, which eventually form *(bone? cartilage?)*, whereas

_____-blasts develop into bone.

    b. Limb buds appear during the *(fifth? seventh? tenth?)* week of development. At this point the skeleton in these buds consists of *(bone? cartilage?)*. Bone begins to form

during week _____.

    c. By week *(six? seven? eight?)* the upper extremity is evident, with defined shoulder, arm, elbow, forearm, wrist, and hand. The lower extremity is also developing but at a slightly *(faster? slower?)* pace.

    d. The notochord develops in the region of the *(head? vertebral column? pelvis?)*. Most of the notochord eventually *(forms vertebrae? disappears?)*. Parts of it persist in the

_____ .

✓ **D2.** Complete this exercise about skeletal changes that occur in the normal aging process.

    a. The amount of calcium in bones *(de? in?)*-creases with age. As a result, bones of the elderly are likely to be *(stronger? weaker?)* than bones of younger persons. This change occurs at a younger age in *(men? women?)*.

This loss of calcium typically begins at about age ___ in females, and by age 70,

about ___% of calcium in bones is lost. In males, calcium loss does not usually begin

until after age _____.

    b. Another component of bones that decreases with age is _____. What is the significance of this change?

    c. Write two or more health practices that can minimize skeletal changes associated with aging.

E. Disorders and medical terminology (pages 178–179)

✓ **E1.** Describe osteoporosis in this exercise.

a. Osteoporotic bones exhibit _____-creased bone mass and _____-creased susceptibility

to fractures. This disorder is associated with a loss of the hormone _____

which normally stimulates osteo-_____ activity.

b. Circle the category of persons at higher risk for osteoporosis in each pair:

Younger persons/older persons   Men/women

Short, thin persons/tall, large-build persons Black persons/white persons

c. List three factors in daily life that will help to prevent osteoporosis. (*Hint:* Focus on diet and activities.)

✓ **E2.** Check your understanding of bone disorders by matching the terms in the box with related descriptions.

Osteoarthritis	**Osteosarcoma**	**Osteomyelitis**

a. Malignant bone tumor: _____
b. Bone infection, for example, caused by *Staphylococcus:* _____
c. Degenerative joint disease caused by destruction of articular cartilage: _____

ANSWERS TO SELECTED CHECKPOINTS: CHAPTER 6

A1. Support, protection, assistance in movement, homeostasis of minerals, hemopoiesis, and fat storage.

A2. Essentially all do. For example, muscles need intact bones for movement to occur; bones are site of blood formation; bones provide protection for viscera of nervous, digestive, urinary, reproductive, cardiovascular, respiratory, and endocrine systems; broken bones can injure integument.

B1. Bone, cartilage, dense connective tissues, epithelium, blood-forming tissues, adipose tissue, and nervous tissue, organ.

B2. Figure LG 6.1A at right.

B3. (a) A. (b) M. (c) P. (d) E.

B4. (a) Osteogenic cells. (b) Osteoblast. (c) Osteocyte. (d) Osteoclast.

B5. (a) Intercellular material. (b) 25, 50. (c) Calcium phosphate, calcium carbonate. (d) Hardness, collagen; tensile strength, collagen fibers. (e) Contains some spaces; provides channels for blood vessels and makes bones lighter weight.

B6. (a) Osteons (Haversian systems); lamella, E. (b) G; F; D. (c) Osteocytes; lacunae, B. (d) C; contain extensions of osteocytes bathed in extra-cellular fluid (ECF); provide routes for oxygen and nutrients to reach osteocytes and for wastes to diffuse

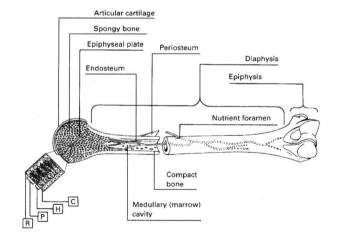

Figure LG 6.1A Diagram of a developing long bone that has been partially sectioned lengthwise.

away. (e) Along the same direction as; change.

B8. (a) Trabeculae, irregular latticework of bony lamellae containing osteocytes surrounded by lacunae and blood vessels. (b) Epiphyses; is also. (c) Only certain areas of; hipbones, ribs, sternum, vertebrae, skull bones, and epiphyses of long bones such as femurs. These areas (such as the sternum) may be biopsied to examine for aplastic anemia or for response of bone marrow to anti-anemia medications. These sites (such as hip-bones) may be utilized for marrow transplant.

C1. (a) Ossification or osteogenesis. (b) Connective, mesenchyme; week. (c) Intramembranous: flat bones of the cranium and the mandibles (lower jaw bones); endochondral: all other bones of the body. (d) Do not.

C2. (a) Mesenchyme; ossification. (b) Osteogenic, osteoblast, osteocyte. (c) Compact, spongy, marrow. (d) Fontanels; growth, protection.

C3. Cartilage. (I) Development: E F. (II) Growth: A B C. (Ill) Primary ossification center: D H I. (IV) Secondary ossification center: G. (V) Articular cartilage and epiphyseal plate: J.

C4. Interstitial, within; appositional, on the cartilage surface.

C5. See Figure LG 6.1A. (a) R. (b) P. (c) H. (d) C.

C6. Epiphysis; diaphysis; diaphysis; shorter than normal.

C7. (a) Line, plate. (b) 18; earlier; shorter. (c) Osteoclasts, osteoblasts.

C8. (a) hGH. (b) IGFs. (c) INS. (d) SH.

C9. (a) Calcium, phosphorus, fluoride, magnesium, iron, and manganese. (b) C, K, B12, and A.

C11. (a) Lysosomes release enzymes that digest the protein collagen; acids dissolve minerals of bone. (b) Osteoblasts lay down new bone. (c) By promoting death of osteoclasts.

C13. (a) Fracture hematoma, hours; inflammation and cell death; osteoclasts, phagocytes, weeks. (b) Procallus; fibroblasts; chondroblasts; weeks. (c) Osteoblasts; months. (d) Remodeling.

C15. (a) See Figure LG 6.3A. (b) AMP; "turns on"; bones, raise. (c) Retain; calcitriol, D. (d) Clasts, resorbed from. (e) Negative, in. (f) Opposite to; deposited in; thyroid, de. (g) Hypo; low, hypo; fragility, lost, hyper. (h) Blood clotting, heart activity, breathing, enzyme function.

C16. (a) Strengthens; weaker. (b) Increased. (c) Osteoblasts, osteoclasts.

D1. (a) Meso, mesenchyme; cartilage, osteo. (b) Fifth; cartilage; six or seven. (c) Eight; slower. (d) Vertebral column; disappears; intervertebral discs.

D2. (a) De; weaker; women; 30, 30; 60. (b) Protein; bones are more brittle and vulnerable to fracture. (c) Weight-bearing exercise; good diet with adequate calcium, vitamin D, and protein; not smoking because smoking constricts blood vessels.

E1. (a) De, in; estrogen; blast. (b) Older; women; short, thin persons; white persons. (c) Adequate intake of calcium, vitamin D, weight-bearing exercise, and not smoking; estrogen replacement therapy (HRT) may be advised for some women.

E2. (a) Osteosarcoma. (b) Osteomyelitis. (c) Osteoarthritis.

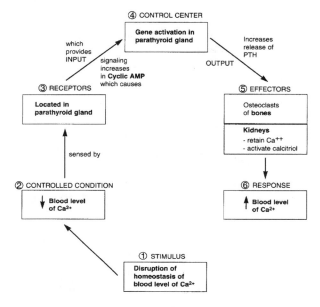

Figure LG 6.3A Negative feedback regulation of calcium.

MORE CRITICAL THINKING: CHAPTER 6

1. Explain how the skeletal system is absolutely vital for basic functions that keep you alive, for example, for breathing, eating, movement, and elimination of wastes.

2. Do you think bone is dynamic or "dead"? State your rationale.

3. Contrast compact bone with spongy bone according to structure and locations.

4. Describe mesenchyme and osteogenic cell roles in formation, growth, and maintenance of skeletal tissues.

5. One process of bone formation involves development of bone from fibrous connective tissue membranes rather than from cartilage. Tell which bones form by this process, and explain how this might be advantageous to the developing human body.

6. Discuss probable effects of these hormonal changes upon bone development and health:
   (a) abnormally high production of sex steroids in a nine-year-old boy
   (b) removal of both ovaries in a 26-year-old woman
   (c) excessive production of parathyroid hormone (PTH) in a 55-year-old woman

MASTERY TEST: CHAPTER 6

Questions 1–15: Circle T (true) or F (false). If the statement is false, change the underlined word or phrase so that the statement is correct.

T F 1. Greenstick fractures occur only in <u>adults</u>.

T F 2. <u>Appositional</u> growth of cartilage or bone means growth in thickness due to action of cells from the perichondrium or periosteum.

T F 3. Calcitriol is an <u>active form of calcium</u>.

T F 4. The epiphyseal plate appears <u>earlier</u> in life than the epiphyseal line.

T F 5. Osteons (Haversian systems) are found in <u>compact bone, but not in spongy bone</u>.

T F 6. Another name for the epiphysis is the <u>shaft of the bone</u>.

T F 7. A compound fracture is defined as one in which the bone <u>is broken into many pieces</u>.

T F 8. Haversian canals run <u>longitudinally (lengthwise) through bone, but perforating (Volkmann's) canals run horizontally across bone</u>.

T F 9. Canaliculi are tiny canals containing <u>blood that nourishes bone cells in lacunae</u>.

T F 10. Compact bone that is of intramembranous origin differs <u>structurally</u> from compact bone developed from cartilage.

T F 11. In a long bone the primary ossification center is located in the <u>diaphysis whereas the secondary center of ossification is in the epiphysis</u>.

T F 12. <u>Osteoblasts</u> are bone-destroying cells.

T F 13. Most bones start out in embryonic life as <u>hyaline cartilage</u>.

T F 14. The layer of compact bone is <u>thicker</u> in the diaphysis than in the epiphysis.

T F 15. <u>Lamellae</u> are small spaces containing bone cells.

Questions 16–20: Arrange the answers in correct sequence.

_____ _____ _____ 16. Steps in marrow formation during endochondral bone formation, from first to last:
  A. Cartilage cells burst, causing intercellular pH to become more alkaline.
  B. pH changes cause calcification, death of cartilage cells; eventually spaces left by dead cells are penetrated by blood vessels.
  C. Cartilage cells hypertrophy with accumulated glycogen.

_____ _____ _____ 17. From most superficial to deepest:
  A. Endosteum
  B. Periosteum
  C. Compact bone

_____ _____ _____ 18. Phases in repair of a fracture, in chronological order:
  A. Fracture hematoma formation
  B. Remodeling
  C. Callus formation

_____ _____ _____ _____ 19. Phases in formation of bone in chronological order:
  A. Mesenchyme cells
  B. Osteocytes
  C. Osteoblasts
  D. Osteogenic

_____ _____ _____ _____ 20. Portions of the epiphyseal plate, from closest to epiphysis to closest to diaphysis:
  A. Zone of resting cartilage
  B. Zone of proliferating cartilage
  C. Zone of calcified cartilage
  D. Zone of hypertrophic cartilage

Questions 21–25: Fill-ins. Write the word or phrase that best completes the statement.

_____ 21. _____ is a term that refers to the shaft of the bone.

_____ 22. _____ is a disorder primarily of older women associated with decreased estrogen level; it is characterized by weakened bones.

_____ 23. _____ is the connective tissue covering over cartilage in adults and also over embryonic cartilaginous skeleton.

_____ 24. The majority of bones formed by intramembranous ossification are located in the _____.

_____ 25. Vitamin _____ is a vitamin that is necessary for absorption of calcium from the gastrointestinal tract, and so it is important for bone growth and maintenance.

ANSWERS TO MASTERY TEST: CHAPTER 6

True-False

1. F.   Children
2. T
3. F.   Active form of vitamin D
4. T
5. T
6. F.   End of the bone (often bulbous)
7. F.   Protrudes through the skin
8. T
9. F.   Parts of osteocytes and fluid from blood vessels in Haversian canals, but not blood itself
10. F.   Only in origin
11. T
12. F.   Osteoclasts
13. T
14. T
15. F.   Lacunae

Arrange

16. C A B
17. B C A
18. A C B
19. A D C B
20. A B D C

Fill-ins

21. Diaphysis
22. Osteoporosis
23. Perichondrium
24. Skull or cranium
25. D (or D3)

FRAMEWORK 7
Axial Skeleton

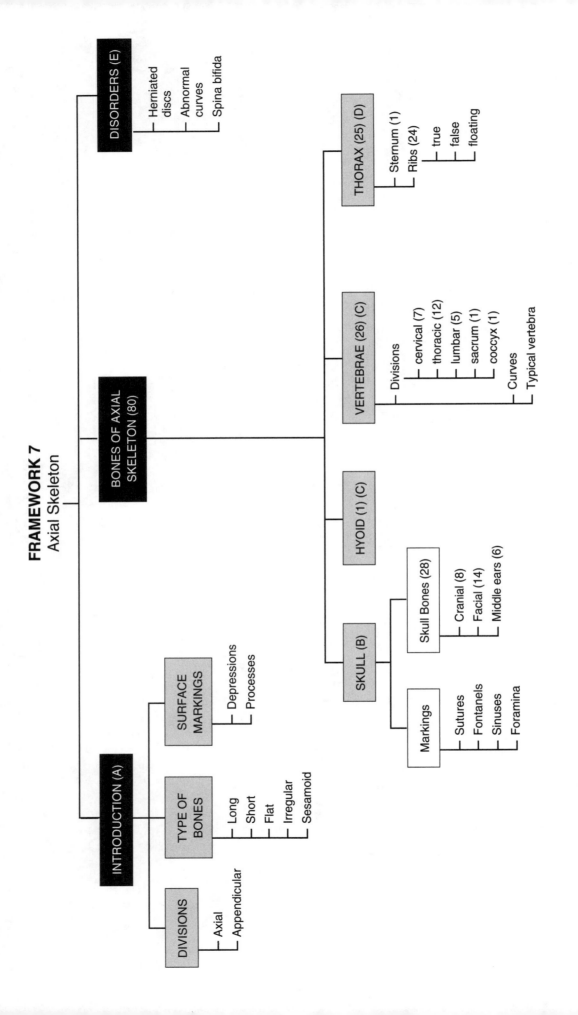

INTRODUCTION (A)

DIVISIONS
- Axial
- Appendicular

TYPE OF BONES
- Long
- Short
- Flat
- Irregular
- Sesamoid

SURFACE MARKINGS
- Depressions
- Processes

BONES OF AXIAL SKELETON (80)

SKULL (B)

Markings
- Sutures
- Fontanels
- Sinuses
- Foramina

Skull Bones (28)
- Cranial (8)
- Facial (14)
- Middle ears (6)

HYOID (1) (C)

VERTEBRAE (26) (C)

Divisions
- cervical (7)
- thoracic (12)
- lumbar (5)
- sacrum (1)
- coccyx (1)

Curves

Typical vertebra

THORAX (25) (D)
- Sternum (1)
- Ribs (24)
  - true
  - false
  - floating

DISORDERS (E)
- Herniated discs
- Abnormal curves
- Spina bifida

Note: numbers in ( ) refer to number of each bone in the body.

C H A P T E R **7**

The Skeletal System: The Axial Skeleton

The 206 bones of the human skeleton are classified into two divisions—axial and appendicu-lar—based on their locations. The axial skeleton is composed of the portion of the skeleton immediately surrounding the axis of the skeleton, primarily the skull bones, vertebral column, and bones surrounding the thorax. Bones of the appendages, or extremities, comprise the appendicular skeleton. Chapter 7 begins with an introduction to types and surface characteris-tics (markings) of all bones and then focuses on the 80 bones of the axial skeleton. Disorders of the vertebral column are also included.

As you begin your study of the axial skeleton, carefully examine the Chapter 7 Frame-work and note relationships among concepts and key terms there. Also refer to the Topic Out-line and Objectives; you may want to check off each objective as you complete it.

TOPIC OUTLINE AND OBJECTIVES

A. Divisions of the skeletal system, types of bones, and bone surface markings

1. Describe how the skeleton is divided into axial and appendicular divisions.
2. Classify bones on the basis of shape and location.
3. Describe the principal surface markings on bones and the functions of each.

B. Skull

4. Name the cranial and facial bones and indicate the number of each.

C. Hyoid bone and vertebral column

5. Describe the relationship of the hyoid bone to the skull.
6. Identify the regions and normal curves of the vertebral column.

D. Thorax

7. Identify the bones of the thorax.

E. Disorders

WORDBYTES

Now become familiar with the language of this chapter by studying each wordbyte, its meaning, and an example of its use within a term. After you study the entire list, self-check your understanding by writing the meaning of each wordbyte on the line. As you continue through the *Learning Guide*, identify (and fill in) additional terms that contain the same wordbyte.

Wordbyte	Self-check	Meaning	Example(s)
annulus	_____	ring	*annulus* fibrosus
cervic-	_____	neck	*cervic*al collar
costa	_____	rib	*costa*l cartilage
cribr-	_____	sieve	*cribr*iform plate
crist-	_____	crest	*crist*a galli; *crist*ae
ethm-	_____	sieve	*ethm*oid bone
infra-	_____	below	*infra*orbital
lambd-	_____	L-shaped	*lambd*oidal suture
lumb-	_____	loin	*lumb*osacral
optic	_____	eye	*optic* foramen
pulp	_____	soft, flesh	nucleus *pulp*osus
sacrum-	_____	sacred bone	*sacr*oiliac
sphenoid-	_____	wedge-shaped	*sphen*oid bone
supra-	_____	above	*supra*orbital

CHECKPOINTS

A. Divisions of the skeletal system, types of bones, and bone surface markings (pages 183–186)

   A1. Explain how the skeletal system is absolutely necessary for vital life functions of eating, breathing, and movement.

✓ **A2.** Describe the bones in the two principal divisions of the skeletal system by completing this exercise. It may help to refer to Figure LG 8.1, page 145.

   a. Bones that lie along the axis of the body are included in the *(axial? appendicular?)* skeleton.

   b. The axial skeleton includes the following groups of bones. Indicate how many bones are in each category.

   _____ Skull (cranium, face)   _____ Vertebrae

   _____ Earbones (ossicles)   _____ Sternum

   _____ Hyoid   _____ Ribs

   c. The total number of bones in the axial skeleton is _____.

   d. The appendicular skeleton consists of bones in which parts of the body?

   e. Write the number of bones in each category. Note that you are counting bones on one side of the body only.

    _____ Left shoulder girdle

    _____ Left upper extremity (arm, forearm, wrist, hand)

    _____ Left hipbone

    _____ Left lower extremity (thigh, kneecap, leg, foot)

   f. There are _____ bones in the appendicular skeleton.

   g. In the entire human body there are _____ bones.

▲ **A3.** Complete the table about the four major and two minor types of bones.

Type of Bone	Structural Features	Examples
a. Long	Slightly curved to absorb stress better	
b.		Bones in wrist and back of feet
c.	Composed of two thin plates of bone	
d. Irregular		
e. Sutural (or wormian)		
f. Small bones in tendons		

A4. In general, what are the functions of surface markings of bones?

A5. Contrast the bone markings in each of the following pairs.
  a. *Tubercle/tuberosity*

  b. *Crest/line*

  c. *Fossa/foramen*

  d. *Condyle/epicondyle*

✓▲**A6.** Write the correct answer from the following list of specific bone markings on the line
  next to its description. The first one has been done for you.

Articular *facet* on vertebra	Maxillary *sinus*
External auditory *meatus*	Optic *foramen*
Greater *trochanter*	Styloid *process*
Head of humerus	Superior orbital *fissure*

  a. Air-filled cavity within a bone, connected to

      nasal cavity __**Maxillary sinus**__
  b. Narrow, cleftlike opening between adjacent
      parts of bone; passageway for blood vessels

      and nerves: _____
  c. Rounded hole, passageway for blood vessels

      and nerves: _____
  d. Tubelike passageway through bone:

      _____

  e. Large, rounded projection above constricted

      neck: _____

  f. Large, blunt projection: _____

  g. Sharp, slender projection: _____

  h. Smooth, flat surface: _____

B. Skull (pages 186–201)

✓ **B1.** The skull consists of two major regions. Describe these in this checkpoint.

a. The _____ consisting of *(6? 8? 14? 16?)* bones. These bones surround

the _____. Located in one pair of the cranial bones (the temporals)
are the tiny bones of the middle ear. Each middle ear contains *(2? 3? 6? 8?)* bones.

b. The remaining bones of the skull are comprised of _____ bones that provide a structural framework for the face. The total number of bones in the cranium, middle ears,

and face is _____ bones.

✓▲**B2.** Answer these questions on the cranium.
a. What is the primary function of the cranium? _____
List several other functions of the cranium.

b. Which bones make up the cranium (rather than the face)? Write the number (1 or 2)
of each bone you name.

O Ethmoid bone
O Frontal bone
O Lacrimal bone
O Mandible bone
O Maxilla bone
O Nasal bone
O Occipital bone
O Parietal bone
O Sphenoid bone
O Temporal bone
O Zygomatic bone

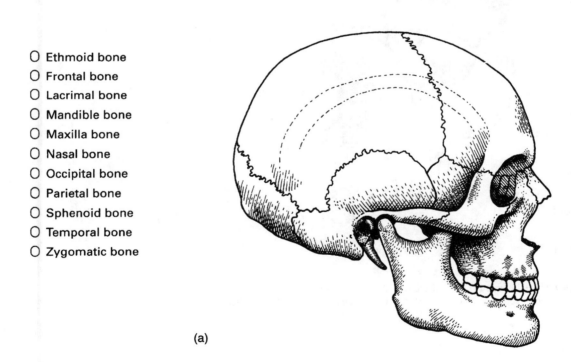

(a)

Figure LG 7.1 Skull bones. (a) Skull viewed from right side. Color and label as directed in Checkpoints
B3, B6, and B9.

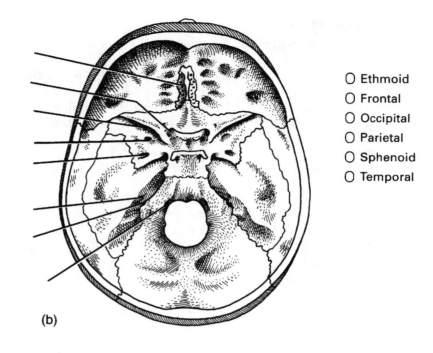

O Ethmoid
O Frontal
O Occipital
O Parietal
O Sphenoid
O Temporal

(b)

O Ethmoid
O Frontal
O Inferior concha
O Lacrimal
O Maxilla
O Nasal
O Sphenoid
O Vomer
O Zygomatic

(c)

Figure LG 7.1 Skull bones. (b) Floor of the cranial cavity. (c) Anterior view of bones that form the right orbit and the nose. Color and label as directed in Checkpoints B3, B6, B7, B9, and B10.

✓▲ **B3.** Color the skull bones on Figure LG 7.1a, b, and c. Be sure to color the corresponding color code oval for each bone listed on the figure.

✓▲ **B4.** Check your understanding of locations and functions of skull bones by matching names of bones in the box with their descriptions below. Use each bone only once.

Ethmoid	**Maxilla**	**Sphenoid**
Frontal	**Nasal**	**Temporal**
Inferior concha	**Occipital**	**Vomer**
Lacrimal	**Palatine**	**Zygomatic (malar)**
Mandible	**Parietal**	

_____ a. This bone forms the lower jaw, including the chin.

_____ b. These are the cheek bones; they also form lateral walls of the orbit of the eye.

_____ c. Tears pass through tiny foramina in these bones; they are the smallest bones in the face.

_____ d. The bridge of the nose is formed by these bones.

_____ e. Organs of hearing (internal part of ears) and mastoid air cells are located in and protected by these bones.

_____ f. This bone sits directly over the spinal column; it contains the foramen through which the spinal cord connects to the brain.

_____ g. The name means "wall." The bones form most of the roof and much of the side walls of the skull.

_____ h. These bones form most of the roof of the mouth (hard palate) and contain the sockets into which upper teeth are set.

_____ i. L-shaped bones form the posterior parts of the hard palate and nose.

_____ j. Commonly called the forehead, it provides protection for the anterior portion of the brain.

_____ k. A fragile bone, it forms much of the roof and internal structure of the nose.

_____ l. It serves as a "keystone" because it binds together many of the other bones of the skull. It is shaped like a bat, with the wings forming part of the sides of the skull and the legs at the back of the nose.

_____ m. This bone forms the inferior part of the septum dividing the nose into two nostrils.

_____ n. Two delicate bones form the lower parts of the side walls of the nose.

✓▲**B5.** Complete the table describing major markings of the skull.

Marking	Bone	Function
a. Greater wings	Sphenoid	
b.		Forms superior portion of septum of nose
c.		Site of pituitary gland
d. Petrous portion		
e.		Largest hole in skull; passageway for spinal cord
f.	(2)	Bony sockets for teeth
g.		Passageway for sound waves to enter ear
h. Condylar process	Mandible	

✓   **B6.** Now label each of the markings listed in the above table on Figure LG 7.1.

✓▲ **B7.** *The Big Picture: Looking Ahead.* Refer to figures on text pages listed in parentheses as you do this activity. The 12 pairs of nerves attached to the brain are called cranial nerves (Figure 14.5 in your text). Holes in the cranium permit passage of these nerves to and from the brain. These nerves are numbered according to the order in which they attach to the brain (and leave the cranium) from I (most anterior) to XII (most posterior). To help you visualize their sequence, label the foramina for cranial nerves in order on the left side of Figure LG 7.1b. Complete the table summarizing these foramina. The first one is done for you.

Number and Name of Cranial Nerve	Location of Opening for Nerve
a. I Olfactory (Figure 16.1)	Cribriform plate of ethmoid bone
b. II Optic (Figure 11.5a)	
c. III Oculomotor (Figure 17.2) IV Trochlear V Trigeminal (ophthalmic branch) VI Abducens	
d. V Trigeminal (maxillary branch)	
e. V Trigeminal (mandibular branch)	
f. VII Facial (Figure 17.2) VIII Vestibulocochlear (Figure 16.16)	
g. IX Glossopharyngeal (Figure 17.2) X Vagus (Figure 17.2) XI Accessory	
h. XII Hypoglossal	

✓▲ **B8.** *For extra review.* Write the names of the markings listed in the box next to the bones in which those markings are located. Write the name of one marking on each line.

Carotid canal	**Mental foramen**	**Superior nasal conchae**
Condylar process	**Olfactory foramina**	**Superior nuchal line**
Crista galli	**Palatine process**	**Superior orbital fissure**
Inferior orbital fissure	**Perpendicular plate**	**Supraorbital margin**
Lesser wings	**Pterygoid processes**	**Zygomatic process**
Mastoid air cells		

a. Ethmoid: _____   _____

                _____   _____

b. Frontal: _____

c. Mandible: _____   _____

d. Maxilla: _____   _____

e. Occipital: _____

f. Sphenoid: _____   _____

                _____

g. Temporal: _____   _____

                _____

✓▲ **B9.** Check your understanding of special features of the skull in this Checkpoint.
Define sutures.

a. Between which two bones is the sagittal suture located? _____
   On Figure LG 7.1a, label the following sutures: *coronal, squamous, lambdoid.*

b. "Soft spots" of a newborn baby's head are known as _____.

   What is the location of the largest one? _____

c. List three functions of fontanels.

d. A good way to test your ability to visualize locations of important skull bones is to try
   to identify bones that form the orbits surrounding and protecting the eyes. On Figure
   LG 7.1c, locate six bones that form the orbit. (*Note:* The tiny portion of the seventh
   bone, the superior tip of the palatine bone, is not visible in the figure.)

✓▲ **B10.** Do this exercise about bony structures related to the nose.

a. Name four bones that contain paranasal sinuses.

   Practice identifying their locations the next time you have a cold!

b. List three functions of paranasal sinuses.

c. Most internal portions of the nose are formed by a bone that includes these
   markings: paranasal sinuses, conchae, and part of the nasal septum. Name the bone.

d. What functions do nasal conchae perform?

e. On Figure LG 7.1c, label the bone that forms the inferior portion of the nasal septum.

   The anterior portion of the nasal septum consists of flexible _____.

✓▲ **B11.** *The Big Picture: Looking Ahead* at details of markings of the skull, circle the correct
   answers in each statement. Refer to figures in Chapters 11, 16, and 23 as you
   complete this Checkpoint.

a. The zygomatic arch (Figure 11.6) is formed of two bones; these are the *(parietal? sphenoid? temporal?)* bones.

b. The styloid process (Figure 11.7) is a penlike projection from the *(parietal? sphenoid?
   temporal?)* bone.

  c. The tear duct (Figure 16.4b) passes through a canal in the *(lacrimal? mental?)* foramen.

  d. The frontal sinus (Figure 23.2b) is located *(superior? inferior?)* to the nasal cavity.

  e. The nasal conchae (Figure 23.2b) are located *(superior to? within?)* the nasal cavity.

  f. The hard palate (Figure 23.2b) is formed anteriorly of the *(hyoid? mandible? maxilla? palatine?)* bone and posteriorly of the *(mandible? maxilla? palatine?)* bone.

▲ **B12.** *Clinical correlation.* Name the skull bone(s) and specific markings likely to be involved in each clinical case.

  a. Ten-year-old Chris suffered from a "black eye" after she was hit by a ball.

    _____   _____

  b. An auto accident left Miji, age 24, with a "deviated septum."

    _____   _____

  c. Susanna, 8 months, was born with a "cleft palate."

    _____   _____

  d. Leonardo, age 40, consulted a pain clinic for chronic "TMJ" pain.

    _____   _____

C. Hyoid bone and vertebral column (pages 201–210)

✓▲ **C1.** Identify the location of the hyoid bone on yourself. Place your hand on your throat and swallow. Feel your larynx move upward? The hyoid sits just *(superior? inferior?)* to the larynx at the level of the mandible.

✓ **C2.** In what way is the hyoid unique among all the bones of the body?

✓▲ **C3.** The five regions of the vertebral column are grouped (A–E) in Figure LG 7.2. Color vertebrae in each region and be sure to select the same color for the corresponding color code oval. Now write on lines next to A–E the number of vertebrae in each region. One is done for you. Also label the first two cervical vertebrae.

✓ **C4.** Note which regions of the vertebral column in Figure 7.2 normally retain an anteriorly concave curvature in the adult: _____ and _____. These are considered *(primary? secondary?)* curvatures. This classification is based upon the fact that these curves *(were present originally during fetal life? are more important?)*.

✓▲ **C5.** On Figure LG 7.3, color the parts of vertebrae and corresponding color code ovals. As you do this, notice differences in size and shape among the three vertebral types in the figure.

✓ **C6.** Identify distinctive features of vertebrae in each region.

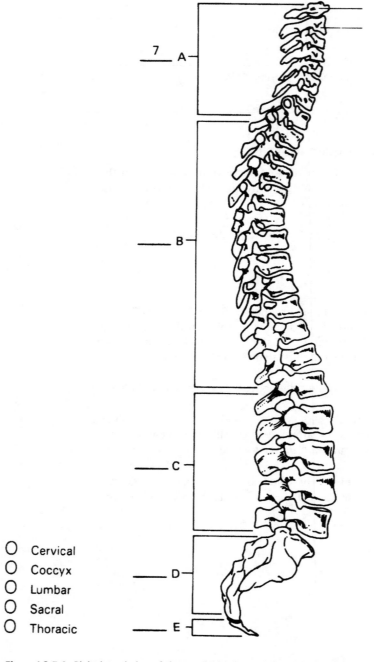

O   Cervical
O   Coccyx
O   Lumbar
O   Sacral
O   Thoracic

Figure LG 7.2 Right lateral view of the vertebral column. Color and label as directed in Checkpoint C3.

✓ **C7.** Describe bones and markings involved in each joint.

C. **Cervical**	**L.** **Lumbar**	**S.** **Sacral**	**T.** **Thoracic**

_____ a. Small body, foramina for vertebral blood vessels in transverse processes

_____ b. Only vertebrae that articulate with ribs

_____ c. Massive body, blunt spinous process and articular processes directed medially or laterally

_____ d. Long spinous processes that point inferiorly

_____ e. Articulate with the two hipbones

_____ f. The atlas and axis are vertebrae in this region

_____ g. Spinous processes in this region include several that are normally bifid

Spinal cord

(a)

(b)

(c)

(d)

O Body
O Facets for articulation with rib
O Intervertebral disc
O Intervertebral foramen
O Lamina
O Pedicle

O Spinous process (spine)
O Superior articular process
O Transverse foramen
O Transverse process
O Vertebral foramen

Figure LG 7.3 Typical vertebrae. (a) Thoracic vertebra, superior view. (b) Thoracic vertebrae, right lateral view (c) Cervical vertebra, superior view. (d) Lumbar vertebra, superior view. Color and label as directed in Checkpoint C5.

✓▲ **C7**. Describe bones and markings involved in each joint.
  a.   Atlanto-occipital joint: _____

  b.   Atlanto-axial (two different joint types): _____

  _____

✓▲ **C8**. *A clinical correlation.* State the clinical significance of these markings.
  a.   Sacral hiatus

  b.   Sacral promontory

  c.   Dens

D. Thorax (pages 210–213)

  D1. Name the structures that compose the thoracic cage. (Why is it called a "cage"?)

✓▲ **D2.** On Figure LG 7.4, color the three parts of the sternum, and label these parts: *sternal angle, suprasternal notch,* and *clavicular notch.*

Figure LG 7.4 Anterior view of the thorax. Color as directed in Checkpoints D2 and D3.

○   Body   ○   Ribs, floating
○   Manubrium   ○   Ribs, true
○   Ribs, false   ○   Xiphoid process

✓▲ **D3.** Complete this exercise about ribs. Refer again to Figure LG 7.4.

    a. There is a total of _____ ribs (_____ pairs) in the human skeleton.

    b. Ribs slant in such a way that the anterior portion of the rib is *(superior? inferior?)* to the posterior end of the rib.

    c. Posteriorly, all ribs articulate with _____. Ribs also pass *(anterior? posterior?)* to and articulate with transverse processes of vertebrae. What is the functional advantage of such an arrangement?

    d. Anteriorly, ribs numbered _____ to _____ attach to the sternum directly by means of

       strips of hyaline cartilage, called _____ cartilage. These ribs are called *(true? false?)* ribs. Color these ribs on Figure LG 7.4, leaving the costal cartilages white.

    e. Ribs 8 to 12 are called _____. Color these ribs a different color, again leaving the costal cartilages white. Do these ribs attach to the sternum? *(Yes? No?)* If so, in what manner?

    f. Ribs _____ and _____ are called "floating ribs." Use a third color for these ribs. Why are they so named?

    g. What function is served by the costal groove?

    h. What occupies intercostal spaces?

✓ **D4.** At what point are ribs most commonly fractured?

E. Disorders (page 213–214)

✓▲ **E1.** Complete this exercise about slipped discs.

    a. The normal intervertebral disc consists of two parts: an outer ring of *(hyaline? elastic?*

       fibro-?) cartilage called _____ and a soft, elastic, inner portion called

       the _____.

    b. Ligaments normally keep discs in alignment with vertebral bodies. What may happen if these ligaments weaken?

c. Why might pain result from a slipped disc?

d. In what part of the vertebral column are slipped discs most common? What symptoms may result from a slipped disc in this region?

E2. Match types of abnormal curvatures of the vertebral column with descriptions below.

K.	Kyphosis	L.	Lordosis	S.	Scoliosis

_____ a. Exaggerated lumbar curvature; "swayback"

_____ c. S- or C-shaped lateral bending; most common abnormal curvature

_____ b. Exaggerated thoracic curvature; "hunchback"

E3. Imperfect union of the vertebral arches at the midline is the condition known as

_____. Why is it crucial that the vertebral foramen be completely surrounded by bone? What problems may result from incomplete closure?

ANSWERS TO SELECTED CHECKPOINTS: CHAPTER 7

A2. (a) Axial. (b) 22 skull, 6 earbones (studied in Chapter 16), 1 hyoid, 26 vertebrae, 1 sternum, 24 ribs. (c) 80. (d) Shoulder girdles, upper extremities, hipbones, lower extremities. (e) 2 left shoulder girdle, 30 left upper extremity, 1 left hipbone, 30 left lower extremity. (f) 126. (g) 206.

A6. (b) Superior orbital fissure. (c) Optic foramen. (d) External auditory meatus. (e) Head of humerus. (f) Greater trochanter. (g) Styloid process. (h) Articular facet on vertebra.

B1. (a) Cranium, 8; brain; 3. (b) 14; 28.

B2. (a) Protection of the brain; attachment for meninges and muscles that move the head;

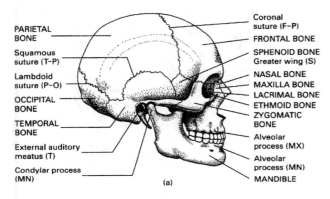

PARIETAL BONE
Squamous suture (T-P)
Lambdoid suture (P-O)
OCCIPITAL BONE
TEMPORAL BONE
External auditory meatus (T)
Condylar process (MN)

Coronal suture (F-P)
FRONTAL BONE
SPHENOID BONE Greater wing (S)
NASAL BONE
MAXILLA BONE
LACRIMAL BONE
ETHMOID BONE
ZYGOMATIC BONE
Alveolar process (MX)
Alveolar process (MN)
MANDIBLE

(a)

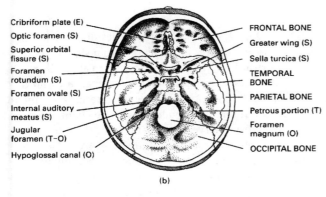

Cribriform plate (E)
Optic foramen (S)
Superior orbital fissure (S)
Foramen rotundum (S)
Foramen ovale (S)
Internal auditory meatus (S)
Jugular foramen (T-O)
Hypoglossal canal (O)

FRONTAL BONE
Greater wing (S)
Sella turcica (S)
TEMPORAL BONE
PARIETAL BONE
Petrous portion (T)
Foramen magnum (O)
OCCIPITAL BONE

(b)

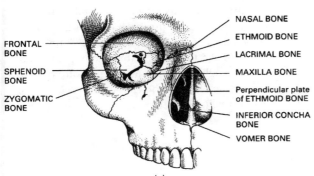

FRONTAL BONE
SPHENOID BONE
ZYGOMATIC BONE

NASAL BONE
ETHMOID BONE
LACRIMAL BONE
MAXILLA BONE
Perpendicular plate of ETHMOID BONE
INFERIOR CONCHA BONE
VOMER BONE

(c)

Figure LG 7.1A Skull bones. (a) Skull viewed from right side. (b) Floor of the cranial cavity. (c) Anterior view of bones that form the right orbit and the nose. Bone labels are capitalized; markings are in lowercase with related bone initial in parentheses.

openings for blood vessels and nerves. (b) Frontal (1), parietals (2), occipital (1), temporals (2), ethmoid (1), and sphenoid (1).

B3. See Figure LG 7.1A.

B4. (a) Mandible. (b) Zygomatic (malar). (c) Lacrimal. (d) Nasal. (e) Temporal. (f) Occipital. (g) Parietal. (h) Maxilla. (I) Palatine. (j) Frontal. (k) Ethmoid. (l) Sphenoid. (m) Vomer. (n) Inferior concha.

B5.

Marking	Bone	Function
a. Greater wings	Sphenoid	**Form part of side walls of skull**
b. **Perpendicular plate**	**Ethmoid**	Forms superior portion of septum of nose
c. **Sella Turcica**	**Sphenoid**	Site of pituitary gland
d. Petrous portion	**Temporal**	**Houses middle ear and inner ear**
e. **Foramen magnum**	**Occipital**	Largest hole in skull; passageway for spinal cord
f. Alveolar processess	(2) **Maxillae and mandible**	Bony sockets for teeth
g. **External auditory meatus**	**Temporal**	Passageway for sound waves to enter ear
h. Condylar process	Mandible	**Articulates with temporal bone (in TMJ)**

B6. See Figure LG 7.1A
B7. See Figure LG 7.1A

Number and Name of Cranial Nerve	Location of Opening for Nerve
a. I Olfactory (Figure 16.1)	Cribriform plate of ethmoid bone
b. II Optic (Figure 11.5a)	**Optic foramen of Sphenoid bone**
c. III Oculomotor (Figure 17.1) IV Trochlear V Trigeminal (ophthalmic branch) VI Abducens	**Superior orbital fissure of sphenoid bone**
d. V Trigeminal (maxillary branch)	**Foramen rotundum of sphenoid bone**
e. V Trigeminal (mandibular branch)	**Foramen ovale of sphenoid bone**
f. VII Facial (Figure 17.2) VIII Vestibulocochlear (Figure 16.16)	**Internal auditory meatus**
g. IX Glossopharyngeal (Figure 17.2) X Vagus (Figure 17.2) XI Accessory	**Jugular foramen**
h. XII Hypoglossal	**Hypoglossal canal**

B8. (a) Crista galli, olfactory foramina, perpendicular plate, superior nasal conchae. (b) Supraorbital margin. (c) Condylar process, mental foramen. (d) Inferior orbital fissure, palatine process. (e) Superior nuchal line. (f) Lesser wings, pterygoid processes, superior orbital fissure. (g) Carotid canal, mastoid air cells, zygomatic process.

B9. (a) Immovable (fibrous) joints between skull bones; between parietal bones; see Figure LG 7.1A. (b) Fontanels; between frontal and parietal bones. (c) They permit change in size and shape of the fetal skull during the birth process and they permit rapid growth of the skull during infancy; the anterior fontanel serves as a clinical landmark. (d) See on Figure LG 7.1cA: frontal, zygomatic, maxilla, lacrimal, ethmoid, and sphenoid.

B10. (a) Frontal, ethmoid, sphenoid, maxilla. (b) Warm and humidify air because air sinuses are lined with mucous membrane; serve as resonant chambers for speech and other sounds; make the skull lighter weight. (c) Ethmoid. (d) Like paranasal sinuses, they are covered with mucous membrane, which warms, humidifies, and cleanses air entering the nose. (e) Vomer; cartilage.

B11. (a) Temporal, zygomatic. (b) Temporal. (c) Lacrimal. (d) Superior. (e) Within. (f) Maxilla, palatine.

B12. (a) Frontal: supraorbital margin. (b) Ethmoid: perpendicular plate; vomer; cartilage (anteriorly). (c) Maxilla bones: palatine processes. (d) Temporal bone: condylar fossa; mandible: condyle.

C1. Superior.

C2. It articulates (forms a joint) with no other hones.

C3. A, cervical (7); B, thoracic (12); C, lumbar (5); D, sacrum (1); E, coccyx (1); C1, atlas; C2, axis.

C4. B (thoracic) and D (sacral); primary; were present originally during fetal life.

C5. See Figure LG 7.3A

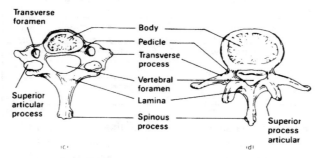

Figure LG 7.3A Typical vertebrae. (a) Thoracic vertebra, superior view. (b) Thoracic vertebrae, right lateral view. (c) Cervical vertebra, superior view. (d) Lumbar vertebra, superior view.

C6. (a) C. (b) T. (c) L. (d) T. (e) S. (f) C. (g) C.

C7. (a) Superior articular facets of atlas-occipital condyles. (b) Inferior articular facets of atlas—superior articular facets of axis; also anterior arch of atlas—dens of axis.

C8. (a) Site used for administration of epidural anesthesia. (b) Obstetrical landmark for measurement of size of the pelvis. (c) In whiplash, it may injure the brainstem.

D2. See Figure LG 7.4A

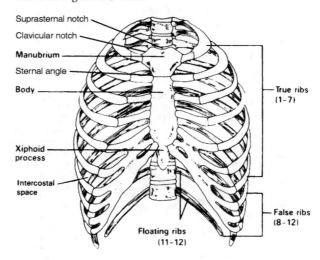

Figure LG 7.4A Anterior view of the thorax.

D3. (a) 24, 12. (b) Inferior. (c) Bodies of thoracic vertebrae; anterior; prevents ribs from slipping posteriorward. (d) 1, 7, costal; true. (e) False; ribs 8–10 attach to sternum indirectly via seventh costal cartilage. (f) 11, 12; they have no anterior attachment to sternum. (g) Provides protective channel for intercostal nerve, artery, and vein. (h) Intercostal muscles. Refer also to Figure LG 7.4A.

D4. Just anterior to the costal angle, especially involving ribs 3 to 10.

E1. (a) Fibro-, annulus fibrosus, nucleus pulposus. (b) Annulus fibrosus may rupture, allowing the nucleus pulposus to protrude (herniate), causing a "slipped disc." (c) Disc tissue may press on spinal nerves. (d) L4 to L5 or L5 to sacrum; pain in posterior of thigh and leg(s) due to pressure on sciatic nerve, which attaches to spinal cord at L4 to S3.

E2. (a) L. (b) K. (c) S.

MORE CRITICAL THINKING: CHAPTER 7

1. Contrast the effects of the following injuries to the skull. Consider what functions of each traumatized bone would be lost, and deduce the relative seriousness in each case.
   a. The right "greater wing" of the sphenoid; the left "lesser wing" of the sphenoid.
   b. The orbit; the foramen magnum. Hint: Which (how many) bone(s) form(s) each?
   c. The mandible; the right maxilla bone
2. One surgical approach to the pituitary gland involves passage of an instrument though the nose, up through the cribriform plate. Which sensation is likely to be impaired after this surgery?
3. The ethmoid bone could be considered as delicate as a French pastry. Look closely at a skull and notice why each of the following parts of this bone is particularly vulnerable:
   (a) Orbital surface
   (b) Superior and middle turbinates (conchae)
   (c) Perpendicular plate
4. Contrast the vertebral column with the spinal column. Then describe functions of the vertebral column, and explain what structural features make the vertebral column strong yet flexible.
5. Consider the structure and functions of the axis and atlas. How do these bones help you to move your head and neck? What parts of these bones (and related joints) are most vulnerable to injury?

MASTERY TEST: CHAPTER 7

Questions 1–10: Circle the letter preceding the one best answer to each question.

1. All of these bones contain paranasal sinuses except:
   A. Frontal   D. Sphenoid
   B. Maxilla   E. Ethmoid
   C. Nasal
2. Choose the one false statement.
   A. There are seven vertebrae in the cervical region.
   B. The cervical region normally exhibits a curve that is slightly concave anteriorly.
   C. The lumbar vertebrae are superior to the sacrum.
   D. Intervertebral discs are located between bodies of vertebrae.
3. The hard palate is composed of _____ bones.
   A. Two maxilla and two mandible
   B. Two maxilla and two palatine
   C. Two maxilla
   D. Two palatine
   E. Vomer, ethmoid, and two temporal
4. The lateral wall of the orbit is formed mostly by which two bones?
   A. Zygomatic and maxilla
   B. Zygomatic and sphenoid
   C. Sphenoid and ethmoid
   D. Lacrimal and ethmoid
   E. Zygomatic and ethmoid

5. Choose the one true statement.
   A. All of the ribs articulate anteriorly with the sternum.
   B. Ribs 8 to 10 are called true ribs.
   C. There are 23 ribs in the male skeleton and 24 in the female skeleton.
   D. Rib 7 is larger than rib 3.
   E. Cartilage discs between vertebrae are called costal cartilages.
6. Which is the largest fontanel?
   A. Frontal   C. Sphenoid
   B. Occipital   D. Mastoid
7. Immovable joints of the skull are called:
   A. Wormian bones   D. Sinuses
   B. Sutures   E. Fontanels
   C. Conchae
8. _____ articulate with every bone of the face except the mandible.
   A. Lacrimal bones   D. Sphenoid bones
   B. Zygomatic bones   E. Ethmoid bones
   C. Maxillae
9. All of these markings are parts of the sphenoid bone except:
   A. Lesser wings   D. Sella turcica
   B. Optic foramen   E. Pterygoid
   C. Crista galli   processes
10. All of these bones are included in the axial skeleton except:
    A. Rib   D. Hyoid
    B. Sternum   E. Ethmoid
    C. Clavicle

Questions 11–15: Arrange the answers in correct sequence.

_____ _____ _____   11. From anterior to posterior:
  A. Ethmoid bone
  B. Sphenoid bone
  C. Occipital bone

_____ _____ _____   12. Parts of vertebra from anterior to posterior:
  A. Vertebral foramen
  B. Body
  C. Lamina

_____ _____ _____   13. Vertebral regions, from superior to inferior:
  A. Lumbar
  B. Thoracic
  C. Cervical

_____ _____ _____   14. From superior to inferior:
  A. Atlas
  B. Axis
  C. Occipital bone

_____ _____ _____   15. From superior to inferior:
  A. Atlas
  B. Manubrium of sternum
  C. Hyoid

Questions 16–20: Circle T (true) or F (false). If the statement is false, change the underlined word or phrase so that the statement is correct.

T F   16. The space between two ribs is called the <u>costal groove</u>.
T F   17. The thoracic and sacral curves are called <u>primary</u> curves, meaning that they retain the original curve of the fetal vertebral column.
T F   18. The jugular vein passes through the same foramen as cranial nerves <u>V, VI, and VII</u>.
T F   19. In general, <u>foramina, meati, and fissures</u> in the skull serve as openings for nerves and blood vessels.
T F   20. The annulus fibrosus portion of an intervertebral disc is a <u>firm ring of fibrocartilage</u> surrounding the nucleus pulposus.

Questions 21–25: Fill-ins. Write the word or phrase that best completes the statement.

_____   21. A finger- or toothlike projection called the dens is part of the

  _____ bone.

_____   22. The ramus, angle, mental foramen, and alveolar processes are all

  markings on the _____ bone.

_____   23. The squamous, petrous, and zygomatic portions are markings on

  the _____ bone.

_____   24. The perpendicular plate, crista galli, and superior and middle conchae are markings found on the bone.

_____   25. The most common type of abnormal curvature of the spine

  is _____ .

ANSWERS TO MASTERY TEST: CHAPTER 7

Multiple Choice

1. C
2. B
3. B
4. B
5. D
6. A
7. B
8. C
9. C
10. C

Arrange

11. A B C
12. B A C
13. C B A
14. C A B
15. A C B

True-False

16. F. Intercostal space
17. T
18. F. IX, X, and XI
19. T
20. T

Fill-Ins

21. Axis (second cervical vertebra)
22. Mandible
23. Temporal
24. Ethmoid
25. Scoliosis

FRAMEWORK 8
The Appendicular Skeleton

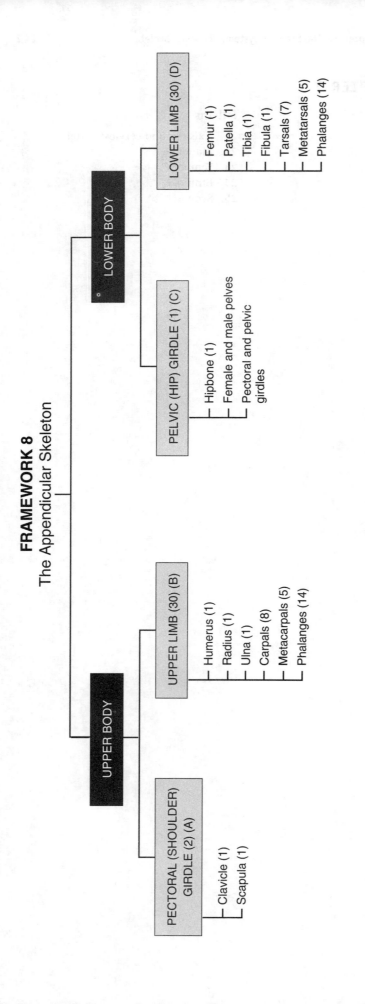

Note: numbers in ( ) refer to number of each bone on *one* side of the body.

The Skeletal System: The Appendicular Skeleton

The appendicular skeleton includes the bones of the limbs, or extremities, as well as the supportive bones of the shoulder and hip girdles. Male and female skeletons exhibit some differences in skeletal structure, noted particularly in bones of the pelvis.

As you begin your study of the appendicular skeleton, carefully examine the Chapter 8 Topic Outline and Objectives and note relationships among concepts and key terms in the Framework.

TOPIC OUTLINE AND OBJECTIVES

A. Pectoral (shoulder) girdle

1. Identify the bones of the pectoral (shoulder) girdle and their principal markings.

B. Upper limb (extremity)

2. Identify the bones of the upper limb and their principal markings.

C. Pelvic (hip) girdle; comparison of female and male pelves

3. Identify the bones of the pelvic girdle and their principal markings.
4. Compare the principal structural differences between female and male pelves.

D. Lower limb (extremity)

5. Identify the bones of the lower limb and their principal markings.

WORDBYTES

Now become familiar with the language of this chapter by studying each wordbyte, its meaning, and an example of its use within a term. After you study the entire list, self-check your understanding by writing the meaning of each wordbyte on the line. As you continue through the *Learning Guide*, identify (and fill in) additional terms that contain the same wordbyte.

Wordbyte	Self-check	Meaning	Example(s)
acrom-	_____	tip	*acrom*ion process
capit-	_____	head	*capit*ulum
cox-	_____	hip	*cox*al bone
genu	_____	knee	*genu* valgum
meta-	_____	after, beyond	*meta*carpal
obtur-	_____	closed up	*obtur*ator foramen
pelv-	_____	basin	*pelv*is
-physis	_____	to grow	public sym*physis*
ram-	_____	branch	inferior *ram*us
semi-	_____	half	*semi*lunar notch
sym-	_____	together	pubic *sym*physis

CHECKPOINTS

A. Pectoral (shoulder) girdle (pages 218–221)

✓▲ **A1.** Do this exercise about the pectoral girdle.
  a. Which bones form the pectoral girdle?

  b. Do these bones articulate (form a joint) with vertebrae or ribs? _____
  c. The pectoral girdle is part of the *(axial? appendicular?)* skeleton. Identify the point (marked by *) on Figure LG 8.1 at which the shoulder girdle articulates with the axial skeleton. Name the two bones forming that joint. Palpate (press and feel) the bones at this joint on yourself.

✓▲ **A2.** *Critical thinking* about the clavicle.
    State two or more functions of the clavicle.

    If the clavicle is fractured, this bone is not set in a cast. Why not?

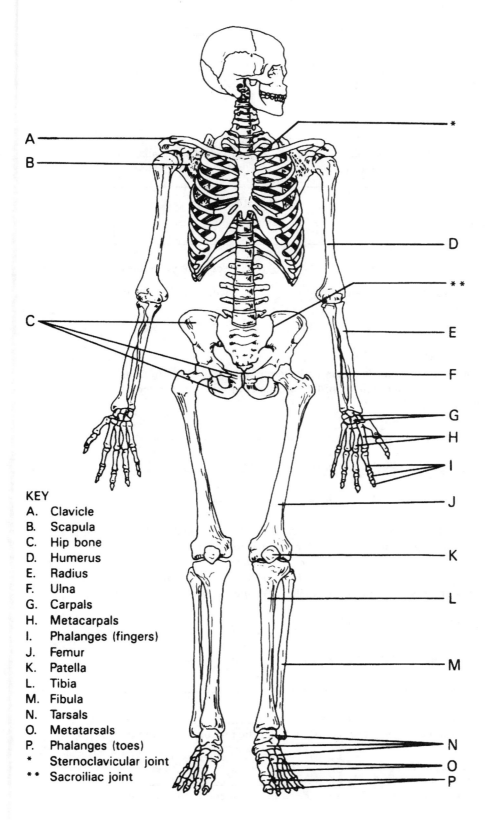

KEY
A. Clavicle
B. Scapula
C. Hip bone
D. Humerus
E. Radius
F. Ulna
G. Carpals
H. Metacarpals
I. Phalanges (fingers)
J. Femur
K. Patella
L. Tibia
M. Fibula
N. Tarsals
O. Metatarsals
P. Phalanges (toes)
* Sternoclavicular joint
** Sacroiliac joint

Figure LG 8.1 Anterior view of the skeleton. Color, label, and answer questions
as directed in Checkpoints A1, B1, C1, D1, D2, and Dg.

✓▲ **A3.** Write the name of the bone that articulates with each of these markings on the scapula.

a. Acromion process: _____

b. Glenoid cavity: _____

c. Coracoid process: _____

✓ **A4.** Study a scapula carefully, using a skeleton or Figure 8.3, in your text. Then match the markings in the box with the descriptions given.

A. **Axillary border**	**M.** **Medial border**
I. **Infraspinatus fossa**	**S.** **Spine**

_____ a. Sharp ridge on the posterior surface   _____ c. Edge closest to the vertebral column

_____ b. Depression inferior to the spine; location   _____ d. Thick edge closest to the arm
of infraspinatus muscle

B. Upper limb (extremity) (pages 221–225)

✓▲ **B1.** List the bone (or groups of bones) in the upper limb from proximal to distal. Indicate how many of each bone there are in one limb. Two are done for you. Refer to Figure LG 8.1 to check your answers.

a. **Humerus** _____ **(1)**   d. _____ **( )**

b. _____ **( )**   e. **Metacarpals** _____ **(5)**

c. _____ **( )**   f. _____ **( )**

✓▲ **B2.** On Figure LG 8.2, select different colors and color each of the markings indicated by ○. Where possible, color markings on both the anterior and posterior views.

✓▲ **B3.** Name the marking that fits each description. Write H, U, or R to indicate whether the marking is part of the humerus (H), ulna (U), or radius (R). One has been done for you.

a. Articulates with glenoid fossa: _____**Head**_____ (_**H**_)

b. Rounded head that articulates with radius: _____ (_____)

c. Posterior depression that receives the olecranon process: _____ (_____)

d. Halfmoon-shaped curved area that articulates with trochlea: _____ (_____)

e. Slight depression in which the head of the radius pivots (at the proximal radioulnar

joint): _____ (_____)

f. Tendon of the biceps brachii muscle attaches here: _____ (_____)

g. Easily palpated (felt) projection on the posterior side of the distal end of the forearm

on the little finger side. _____ (_____)

h. Projection on the medial side of the elbow; sensitive to touch since ulnar nerve passes

through here: _____ (_____)

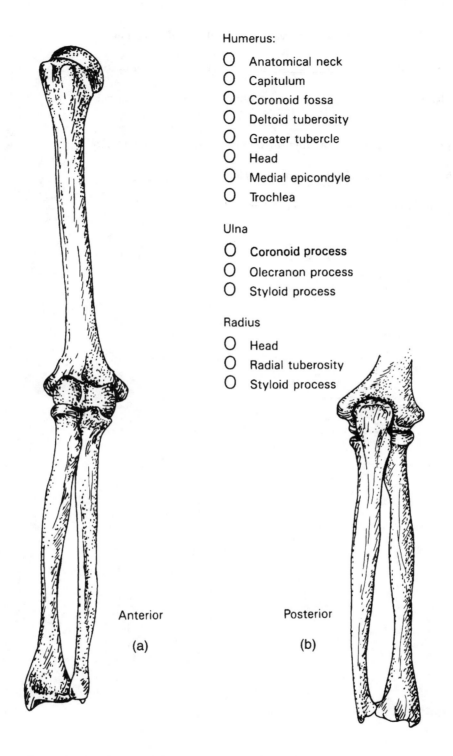

Humerus:

O Anatomical neck
O Capitulum
O Coronoid fossa
O Deltoid tuberosity
O Greater tubercle
O Head
O Medial epicondyle
O Trochlea

Ulna

O **Coronoid process**
O Olecranon process
O Styloid process

Radius

O Head
O Radial tuberosity
O Styloid process

Anterior

(a)

Posterior

(b)

Figure LG 8.2 Right upper limb. (a) Anterior view. (b) Posterior view.
Color and label as directed in Checkpoint B2.

✓▲**B4.** Answer these questions about the wrist.

    a. Wrist bones are called _____. There are *(5? 7? 8? 14?)* of them in
       each wrist.

    b. The wrist bone most subject to fracture is the bone named _____
       located just distal to the *(radius? ulna?).*

c. *Carpal tunnel syndrome*, due to repetitive wrist movements, involves inflammation of

hand muscle tendons and the _____ that are wrapped by a connective

tissue sheathe, the flexor _____, in a "tunnel" formed by four carpal
bones.

✓▲ **B5.** If you wear a ring on your "ring finger," the ring surrounds the *(proximal? distal? mid-*

dle?) phalanx number *(II? IV?)*. The *pollex* is a term that refers to _____.

▲ **B6.** Trace an outline of your hand. Draw in and label all bones.

C. Pelvic (hip) girdle; comparison of female and male pelves (pages 225–231)

✓▲ **C1.** Describe the pelvic bones in this exercise.
   a. Name the bones that form the pelvic girdle. _____

   Which bones form the pelvis? _____
   b. Which of these bones is/are part of the axial skeleton?

   c. Locate the point (**) on Figure LG 8.1 at which the pelvic girdle portion of the appendicular skeleton articulates with the axial skeleton. Name the bones and

   markings involved in that joint. _____ and _____

✓▲ **C2.** Answer the following questions about coxal bones.
   a. Each coxal (hip) bone originates as three bones that fuse early in life. These bones

   are the _____, _____, and _____.

   At what location do the bones fuse? _____

   b. The largest of the three bones is the _____.
      A ridge along the superior border is called the iliac crest. Locate this on yourself.

   c. The iliac crest ends anteriorly as the _____ spine (known by the

   acronym ASIS). A posterior marking known as the _____ spine (or PSIS) causes a dimpling of the skin just lateral to the sacrum, which can be used as a landmark for administering hip injections accurately.

✓▲ **C3.** Complete the table about markings of the coxal bones.

Marking	Location on Coxal Bone	Function
a. Greater sciatic notch		
b.		Supports most of body weight in sitting position
c.		Fibrocartilaginous joint between two coxal bones
d.		Socket for head of femur
e. Obturator oramen	Large foramen surrounded by pubic and ischial rami and acetabulum	

○ Coccyx   ○ Ischium   ○ Sacrum

○ Ilium   ○ Pubis

Figure LG 8.3 Right anterolateral view of the pelvis. Color, label, and answer questions as directed in Checkpoints C4 and C6.

✓ **C4.** Refer to Figure LG 8.3 and answer these questions about pelvic markings.
  a. Trace with your pencil, and then label, the brim of the pelvis on the figure. It is a relatively *(smooth? irregular?)* line that demarcates the *(superior? inferior?)* border of the lesser (true) pelvis. The pelvic brim encircles the pelvic *(inlet? outlet?)*.
  b. The bony border of the pelvic outlet is *(smooth? irregular?)*. Why is the pelvic outlet so named?

  C5. Contrast the two principal parts of the pelvis. Describe their locations and name the structures that compose them.
  a. Greater (false) pelvis

b. Lesser (true) pelvis

✓▲ **C6.** Color the pelvic structures indicated by color code ovals on Figure LG 8.3. Also label the following markings: *acetabulum, anterior superior iliac spine, greater sciatic notch, iliac crest, inferior pubic ramus, ischial spine, ischial tuberosity, obturator foramen, sacral promontory,* and *symphysis pubis.*

C7. State several characteristics of the female pelvis that make it more suitable for childbirth than the male pelvis.

✓▲ **C8.** Identify specific differences in pelvic structure in the two sexes by placing M before characteristics of the male pelvis and F before structural descriptions of the female pelvis.

_____ a. Shallow greater pelvis

_____ b. Heart-shaped inlet

_____ c. Pubic arch greater than 90° angle

_____ d. Acetabulum small

_____ e. Pelvic inlet comparatively small

✓▲ **C9.** *The Big Picture: Looking Ahead.* Refer to figures in Chapter 11 of your text as you check your understanding of the appendicular skeleton in this Checkpoint.
   a. Contrast Figures 11.12 and 11.13. In each figure, find the coccyx and the ischial tuberosities. Which pelvis has a relatively wider distance between ischial tuberosities? *(Female? Male?)* Also notice that *(bone? muscle?)* forms most of the floor of the human pelvis.
   b. The ischial tuberosities are points of attachment of several large muscles (Figure 11.20c and Exhibit 11.18). Name two or more muscles attached to these bony markings of the hip bone.

   c. Which muscles cover the posterior surface of the scapula, padding this thin, hard bone (Figure 11.15c)?

   The _____-spinatus muscle lies superior to the spine of the scapula,

   whereas the _____-spinatus and teres _____ are located inferior to the spine.
   d. Name two muscles that are anchored into the comma-shaped coracoid process of the

   scapula: the pectoralis *(major? minor?)* (Figure 11.14a) and the _____ brachii (Figure 11.16a). These two muscles both lie closer to the *(anterior? posterior?)* surface of the body.
   e. Figure 11.16a shows that the distal tendon of the biceps brachii attaches to the anterior of the *(humerus? radius? ulna?).*

C10. *Critical thinking.* Explain how structural differences in the pectoral and pelvic girdles account for differences in support and mobility of the body.

D. Lower limb (extremity) (pages 231–237)

✓▲ **D1.** Refer to Figure LG 8.1 and list the bones (or groups of bones) in the lower limb from proximal to distal. Indicate how many of each bone there are. One is done for you.

a. **Femur** _____ **(1)**   e. _____ ( )

b. _____ ( )   f. _____ ( )

c. _____ ( )   g. _____ ( )

d. _____ ( )

D2. Contrast the size, location, and names of the bones of the upper and lower limbs by coloring the bones on Figure LG 8.1 as follows. Color on one side of the figure only.

Humerus and femur (red)   Carpals and tarsals (blue)

Patella (brown)   Metacarpals and metatarsals (orange)

Ulna and tibia (green)   Phalanges (purple)

Radius and fibula (yellow)

✓▲ **D3.** Circle the term that correctly indicates the location of these parts of the lower limb.
a. The head is the *(proximal? distal?)* epiphysis of the femur.
b. The greater trochanter is *(lateral? medial?)* to the lesser trochanter.
c. The intercondylar fossa is on the *(anterior? posterior?)* surface of the femur.
d. The tibial condyles are more *(concave? convex?)* than the femoral condyles.
e. The lateral condyle of the femur articulates with the *(fibula? lateral condyle of the tibia?)*.
f. The tibial tuberosity is *(proximal? distal?)* to the patella.
g. The tibia is *(medial? lateral?)* to the fibula.
h. The outer portion of the ankle is the *(lateral? medial?)* malleolus, which is part of the *(tibia? fibula?)*.

✓▲ **D4.** Complete the Checkpoint about the patella.
a. This bone is located on the *(anterior? posterior?)* aspect of the knee joint.
b. The patella is a sesamoid bone. Define sesamoid bone.

c. What are its functions?

d. What function does the patellar ligament serve?

e. Does the patella articulate with the fibula? _____

f. "Runner's knee" is a common term for _____ _____
    syndrome, in which the patella fails to track (glide) in the groove between the

    _____ of the femur.

D5. *A clinical correlation* about the lower extremity.
a. The angle of the femur (between hip and knee joints) is known as the angle of

    _____ since the knees are normally more *(medial? lateral?)* than the
    hips. This angle is normally *(larger? smaller?)* in females because hips are broader
    in females.
b. Contrast *genu valgum* with *genu varum.*

✓▲ **D6.** The tarsal bone that is most superior in location (and that articulates with the tibia

    and fibula) is the _____. The largest and strongest of the tarsals is the

    _____.

✓▲ **D7.** Answer these questions about the arch of the foot.
a. How is the foot maintained in an arched position?

b. Locate each of these arches on your own foot. Refer to Figure 8.17 of your text. *(For
    extra review:* List the bones that form each arch.)
    Longitudinal: medial side

    Longitudinal: lateral side

    Transverse

c. What causes flatfoot?

✓▲ **D8.** Now that you have seen all of the bones of the appendicular skeleton, complete this table relating common and anatomical names of bones.

Common name	Anatomical term
a. Shoulder blade	
b.	Pollex
c. Collarbone	
d. Heel bone	
e.	Olecranon process
f. Kneecap	
g.	Tibial crest
h. Toes	
i. Palm of hand	
j. Wrist bones	

D9. *For extra review.* Label all bones marked with label lines on Figure LG 8.1.

ANSWERS TO SELECTED CHECKPOINTS: CHAPTER 8

A1. (a) Two clavicles and two scapulas. (b) No. (c) Appendicular; clavicles, manubrium of sternum.

A2. The clavicle serves as a thick pencil-like wedge that prevents the arm from collapsing inward towards the sternum; serves as attachment point for muscles. Totally immobilizing the clavicle would limit movements of breathing.

A3. Clavicle. (b) Humerus. (c) None; tendons attach here.

A4. (a) S. (b) I. (c) M. (d) A.

B1. (a) Humerus, 1. (b) Ulna, 1. (c) Radius, 1. (d) Carpals, 8. (e) Metacarpals, 5. (f) Phalanges, 14.

B2. See Figure LG 8.2A

B3. (b) Capitulum (H). (c) Olecranon fossa (H). (d) Trochlear (semilunar) notch (U). (e) Radial notch (U). (f) Radial tuberosity (R). (g). Styloid process (U). (h) Medial epicondyle (H).

B4. (a) Carpals, 8. (b) Scaphoid, radius. (c) Median, retinaculum.

B5. Proximal, IV; thumb.

Figure LG 8.2A Right upper limb. (a) Anterior view. (b) Posterior view.

C1. (a) Two coxal (hip) bones; hip bones plus sacrum and coccyx. (b) Sacrum and coccyx. (c) Sacrum (right and left auricular surfaces) and hip bone (auricular surface).

C2. (a) Ilium, ischium, pubis; acetabulum (socket for the head of the femur). (b) Ilium. (c) Anterior superior iliac spine (ASIS); posterior superior iliac spine (PSIS).

C3. See Table at right.

C4. See Figure LG 8.3A. (a) Smooth, superior, inlet. (b) Irregular; feces, urine, semen, menstrual flow and baby (at birth) exit via openings in the muscular floor attached to bony outlet.

C6. See Figure LG 8.3A.

C8. (a) F (b) M. (c) F. (d) F. (e) M.

C9. (a) Female; muscle. (b) The three heads of the hamstrings (semimembranosus, semitendinosus, and biceps femoris), as well as the adductor magnus, quadratus femoris, and inferior gemellus. (c) Supra, infra, minor. (d) Minor biceps: anterior. (e) Radius.

D1. (a) Femur, 1. (b) Patella, 1. (c) Tibia, 1. (d) Fibula, 1. (e) Tarsals, 7. (f) Metatarsals, 5. (g) Phalanges, 14.

D3. (a) Proximal. (b) Lateral. (c) Posterior. (d) Concave. (e) Lateral condyle of the tibia. (f) Distal. (g) Medial. (h) Lateral, fibula.

D4. (a) Anterior. (b) Sesame seed-shaped bone that develops within a tendon. (c) Protects the joint and its tendons from wear and tear; changes the direction of pull of a tendon, which provides mechanical advantage and some stability at the joint. (d) Attaches the patella to the tibial tuberosity. (e) No. (f) Patellofemoral stress; condyles.

D6. Talus; calcaneus.

D8. (a) Scapula. (b) Thumb. (c) Clavicle. (d) Calcaneus. (e) Elbow. (f) Patella. (g) Shinbone. (h) Phalanges. (i) Metacarpals. (j) Carpals.

Marking	Location on Coxal Bone	Function
a. Greater sciatic notch	Inferior to posterior inferior iliac spine	Sciatic nerve passes inferior to notch
b. Ischial tuberosity	Posterior and inferior to obturator foramen	Supports most of body weight in sitting position
c. Symphysis pubis	Most anterior portion of pelvis	Fibrocartilaginous joint between two coxal bones
d. Acetabulum	At junction of ileum, ischium, and pubis	Socket for head of femur
e. Obturator oramen	Large foramen surrounded by pubic and ischial rami and acetabulum	Blood vessels and nerves pass through

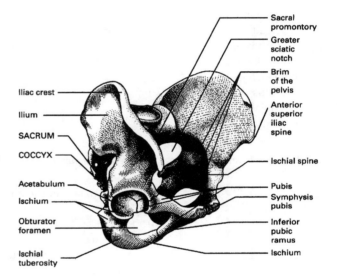

Figure LG 8.3A Right anterolateral view of the pelvis. Bone labels are capitalized; markings are in lowercase.

MORE CRITICAL THINKING: CHAPTER 8

1. Which joint is the point (bilaterally) at which the entire upper appendage attaches to the axial skeleton? At which joint (bilaterally) does the entire lower extremity attach to the axial skeleton? Speculate about what additional structures help to prevent the appendages from falling away from the axial skeleton.

2. Name the bones with which the scapula articulates. Based on the shape of bones in each joint, which joint appears to be most stable?

3. A fractured elbow can involve disruption of any/all of three joints. Describe these.

4. Identify functional differences that can be attributed to differences in location and structure of the carpal and tarsal bones.

MASTERY TEST: CHAPTER 8

Questions 1–6: Circle the letter preceding the one best answer to each question.

1. The point at which the upper part of the appendicular skeleton is joined to (articulates with) the axial skeleton is at the joint between:
   A. Sternum and ribs
   B. Humerus and clavicle
   C. Scapula and clavicle
   D. Scapula and humerus
   E. Sternum and clavicle

2. All of the following are markings on the femur *except*.
   A. Acetabulum   D. Greater trochanter
   B. Head   E. Intercondylar fossa
   C. Condyles

3. All of the following are bones in the lower limb *except:*
   A. Talus   D. Ulna
   B. Tibia   E. Fibula
   C. Calcaneus

4 Which structures are on the posterior surface of the upper limb (in anatomical position)?
   A. Radial tuberosity and lesser tubercle
   B. Trochlea and capitulum
   C. Coronoid process and coronoid fossa
   D. Olecranon process and olecranon fossa

5. The humerus articulates with all of these bones *except:*
   A. Ulna   C. Clavicle
   B. Radius   D. Scapula

6. Choose the *false* statement.
   A. The capitulum articulates with the head of the radius.
   B. The medial and lateral epicondyles are located at the distal ends of the tibia and fibula.
   C. The coronoid fossa articulates with the ulna when the forearm is flexed elbow joint.
   D. The trochlea articulates with the trochlear notch of the ulna.

Questions 7–11: Arrange the answers in correct sequence.

_____ _____ _____   7. According to size of the bones, from largest to smallest:
   A. Femur
   B. Ulna
   C. Humerus

_____ _____ _____   8. Parts of the humerus, from proximal to distal:
   A. Anatomical neck
   B. Surgical neck
   C. Head

_____ _____ _____   9. From proximal to distal:
   A. Phalanges
   B. Metacarpals
   C. Carpals

_____ _____ _____ 10. From superior to inferior:
   A. Lesser pelvis
   B. Greater pelvis
   C. Pelvic brim

_____ _____ _____ 11. Markings on hipbones in anatomical position, from superior to inferior:
   A. Acetabulum
   B. Ischial tuberosity
   C. Iliac crest

Questions 12–20: Circle T (true) or F (false). If the statement is false, change the underlined word or phrase so that the statement is correct.

T F   12. Another name for the true pelvis is the <u>greater</u> pelvis.
T F   13. The scapulae <u>do</u> articulate with the vertebrae.
T F   14. The olecranon process is a marking on the <u>ulna, and the olecranon fossa is a marking on the humerus.</u>
T F   15. The female pelvis is <u>deeper and more heart-shaped</u> than the male pelvis.
T F   16. The greater tubercle of the humerus is <u>lateral</u> to the lesser tubercle.
T F   17. There are <u>14 phalanges in each hand and also in each foot.</u>
T F   18. The fibula articulates with the <u>femur, tibia, talus, and calcaneus.</u>
T F   19. The organs contained within the right iliac, hypogastric, and left iliac portions (ninths) of the abdomen are located in the <u>true</u> pelvis.
T F   20. The total number of bones in one upper limb (including arm, forearm, wrist, hand, and fingers, but excluding shoulder girdle) is <u>29</u>.

Questions 21–25: Fill-ins. Write the word or phrase that best completes the statement.

_____ 21. In about three-fourths of all carpal bone fractures, only the _____ bone is involved.

_____ 22. The _____ is the thinnest bone in the body compared to its length.

_____ 23. The point of fusion of the three bones forming the hip bone is the _____.

_____ 24. The kneecap is the common name for the _____.

_____ 25. The bone that forms the heel is the _____.

ANSWERS TO MASTERY TEST: CHAPTER 8

Multiple Choice
1. E
2. A
3. D
4. D
5. C
6. B

Arrange
7. A C B
8. C A B
9. C B A
10. B C A
11. C A B

True-False
12. F. Lesser
13. F. Do not
14. T
15. F. Shallower and more oval
16. T
17. T
18. F. Tibia and talus only
19. F. False
20. F. 30

Fill-ins
21. Scaphoid
22. Fibula
23. Acetabulum
24. Patella
25. Calcaneus

FRAMEWORK 9
Joints

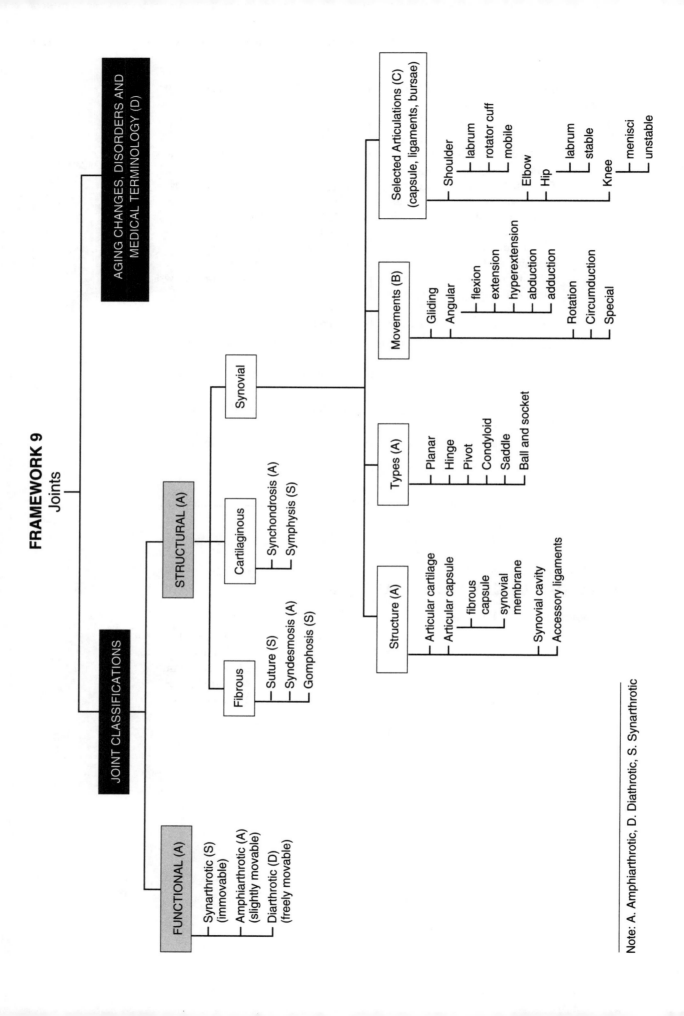

Note: A. Amphiarthrotic, D. Diathrotic, S. Synarthrotic

CHAPTER **9**

Joints

In the last three chapters you have learned a great deal about the 206 bones in the body. Separated, or disarticulated, these bones would constitute a pile as disorganized as the rubble of a ravaged city deprived of its structural integrity. Fortunately, bones are arranged in precise order and held together in specific conformations—articulations or joints—that permit bones to function effectively. Joints may be classified by function (how movable) or by structure (the type of tissue forming the joint). Synovial joints are most common; their structure, movements, and types will be discussed. A detailed examination of four important synovial joints (shoulder, elbow, hip, and knee) is also included. Joint disorders such as arthritis are usually not life threatening but plague much of the population, particularly the elderly. An introduction to joint disorders completes this chapter.

As you begin your study of articulations, carefully examine the Chapter 9 Topic Outline and Objectives and note relationships among concepts and key terms in the Framework.

TOPIC OUTLINE AND OBJECTIVES

A. Joint classifications

1. Describe the structural and functional classifications of joints.
2. Describe the structure and functions of the three types of fibrous joints.
3. Describe the structure and functions of the two types of cartilaginous joints.
4. Describe the structure of synovial joints.
5. Describe the six subtypes of synovial joints.

B. Types of movements at synovial joints

6. Describe the types of movements that can occur at synovial joints.

C. Selected joints of the body

7. Become familiar with the structure and function of the shoulder, elbow, hip, and knee joints.
8. Describe six factors that influence the type of movement and range of motion possible at a synovial joint.

D. Aging and joints; disorders and medical terminology

9. Explain the effects of aging on joints.

WORDBYTES

Now become familiar with the language of this chapter by studying each wordbyte, its meaning, and an example of its use within a term. After you study the entire list, self-check your understanding by writing the meaning of each wordbyte on the line. As you continue through the *Learning Guide*, identify (and fill in) additional terms that contain the same wordbyte.

Wordbyte	Self-check	Meaning	Example(s)
amphi-	_____	both	*amphi*arthrotic
arthr-	_____	joint	osteo*arthr*itis
articulat-	_____	joint	*articulat*ion
cruci-	_____	cross	*cruci*ate
gompho-	_____	bolt, nail	*gompho*sis
-itis	_____	inflammation	osteoarthr*itis*
ov-	_____	egg	syn*ov*ial fluid
sutur-	_____	seam	coronal *suture*
syn-	_____	together	*syn*arthrotic
syndesmo-	_____	band or ligament	*syndesmo*sis

CHECKPOINTS

A. Joint classifications (pages 241–247)

✓ **A1.** Define the term *joint (articulation)*.

State two criteria used to classify joints according to structure.

✓ **A2.** Name three classes of joints based on structure.

✓ **A3.** Fill in the blanks below to name three classes of joints according to the amount of movement they permit.

a. Synarthrosis: _____

b. _____: slightly movable

c. _____: freely movable

✓▲ **A4.** Describe fibrous joints by completing this exercise.
a. Fibrous joints *(have? lack?)* a joint cavity. They are held together by

_____ connective tissue.

b. One type of fibrous joint is a _____ found between skull bones.

Such joints are *(freely? slightly? im-?)* movable, or _____-arthrotic. Replacement
of a fibrous suture with bony fusion is known as a *(synchondrosis? synostosis?)*.

c. Which of the following is a site of a gomphosis?
    A. Joint at distal ends of tibia and fibula
    B. Epiphyseal plate
    C. Attachment of tooth by periodontal ligament to tooth socket in maxilla or mandible

d. The distal end of the tibia/fibula joint is a fibrous joint. It is *(more? less?)* mobile

than a suture and is therefore _____ -arthrotic. It is called a _____.

✓▲ **A5.** Describe cartilaginous joints by completing this exercise.

a. Synchondroses involve *(hyaline? fibrous?)* cartilage between regions of bone. An

example is the _____ cartilage between diaphysis and epiphysis of
a growing bone. These joints are *(immovable? somewhat movable?)*. This cartilage
(persists through life? is replaced by bone during adult life?), forming a *(symphysis? syndesmosis? synostosis?)*.

b. Fibrocartilage is present in the type of joint known as a _____.

These joints permit some movement and so are called _____-arthrotic.

Two locations of symphyses are _____ and _____.

✓ **A6.** What structural features of synovial joints make them more freely movable than
fibrous or cartilaginous joints?

✓▲ **A7.** On Figure LG 9.1, color the indicated structures.

O Articular cartilage   O Periosteum

O Articulating bone   O Synovial (joint) cavity

O Fibrous capsule   O Synovial membrane

Figure LG 9.1 Structure of a generalized synovial
joint. Color as directed in Checkpoint A7.

✓▲ **A8.** Select the parts of a synovial joint (listed in the box) that fit the descriptions below.

A.	**Articular cartilage**	SF. **Synovial fluid**
F.	**Fibrous capsule**	SM. **Synovial membrane**
L.	**Ligaments**	

_____ a. Hyaline cartilage that covers ends of articulating bones but does not bind them together

_____ b. With the consistency of uncooked egg white or oil, it lubricates the joint and nourishes the avascular articular cartilage

_____ c. Connective tissue membrane that lines synovial cavity and secretes synovial fluid

_____ d. Parallel fibers in some fibrous capsules; bind bones together

_____ e. Together these form the articular capsule (two answers)

_____ f. Provides both flexibility and tensile strength

_____ g. Reduces friction at the joint

_____ h. Contains phagocytes that remove debris and microbes from joint

✓▲ **A9.** Describe the structure, function, and location of the following structures that are associated with synovial joints.
  a. Intracapsular ligaments

  b. Extracapsular ligaments

  c. Articular discs (menisci)

A10. Do this activity about "torn cartilage" of the knee.
  a. In this condition, the injured tissues are *(articular cartilages? menisci?)*.
  b. What functions are normally served by these cartilages?

  c. How may these cartilages be repaired with minimal destruction to other tissues?

✓ **A11.** A *(sprain? strain?)* is an overstretching of a muscle, whereas a *(sprain? strain?)* involves a more serious injury to joint structures. The joint most often sprained is the *(ankle? knee?)* joint.

▲ **A12.** Complete Table LG 9.1 on six subtypes of synovial joints. Select answers from lists in the box. The first one is done for you. *Note: omit the third column (types of movement) until Checkpoint B4.*

Subtype of synovial joint	Planes of movement	Types of movement	Examples of joints
Ball-and-socket	Biaxial	ABD. Abduction	List name of joint,
Condyloid (ellipsoidal)	Monoaxial	ADD. Adduction	articulating bones
Hinge	Triaxial	CIR. Circumduction	with specific
Pivot		E. Extension	markings at points
Planar		F. Flexion	of articulation.
Saddle		ROT. Rotation	

Table LG 9.1 Subtypes of synovial joints.

Subtype of Synovial Joint	Planes of Movement	Types of Movement	Examples
a. Ball-and-socket	Triaxial	F, E, ADD, ABD, CIR, ROT	1. Shoulder: scapula (glenoid cavity)-humerus (head) 2. Hip: hip bone (acetabulum)-femur (head)
b. Condyloid (ellipsoidal)			Wrist (radius-carpals)
c.			Carpal (trapezium)-first metacarpal (base)
d. Hinge		F, E	1. Knee: femur (condyles)-tibia (condyles)
e.	Monoaxial	ROT	
f.	(Omit)	Gliding	

Table LG 9.1 Subtypes of synovial joints. Complete according to directions in Checkpoints A12 and B4.

✓▲ **A13.** *For extra review* of types of synovial joints, choose the type of joint that fits the description. (Answers may be used more than once.)

B.	**Ball-and-socket**	**Piv.**	**Pivot**
C.	**Condyloid**	**Pla.**	**Planar**
H.	**Hinge**	**S.**	**Saddle**

_____ a. Monaxial joint; only rotation possible

_____ b. Examples include atlas-axis joint and joint between head of radius and radial notch at proximal end of ulna

_____ c. Triaxial joint, allowing movement in all three planes

_____ d. Hip and shoulder joints

_____ e. Spoollike (convex) surface articulated with concave surface, for example, elbow, ankle, and joints between phalanges

_____ f. One type of monoaxial joint in which only flexion and extension are possible

_____ g. Found in joints at sternoclavicular and claviculoscapular joints

_____ h. Thumb joint located between metacarpal proximal to thumb and carpal bone (trapezium)

_____ i. Biaxial joints (two answers)

A14. Contrast *bursae* and *tendon sheaths* according to structure and locations.

B. Types of movements at synovial joints (pages 247–253)

✓▲ **B1.** From the terms listed in the box, choose the one that fits the type of movement in each case. Not all answers will be used.

Abd.	**Abduction**	**E.**	**Extension**	**LF.**	**Lateral flexion**
Add.	**Adduction**	**F.**	**Flexion**	**O.**	**Opposition**
C.	**Circumduction**	**G.**	**Gliding**	**P.**	**Plantar flexion**
D.	**Dorsiflexion**	**I.**	**Inversion**	**R.**	**Rotation**

_____ a. Decrease in angle between anterior surfaces of bones (or between posterior surfaces at knee and toe joints)

_____ b. Simplest kind of movement that can occur at a joint; no angular or rotary motion involved; example: ribs moving against vertebrae

_____ c. State of entire body when it is in anatomical position

_____ d. Movement away from the midline of the body

_____ e. Movement of a bone around its own axis

_____ f. Position of foot when heel is on the floor and rest of foot is raised

✓▲ **B2.** Perform the action described. Then write in the name of the type of movement.

 a. Describe a cone with your arm, as if you were winding up to pitch a ball.

 The movement at your shoulder joint is called _____.

 b. Stand in anatomical position (palms forward). Turn your palms backward. This action

 is called _____.

 c. Move your fingers from "fingers together" to "fingers apart" position. This action is

 _____ of fingers.

 d. Raise your shoulders, as if to shrug them. This movement is called

 _____ of the shoulders.

 e. Stand on your toes. This action at the ankle joint is called _____.

 f. Grasp a ball in your hand. Your fingers are performing the type of movement called

 _____.

 g. Sit with the soles of your feet pressed against each other. In this position, your feet

 are performing the action called _____.

 h. Thrust your jaw outward (gently!). This action is _____ of the mandible.

 i. Move your thumb across your palm to touch your fingertips. _____

 j. Move to the right or left at the waist, as if to slide your fingertips down the lateral aspect of your leg. _____

✓▲ **B3.** Identify the kinds of movements shown in Figure LG 9.2. Write the name of the movement below each figure. Use the following terms: *abduction, adduction, extension, flexion,* and *hyperextension.*

✓ **B4.** Now complete the "types of movement" column in Table LG 9.1. (See page LG 163).

C. Selected joints of the body (pages 253–263)

✓▲ **C1.** Check your understanding of structure, function, and disorders of the shoulder joint in this Checkpoint.

 a. The shoulder has *(greater? less?)* freedom of movement than any other joint in the

 body. This joint is said to be _____-axial. What structural features of the joint account for this?

 b. The joint has several ligaments supporting the capsule. Which ligament is broad and strengthens the upper part of the articular capsule? *(Coraco-? Gleno-? Transverse)* humeral ligament.

 c. The glenoid labrum is composed of *(bone? fat? fibrocartilage?)*. How does the labrum stabilize the shoulder joint?

 d. Four bursae provide some cushioning and reduce friction at the shoulder joint. These bursae are located deep to bone or muscle tissue in this joint. Name them:

 sub-_____ , sub-_____ , sub-_____ ,

 and sub-_____ .

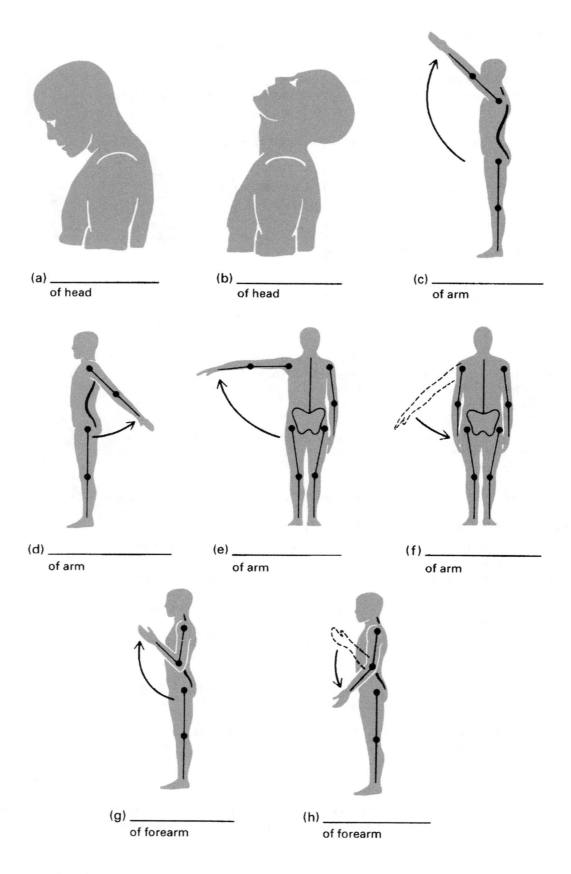

(a) _____
of head

(b) _____
of head

(c) _____
of arm

(d) _____
of arm

(e) _____
of arm

(f) _____
of arm

(g) _____
of forearm

(h) _____
of forearm

Figure LG 9.2 Movements at synovial joints. Answer questions
as directed in Checkpoint B3 and in Chapter 11.

(i) _____
　　　　of thigh

(j) _____
　　　　of thigh

(k) _____
　　　　of leg

(l) _____
　　　　of leg

(m) _____
　　　　of thigh

(n) _____
　　　　of thigh

(o) _____
　　　　of thigh

(p) _____
　　　　of thigh

e. What is the function of the rotator cuff?

What happens if the rotator cuff fails to do its job?

What tissues comprise the rotator cuff?

✓▲ **C2.** Contrast the hip joint with the shoulder joint in this exercise.
  a. The hip joint exhibits *(more? less?)* stability and *(more? less?)* range of motion than the shoulder joint. State three factors that contribute to the strength of this joint.

  b. Name three ligaments that limit hyperextension at the hip joint:

_____-femoral, _____-femoral, and _____-femoral. What limits adduction at this joint?

✓▲ **C3.** Complete this exercise describing the knee (tibiofemoral) joint. Consult Figure 9.14 in your text.
  a. The knee joint actually consists of three joints:
      1. Between the femur and the *(patella? fibula?)*
      2. Between the lateral condyle of the femur, *(medial? lateral?)* meniscus, and the lateral condyle of the *(tibia? fibula?)*
      3. Between the medial condyle of the femur, *(medial? lateral?)* meniscus, and the medial condyle of the *(tibia? fibula?)*
  b. The first joint (1) described above is a *(modified hinge? pivot? planar?)* joint. The other

two joints (2 and 3) are _____ joints.
  c. The knee (tibiofemoral) joint *(does? does not?)* include a complete capsule uniting the two bones. This factor contributes to the relative *(strength? weakness?)* of this joint.
  d. The medial and lateral patellar retinacula and patellar ligament are both tissues that serve as insertions of the *(hamstrings? quadriceps femoris?)* muscles.
  e. A number of other ligaments provide strength and support. Identify whether the following ligaments are extracapsular (E) or intracapsular (I).

      1. Arcuate popliteal ligament: _____
      2. Anterior and posterior cruciate ligaments: _____
      3. Tibial (medial) and fibular (lateral) collateral ligaments: _____
  f. Two cartilages, much like two C-shaped stadiums facing one another, each making an incomplete circle, are located between condyles of the femur and tibia. These

articular discs are called _____. What is their function?

  g. There are *(no? several?)* bursae associated with the knee joint. Inflammation of bursae

is called _____.

✓▲**C4.** Match the names of joints with descriptions provided.

A.	Ankle joint	S.	Shoulder
Ao.	Atlanto-occipital joint	TMJ.	Temporomandibular joint
E.	Elbow joint	W.	Wrist
H.	Hip (coxal) joint		

_____ a. Talocrural joint; capable of plantar flexion and dorsiflexion

_____ b. Glenohumoral joint

_____ c. A synovial joint with hinge and gliding actions; moves the lower (but not upper) jaw bone within the mandibular fossa of the temporal bone

_____ d. Radiocarpal joint

_____ e. Joint between the skull and the first cervical vertebra; condyloid joint

_____ f. Joint between the humerus (trochlea) and the trochlear (semilunar) notch of the ulna as well as the capitulum of humerus with head of the radius

_____ g. Joint between the head of the femur and acetabulum of the coxal bone

✓ **C5.** *The Big Picture: Looking Back and Looking Ahead.* The design of synovial joints permits free movement of bones. However, if bones moved too freely, they would potentially move right out of their joint cavities (dislocation). A number of factors modify movement at synovial joints. Identify these factors as you refer to figures and text in your textbook. Use key factors listed in the box to fill in lines with*.

Bones	Hormones	Muscle	Ligaments	Soft parts

a. Shape of articulating *_____ affects movement at joint, for example, at shoulder (Figure 8.4) and hip joints (Figure 8.10). Which joint has a deeper socket, contributing to increased stability and decreased mobility? *(Shoulder? Hip?)*

b. Position and tautness of *_____, such as the anterior cruciate (Figure 9.14), strengthen the joint and limit excessive knee movement. In about _____% of all serious knee injuries, this ligament is damaged. It normally forms a cross ("cruci") with the _____ cruciate ligament.

c. Figure 11.16, page 343 shows _____ located on the anterior and posterior of the arm. The *(biceps? triceps?)* lies on the anterior of the arm. When this muscle contracts, the arm moves forward (flexion). The triceps, located on the *(anterior? posterior?)* of the arm, must stretch during this movement; because it has a limited ability to stretch, the triceps *(enhances? limits and opposes?)* flexion of the arm. In fact, the triceps is said to be an *antagonist* to the biceps.

d. Bend (flex) your own forearm (Figure 11.1 page 304). Try to place your wrist directly on your shoulder. If this is not possible for you, it is probably due to the meeting (apposition) of _____ of your forearm against those covering your arm. Muscles are "soft parts." Name several others.

e. Another factor affecting movement at joints is *_____ such as relaxin

(Figure 28.25, page 1002). Made by ovaries and the _____ toward the end of pregnancy, this hormone helps to relax the anterior joint between hipbones,

known as the _____ _____.

D. Aging and joints; disorders and medical terminology (pages 263–264).

✓ **D1.** Fill in blanks to address these question about aging effects on joints.

a. With aging, the volume of synovial fluid tends to _____-crease.

b. Articular cartilage tends to _____-crease in thickness.

c. Ligaments are likely to _____-crease in length and flexibility.

d. One type of arthritis, namely _____, is at least partly age-related.

✓ **D2.** How are *arthritis* and *rheumatism* related? Choose the correct answer.
   A. Arthritis is a form of rheumatism.
   B. Rheumatism is a form of arthritis.

✓ **D3.** List several forms of arthritis. Name one symptom common to all forms of this ailment.

✓ **D4.** Contrast *rheumatoid arthritis* with *osteoarthritis* in Table LG 9.2. Write *Yes, No, Large,* or *Smaller* for answers.

Table LG 9.2 Comparison of major types of arthritis.

	Rheumatoid Arthritis	Osteoarthritis
a. Is known as "wear-and-tear" arthritis		
b. Which types of joints are most affected?		
c. Is this an inflammatory, autoimmune condition?		
d. Is synovial membrane affected?		
e. Does articular cartilage degenerate?		
f. Does fibrous tissue join bone ends?		
g. Is movement limited?		

✓ **D5.** Complete this exercise describing disorders involving articulations.

a. Gouty arthritis is a condition due to an excess of _____ blood leading

to deposit of _____ in the joints. Gout can also involve damage to

joints or to organs such as the _____ because crystals may be
deposited there also.

b. An acute chronic inflammation of a bursa is called _____.

c. *Luxation,* or _____, is displacement of a bone from its joint with
tearing of ligaments.

d. Pain in a joint is known as _____.

ANSWERS TO SELECTED CHECKPOINTS: CHAPTER 9

A1. Point of contact between bones, between bone
and cartilage, and between teeth and bone; the
presence or absence of a space (synovial cavity)
between bones, and the type of connective tissue
that binds the bones together.

A2. Fibrous, cartilage, and synovial.

A3. (a) Immovable. (b) Amphiarthroses. (c) Diarth-
roses.

A4. (a) Lack; fibrous. (b) Suture; im-, syn; synostosis.
(c) C. (d) More, amphi; syndesmosis.

A5. (a) Hyaline; epiphyseal; immovable; is replaced by
bone during adult life, synostosis. (b) Symphysis;
amphi; discs between vertebrae, symphysis pubis
between hipbones.

A6. The space (synovial cavity) between the articulating bones and the absence of tissue between those
bones (which might restrict movement) make the
joints more freely movable. The articular capsule
and ligaments also contribute to free movement.

A7.

Figure LG 9.1A Structure of a generalized synovial joint.

A8. (a) A. (b) SF. (c) SM. (d) L. (e) SM and F. (f) F
(and L, somewhat). (g) SF. (h) SF.

A11. Strain, sprain; ankle.

A12. See Table LG 9.1A.

Table LG 9.1 Subtypes of synovial joints.

Subtype of Synovial Joint	Planes of Movement	Types of Movement	Examples
a. Ball-and-socket	Triaxial	F, E, ADD, ABD, CIR, ROT	1. Shoulder: scapula (glenoid cavity)-humerus (head) 2. Hip: hip bone (acetabulum)-femur (head)
b. Condyloid (ellipsoidal)	**Biaxial**	**F, E, ADD, ABD, CIR**	Wrist (radius-carpals)
c. Saddle	**Biaxial**	**F, E, ABD, ADD, CIR**	Carpal (trapezium)-first metacarpal (base)
d. Hinge	**Monoaxial**	F, E	1. Knee: femur (condyles)-tibia (condyles) 2. **Elbow: humerus (trochlea)-ulna (trochlear [semilunar] notch)** 3. Ankle: tibia and fibula-talus. 4. Phalanges 5. Atlanto-occipital: atlas-occipital bone (condyles)
e. **Pivot**	Monoaxial	ROT	1. Atlas-axis (dens) 2. Radius (head)-ulna (radial notch)
f. Planar	—	Gliding	1. Ribs (heads, tubercles)-vertebrae (bodies, transverse processes) 2. Clavicle-sternum (manubrium), Clavicle-scapula (acromiom) 3. Intercarpal and intertarsal joints

Table LG 9.1A Subtypes of synovial joints.

A13. (a) Piv. (b) Piv. (c) B. (d) B. (e) H. (f) H. (g) Pla.
(h) S (i) C, S.

B1. (a) F. (b) G. (c) E. (d) Abd. (e) R. (f) D.

B2. (a) Circumduction. (b) Pronation. (c) Abduction.
(d) Elevation. (e) Plantar flexion. (f) Flexion.
(g) Inversion. (h) Protraction. (i) Opposition.
(j) Lateral flexion.

B3. (a) Flexion. (b) Hyperextension. (c) Flexion.
(d) Hyperextension. (e) Abduction. (f) Adduction.
(g) Flexion. (h) Extension. (i) Flexion, while leg
also slightly flexed. (j) Hyperextension, while leg
extended. (k) Flexion, while thigh extended.
(1) Extension, with thigh flexed. (m) Flexion.
(n) Hyperextension. (o) Abduction. (p) Adduction.

B4. See Table LG 9.1A.

C1. (a) Greater; tri; loose articular capsule and large, shallow glenoid cavity/socket for the head of the humerus. (b) Coraco-. (c) Fibrocartilage; increases the depth of the glenoid cavity. (d) Acromi-, coracoid, deltoid, scapular. (e) Helps to hold the humerus in the glenoid cavity; subluxation (dislocation) of the humerus out of the shoulder socket; four muscles and their tendons: supraspinatus, infraspinatus, teres minor, and subscapularis.

C2. (a) More, less; strong accessory ligaments and muscles around the joint capsule, as well as the fit of the femur into the deep acetabular socket. (b) Ilio, pubo, ischio; contact with the opposite thigh, as well as ligamentous tension.

C3. (a) (1) Patella, (2) lateral, tibia, (3) medial, tibia. (Note that neither the femur nor the patella articulates with the fibula.) (b) Planar; modified hinge. (c) Does not; weakness. (d) Quadriceps femoris. (e1) I; (e2) E, (e3) I. (f) Menisci; provide some stability to an otherwise unstable joint. (g) Several; bursitis.

C4. (a) A. (b) S. (c) TMJ. (d) W. (e) Ao. (f) E. (g) H.

C5. (a) Bones; hip. (b) Ligaments; 70; posterior (c) Muscles; biceps; posterior; limits and opposes. (d) Soft parts; skin, fat, bursae, muscles and blood vessels. (e) Hormones; placenta, pubic symphysis.

D1. (a–b) in. (c) De. (d) Osteoarthritis.

D2. A.

D3. Rheumatoid arthritis (RA), osteoarthritis (OA), and gouty arthritis; pain.

D4.

Table LG 9.2 Comparison of major types of arthritis.

	Rheumatoid Arthritis	**Osteoarthritis**
a. Is known as "wear-and-tear" arthritis	No	Yes
b. Which types of joints are affected first?	Smaller	Larger
c. Is this an inflammatory autoimmune condition?	Yes	No
d. Is synovial membrane affected?	Yes	No
e. Does articular cartilage degenerate?	Yes	Yes
f. Does fibrous tissue join bone ends?	Yes	No
g. Is movement limited?	Yes	Yes

Table 9.2 Comparison of major types of antritis.

D5. (a) Uric acid, sodium urate; kidneys. (b) Bursitis. (c) Dislocation. (d) Arthralgia.

MORE CRITICAL THINKING: CHAPTER 9

1. Synovial joints do contain cartilage. Exactly where is it located? Is it vascular? How does synovial fluid help with this problem?

2. Explain what prevents hyperextension at joints such as the elbow and knee. Speculate about what causes "double-jointedness," in which these joints do exhibit some hyperextension.

3. What is the rotator cuff? What is likely to cause a "torn rotator cuff," and what are likely to be the consequences of that injury?

4. Which joint is usually more mobile? Hip or shoulder? Describe several anatomical factors that account for this difference.

MASTERY TEST: CHAPTER 9

Questions 1-13: Circle T (true) or F (false). If the statement is false, change the underlined word or phrase so that the statement is correct.

T F   1. A fibrous joint is one in which there is <u>no joint cavity and bones are held together by fibrous connective tissue</u>.

T F   2. <u>Sutures, syndesmoses, and symphyses</u> are examples of fibrous joints.

T F   3. Flexion and extension are movements that both occur in the <u>sagittal</u> plane.

T F   4. <u>Ball-and-socket, planar, pivot, and ellipsoidal joints</u> are all diarthrotic joints.

T F   5. All fibrous joints <u>are synarthrotic and all cartilaginous joints are amphiarthrotic</u>.

T F   6. In synovial joints, synovial membranes <u>cover the surfaces of articular cartilages</u>.

T F   7. Bursae are <u>saclike structures that reduce friction</u> at joints.

T F   8. Abduction of fingers occurs around an imaginary line through the <u>middle finger</u>, whereas abduction of the toes occurs around an imaginary line through the <u>second toe</u>.

T F   9. When your arm is in the supine position, your radius and ulna are <u>parallel (not crossed)</u>.

T　F　　10. When you touch your toes, the major action you perform at your hip joint is called <u>hyperextension</u>.

T　F　　11. Abduction is movement <u>away from</u> the midline of the body.

T　F　　12. The elbow, knee, and ankle joints are all <u>hinge</u> joints.

T　F　　13. Joints that are relatively stable (such as hip joints) tend to have <u>more</u> mobility than joints that are less stable (such as shoulder joints).

Questions 14-15: Arrange the answers in correct sequence.

_____ _____ _____ 14. From most mobile to least mobile:

    A. Amphiarthrotic

    B. Diarthrotic

    C. Synarthrotic

_____ _____ _____ 15. Stages in rheumatoid arthritis, in chronological order:

    A. Articular cartilage is destroyed and fibrous tissue joins exposed bones.

    B. The synovial membrane produces pannus which adheres to articular cartilage.

    C. Synovial membrane becomes inflamed and thickened, and synovial fluid accumulates.

Questions 16-20: Circle the letter preceding the one best answer to each question.

16. The movement of circumscribing a cone with the humerus in the glenoid fossa is an example of the type of movement called:

    A. Supination   C. Circumduction

    B. Pronation   D. Internal rotation

17. Which structure is extracapsular in location?

    A. Meniscus

    B. Posterior cruciate ligament

    C. Synovial membrane

    D. Tibial collateral ligament

18. Which joint is amphiarthrotic and cartilaginous?

    A. Symphysis   C. Syndesmosis

    B. Synchondrosis   D. Synostosis

19. All of these structures are associated with the knee joint *except:*

    A. Glenoid labrum

    B. Patellar ligament

    C. Infrapatellar bursa

    D. Medial meniscus

    E. Fibular collateral ligament

20. A suture is found between:

    A. The two pubic bones

    B. The two parietal bones

    C. Radius and ulna

    D. Diaphysis and epiphysis

    E. Tibia and fibula (distal ends)

Questions 21-25: Fill-ins. Write the word or phrase that best completes the statement.

_____ 21. _____ is the forcible wrenching or twisting of a joint with partial rupture of it, but without dislocation.

_____ 22. The action of pulling the jaw back from a thrust-out position so that it becomes in line with the upper jaw is the movement

called _____.

_____ 23. Another name for a freely movable joint is _____.

_____ 24. The type of joint between the atlas and axis and also between

proximal ends of the radius and ulna is a _____ joint.

_____ 25. Synovial fluid normally has the consistency resembling

_____.

ANSWERS TO MASTERY TEST: CHAPTER 9

True-False

1. T
2. F. Sutures, syndesmoses, and gomphoses
3. T
4. T
5. F. Either synarthrotic or amphiarthrotic, and the same is true of cartilaginous
6. F. Do not cover surfaces of articular cartilages. which may be visualized as floor and ceiling of a room; but synovial membranes do line the rest of the inside of the joint cavity (much like wallpaper covering the four walls of the room).
7. T
8. T
9. T
10. F. Flexion
11. T
12. T. (Or modified hinge joints)
13. F. Less

Arrange

14. B A C
15. C B A

Multiple Choice

16. C
17. D
18. B
19. A
20. B

Fill-ins

21. Sprain
22. Retraction
23. Diarthrotic
24. Pivot (or synovial or diarthrotic)
25. Egg white

FRAMEWORK 10
Muscle Tissue

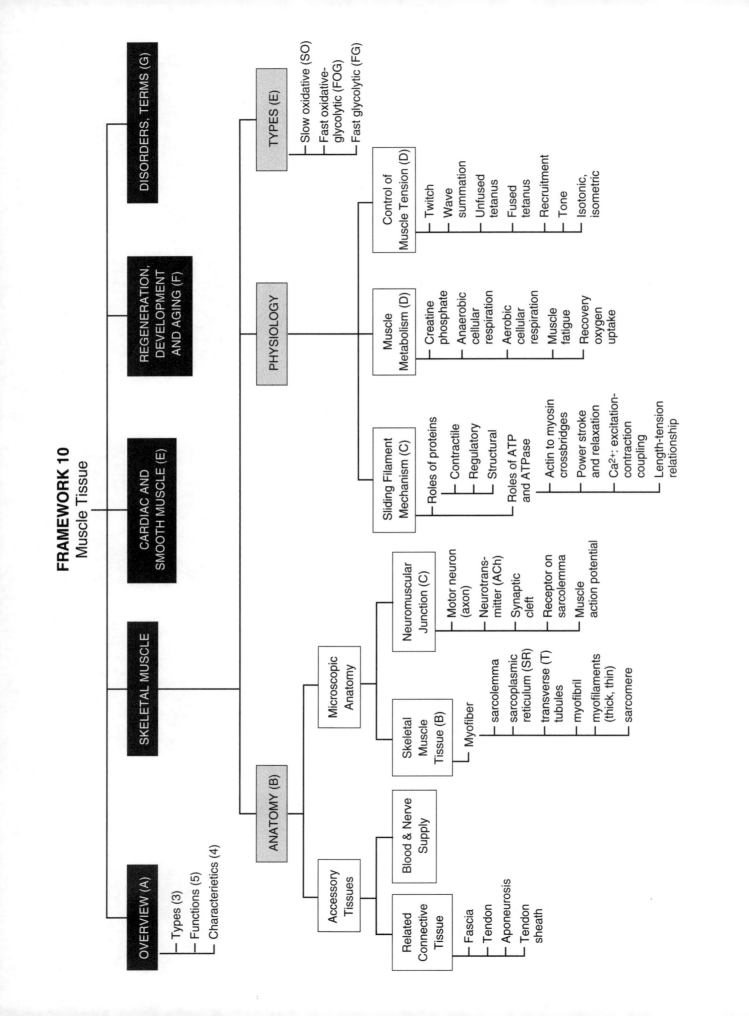

CHAPTER 10

Muscle Tissue

Muscles are making it possible for you to read this paragraph: facilitating movement of your eyes and head, maintaining your posture, producing heat to keep you comfortable. Muscles do not work in isolation, however. Connective tissue binds muscle cells into bundles and attaches muscle to bones. Blood delivers the oxygen and nutrients for muscle work, and nerves initiate a cascade of events that culminate in muscle contraction.

A tour of the mechanisms of muscle action finds the visitor caught in the web of muscle protein filaments sliding back and forth as the muscle contracts and relaxes. Neurotransmitters, as well as calcium and power-packing ATP, serve as chief regulators. Muscles are observed contracting in a multitude of modes, from twitch to treppe and isotonic to isometric. It may be surprising to hear a tour guide's announcement that most muscle cells present now were actually around at birth. Very few muscle cells can multiply, but fortunately muscle cells can greatly enlarge, as with exercise. With aging, cell numbers decline and so do muscle strength and endurance. Certain disorders decrease muscle function and debilitate even in very early years of life.

As you begin your study of muscle tissue, carefully examine the Chapter 10 Topic Outline and Objectives; check off each one as you complete it. To organize your study of muscle tissue, glance over the Chapter 10 Framework now. Be sure to refer to the Framework frequently and note relationships among key terms in each section.

TOPIC OUTLINE AND OBJECTIVES

A. Overview of muscle tissue

1. Correlate the three types of muscle tissue with their functions and special properties.

B. Skeletal muscle tissue

2. Explain the relation of connective tissue components, blood vessels, and nerves to skeletal muscles.
3. Describe the microscopic structure of a skeletal muscle fiber.

C. Contraction and relaxation of skeletal muscle fibers

4. Outline the steps involved in the sliding filament mechanism of muscle contraction.
5. Describe how muscle action potentials arise at neuromuscular junctions.

D. Muscle metabolism; control of muscle tension

6. Describe the reactions by which muscle fibers produce ATP.
7. Describe the structure and function of a motor unit.
8. Explain the phases of a twitch contraction.
9. Describe how frequency of stimulation affects muscle tension.

E. Types of skeletal muscle fibers; cardiac and smooth muscle

10. Compare the structure and function of the three types of skeletal muscle fibers.
11. Describe the main structural and functional characteristics of cardiac muscle tissue.
12. Describe the main structural and functional characteristics of smooth muscle tissue.

F. Regeneration of muscle tissue; develop mental anatomy and aging of muscles

13. Explain how muscle fibers regenerate.
14. Describe the development of the muscular system.
15. Explain how aging affects skeletal muscle.

G. Disorders and medical terminology

WORDBYTES

Now become familiar with the language of this chapter by studying each wordbyte, its meaning, and an example of its use within a term. After you study the entire list, self-check your understanding by writing the meaning of each wordbyte on the line. As you continue through the *Learning Guide*, identify (and fill in) additional terms that contain the same wordbyte.

Wordbyte	Self-check	Meaning	Example(s)
a-	_____	not	*a*trophy
-algia	_____	pain	fibromy*algia*
apo-	_____	from	*apo*neurosis
dys-	_____	bad, difficult	muscular *dys*trophy
endo-	_____	within	*endo*mysium
-graph	_____	to write	electromyo*graph*y
hyper-	_____	above, excessive	*hyper*trophy
iso-	_____	equal	*iso*metric
-lemma	_____	rind, skin	sarco*lemma*
-metric	_____	measure, length	iso*metric*
myo-	_____	muscle	*myo*fiber, *myo*sin
mys-	_____	muscle	epi*mys*ium
peri-	_____	around	*peri*mysium
sarc-	_____	flesh, muscle	*sarc*olemma
tetan-	_____	rigid, tense	*tetan*us
titan	_____	gigantic	*tit*in
-tonic	_____	tension	iso*tonic*
tri-	_____	three	*tri*ad
troph-	_____	nourishment	muscular dys*troph*y

CHECKPOINTS

A. Overview of muscle tissue (pages 268–269)

✓ **A1.** Match the muscle types listed in the box with descriptions below.

C. Cardiac	**Sk.** Skeletal	**Sm.** Smooth

_____ a. Involuntary muscle found in blood vessels and intestines

_____ b. Involuntary striated muscle

_____ c. Striated muscle attached to bones

_____ d. The only type of muscle that is voluntary

_____ e. Involuntary muscle in arrector pili muscles that causes "goose bumps"

✓ **A2.** Describe five functions of muscles that are important for maintenance of homeostasis.

a. _____ due to action of muscles pulling on bones.

b. Movement of substances within the body, such as:

1. _____ within blood vessels due to pumping of heart muscle,

2. _____ within the digestive system,

3. _____ through the urinary system, and

4. _____ within the reproductive system.

c. _____ of body position, and maintenance of stored fluids within

organs such as _____.

d. Thermogenesis, which is generation of _____ by muscles; an

example is the warming effect of _____ during cold weather.

e. Ringlike bands of smooth muscle known as _____ control outflow of substances from organs, such as urine from the urinary bladder, or feces from the rectum.

✓ **A3.** Match the characteristics of muscles listed in the box with descriptions below.

Contractility	**Elasticity**	**Excitability**	**Extensibility**

a. Ability of muscle to stretch without damaging the tissue _____

b. Tendency of stretched or contracted muscle to return to its original shape

_____

c. Ability of muscle cells and nerve cells (neurons) to respond to stimuli by producing action potentials; irritability

_____

d. Ability of muscle tissue to shorten and thicken in response to action potential

_____

✓ **A4.** Contrast isometric and isotonic contractions by doing this exercise.

    a. A contraction in which a muscle shortens while tension (tone) of the muscle remains constant is known as an *(isometric? isotonic?)* contraction.

    b. In an isometric contraction the muscle length *(shortens? stays about the same?)* and tension of the muscle *(increases? stays the same?)*.

B. Skeletal muscle tissue (pages 269–276)

✓ **B1.** Contrast two kinds of fascia by indicating which of the following are characteristics of superficial fascia (S) or deep fascia (D).

    _____ a. Located immediately under the skin (subcutaneous)

    _____ b. Composed of dense connective tissue that extends inward to surround and compartmentalize muscles

    _____ c. A route for nerves and blood vessels to enter muscles; contains much fat, so provides insulation and protection

✓ **B2.** Arrange the following terms (connective tissue) in correct sequence according to the amount of muscle surrounded: *endomysium, epimysium, perimysium.*

_____ → _____ → _____
    (entire muscle)   (bundle [fascicle] of muscle fibers) (individual muscle fiber)

For extra review. Color these three connective tissues on Figure LG 10.la. Label a *fascicle.*

✓ **B3.** The connective tissues surrounding and compartmentalizing muscle fascicles join

together to attach the muscle into the _____ covering over bone. A(n) *(aponeurosis? tendon?)* resembles a cord attaching the tapered end of a muscle

into the covering over bone, whereas a(n) _____ is a broad, flat structure attaching a broad muscle into periosteum.

✓ **B4.** Summarize the nerve and blood supply to skeletal muscle in this Checkpoint.

    a. Skeletal muscle receives its nerve supply by cells known as _____.

       Branches (called _____) of these nerve cells supply nerve impulses

       to several muscle fibers (cells). The ends (or synaptic end _____) of

       the nerve cells lie close to muscle cells at NMJs or _____ junctions. One will be considered in more detail in Checkpoint C7.

    b. State two or more functions of blood vessels that supply muscles.

B5. Describe the roles of *myoblasts* in muscle cell development.

Do myoblasts multiply after birth? _____

O Endomysium
O Epimysium
O Perimysium
O Tendon

Muscle cell
(myofiber)

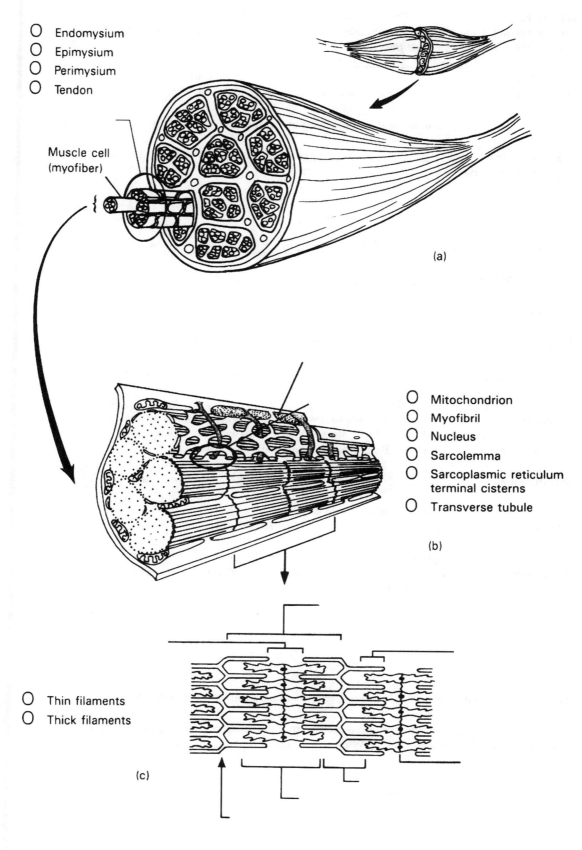

(a)

O Mitochondrion
O Myofibril
O Nucleus
O Sarcolemma
O Sarcoplasmic reticulum
terminal cisterns
O Transverse tubule

(b)

O Thin filaments
O Thick filaments

(c)

Figure LG 10.1 Diagram of skeletal muscle. (a) Skeletal muscle cut to show cross section and longitudinal section with connective tissues. (b) Section of one muscle cell (myofiber). (c) Detail of sarcomere of muscle cell. Color and label as indicated in Checkpoints B2, B6, and B10.

✓ **B6.** Refer to Figure LG 10.1, and do this exercise about muscle structure.

    a. Arrange the following terms in correct order from largest to smallest in size: *filaments (thick or thin), myofibrils, muscle fiber (cell)*.

    _____ → _____ → _____

          (largest)   (smallest)

    b. The plasma membrane of the skeletal muscle fiber (cell) is known as the

    _____. Continuous with this membrane are thousands of tiny

    structures called _____ (_____) tubules. What major function do these structures serve?

    c. The cytoplasm of the skeletal muscle cell is called its _____. Two

    major chemicals within it are _____, which can be broken down to

    form glucose and _____, which gives muscle its red color and binds

    the gas _____. Both of these chemicals are needed for

    _____ production, which occurs in the myriad of organelles named

    _____ within the muscle.

    d. Sarcoplasmic reticulum (SR) is comparable to *(endoplasmic reticulum? ribosomes? mitochondria?)* in nonmuscle cells. *(Ca²⁺? K⁺?)* stored in SR is released to sarcoplasm as the trigger for muscle contraction. Each T tubule along with the dilated

    ends (terminal _____) of SR on both sides of it are known as a

    _____. Label the triad circled on Figure LG 10.1b.

    e. Color all structures indicated by color code ovals in Figured LG 10.1.

✓ **B7.** *A clinical correlation.* Complete this activity on changes in muscle size.

    a. Hypertrophy refers to increase in *(number? size)* of muscle fibers. State two or more factors that may lead to hypertrophy.

    b. When the nerve supply of a muscle is disrupted, the muscle is likely to experience

    _____ atrophy. Within 6 months to 2 years, the muscle is likely to be reduced in size by *(25% 50% 75%)*. What type of tissue replaces the muscle?

    _____.

    c. State two examples of causes of disuse atrophy.

✓ **B8.** Match the correct term from the list in the box with its description below.

A band	M line	Sarcoplasm	Zone of overlap
I band	Sarcomere	Z disc	

a. Cytoplasm of muscle cell _____

b. Extends from Z disc to Z disc _____

c. Dark area in striated muscle; contains thick and thin filaments _____

d. Light area on either side of Z disc; location of thin filaments only _____

e. Located in the center of the H zone, which is the exact middle of the sarcomere.

f. Located in the center of the I band; appear damaged in electron micrographs of

   severely exercised muscle _____.

g. Located at both ends of each A band; consists of six thin filaments arranged around

   each thick filament _____.

✓ **B9.** *A clinical correlation.* Describe possible effects of intensive exercise upon muscles by filling in the blanks in this activity.

a. The acronym DOMS refers to D _____ O _____

   M _____ S _____.

b. _____ and _____ are two proteins normally present within healthy muscle cells. Increased blood levels of these proteins may be correlated to the degree of muscle cell damage, as these chemicals move out of injured cells and into blood.

c. DOMS is likely to occur within _____ to _____ hours after strenuous exercise.

✓ **B10.** After you study Figures 10.4 and 10.6 in your text, describe the arrangement of the three categories of proteins in myofibrils.

a. Using the leader lines provided on Figure LG 10.1, label the following structures: *sarcomere, A band, I band, H zone, M line, Z disc, zone of overlap.*

b. Actin and myosin are the major components of thin and thick filaments, and they

   are both *(contractile? regulatory? structural?)* proteins. _____ and

   _____ are two regulatory proteins.

c. Each *(thick? thin?)* filament is composed of about 300 myosin molecules. Each molecule is shaped like two intertwined *(footballs? golf clubs?)*. The ends of the "handles" point toward the *(M lines? Z discs?)*. The rounded head of the "club" is called a

   _____ and it attaches to *(actin? troponin?)* in one of the six thin filaments as muscles begin contraction.

d. Thin filaments are anchored at *(M lines? Z discs?)* and *(do? do not?)* extend into H zones. These filaments are composed mostly of *(actin? myosin?)* molecules that are twisted into a helix. Thin filaments also contain two other proteins. The protein that covers myosin-binding sites in relaxed muscle is called *(tropomyosin? troponin?)*,

   whereas tropomyosin is held in place by the regulatory protein _____.

e. A third category of muscle proteins includes about 12 _____ proteins. One of these proteins, titin, is named so because of its *(huge? tiny?)* size. Titin functions to anchor *(thick? thin?)* filaments between two parts of the sarcomere,

namely _____ and _____. Titin contributes greatly to the *(tensile strength? elasticity?)* of muscles.

f. Which structural protein forms the M line? _____ Which one reinforces connections between the thin filaments, sarcolemma, and fibrous con-

nective tissue surrounding muscles? _____

C. Contraction and relaxation of skeletal muscle fibers (pages 276–283)

✓ **C1.** Summarize one theory of muscle contraction in this Checkpoint.

a. In order to effect muscle shortening (or contraction), heads (cross bridges) of

_____ filaments pull on molecules of thin filaments, moving them towards the *(M line? Z discs?)*.

b. As a result, *(thick? thin?)* filaments move toward the center of the sarcomere. Because the thick and thin filaments *(shorten? slide?)* to decrease the length of the

sarcomere, this theory of muscle contraction is known as the _____ theory. During this sliding action, the lengths of the thick and thin filaments *(do? do not?)* shorten.

✓ **C2.** Check your understanding of events that occur in muscle contraction and relaxation by completing this Checkpoint.

a. To initiate a muscle contraction, calcium ions (Ca^{2+}) are released from storage areas within the *(mitochondria? sarcoplasmic reticulum? T tubules?)*. (More about this process below in Checkpoint C3.)

b. The role of calcium ions is to bind to the protein named _____ and therefore to cause the troponin-tropomyosin complex to pull away from sites on *(actin?*

myosin?) where "heads" of the other contractile protein (_____) must bind.

c. Myosin heads have ATP attached to them at this point. The enzyme

_____ on myosin breaks down ATP into its products:

_____ and _____, which are still attached to myosin. The energy released by this reaction permits myosin heads to attach to

binding sites on _____.

d. Release of the phosphate group triggers the "_____ stroke," in which the myosin head rotates, releases *(ATP? ADP?)*, and slides the thin filament towards

the center of the _____.

e. Recall that each thick filament includes _____ myosin molecules *(Hint:* Look back at Checkpoint B.10c). Each of these myosin molecules undergoes power strokes at a

rate of about _____ times per second.

f. More ATP is then needed to bind to the myosin head and detach it from

_____. Splitting of ATP by ATPase located on *(actin? myosin?)* again activates myosin and returns it to its original position, where it is ready to bind to

another _____-binding site on an actin further along the thin filament.

g. Repeated power strokes slide actin filaments *(toward or even across? away from?)* the H zone and M line and so shorten the sarcomere (and entire muscle). If this process is compared to running on a treadmill, *(actin? myosin?)* plays the role of the runner staying in one place, whereas the treadmill that slides backward (toward H zones) is comparable to the *(thick? thin?)* sliding filament.

✓ **C3.** Describe the significance of intracellular locations of calcium and also the availability of ATP in regulation of muscle contraction and relaxation.

a. In a relaxed muscle, the concentration of calcium ions (Ca^{2+}) is *(high? low?)* in sarcoplasm but very high inside of SR. This concentration gradient is maintained in relaxed muscle by two mechanisms:

  1) *(Diffusion? Active transport?)* across the SR membrane.

  2) To enhance this effect, proteins named _____ in SR membranes bind to the Ca^{2+} that enters SR, concentrating it inside SR *(10? 100? 10,000)* times over the level in the sarcoplasm. As a consequence, enormous amounts of calcium ions stand waiting in the SR for the next signal for the muscle to contract.

b. The signal to begin a muscle contraction is stimulation of a muscle by a nerve, which spreads an action potential across the sarcolemma and across the

  _____ to SR. As a result, Ca^{2+} release _____ in the SR membrane will open up, allowing Ca^{2+} to flow out of SR into

  _____ surrounding thick and thin filaments. These ions are then available to exert their impact on regulatory proteins (troponin-tropomyosin) to initiate muscle contraction (review Checkpoint C2b).

c. When nerve stimulation ceases, the muscle relaxes as SR channels are no longer signaled to release Ca^{2+} into the cytosol. In fact, Ca^{2+} are sequestered back into SR by mechanisms just described. With such a low level of Ca^{2+} now in the sarcoplasm surrounding myofilaments, the troponin-tropomyosin complex once again blocks

  binding sites on _____. As a result, thick and thin filaments detach,

  slip back into normal position, and the muscle is said to _____.

d. The movement of Ca^{2+} back into sarcoplasmic reticulum (C3a) is *(an active? a passive?)* process. After death, a supply of ATP *(is? is not?)* available, so an active transport process cannot occur. Explain why the condition of rigor mortis results.

C4. *For extra review.* Sequence the events in excitation-contraction coupling by writing the letters of the five answers in correct chronological order.

_____ _____ _____ _____ _____

A. The troponin-tropomyosin complex is moved away from actin; myosin's ATP is hydrolyzed to ADP and phosphate, which stay attached to myosin.

B. The thin filament slides towards the center of the sarcomere as the muscle shortens; another ATP binds to myosin, releasing myosin from actin so the cycle can repeat.

C. Nerve impulse occurs; action potential spreads across sarcolemma and T-tubules.

D. Calcium release channels release Ca^{2+} from SR to cytoplasm; Ca^{2+} complexes with troponin.

E. The energized myosin head can now attach to actin; phosphate is released from myosin, triggering rotation of myosin with release of ADP and the "power stroke."

✓ **C5.** Do this activity about changes in the force of muscle contraction.

a.   When a muscle is stretched to its optimal length, there is *(much? little or no?)* overlap of myosin cross bridges on thick filaments with actin on thin filaments. In this case, a muscle demonstrates *(minimal? maximal?)* force of contraction. When a skeletal muscle is stretched excessively (such as to 175% of its optimal length), then *(many? no?)* myosin cross bridges can bind to actin.

b.   As a general rule, the more a muscle is stretched (within limits), the *(stronger? weaker?)* the contraction.

C6. Muscle tension that ultimately moves an entire body part (such as the arm) is of two types. One type is generated by actin and myosin; this is *(active? passive?)* tension. What accounts for passive tension?

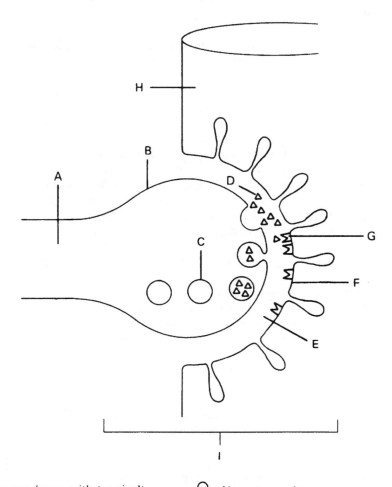

O   Motor neuron (axon with terminal)   O   Neurotransmitter
O   Synaptic vesicle   O   Skeletal muscle myofiber (cell)

Figure LG 10.2 Diagram of a neuromuscular junction (motor end plate). Refer to Checkpoints C7 and C8.

✓ **C7.** Do this activity about the nerve supply of skeletal muscles.

a.   The combination of a neuron plus the muscle fibers (cells) it innervates is called a

_____. Motor neurons form at least two, and possibly thousands of, branches called _____, each of which supplies an individual skeletal muscle fiber (cell). When the motor neuron "fires," *(just one muscle fiber supplied by one axon? all muscle fibers within that motor unit?)* will be stimulated to contract.

b. Refer to Figure LG 10.2, in which we zoom in on the portion of a motor unit at which a branch of one neuron stimulates a single muscle fiber (cell). A nerve impulse

travels along an axon terminal (at letter _____) toward one muscle fiber (letter

_____). The axon is enlarged at its end into a synaptic _____ (at letter B).

c. The nerve impulse causes synaptic vesicles (letter _____) to fuse with the plasma membrane of the axon. Next the vesicles release the neurotransmitter (letter

_____) named _____ into the synaptic cleft (letter _____)

d. The region of the muscle fiber membrane (sarcolemma) close to the axon terminals is called a *(motor end plate? neuromuscular junction [NMJ]?)*. This site contains

specific _____ (letter G) that recognize and bind to ACh. A typical motor end plate contains about *(30-40? 3000-4000? 30-40 million?)* ACh receptors.

e. The effect of ACh is to cause Na⁺ channels in the sarcolemma to *(open? close?)* so that Na⁺ enters the muscle fiber. As a result, an action potential is initiated, leading to *(contraction? relaxation?)* of the muscle fiber. The diagnostic technique of recording

such electrical activity in muscle cells is called _____.

f. The combination of the axon terminals and the motor end plate is known as a

_____. At only one site along a muscle fiber (at about its middle) does a nerve approach a muscle cell to innervate it. In other words, there is/are usually

(only one? many?) neuromuscular junction(s), labeled _____ on the figure, for each muscle fiber (cell). (In Chapter 12 we will consider regions similar to NMJ's, but where one neuron meets another neuron. These sites are known as

_____.)

C8. *For extra review.* Color the parts of Figure LG 10.2 indicated by color code ovals.

✓ **C9.** Complete this exercise about pharmacology of the neuromuscular junction. Match the names of chemicals in the box with descriptions below. Short lines following each description will be used in Checkpoint C10.

AChE.	Acetylcholinesterase (AChE)	CUR.	Curare
CBT.	*Clostridium botulinum* toxin	NEO.	Neostigmine

_____ a. Synaptic cleft enzyme that breaks down ACh into acetyl and choline. _____

_____ b. Muscle relaxant that acts by blocking ACh receptors on motor end plate. _____

_____ c. Anticholinesterase medication used for patients with myasthenia gravis. _____

_____ d. Blocks release of ACh. _____

✓ **C10.** Now write arrows on lines in Checkpoint C9 to indicate whether each of the chemicals described there increases (↑) or decreases (↓) muscle stimulation.

D. Muscle metabolism; control of muscle tension (pages 283–288)

✓ **D1.** Complete this exercise about energy sources for muscle contraction.

a. Breakdown of *(ADP? ATP?)* provides the energy muscles use for contraction. Recall from Checkpoint C2c that ATP is attached to *(actin? myosin?)* cross bridges and so is available to energize the power stroke. Complete the chemical reaction showing ATP breakdown.

ATP →

b. ATP must be regenerated constantly. One method involves use of ADP and energy from food sources. Complete that chemical reaction.

ADP +

In essence, ADP is serving as a transport vehicle that can pick up and drop off energy stored in an extra high energy phosphate bond (~ P).

c. But ATP is used for other cell activities such as _____. To assure adequate energy for muscle work, muscle cells contain an additional molecule for

transporting high energy phosphate; this is _____. Complete the reaction in Figure LG 10.3 showing how the creatine phosphate and ADP transport "vehicles" can meet and transfer the high phosphate "trailer" so that more ATP is formed for muscle work.

creatine phosphate +   ADP   _____ + _____

Figure LG 10.3 High energy molecules of the creatine phosphate system: the (.-P) "trailer" tradeoff. Complete figure as indicated in Checkpoint D11c, d.

d. How is creatine phosphate regenerated?
Show this on Figure LG 10.3. What enzyme catalyzes this reaction?_____.
Which is more plentiful in the sarcoplasm? *(ATP? Creatine phosphate?)*.

e. ATP and creatine phosphate together provide only enough energy to power muscle activity for about *(an hour? 10 minutes? 15 seconds?)*. After that, muscles turn first to *(aerobic? anaerobic?)* pathways and later to *(aerobic? anaerobic?)* pathways. Complete Figure 10.4 to show how muscles get energy via these pathways.

f. During strenuous exercise an adequate supply of oxygen may not be available to

muscles. They must convert pyruvic acid to _____ acid by an *(aerobic? anaerobic?)* process. Excessive amounts of lactic acid in muscle tissue con-

tribute to some muscle _____. Name several types of cells that can use lactic acid to form ATP.

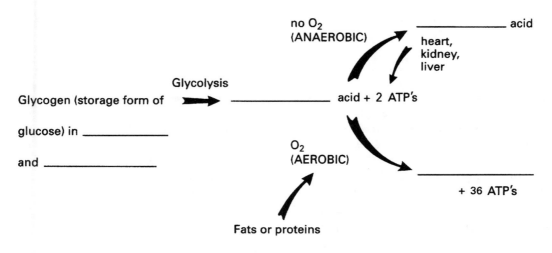

no O$_2$
(ANAEROBIC)   _____ acid

                              heart,
                              kidney,
                              liver

          Glycolysis
Glycogen (storage form of ➤ _____ acid + 2 ATP's

glucose) in _____

                    O$_2$
and _____   (AEROBIC)

                                              _____

                                              + 36 ATP's

          Fats or proteins

Figure LG 10.4 Anaerobic and aerobic energy sources for muscles.
Complete figure as indicated in Checkpoint D1.

g. One substance in muscle that stores oxygen until oxygen is needed by mitochondria

    is _____. This protein is structurally somewhat like the

    _____ -globin molecule in blood that also binds to and stores oxygen.

    Both of these molecules have a _____ color that accounts for the
    color of blood and also of red muscle.

h. During exercise lasting more than 10 minutes, more than 90% of ATP is provided by
    the *(anaerobic? aerobic?)* breakdown of pyruvic acid (See Figure LG 10.4.).

D2. List several other factors besides the buildup of lactic acid that contribute to the
following effects of exercise:
a. Muscle fatigue

b. Need for recovery oxygen uptake

✓ **D3.** *For extra review.* Complete this Checkpoint to identify metabolic processes used to
supply energy for maximal exercise of varying durations. Select answers from those
in the box.

AER.	**Aerobic cellular respiration**	**CP.**	**Creatine phosphate**
ANAER.	**Anaerobic cellular respiration**		

_____ a. 15-30 sec (100-meter dash)

_____ b. 30-40 sec (300-meter race); carbohydrate
        loading prior to an athletic event helps
        prepare muscles by maximizing glycogen
        storage there; does not require O$_2$

_____ c. Longer than 10 minutes (2 miles to 26-
        mile marathon); "fuel" sources include
        glucose, fatty acids, and amino acids; does
        require O$_2$

✓ **D4.** List three or more factors that can decrease the strength of a muscle contraction.

✓ **D5.** In most cases a single nerve cell (neuron) innervates *(one muscle fiber [cell]? an aver-age of 150 muscle fibers [cells]?)*. An example of a motor unit that is likely to consist of just two or three muscle fibers (cells) precisely innervated by one neuron is *(laryngeal muscles controlling speech? the calf [gastrocnemius] muscles controlling walking?)*

✓ **D6.** Match the terms in the box with definitions below. One answer will be used twice; one will not be used.

C.	**Contraction period**		**Ref.**	**Refractory period**
L.	**Latent period**		**Rel.**	**Relaxation period**
M.	**Myogram**		**T.**	**Twitch**

_____ a. Rapid, jerky response to a single action potential by a motor neuron

_____ b. Recording of a muscle contraction

_____ c. Period between application of a stimulus and start of a contraction; Ca^{2+} is being released from the SR during this time

_____ d. Period when a muscle is not responsive to a stimulus

_____ e. Active transport of Ca^{2+} back into the SR is occurring

_____ f. A short period (about 0.005 second) in skeletal muscle and a long one (about 0.30 second) in cardiac muscle

✓ **D7.** Choose the type of contraction from the box that fits each descriptive phrase.

F.	**Fused tetanus**	**U.**	**Unfused tetanus**	**W.**	**Wave summation**

_____ a. Sustained contraction due to stimulation at a rate of 20–30 stimuli per second

_____ b. Sustained contraction due to stimulation at a rate of 80–100 stimuli per second

_____ c. Stimuli arriving at different times cause larger contractions

_____ d. Smooth, sustained voluntary muscle contractions are due primarily to out-of-syn-chrony _____ in different motor units

✓ **D8.** Contrast effects of different types of training in this Checkpoint.
  a. Repeated isometric contractions (such as weight lifting) rely on *(aerobic? anaerobic?)* ATP production. Repeated short bouts of exercise such as sprints also depend on *(aer-obic? anaerobic?)* ATP production.
  b. Repeated isotonic contractions (such as jogging or dancing) rely on *(aerobic? anaero-bic?)* ATP production. *(Aerobic? Anaerobic?)* workouts as in endurance training or prolonged jogging are more likely to increase blood vessels to skeletal and heart mus-cle to provide needed oxygen.

  D9. Contrast concentric isotonic contraction with eccentric isotonic contraction.

E. Types of skeletal muscle fibers; cardiac and smooth muscle (pages 288–293)

✓ **E1.** Refer to the table below and contrast the three types of skeletal muscle fibers by doing this activity. You may find it helpful to refer back to Checkpoints D3 and D8 also.

Type and Rate of Contraction	Color	Myoglobin Concentration	Mitochondria and Blood Vessels	Source of ATP	Fatigue Easily?
SO. Slow oxidative	Red	High	Many	Aerobic	No
FOG. Fast oxidative-glycolytic	Red	Very high	Many	Aerobic	Moderate
FG. Fast glycolytic	White	Low	Few	Anaerobic	Yes

a. Postural muscles (as in neck and back) are used *(constantly? mostly for short bursts of energy?)*. Therefore it is appropriate that they contract at a *(fast? slow?)* rate. Such muscle tissue consists mostly of slow red fibers that *(do? do not?)* fatigue easily. These appear red because they have *(much? little?)* myoglobin and *(many? few?)* blood vessels. They have *(many? few?)* mitochondria and therefore can depend on *(aerobic? anaerobic?)* metabolism for energy. These muscles are classified as type *(SO? FOG? FG?)*.

b. Muscles of the arms are used *(constantly? mostly for short bursts of energy?)* as in lifting and throwing. Therefore they must contract at a *(fast? slow?)* rate. The arms consist mainly of fast twitch white fibers. These respond rapidly to nerve impulses and contract *(fast? slowly?)*. They *(do? do not?)* fatigue easily because they are designed for the relatively inefficient processes of *(aerobic? anaerobic?)* metabolism. These muscles are classified as *(SO? FOG? FG?)*.

c. Endurance exercises tend to enhance development of the more efficient (aerobic)

   _____ fibers, whereas weight lifting, which requires short bursts of

   energy, tends to develop _____ fibers.

d. Most skeletal muscles, such as the biceps brachii, are *(all of one type? mixtures of types SO, FOG, and FG?)*. The muscles of any one motor unit are *(all of one type? mixtures of types SO, FOG, and FG?)*.

e. Arrange the three types of fibers in the sequence in which they are "called to action":

   1. _____ → 2. _____ → 3. _____
      (weak contraction)   (stronger contraction)   (maximal contraction)

f. Endurance training (such as running or swimming) tends to cause transformation

   of some _____.
   A. Fast oxidative-glycolytic (FOG) into fast glycolytic (FG) fibers
   B. Fast glycolytic (FG) into fast oxidative-glycolytic (FOG) fibers

E2. State the desired effects as well as disadvantages of uses of anabolic steroids.

✓ **E3.** Compare the structure and function of skeletal and cardiac muscle by completing this table. State the significance of characteristics that have asterisks (*).

Characteristic	Skeletal Muscle	Cardiac Muscle
a. Number of nuclei per myofiber	Several (multinucleate)	
b. Number of mitochondria per myofiber (More or Fewer)*		
c. Striated appearance due to alternated actin and myosin myofilaments (Yes or No)*	Yes	
d. Presence of intercalated discs (Yes or No)*		
e. Nerve stimulation with release of ACh required for contraction (Yes or No)*		
f. Duration of muscle contraction (Long or Short)*		

E4. Compare the two types of smooth muscle.

Muscle Type	Structure	Spread of Stimulus	Locations
a. Visceral		Impulse spreads and causes contraction of adjacent fibers	
b.			Large blood vessels and airways, iris and ciliary muscles of eye, arrector pili muscles

✓ **E5.** Contrast smooth muscle with the muscle tissue you have already studied by circling the correct answers in this paragraph.

Smooth muscle fibers are *(cylinder shaped? tapered at each end?)* with *(several nuclei? one nucleus?)* per cell. They *(do? do not?)* contain actin and myosin. However, due to the irregular arrangement of these filaments, smooth muscle tissue appears *(striated? nonstriated or "smooth"?)*. In general, smooth muscle contracts and relaxes more *(rapidly? slowly?)* than skeletal muscle does, and smooth muscle holds the contraction for a *(shorter? longer?)* period of time than skeletal muscle does. This difference in smooth muscle is due at least partly to the fact that smooth muscle has *(no? many?)* transverse tubules to transmit calcium into the muscle fiber and has a *(large? small?)* amount of SR.

✓ **E6.** Smooth muscle *(is? is not?)* normally under voluntary control. List three chemicals released in the body or other factors that can also lead to smooth muscle contraction or relaxation.

✓ **E7.** *For extra review.* Of the three muscle types, select the correct muscle type from the box that matches the description below.

C. Cardiac	**Sk. Skeletal**	**Sm. Smooth**

_____ a. Starts most slowly and lasts much longer

_____ b. Has no transverse tubules and only scanty sarcoplasmic reticulum

_____ c. Has no striations because thick and thin filaments are not arranged in a regular pattern

_____ d. Has intermediate filaments attached to dense bodies (comparable to Z discs)

_____ e. Calmodulin, the regulatory protein in this muscle, activates an enzyme called myosin light chain kinase

_____ f. These muscles demonstrate the stress-relaxation response

_____ g. Spindle-shaped fibers with one nucleus per cell

_____ h. Unbranched, cylindrical cells; multi-nucleate

_____ i. Has fastest speed of contraction

_____ j. Stimulated only by the neurotransmitter acetylcholine (ACh)

_____ k. Includes three types: SO, FOG, and FG

_____ l. Branched, cylindrical cells; usually have one nucleus per cell

_____ m. Found only in the heart

_____ n. Striated, involuntary muscle

_____ o. Contains intercalated discs containing desmosomes and gap junctions

_____ p. Has least amount of SR and lacks T tubules

E8. *The Big Picture: Looking Back.* For additional review of the three muscle types, look back at LG Chapter 4, Checkpoints El-E3, and Figure LG 4.1, M-O, page LG 71.

F. Regeneration of muscle tissue; developmental anatomy and aging of muscles (pages 293–296)

✓ **F1.** Arrange the three types of your muscle tissue in correct sequence according to ability (from most to least) to regenerate during your lifetime.

_____ _____ _____   A. Heart muscle

                           B. Biceps muscle

                           C. Muscle of an artery or intestine

✓ **F2.** Both skeletal and cardiac muscle tissue increases in size by *(hyperplasia? hypertrophy?)*, which means increase in *(number? size?)* of cells. If either of these tissues is damaged,

fibrosis, which means replacement of muscle tissue with _____ tissue, is likely to occur.

F3. Describe the roles of the following cells in regeneration of muscle tissue:
a. Satellite cells

b. Pericytes

✓ **F4.** Complete this exercise about the development of muscles.
a. Which of the three types of muscle tissue develop from mesoderm? *(Skeletal? Cardiac? Smooth?)*
b. Part of the mesoderm forms columns on either side of the developing nervous

system. This tissue segments into blocks of tissue called _____. The first pair of somites forms on day *(10? 20? 30?)* of gestation. By day 30 a total of _____ pairs of somites are present.
c. Which part of a somite develops into vertebrae? *(Myo-? Derma-? Sclero-?)* derm. The

dermis of the skin and other connective tissues is formed from _____

-tomes. while most skeletal muscles develop from _____-tomes.

F5. With aging, muscles experience a relative increase in *(SO? FOG? FG?)* fibers. Discuss possible reasons for this change.

G. Disorders and medical terminology (pages 296–297)

G1. Write the correct medical term related to muscles after its description.

a. Muscle or tendon pain and stiffness, such as "charley horse": _____

b. Increased muscle tone, rigidity, or spasticity: _____

c. Inherited, muscle-destroying disease causing atrophy of muscles: _____

d. A muscle tumor: _____
e. An involuntary, brief twitch of a muscle that is ordinarily under voluntary control,

such as in the eyelid: _____
f. An involuntary, brief twitch of an entire motor unit that is visible under the skin;

may be seen in MS or ALS patients: _____

✓ **G2.** Describe myasthenia gravis in this exercise.
a. In order for skeletal muscle to contract, a nerve must release the chemical

_____ at the myoneural junction. Normally ACh binds to

_____ on the muscle fiber membrane.

b. It is believed that a person with myasthenia gravis produces _____ that bind to these receptors, making them unavailable for ACh binding. Therefore, ACh *(can? cannot?)* stimulate the muscle, and it is weakened.

c. One treatment for this condition employs _____ drugs, which enhance muscle contraction by permitting the ACh molecules that *do* bind to act longer (and not be destroyed by AChE).

d. Other treatments include _____-suppressants, which decrease the

    patient's antibody production, or _____ , which segregates the patient's harmful antibodies.

G3. Define each of these types of abnormal muscle contractions.
a. Spasm

b. Fibrillation

c. Tremor

ANSWERS TO SELECTED CHECKPOINTS

A1. (a) Sm. (b) C. (c) Sk. (d) Sk. (e) Sm.
A2. (a) Motion or movement. (b1) Blood, (b2) food, bile, enzymes, or wastes, (b3) urine, (b4) ova and sperm or eggs. (c) Stabilization, stomach or bladder. (d) Heat, shivering. (e) Sphincters.
A3. (a) Extensibility. (b) Elasticity. (c) Excitability. (d) Contractivity
A4. (a) Isotonic. (b) Stays about the same, increases.
B1. (a) S. (b) D. (c) S.
B2. Epimysium → perimysium → endomysium. See Figure LG 10.1A
B3. Periosteum; tendon, aponeurosis.
B4. (a) Somatic motor neurons; axons; bulbs, neuromuscular. (b) Provide nutrients and oxygen for ATP production, and remove heat and wastes.
B6. (a) Muscle fiber (cell) → myofibrils → filaments (thick or thin). (b) Sarcolemma; T (transverse); they transmit action potentials from the sarcolemma throughout the muscle fiber so that all parts of it contract in unison. (c) Sarcoplasm; glycogen, myoglobin, oxygen; ATP, mitochondria. (d) Endoplasmic reticulum; Ca^{2+}; cisternae , triad; see Figure LG 10.1A. (e) See Figure LG 10.1A.

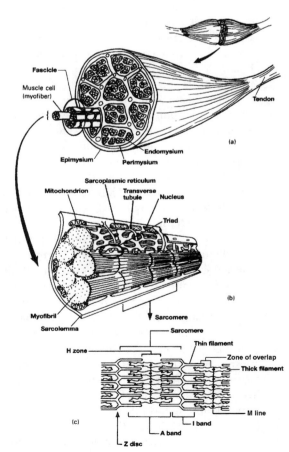

Figure LG 10.1A Diagram of skeletal muscle. (a) Skeletal muscle cut to show cross section and longitudinal section with connective tissues. (b) Section of one muscle cell (myofiber). (c) Detail of sarcomere of muscle cell.

B7. (a) Size; exercise and good nutrition; the heart can also hypertrophy when it must pump hard to move blood out into a high pressure system (in a person with high blood pressure).(b) Denervation; 75; fibrous connective. (c) Immobilization of the extremity by a cast or other immobilizer, lack of exercise; being bedridden (for example, in a coma).

B8. (a) Sarcoplasm. (b) Sarcomere. (c) A band. (d) I band. (e) M line. (f) Z disc. (g) zone of overlap.

B9. (a) Delayed onset muscle soreness. (b) Myoglobin, creatine kinase. (c) 12, 48.

B10. (a) See Figure LG 10.1A. (b) Contractile; troponin, tropomyosin. (c) Thick; golf clubs; M lines; myosin head or crossbridge, actin. (d) Z discs, do not; actin; tropomyosin, troponin. (e) Structural; huge (titanic); thick, Z disc, M line; elasticity. (f) Myomesin; dystrophin.

C1. (a) Myosin, actin, M line. (b) Thin; slide, sliding-filament; do not.

C2. (a) Sarcoplasmic reticulum (SRI). (b) Troponin, actin, myosin. (c) ATPase, ADP, phosphate; actin. (d) Power, ADP, sarcomere. (e) 600 (2 per myosin x 300 myosin molecules per thick filament); five. (f) Actin; myosin, myosin. (g) Toward or even across; myosin, thin.

C3. (a) Low; (1) Active transport; (2) Calsequestrin, 10,000. (b) T tubules; channels, sarcoplasm. (c) Actin; relax. (d) An active; is not; myosin cross bridges stay attached to actin and muscles remain in a state of partial contraction (rigor mortis) for about a day.

C4. C D A E B.

C5. (a) Much; maximal; no. (b) Stronger.

C6. Active; elastic components including titin, endomysium, perimysium, epimysium, and tendons.

C7. (a) Motor unit; synaptic end bulbs; all muscle fibers within that motor unit. (b) A, H; end bulb. (c) C; D, acetylcholine (ACh), E. (d) Motor end plate; receptors; 30-40 million. (e) Open; contraction; electromyography (EMG). (f) Neuromuscular junction (NMJ); only one. I; synapses.

C9-10. (a) AChE; ↓. (b) CUR; ↓. (c) NEO; ↑. (d) CBT; ↓.

D1. (a) ATP; myosin; ATP → ADP + P + energy. (b) ADP + P + energy (from foods) → ATP. (c) Active transport; creatine (that combines with phosphate to form creatine phosphate; creatine phosphate + ADP → creatine + ATP. (d) Creatine phosphate + ADP ← creatine + ATP (draw arrow to LEFT in Figure LG 10.3); creatine kinase (CK); creatine phosphate. (e) 15 seconds; anaerobic; aerobic; see Figure LG 10.4A. (f) Lactic, anaerobic; fatigue; heart, liver, kidney; see Figure LG 10.4A. (g) Myoglobin; hemo; red. (h) Aerobic.

D3. (a) CP. (b) ANAER. (c) AER.

D4. Decrease in frequency of stimulation by neurons, shorter length muscle fibers just prior to contrac-tion, fewer and smaller motor units recruited, lack of nutrients, lack of oxygen. and muscle fatigue.

D5. An average of 150 muscle fibers [cells]; laryngeal muscles controlling speech.

D6. (a) T. (b) M. (c) L. (d) Ref. (e) Rel. (f) Ref.

D7. (a) U. (b) F. (c) W. (d) U.

D8. (a) Anaerobic; anaerobic. (b) Aerobic; aerobic.

E1. (a) Constantly; slow; do not; much, many; many, aerobic; SO. (b) Mostly for short bursts of energy; fast; fast; do, anaerobic; FG. (c) FOG, FG. (d) Mixtures of types SO, FOG, and FG; all of one type. (e1) SO, (e2) FOG, (e3) FG. (f) B.

E3.

Characteristic	Skeletal Muscle	Cardiac Muscle
a. Number of nuclei per myofiber	Several (multinucleate)	One
b. Number of mitochondria per myofiber (More or Fewer)*	Fewer	More, because heart muscle requires constant generation of energy
c. Striated appearance due to alternated actin and myosin myofilaments (Yes or No)*	Yes	Yes
d. Presence of intercalated discs (Yes or No)*	No	Yes, allow muscle action potentials to spread from one cardiac muscle fiber to another
e. Nerve stimulation with release of ACh required for contraction (Yes or No)*	Yes	No, so heart can contract without nerve stimulation, but nerves can increase or decrease heart rate
f. Duration of muscle contraction (Long or Short)*	Short (0.01 sec.–0.100 sec)	10–15 times longer so heart can pump blood out of heart

E5. Tapered at each end, one nucleus; do; nonstriated or "smooth"; slowly, longer; no, small.

E6. Is not; hormones, pH or temperature changes, O_2 and CO_2 levels, and certain ions.

E7. (a–g) Sm. (h–k) Sk. (l–o) C. (p) Sm.

F1. C B A

F2. Hypertrophy, size; fibrous (or connective or scar).

F4. (a) All three: skeletal, cardiac, and smooth. (b) Somites; 20; 44. (c) Sclero-; derma-, myo-.

G1. (a) Myalgia. (b) Hypertonia (c) Muscular dystrophy. (d) Myoma. (e) Tic. (f) Fasciculation.

G2. (a) Acetylcholine (ACh); receptors. (b) Antibodies; cannot. (c) Anticholinesterase (antiAChE). (d) Immuno-, plasmapheresis.

Figure LG 10.4A Anaerobic and aerobic energy sources for muscles.

MORE CRITICAL THINKING: CHAPTER 10

1. Explain why stretching a muscle before exercising can maximize muscle contraction. Then explain why *over*stretched heart muscle (as in the condition heart failure) may decrease the effectiveness of the heart as a pump.
2. Discuss results of research on the potential value of using creatine as a performance-enhancing supplement.
3. Discuss the pros and cons of anabolic steroids as you would with a friend who is considering using these chemicals.
4. Explain how the microscopic structure of cardiac muscle fibers accounts for its continuous, rhythmic activity.
5. Describe several reasons why smooth muscle contracts more slowly than other muscle types.
6. Write a rationale for why endurance and strength training programs are effective for maintaining healthy musculature in elderly persons.

MASTERY TEST: CHAPTER 10

Questions 1-9: Circle T (true) or F (false). If the statement is false, change the underlined word or phrase so that the statement is correct.

T   F   1. Tendons are <u>cords of connective tissue, whereas aponeuroses are broad, flat bands</u> of connective tissue.

T   F   2. During contraction of muscle <u>both A bands and I bands</u> get shorter.

T   F   3. In general, <u>red (rather than white)</u> skeletal muscle is more suited for endurance exercise than for short bursts of energy.

T   F   4. <u>Elasticity</u> is the ability of a muscle to be stretched or extended.

T   F   5. Myasthenia gravis is caused by <u>an excess of acetylcholine production at the myoneural junction</u>.

T   F   6. Muscle fibers remain relaxed if there are <u>few calcium ions in the sarcoplasm</u>.

T   F   7. <u>Most myoblasts do</u> continue to divide to form new skeletal muscle cells throughout life.

T   F   8. Muscles that are used mostly for quick bursts of energy (such as those in the arms) contain large numbers of <u>fast glycolytic (FG) fibers</u>.

T   F   9. In a relaxed muscle fiber <u>thin and thick filaments overlap to form the zone of overlap</u>.

Questions 10-12: Arrange the answers in correct sequence.

_____ _____ _____   10. Number of transverse tubules from greatest to least per muscle cell in these types of muscle:

   A. Smooth

   B. Skeletal

   C. Cardiac

_____ _____ _____   11. According to the amount of muscle tissue they surround, from most to least:

   A. Perimysium

   B. Endomysium

   C. Epimysium

_____ _____ _____   12. From largest to smallest:

   A. Myofibril

   B. Filament

   C. Muscle fiber (myofiber or cell)

Questions 13-20: Circle the letter preceding the one best answer to each question.

13. All of the following molecules are parts of thin filaments *except:*
    A. Actin
    B. Myosin
    C. Tropomyosin
    D. Troponin

14. Which of the following answers includes only structural (not contractile or regulatory) proteins of muscle?
    A. Actin and myosin
    B. Titin, nebulin, and myomesin
    C. Troponin and tropomyosin
    D. Actin, troponin, and tropomyosin

15. Choose the one statement that is *false:*
    A. The A band is darker than the I band.
    B. Thick filaments reach the Z disc in relaxed muscle.
    C. The H zone contains thick filaments, but not thin ones.
    D. The I band is shorter in contracted muscle than in relaxed muscle.

16. All of the following terms are correctly matched with descriptions *except:*
    A. Denervation atrophy—wasting of a muscle to one-quarter of its size within two years of loss of nerve supply to a muscle
    B. Disuse atrophy—decrease of muscle mass in a person who is bedridden or who has a cast on
    C. Muscular hypertrophy—increase in size of a muscle by increase in the number of muscle cells
    D. Cramp—painful, spasmodic contraction

17. Which statement about muscle physiology in the relaxed state is *false?*
    A. Myosin cross bridges are bound to ATP.
    B. Calcium ions are stored in sarcoplasmic reticulum.
    C. Myosin cross bridges are bound to actin.
    D. Tropomyosin-troponin complex is bound to actin.

18. All of the following answers match a chemical with its correct description *except:*
    A. Neurotransmitter released by nerve at a neuromuscular junction: ACh
    B. Enzyme that destroys ACh within the synaptic cleft: AChE
    C. A contractile protein: tropomyosin
    D. Chemical that binds to and changes shape of troponin-tropomyosin complex, exposing myosin-binding site on actin: Ca^{2+}

19. Choose the *false* statement about cardiac muscle.
    A. Cardiac muscle is involuntary.
    B. Cardiac muscle typically has one nucleus per fiber.
    C. Cardiac muscle cells are called cardiac muscle fibers.
    D. Cardiac fibers are separated by intercalated discs.
    E. Cardiac fibers are tapered at each end with no striations.

20. Most voluntary movements of the body are results of _____ contractions.
    A. Isometric
    B. Fibrillation
    C. Twitch
    D. Unfused tetanus
    E. Fused tetanus

Questions 21-25: Fill-ins. Complete each sentence with the word or phrase that best fits.

_____ 21. _____ muscle cells are nonstriated and tapered at each end with one nucleus per cell.

_____ 22. Cardiac muscle remains contracted longer than skeletal muscle does because _____ is slower in cardiac than in skeletal muscle.

_____ 23. DOMS is an acronym that means _____.

_____ 24. _____ is the transmitter released from synaptic vesicles of axons supplying skeletal muscle.

_____ 25. ADP + creatine phosphate → _____ (Write the products.)

ANSWERS TO MASTERY TEST: CHAPTER 10

True-False

1. T
2. F. I bands but not A bands
3. T
4. F. Extensibility
5. F. An autoimmune disorder in which antibodies are produced that bind onto receptors on the sarcolemma
6. T
7. Few myoblasts (those called satellite cells) do
8. T
9. T

Arrange

10. B C A
11. C A B
12. C A B

Multiple Choice

13. B
14. A
15. B
16. C
17. C
18. C
19. E
20. D

Fill-Ins

21. Smooth
22. Passage of calcium ions from the extracellular fluid through sarcolemma
23. Delayed onset muscle soreness
24. Acetylcholine
25. ATP + creatine

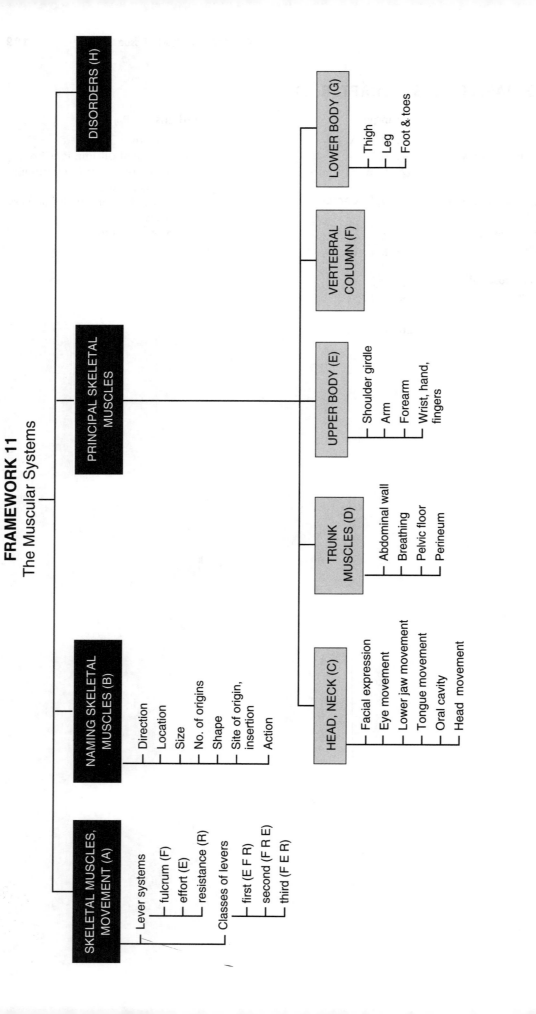

FRAMEWORK 11
The Muscular Systems

SKELETAL MUSCLES, MOVEMENT (A)
- Lever systems
  - fulcrum (F)
  - effort (E)
  - resistance (R)
- Classes of levers
  - first (E F R)
  - second (F R E)
  - third (F E R)

NAMING SKELETAL MUSCLES (B)
- Direction
- Location
- Size
- No. of origins
- Shape
- Site of origin, insertion
- Action

PRINCIPAL SKELETAL MUSCLES

DISORDERS (H)

HEAD, NECK (C)
- Facial expression
- Eye movement
- Lower jaw movement
- Tongue movement
- Oral cavity
- Head movement

TRUNK MUSCLES (D)
- Abdominal wall
- Breathing
- Pelvic floor
- Perineum

UPPER BODY (E)
- Shoulder girdle
- Arm
- Forearm
- Wrist, hand, fingers

VERTEBRAL COLUMN (F)

LOWER BODY (G)
- Thigh
- Leg
- Foot & toes

C H A P T E R **11**

The Muscular System

More than 600 different muscles attach to the bones of the skeleton. Muscles are arranged in groups that work much as a symphony, with certain muscles quiet while others are performing. The results are smooth, harmonious movements, rather than erratic, haphazard discord.

Much information can be gained by a careful initial examination of two aspects of each muscle: its name and location. A muscle name may offer such clues as size or shape, direction of fibers, or points of attachment of the muscle. A look at the precise location of the muscle and the joint it crosses—combined with logic—can usually lead to a correct understanding of action(s) of that muscle.

As you begin your study of the muscular system, carefully examine the Chapter 11 Topic Outline and Objectives; check off each one as you complete it. To organize your study of the muscular system, glance over the Chapter 11 Framework now. Be sure to refer to the Framework frequently and note relationships among key terms in each section.

TOPIC OUTLINE AND OBJECTIVES

A. How skeletal muscles produce movement

1. Describe the relationship between bones and skeletal muscles in producing body movements.
2. Define lever and fulcrum, and compare the three types of levers on the basis of location of the fulcrum, effort, and resistance.
3. Identify the various arrangements of muscle fibers in a skeletal muscle, and relate the arrangements to strength of contraction and range of motion.
4. Explain how the prime mover, antagonist, synergist, and fixator in a muscle group work together to produce movement.

B. How skeletal muscles are named

5. Explain seven characteristics used in naming skeletal muscles.

C. Principal skeletal muscles of the head and neck

6. Describe the origin, insertion, action, and innervation of the muscles of facial expression.
7. Describe the origin, insertion, action, and innervation of the extrinsic muscles of the eyeballs.

8. Describe the origin, insertion, action, and innervation of the muscles that move the eyeballs.
9. Describe the origin, insertion, action, and innervation of the extrinsic muscles of the tongue.
10. Describe the origin, insertion, action, and innervation of the muscles of the floor of the oral cavity.
11. Describe the origin, insertion, action, and innervation of the muscles that move the head.

D. Principal skeletal muscles that act on the abdominal wall, muscles used in breathing, muscles of the pelvic floor and perineum

12. Describe the origin, insertion, action, and innervation of the muscles that act on the abdominal wall.
13. Describe the origin, insertion, action, and innervation of the muscles used in breathing.
14. Describe the origin, insertion, action, and innervation of the muscles of the pelvic floor.
15. Describe the origin, insertion, action, and innervation of the muscles of the perineum.

E. Principal skeletal muscles that move the shoulder girdle and upper limb

16. Describe the origin, insertion, action, and innervation of the muscles that move the pectoral girdle.
17. Describe the origin, insertion, action, and innervation of the muscles that move the humerus.
18. Describe the origin, insertion, action, and innervation of the muscles that move the radius and ulna.
19. Describe the origin, insertion, action, and innervation of the muscles that move the wrist, hand, and digits.
20. Describe the origin, insertion, action, and innervation of the intrinsic muscles of the hand.

F. Principal skeletal muscles that move the vertebral column

21. Describe the origin, insertion, action, and innervation of the muscles that move the vertebral column.

G. Principal skeletal muscles that move the lower limb

22. Describe the origin, insertion, action, and innervation of the muscles of the femur.
23. Describe the origin, insertion, action, and innervation of the muscles that act on the femur, tibia, and fibula.
24. Describe the origin, insertion, action, and innervation of the muscles that move the foot and toes.
25. Describe the origin, insertion, action, and innervation of the intrinsic muscles of the foot.

H. Disorders

WORDBYTES

Now become familiar with the language of this chapter by studying each wordbyte, its meaning, and an example of its use within a term. After you study the entire list, self-check your understanding by writing the meaning of each wordbyte on the line. As you continue through the *Learning Guide*, identify (and fill in) additional terms that contain the same wordbyte.

Wordbyte	Self-check	Meaning	Example(s)
bi-	_____	two	*bi*ceps
brachi-	_____	arm	*brachi*alis
brev-	_____	short	flexor digitorum *brev*is
bucc-	_____	mouth, cheek	*bucc*inator
cap-, -ceps	_____	head	*cap*itis, tri*ceps*
-cnem-	_____	leg	gastro*cnem*ius
delt-	_____	Greek D Δ	*delt*oid
gastro-	_____	stomach, belly	*gastro*cnemius
genio-	_____	chin	*genio*glossus
glossus-	_____	tongue	*gloss*ary, stylo*glossus*
glute-	_____	buttock	*glute*us medius
grac-	_____	slender	*grac*ilis
-issimus	_____	the most	lat*issimus* dorsi
maxi-	_____	large	gluteus *maxi*mus
mini-	_____	small	gluteus *mini*mus

Wordbyte	Self-check	Meaning	Example(s)
or-	_____	mouth	orbicularis *oris*
rect-	_____	straight	*rect*us femoris
sartor-	_____	tailor	*sartor*ius
serra-	_____	toothed, notched	*serra*tus anterior
teres-	_____	round	*teres* major
tri-	_____	three	*tri*ceps femoris
vast-	_____	large	*vast*us lateralis

CHECKPOINTS

A. How skeletal muscles produce movement (pages 302–307)

✓ **A1.** What structures constitute the *muscular system*?

✓▲**A2.** Refer to Figure LG 11.1 and consider flexion of your own forearm as you do this learning activity.
   a. In flexion your forearm serves as a rigid rod, or _____, which moves

about a fixed point, called a _____ (your elbow joint, in this case).
   b. Hold a weight in your hand as you flex your forearm. The weight plus your forearm serve as the *(effort? fulcrum? resistance?)* during this movement.
   c. The effort to move this resistance is provided by contraction of a _____. Note that if you held a heavy telephone book in your hand while your forearm was flexed, much

more _____ by your arm muscles would be required.
   d. In Figure LG 11.1 identify the exact point at which the muscle causing flexion attaches to the forearm. It is the *(proximal? distal?)* end of the *(humerus? radius? ulna?)*. Write an E and an I on the two lines next to the arrow at that point in the figure. This indicates that this is the site where the muscle exerts its effort (E) in the lever system, and it is also the insertion (I) end of the muscle. (More about insertions in a minute.)

Figure LG 11.1 The lever-fulcrum principle is illustrated by flexion of the forearm. Complete the figure as directed in Checkpoint A2.

   e. Each skeletal muscle is attached to at least two bones. As the muscle shortens, one bone stays in place and so is called the *(origin? insertion?)* end of the muscle. What

bone in the figure appears to serve as the origin bone? _____

Write O on that line at that point in the figure.

  f. Now label the remaining arrows in Figure LG 11.1: F at fulcrum and R at resistance. This is an example of a *(first? second? third?)* -class lever.

▲ **A3.** Complete the table about levers. KEY: E. Effort, F. Fulcrum, and R. Resistance.

Class	Analogy of EFR	Arrangement	Mechanical advantage: (Yes, No). Explain.	Example in human body
a. First				Head and neck: atlanto-occipital joint
b.	Wheelbarrow	F R E		
c.	(Omit)		No since E is closer (than R) to F	

✓ **A4.** *For extra review.* Check your understanding of lever systems in the body in this Checkpoint.

  a. Refer to Figure 11.2a in your text. Now hyperextend your head as if to look at the sky. The weight of your face and jaw serves as *(E? F? R?)*, while your neck muscles

    provide *(E? F? R?)*. The fulcrum is the joint between the _____ and the

    _____ bones. This is an example of a _____ class lever.

  b. Mechanical advantage (leverage) means that a *(more? less?)* forceful effort can move a *(more? less?)* forceful resistance. Mechanical advantage is provided by having the resistance always *(closer to? farther from?)* the fulcrum than the effort. Which class of lever provides more mechanical advantage?
    A. Second-class lever   B. Third-class lever

✓▲ **A5.** Correlate fascicular arrangement with muscle power and range of motion of muscles.

  a. A muscle with *(many? long?)* fibers will tend to have great strength. An example is the *(parallel? pennate?)* arrangement.

  b. A muscle with *(many? long?)* fibers will tend to have great range of motion (but less power). An example is the *(parallel? pennate?)* arrangement.

✓ **A6.** Refer again to Figure LG 11.1 and do this exercise about how muscles of the body work in groups.

  a. The muscle that contracts to cause flexion of the forearm is called a _____ .

    An example of a prime mover in this action would be the _____ muscle.

  b. The triceps brachii must relax as the biceps brachii flexes the forearm. The triceps is an extensor. Because its action is opposite to that of the biceps, the triceps is called *(a synergist? an agonist? an antagonist?)* of the biceps.

  c. What would happen if the flexors of your forearm were functional, but not the antagonistic extensors?

  d. What action would occur if both the flexors and extensors contracted simultaneously?

e.　Muscles that assist or cooperate with the prime mover to cause a given action are

known as _____.

B.　How skeletal muscles are named (page 307)

✓ **B1.** Review the Wordbyte section above. Match the names of the following muscles with
their meanings.

A.　Large muscle of the buttock region	**D.　The broadest muscle of the back**
B.　Belly-shaped muscle in leg	**E.　Large muscle in medial thigh area**
C.　Thigh muscle with four origins	**F.　Muscle that raises the upper lip**

_____ a.　Latissimus dorsi　　　　　　_____ d.　Gastrocnemius

_____ b.　Vastus medialis　　　　　　_____ e.　Gluteus maximus

_____ c.　Quadriceps femoris　　　　　_____ f.　Levator labii superioris

✓▲ **B2.** As you study the names of muscles, you will find that most of them provide a good
description of the muscle. For each of the following, indicate the type of clue that
each part of the name gives. The first one is done for you.

A.　Action	**N.　Number of heads or origins**
D.　Direction of fibers	**P.　Points of attachment of origin and insertion**
L.　Location	**S.　Size or shape**

__**D, L**__ a.　Rectus abdominis　　　　　_____ d.　Sternocleidomastoid

_____ b.　Flexor carpi ulnaris　　　　_____ e.　Adductor longus

_____ c.　Biceps brachii

C.　Principal skeletal muscles of the head and neck (pages 307–324)

✓▲ **C1.** After studying Exhibit 11.1 (pages 311-312) in your text, check your understanding of
the muscles of facial expression. Write the name of the muscle that answers each
description. Locate muscles in Figure LG 11.2a O-R. Cover the key and write the
name of each facial muscle next to its lettered leader line.

a.　Allows you to show surprise by raising your eyebrows and forming horizontal

forehead wrinkle: _____

b.　Muscle surrounding opening of your mouth; allows you to use your lips in kissing

and in speech: _____

c.　Muscle for smiling and laughing because it draws the outer portion of the mouth

upward and outward: _____

d.　Circular muscle around eye; closes eye: _____

✓ **C2.** Essentially all of the muscles controlling facial expression receive nerve impulses via

the _____ nerve, which is cranial nerve *(III? V? VII?)*.

O   Flexors
O   Extensors

KEY

A.   Temporalis
B.   Masseter
C.   Trapezius
D.   Pectoralis minor
E.   Serratus anterior
F.   Intercostals
G.   Rectus abdominis
H.   Internal oblique
I.   Transversus abdominis
J.   Adductor group
K.   Gracilis
L.   Peroneus longus
M.   Flexor digitorum longus
N.   Tibialis anterior
O.   Frontalis
P.   Orbicularis oculis
Q.   Zygomaticus major
R.   Orbicularis oris
S.   Sternocleidomastoid
T.   Deltoid
U.   Pectoralis major
V.   Biceps brachii
W.   Brachialis
X.   External oblique
Y.   Gluteus medius
Z.   Tensor fasciae latae
AA.   Sartorius
BB.   Rectus femoris
CC.   Vastus lateralis
DD.   Vastus medialis
EE.   Gastrocnemius
FF.   Soleus

(a)

Figure LG 11.2 Major muscles of the body. Some deep muscles are shown on the left side of the figure. All muscles on the right side are superficial. Label and color as directed. (a) Anterior view. (b) Posterior view.

○ Flexors
○ Extensors

A
B
C
D
E
F
G
H
I
J
K

L
M
N
O
P
Q
R
S
T

(b)

KEY

A.   Levator scapulae
B.   Rhomboideus minor
C.   Rhomboideus major
D.   Supraspinatus
E.   Infraspinatus
F.   Teres major
G.   Erector spinae
H.   Adductors
I.   Vastus lateralis
J.   Biceps femoris
K.   Semimembranosus and
       semitendinosus
L.   Trapezius
M. Deltoid
N.   Triceps brachii
O.   Latissimus dorsi
P.   External oblique
Q.   Gluteus medius
R.   Gluteus maximus
S.   Gastrocnemius
T.   Soleus

✓▲ **C3.** Complete this exercise about muscles that move the eyeballs.
  a. Why are these muscles called *extrinsic* eyeball muscles?

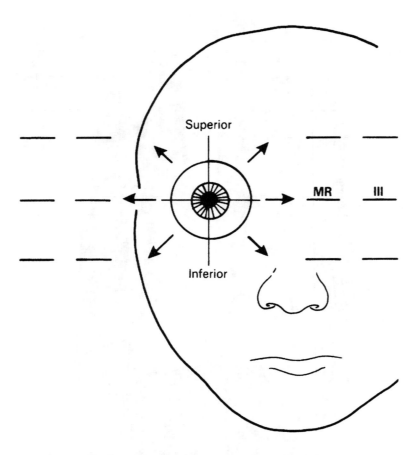

Figure LG 11.3 Right eyeball with arrows indicating directions of eye movements produced by the extrinsic eye muscles. Label as directed in Checkpoint C3.

  b. Label Figure LG 11.3 with initials of the names of the six extrinsic eye muscles listed in the box to indicate *direction of movement* of each muscle. One has been done for you.

IO.	Inferior oblique	MR.	Medial rectus
IR.	Inferior rectus	SO.	Superior oblique
LR.	Lateral rectus	SR.	Superior rectus

  c. Now write next to each muscle the number of the cranial nerve innervating that muscle. One has been done for you.
  d. Note that most of the extrinsic eye muscles are innervated by cranial nerve *(Ill? IV? VI?)*, as indicated by the name of that nerve *(oculomotor)*. Cranial nerve VI, the *abducens*, supplies the muscles that cause most *(medial? lateral?)* movement, or abduction, of the eye. [*Hint:* Remember the chemical symbol for sulfate (SO4) to remind yourself that the remaining muscle, superior oblique (SO), is supplied by the fourth cranial nerve.]

✓ **C4.** *For extra review* of actions of eye muscles, work with a study partner. One person moves the eyes in a particular direction; the partner then names the eye muscles used for that action. *Note:* Will both eyes use muscles of the same name? For example, as you look to your right, will the lateral rectus muscles attached to both eyes contract?

✓▲ **C5.** Place your index finger and thumb on the origin and insertion of each of the muscles that move your lower jaw. (Refer to Exhibit 11.3 and Figure 11.6 in the text, for help.) Then do this learning activity.
   a. Two large muscles help you to close your mouth forcefully, as in chewing. Both of these act by *(lowering the maxilla? elevating the mandible?)*. The

   _____ covers your temple, and the _____ covers the ramus of the mandible.
   b. Most of the muscles involved in chewing are ones that help you to *(open? close?)* your mouth. Think about this the next time you go to the dentist and try to hold

   your mouth open for a long time, with only your _____ muscles to force your mouth wide open.
   c. The muscles that move the lower jaw and so aid in chewing are innervated by

   cranial nerve *(III? V? VII?)*, known as the _____ nerve.
   d. Refer to Figure LG 11.2a. Muscles *A* and *B* are used primarily for *(facial expressions? chewing?)*, whereas muscles *0, P, Q,* and *R* are used mainly for *(facial expressions? chewing?)*.

✓ **C6.** Actions such as opening your mouth for eating or speaking and also swallowing require integrated action of a number of muscles. You have already studied some that allow you to open your mouth (such as masseter and temporalis). We will now consider muscles of the tongue and the floor of the oral cavity.
   Identify locations of three bony structures relative to your tongue. Tell whether each is anterior or posterior and superior or inferior to your tongue. (For help refer to Figure 11.7.)
   a. Styloid process of temporal bone _____

   b. Hyoid bone _____

   c. Chin (anterior portion) _____

✓▲ **C7.** Remembering the locations of these bony points and that muscles move a structure (tongue) by pulling on it, determine the direction that the tongue is pulled by each of these muscles.
   a. Styloglossus: _____

   b. Hyoglossus: _____

   c. Genioglossus: _____

✓▲ **C8.** The three muscles listed in Checkpoint C7 are innervated by cranial nerve _____;

   the name _____ nerve indicates that this pair of cranial nerves supplies the region "under the tongue."

✓▲ **C9.** Match groups of muscles in each answer with the descriptions below.

> **A. Styloglossus, hyoglossus, and genioglossus**
>
> **B. Stylohyoid, mylohyoid, and geniohyoid**

_____ a. Located superior to the hyoid (supra-hyoid), these form the floor of the oral cavity. (Recall that the hyoid is directly posterior to the inferior border of the U-shaped mandible.)

_____ b. Muscles permitting tongue movements.

✓ **C10.** Using a mirror, find the origin and insertion of your left sternocleidomastoid muscle. (See Figure 11.9 in your text.) The muscle contracts when you pull your chin down and to the right; this diagonal muscle of your neck will then be readily located. Note that the left sternocleidomastoid pulls your face toward the (same? opposite?) side. It also (flexes? extends?) the head.

✓ **C11.** Alternately flex and extend vertebrae of your neck by looking at your toes and then toward the sky. As you look at the sky, you are (flexing? extending and then hyper-extending?) your head and cervical vertebrae. On which surface of the neck would you expect to find extensors of the head and neck (Anterior? Posterior?)? Note that these muscles are the most superior muscles of the columns of extensors of the verte-brae. Find them on Figure 11.19. Now name three of these muscles.

✓ **C12.** _The Big Picture: Looking Ahead._ Refer to Figure 14.5 of your text, as well as Table 14.2, page 476–479 and summarize head and neck muscle innervation by cranial nerves. Write the correct cranial nerve numbers next to the names of muscles innervated. Select from answers in the box. All except one answer will be used.

> | III. | Oculomotor | VII. | Facial |
> | IV. | Trochlear | IX. | Glossopharyngeal |
> | V.(mand) | Trigeminal (mandibular branch) | XI. | Spinal accessory |
> | VI. | Abducens | XII. | Hypoglossal |

_____ a. Most extrinsic eye muscles

_____ b. Superior oblique eye muscle

_____ c. Lateral rectus eye muscle

_____ d. Upper eyelid (raises it)

_____ e. Muscles of mastication (chewing), such as temporalis and masseter

_____ f. Muscles that move the tongue

_____ g. Muscles of facial expression, such as smiling, laughing, and frowning, as well as muscles used in speaking, pouting, or sucking

_____ h. Sternocleidomastoid muscle that causes flexion of the head

D. Principal skeletal muscles that act on the abdominal wall, muscles used in breathing, muscles of the pelvic floor and perineum (pages 325–333)

✓▲ **D1.** Each half of the abdominal wall is composed of *(two? three ? four?)* muscles. Describe these in the exercise below.

  a. Just lateral to the midline is the rectus abdominis muscle. Its fibers are *(vertical? horizontal?),* attached inferiorly to the _____ and superiorly to the _____. Contraction of this muscle permits *(flexion? extension?)* of the vertebral column.

  b. List the remaining abdominal muscles that form the sides of the abdominal wall from most superficial to deepest.

    _____ → _____ → _____
       most superficial   deepest

  c. Do all three of these muscles have fibers running in the same direction? *(Yes? No?)* Of what advantage is this?

✓▲ **D2.** Answer these questions about muscles used for breathing.

  a. The diaphragm is _____-shaped. Its oval origin is located _____ _____. Its insertion is not into bone, but rather into dense connective tissue forming the roof of the diaphragm; this tissue is called the _____.

  b. Contraction of the diaphragm flattens the dome, causing the size of the thorax to *(increase? decrease?),* as occurs during *(inspiration? expiration?).*

  c. The name intercostals indicates that these muscles are located _____. Which set is used during expiration? *(Internal? External?)*

D3. *For extra review.* Cover the key to Figure LG 11.2a and write labels for muscles F, G, H, I, and X.

▲ **D4.** Look at the inferior of the human pelvic bones on a skeleton (or refer to Figure 8.9 in your text). Note that a gaping hole (outlet) is present. Pelvic floor muscles attach to the bony pelvic outlet. Name these muscles and state functions of the pelvic floor.

✓▲ **D5.** Fill in the blanks in this paragraph. *Diaphragm* means literally _____ *(dia-)* _____ *(-phragm).* You are familiar with the diaphragm that separates the _____ from the abdomen. The *pelvic diaphragm* consists of all of the muscles of the _____ floor plus their fasciae. It is a "wall" or barrier between the inside and outside of the body. Name the largest muscle of the pelvic floor. _____

E. Principal skeletal muscles that move the shoulder girdle and upper limb (pages 334–352)

✓▲ **E1.** From the list in the box, select the names of all muscles that move the shoulder girdle as indicated. Write one answer on each line provided.

LS.	Levator scapulae	SA.	Serratus anterior
RMM.	Rhomboideus major and minor	T.	Trapezius
PM.	Pectoralis minor		

Superiorly (elevation) _____ _____ _____

Inferiorly (depression) _____ _____

Toward vertebrae (adduction) _____ _____

Away from vertebrae (abduction) _____

✓ **E2.** Note on Figure LG 11.2a and b that the only two muscles listed in Checkpoint El

that are superficial are the _____ and a small portion of the

_____. The others are all deep muscles.

✓ **E3.** On Figure LG 11.2a and b identify the pectoralis major, deltoid, and latissimus dorsi muscles. All three of these muscles are *(superficial? deep?)*. They are all directly involved with movement of the *(shoulder girdle? humerus? radius/ulna?)*

Points of origin or insertion		Actions (of humerus)	
C.	Clavicle	Ab.	Abducts
H.	Humerus	Ad.	Adducts
I.	Ilium	EH.	Extension, hyperextension
RC.	Ribs or costal cartilages	F.	Flexion
Sc.	Scapula		
St.	Sternum		
VS.	Vertebrae and sacrum		

✓▲ **E4.** Fill in letters for *all points of origin and insertion* and *all actions* that apply for each muscle. Numbers in parentheses indicate the number of answers required in each case.

Muscles	Points of Origin or Insertion	Actions (of Humerus)
a. Pectoralis major	_____ (4)	_____ (2)
b. Deltoid	_____ (3)	_____ (3)
c. Latissimus dorsi	_____ (4)	_____ (2)

✓▲ **E5.** *The Big Picture: Looking Back.* Combine your knowledge of muscle actions with your knowledge of movements at joints from Chapter 9. Return to Figure LG 9.2, page LG 166. Write the name(s) of one or two muscles that produce each of the actions (a)-(h). Write muscle names next to each figure.

✓▲ **E6.** Complete the table describing three muscles that move the forearm.

Muscle name	Origin	Insertion	Action on Forearm
a.		Radial tuberosity	
b. Brachialis			
c.			Extension

✓▲ **E7.** Complete this exercise about muscles that move the wrist and fingers.
   a. Examine your own forearm, palm, and fingers. There is more muscle mass on the *(anterior? posterior?)* surface. You therefore have more muscles that can *(flex? extend?)* your wrist and fingers.
   b. Locate the flexor carpi ulnaris muscle on Figure 11.17 your text. What action does it

   have other than flexion of the wrist? _____ What muscles would you expect to abduct the wrist?

   c. What is the difference in location between flexor digitorum superficialis and flexor digitorum profundus?

   d. What muscle helps you to point (extend) your index finger?

✓▲ **E8.** For review of more details of the structures of the hand, do this exercise.
   a. The *(extensor retinaculum? flexor retinaculum?)* is located over the palmar surface of the hand. This ligament secures the position of a number of long tendons of the

   wrist and digits, as well as the _____ nerve. Inflammation of these tissues by repetitive wrist flexion may lead to carpal tunnel syndrome.
   b. The flexor pollicis longus is a muscle that flexes the *(thumb? middle finger? little finger?)*.
   c. Which action of the thumb is movement of the thumb medially across the palm? *(Hint:* Refer to Exhibit 11.18.) *(Abduction? Adduction? Extension? Flexion? Opposition?)* Which action allows you to touch your thumb to the tip your of your little finger?

   _____
   d. Extensor digitorum muscles extend the *(wrist? fingers?)*.

e. Thenar and hypothenar muscles are *(extrinsic? intrinsic?)* muscles of the hand. Thenar muscles form the pronounced muscle mass of the hand that moves the

(thumb? little finger?), whereas hypothenar muscles move the _____.

✓ **E9.** *For extra review* of muscles that move the upper limbs, write the name of one or more muscles that fit these descriptions.

a. Covers most of the posterior of the humerus: _____

b. Turns your hand from palm down to palm up position: _____

c. Originates from upper eight or nine ribs; inserts on scapula; moves scapula laterally:

_____

d. Used when a baseball is grasped: _____

e. Antagonist to serratus anterior: _____

f. Largest muscle of the chest region; used to throw a ball in the air (flex humerus) and

to adduct arm: _____

g. Raises or lowers scapula, depending on which portion of the muscle contracts:

_____

h. Controls action at the elbow for a movement such as the downstroke in hammering

a nail _____

i. Hyperextends the humerus, as in doing the "crawl" stroke in swimming or exerting

a downward blow; also adducts the humerus: _____

F. Principal skeletal muscles that move the vertebral column (pages 353–357)

✓▲ **F1.** Describe the muscles that comprise the sacrospinalis. Locate and label on Figure LG 11.2b.

a. The sacrospinalis muscle is also called the _____.

b. The muscle consists of three groups: _____ (lateral),

_____ (intermediate), and _____ (medial).

c. In general, these muscles have attachments between _____.

d. They are *(flexors? extensors?)* of the vertebral column, and so are *(synergists? antagonists?)* of the rectus abdominis muscles.

✓ **F2.** *Critical thinking.* Explain why it is common for women in their final weeks of pregnancy to experience frequent back pains.

✓ **F3.** Choose the correct origin and insertion of the scalene muscles:

A. Ribs—iliac crest

B. Cervical vertebrae—occipital and temporal bones

C. Cervical vertebrae—first two ribs

D. Thoracic vertebrae—sacrum

G. Principal skeletal muscles that move the lower limb (pages 358–373)

G1. Cover the key in Figure LG 11.2a and b and identify by size, shape, and location the major muscles that move the lower limb.

✓▲ **G2.** Now match muscle names in the box with their descriptions below.

Ad.	**Adductor group**	**Ham.**	**Hamstrings**
Gas.	**Gastrocnemius**	**Il.**	**Iliopsoas**
GMax.	**Gluteus maximus**	**QF.**	**Quadriceps femoris**
GMed.	**Gluteus medius**	**Sar.**	**Sartorius**

_____ a. Consists of four heads: rectus femoris and three vastus muscles (lateralis, medialis, and intermedius)

_____ b. This muscle mass lies in the posterior (flexor) compartment of the thigh; antagonist to quadriceps femoris

_____ c. Attached to lumbar vertebrae, anterior of ilium, and lesser trochanter, it crosses anterior to hip joint

_____ d. Large muscle mass of the buttocks; antagonist to the iliopsoas

_____ e. The only one of these muscles located in the leg (between knee and ankle), it forms the "calf;" attaches to calcaneus by "Achilles tendon"

_____ f. Forms the medial compartment of the thigh; moves femur medially

_____ g. Crossing the femur obliquely, it moves lower extremity into "tailor position"

_____ h. Located posterior to upper, outer portion of the ilium, it forms a preferred site for intramuscular (IM) injections

✓▲ **G3.** Complete the table by marking an X below each action produced by contraction of the muscles listed. (Some muscles will have two or three answers.)

Key to actions in table: Ab. Abduct; Ad. Adduct; EH, extend or hyperextend; F, flex

	Movements of Thigh (Hip Joint)				Movement of Leg (Knee)	
	Ab	Ad	EH	F	EH	F
a. Iliopsoas						
b. Gluteus maximus and medius						
c. Adductor mass						
d. Tensor fasciae latae						
e. Quadriceps femoris						
f. Hamstrings						
g. Gracilis						
h. Sartorius						

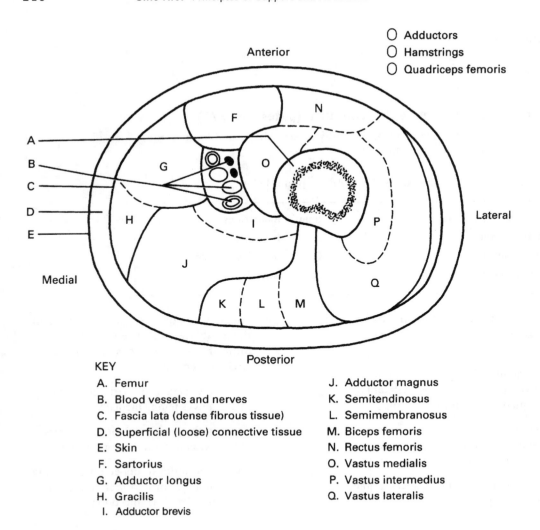

○ Adductors
○ Hamstrings
○ Quadriceps femoris

KEY

A. Femur
B. Blood vessels and nerves
C. Fascia lata (dense fibrous tissue)
D. Superficial (loose) connective tissue
E. Skin
F. Sartorius
G. Adductor longus
H. Gracilis
I. Adductor brevis

J. Adductor magnus
K. Semitendinosus
L. Semimembranosus
M. Biceps femoris
N. Rectus femoris
O. Vastus medialis
P. Vastus intermedius
Q. Vastus lateralis

Figure LG 11.4 Cross section of the right thigh midway between hip and knee joints. Structures A to E are nonmuscle; F to Q are skeletal muscles of the thigh. Color as directed in Checkpoint G4.

✓▲ **G4.** Refer to Figure LG 11.4.
  a. Locate the major muscle groups of the right thigh in this cross section. Color the groups as indicated by color code ovals on the figure.
  b. Cover the key to Figure LG 11.4 and identify each muscle. Relate position of muscles in this figure to views of muscles in Figure LG 11.2a and b.

✓ **G5.** *The Big Picture: Looking Back.* What muscles cause the actions i-p shown in Figure LG 9.2 on page LG 167. Write the muscle names next to each diagram.

✓ **G6.** Perform these actions of your foot and toes. Feel which muscles are contracting. Then match names of actions with descriptions.

DF.	**Dorsiflex**		**F.**	**Flex toes**
Ev.	**Evert foot**		**In.**	**Invert foot**
Ex.	**Extend toes**		**PF.**	**Plantar flex**

_____ a. Jump, as if to touch ceiling

_____ b. Walk around on your heels

_____ c. Curl toes down

_____ d. Lift toes upward away from floor

_____ e. Move sole of foot medially

_____ f. Move sole of foot laterally

✓▲ **G7.** To review details of leg muscles that move the foot, complete this table. Note that muscles within a compartment tend to have similar functions. Muscles with * flex or extend only the great toe (not all toes). Use the same key for foot and toe actions as in the box for Checkpoint G6.

	DF PF	In Ev	F Ex
Posterior compartment: a. Gastrocnemius and soleus b. Tibialis posterior c. Flexor digitorum longus d. Flexor hallucis longus*			
Lateral compartment: e. Peroneus (longus and brevis)			
Anterior compartment: f. Tibialis anterior g. Extensor digitorum longus h. Extensor hallucis longus*			

✓ **G8.** *For extra review* of all muscles, color flexors and extensors using color code ovals on Figure LG 11.2a and b. This activity will allow you to see on which sides of the body most muscles with those actions are located. Omit plantar flexors and dorsiflexors. Note that some muscles are flexors *and* extensors-at different joints. (See Mastery Test question 23 also.)

H. Disorders (page 374)

✓ **H1.** Answer these questions about running injuries.
   a. The most common site of injury for runners is the *(calcaneal tendon? groin? hip? knee?)*.
   b. *Patellofemoral stress syndrome* is a technical term for _____. Briefly describe this problem.

   c. *Shinsplint syndrome* refers to soreness along the *(patella? tibia? fibula?)*.
   d. Write several suggestions you might make to a beginning runner to help to avoid runners' injuries.

   e. Initial treatment of sports injuries usually calls for *RICE* therapy. To what does RICE therapy refer? R _____ I _____

   C _____ E _____

H2. Write three or more possible causes of compartment syndrome. Then explain how this syndrome may lead to contractures.

ANSWERS TO SELECTED CHECKPOINTS: CHAPTER 11

A1. Skeletal muscle tissues and connective tissues.

A2. (a) Lever, fulcrum. (b) Resistance. (c) Muscle; effort. (d) Proximal, radius. (e) Origin, scapula. (f) See Figure LG 11.1A; third.

A3.

Figure LG 11.1A The lever-fulcrum principle is illustrated by flexion of the forearm.

Class	Analogy of EFR	Arrangement	Mechanical advantage: (Yes, No). Explain.	Example in human body
a. First	Seesaw	E F R	Yes if R is closer to F / No if E is closer to F	Head and neck: atlanto-occipital joint
b. Second	Wheelbarrow	F R E	Yes since R is closer (than E) to F	Ankle: gastro-cnemius to calcaneus
c. Third	(Omit)	F E R	No since E is closer (than R) to F	Elbow joint: biceps brachii

A4. (a) R, E; atlas, occipital; first. (b) Less, more, closer to; A.

A5. (a) Many; pennate. (b) Long; parallel.

A6. (a) Prime mover (agonist); biceps brachii. (b) An antagonist. (c) Your forearm would stay in the flexed position. (d) None: each opposing muscle would negate the action of the other. (e) Synergists.

B1. (a) D. (b) E. (c) C. (d) B. (e) A. (f) F.

B2. (b) A, P, L. (c) N, L. (d) P. (e) A, S.

C1. (a) Frontalis portion of epicranius. (b) Orbicularis oris. (c) Zygomaticus major. (d) Orbicularis oculi.

C2. Facial, VII.

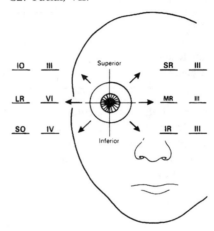

Figure LG 11.3A Right eyeball with arrows indicating directions of eye movements produced by the extrinsic eye muscles.

C3. (a) They are outside of the eyeballs, not intrinsic like the iris. (b) and (c) See Figure LG 11.3A. (d) III; lateral.

C4. No. The left eye contracts its medial rectus, while the right eye uses its lateral rectus and exerts some tension upon both oblique muscles.

C5. (a) Elevating the mandible; temporalis, masseter. (b) Close; lateral pterygoid. (c) V, trigeminal (d) Chewing, facial expression.

C6. (a) Posterior and superior. (b) Inferior. (c) Anterior and inferior.

C7. Same answers as for C6.

C8. XII, hypoglossal.

C9. (a) B. (b) A.

C10. Opposite flexes.

C11. Extending and then hyperextending; posterior; longissimus capitis, semispinalis capitis, and splenius capitis.

C12. (a) III. (b) IV: remember SO4. (c) VI. (d) III. (e) V (mand). (f) XII. (g) VII. (h) XI.

D1. Four. (a) Vertical, pubic crest and symphysis pubis, ribs 5 to 7; flexion. (b) External oblique → internal oblique → transversus abdominis. (c) No; strength is provided by the three different directions.

D2. (a) Dome; around the bottom of the rib cage and on lumbar vertebrae; central tendon. (b) Increase, inspiration. (c) Between ribs (costa); internal.

D5. Across, wall; thorax; pelvic; levator ani (pubococcygeus and iliococcygeus).

E1. (a) LS, RMM, upper fibers of T. (b) PM, lower fibers of T. (c) RMM, T. (d) SA.

E2. Trapezius, serratus anterior.

E3. Superficial; humerus.

E4. (a) C, St, RC, H; Ad, F. (b) C, H, Sc; Ab, EH (posterior fibers), F (anterior fibers). (c) I, RC, VS, H; Ad, EH.

E5. (a) Sternocleidomastoid. (b) Three capitis muscles. (c) Pectoralis major, coracobrachialis, deltoid (anterior fibers), biceps brachii. (d) Latissimus dorsi, teres major, deltoid (posterior fibers), triceps brachii. (e) Deltoid and supraspinatus. (f) Pectoralis major, latissimus dorsi, teres major. (g) Biceps brachii, brachialis, brachioradialis. (h) Triceps brachii.

E6.

Muscle name	Origin	Insertion	Action on Forearm
a. Biceps brachii	Scapula (2 sites)	Radial tuberosity (Anterior)	Flexion, supination (*Note:* also flexes humerus)
b. Brachialis	Anterior humerus	Ulna (coronoid process)	Flexion
c. Triceps brachii	Scapula and 2 sites on posterior of humerus	Posterior of Ulna (olecranon)	Extension (also extension of humerus)

E7. (a) Anterior; flex. (b) Adducts wrist; those lying over radius, such as flexor and extensor carpi radialis. (c) Superficialis is more superficial, and profundus lies deep. (d) Extensor indicis.

E8. (a) Flexor retinaculum; median. (b) Thumb. (c) Flexion; opposition. (d) Fingers (digits). (e) Intrinsic; thumb, little finger.

E9. (a) Triceps brachii. (b) Supinator and biceps brachii. (c) Serratus anterior. (d) Flexor digitorum superficialis and profundus. (e) Trapezius. (f) Pectoralis major. (g) Trapezius. (h) Triceps brachii. (i) Latissimus dorsi.

F1. (a) Erector spinae. (b) Iliocostalis, longissimus, spinalis. (c) Hipbone (ilium), ribs and vertebrae. (d) Extensors, antagonists.

F2. Extra weight of the abdomen demands extra support (contraction) by sacrospinalis muscles.

F3. C.

G2. (a) QF. (b) Ham. (c) Il. (d) GMax. (e) Gas. (f) Ad. (g) Sar. (h) Gmed.

G3.

	Movements of Thigh (Hip Joint)				Movement of Leg (Knee)	
	Ab	Ad	EH	F	EH	F
a. Iliopsoas				X		
b. Gluteus maximus and medius			X			
c. Adductor mass		X		X		
d. Tensor fasciae latae	X			X		
e. Quadriceps femoris				X	X	
f. Hamstrings			X			X
g. Gracilis		X				X
h. Sartorius	X			X		X

G4. (a) Adductors: G H I J; Hamstrings: K L M; Quadriceps: N O P Q. (b) See key to Figure LG 11.4.

G5. (i) Iliacus + psoas (iliopsoas), rectus femoris, adductors, sartorius; (j) Gluteus maximus, hamstrings, adductor magnus (posterior portion); (k) Hamstrings, gracilis, sartorius, gastrocnemius; (l) Quadriceps femoris; (m) Same as i but bilateral; (n) Same as j but bilateral; (o) Tensor fasciae latae, gluteus (medius and minimus), piriformis, superior and inferior gemellus, and obturator internus; (p) Adductors (longus, magnus, and brevis), pectineus, quadratus femoris, and gracilis.

G6. (a) PF. (b) DF. (c) F. (d) Ex. (e) In. (f) Ev.

G7.

	DF	PF	In	Ev	F	Ex
Posterior compartment:						
a. Gastrocnemius and soleus		X				
b. Tibialis posterior		X	X			
c. Flexor digitorum longus		X	X		X	
d. Flexor hallucis longus*		X	X		X	
Lateral compartment:						
e. Peroneus (longus and brevis)		X		X		
Anterior compartment:						
f. Tibialis anterior	X		X			
g. Extensor digitorum longus	X			X		X
h. Extensor hallucis longus*	X			X		X

G8. Figure LG 11.2a: flexors: G, J, K, S, T (anterior portion), U, V, W, AA, BB, EE; extensors: BB, CC, DD. Figure LG 11.2b: flexors: J, K, S; extensors: F, G, H, I, J, K, L, M (posterior fibers), N, O, R.

H1. (a) Knee. (b) "Runner's knee"; the patella tracks (glides) laterally, causing pain. (c) Tibia. (d) Replace worn shoes with new, supportive ones; do stretching and strengthening exercises; build up gradually; get proper rest; consider other exercise if prone to leg injuries because most (70%) runners do experience some injuries. (e) Rest, ice, compression (such as by elastic bandage), and elevation (of the injured part).

MORE CRITICAL THINKING: CHAPTER 11

1. Explain the kinds of information conveyed about muscles by examining muscle names. Give at least six examples of muscle names that provide clues to muscle locations, actions, size or shape, direction of fibers, numbers of origins, or points of attachment.
2. Identify types of movements required at shoulder, elbow, and wrist as you perform the action of tossing a tennis ball upward. Indicate groups of muscles that must contract and relax to accomplish this action.
3. Perform the actions of abducting and adducting your right thigh. Contrast the amount of muscle mass contracting to effect each of these movements. Name several muscles contracting as you carry out each action.

MASTERY TEST: CHAPTER 11

Questions 1-2: Arrange the answers in correct sequence.

_____ _____ _____ 1. Abdominal wall muscles, from superficial to deep:
    A. Transversus abdominis
    B. External oblique
    C. Internal oblique

_____ _____ _____ 2. From superior to inferior in location:
    A. Sternocleidomastoid
    B. Pelvic diaphragm
    C. Diaphragm and intercostal muscles

Questions 3-12: Circle T (true) or F (false). If the statement is false, change the underlined word or phrase so that the statement is correct

T F   3. In extension of the thigh the hip joint serves as the <u>fulcrum (F)</u>, while the <u>hamstrings and gluteus maximus</u> serve as the effort (E).
T F   4. The wheelbarrow and gastrocnemius are both examples of the action of <u>second</u>-class levers.
T F   5. The hamstrings are <u>antagonists</u> to the quadriceps femoris.
T F   6. The name deltoid is based on the <u>action</u> of that muscle.
T F   7. The most important muscle used for normal breathing is the <u>diaphragm</u>.
T F   8. The insertion end of a muscle is the attachment to the bone that <u>does move</u>.
T F   9. In general, adductors (of the arm and thigh) are located more on the <u>medial</u> than on the <u>lateral</u> surface of the body.
T F   10. Both the <u>pectoralis major and latissimus dorsi muscles extend</u> the humerus.
T F   11. The capitis muscles (such as splenius capitis) are <u>extensors</u> of the head and neck.
T F   12. The biceps brachii and biceps femoris are both muscles with <u>two heads of origin located on the arm</u>.

Questions 13-20: Circle the letter preceding the one best answer to each question.

13. All of these muscles are located in the lower limb *except:*
    A. Hamstrings   D. Deltoid
    B. Gracilis   E. Peroneus longus
    C. Tensor fasciae latae

14. All of these muscles are located on the anterior of the body *except:*
    A. Tibialis anterior   D. Pectoralis major
    B. Rectus femoris   E. Rectus abdominis
    C. Sacrospinalis

15. Contraction of all of the following muscles causes extension of the leg *except:*
    A. Rectus femoris   D. Vastus intermedius
    B. Biceps femoris   E. Vastus lateralis
    C. Vastus medialis

16. All of these muscles are attached to ribs *except:*
    A. Serratus anterior   D. Iliocostalis
    B. Intercostals   E. External oblique
    C. Trapezius

17. All of these muscles are directly involved with movement of the scapulae except:
    A. Levator scapulae   D. Rhomboideus major
    B. Pectoralis major   E. Serratus anterior
    C. Pectoralis minor
18. All of these muscles have attachments to the hip bones *except:*
    A. Adductor muscles
       (longus, magnus, brevis)
    B. Biceps femoris   D. Vastus medialis
    C. Rectus femoris   E. Latissimus dorsi

19. The masseter and temporalis muscles are used for:
    A. Chewing   D. Depressing tongue
    B. Pouting   E. Elevating tongue
    C. Frowning
20. All of these muscles are used for facial expression *except:*
    A. Zygomaticus major   D. Rectus abdominis
    B. Orbicularis oculi   E. Mentalis
    C. Platysma

Questions 21-25: Fill-ins. Refer to Figure LG 11.2a and b on pages LG 206–207. Write the word or phrase or key letters of muscles that best complete the statement or answer the questions.

_____   21. Muscles G, S, U, V, and W on Figure LG 11.2a all have in common the fact that they carry out the action of _____.

_____   22. Muscles G, N, 0, and R on Figure LG 11.2b all have in common the fact that they carry out the action _____.

_____   23. If you colored all flexors red and all extensors blue on these two figures, the view of the _____ surface of the body would appear more blue.

_____   24. Choose the letters of all of the muscles listed below that would contract as you raise your left arm straight in front of you, as if to point toward a distant mountain: Figure LG 11.2a: D T U V; Figure LG 11.2b: N O P.

_____   25. Choose the letters of all of the muscles listed below that would contract as you raise your knee and extend your leg straight out in front of you, as if you are starting to march off to the distant mountain: Figure LG 11.2a: AA BB CC DD; Figure LG 11.2b; I J K R.

ANSWERS TO MASTERY TEST: CHAPTER 11

Arrange
1. B C A
2. A C B

True-False
3. T
4. T
5. T
6. F. Shape
7. T
8. T
9. T
10. F. The latissimus dorsi extends, but the pectoralis major flexes.
11. T
12. F. Two heads of origin; but origins of biceps brachii are on the scapula, and origins of biceps femoris are on ischium and femur.

Multiple Choice
13. D
14. C
15. B
16. C
17. B
18. D
19. A
20. D

Fill-ins
21. Flexion
22. Extension
23. Posterior
24. Figure LG 11.2a: T (anterior fibers) U V; Figure LG 11.2b: none.
25. Figure LG 11.2a: AA BB CC DD; Figure LG 11.2b: I.

FRAMEWORK 12
Nervous Tissue

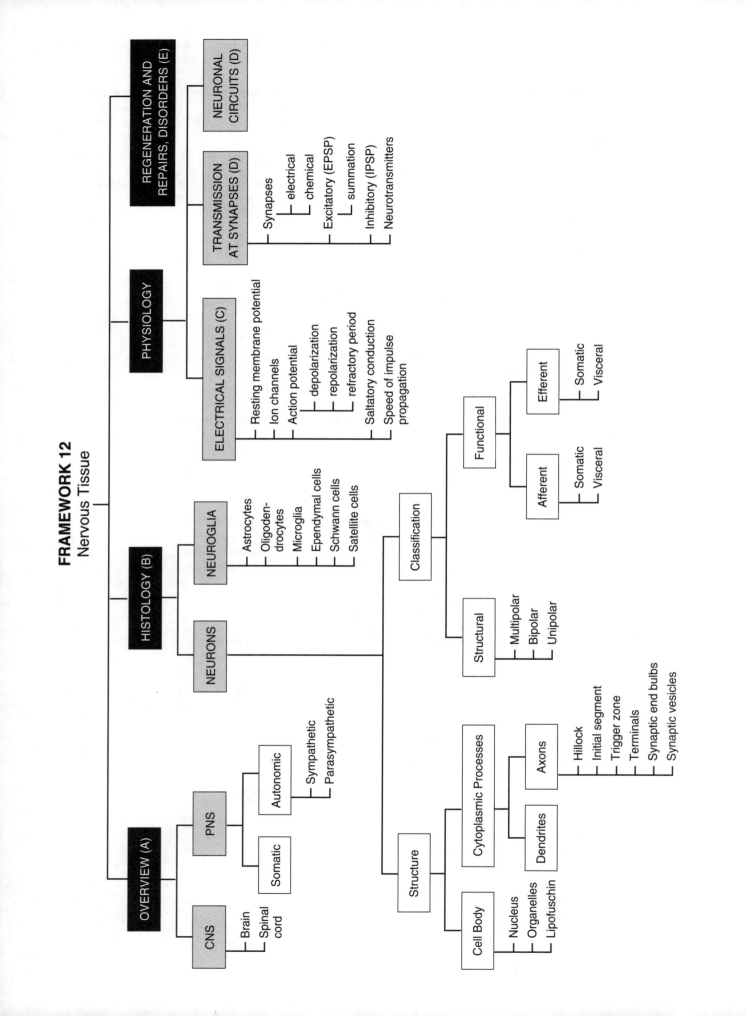

REGENERATION AND REPAIRS, DISORDERS (E)

NEURONAL CIRCUITS (D)

TRANSMISSION AT SYNAPSES (D)
- Synapses
  - electrical
  - chemical
- Excitatory (EPSP)
  - summation
- Inhibitory (IPSP)
- Neurotransmitters

PHYSIOLOGY

ELECTRICAL SIGNALS (C)
- Resting membrane potential
- Ion channels
- Action potential
  - depolarization
  - repolarization
  - refractory period
- Saltatory conduction
- Speed of impulse propagation

HISTOLOGY (B)

NEUROGLIA
- Astrocytes
- Oligoden-drocytes
- Microglia
- Ependymal cells
- Schwann cells
- Satellite cells

NEURONS

Classification
- Structural
  - Multipolar
  - Bipolar
  - Unipolar
- Functional
  - Afferent
    - Somatic
    - Visceral
  - Efferent
    - Somatic
    - Visceral

Structure
- Cell Body
  - Nucleus
  - Organelles
  - Lipofuschin
- Cytoplasmic Processes
  - Dendrites
  - Axons
    - Hillock
    - Initial segment
    - Trigger zone
    - Terminals
    - Synaptic end bulbs
    - Synaptic vesicles

OVERVIEW (A)

CNS
- Brain
- Spinal cord

PNS
- Somatic
- Autonomic
  - Sympathetic
  - Parasympathetic

Nervous Tissue

Two systems—nervous and endocrine—are responsible for regulating our diverse body functions. Each system exerts its control with the help of specific chemicals, namely neurotransmitters (nervous system) and hormones (endocrine system). In Unit III both systems of regulation will be considered, starting with the tissue of the nervous system.

Nervous tissue consists of two types of cells: neurons and neuroglia. The name neuroglia (*glia* = glue) offers a clue as to function of these cells: They bind, support, and protect neurons. Neurons perform the work of transmitting nerve impulses. These long and microscopically slender cells sometimes convey information several feet along a single neuron. Their function relies on an intricate balance between ions (Na^+ and K^+) found in and around nerve cells. Neurons release neurotransmitters that bridge the gaps between adjacent neurons and at nerve—muscle or nerve—gland junctions. Analogous to a complex global telephone system, the nervous tissue of the human body boasts a design and organization that permits accurate communication, coordination, and integration of virtually all thoughts, sensations, and movements.

As you begin your study of nervous tissue, carefully examine the Chapter 12 Topic Outline and Objectives; check off each one as you complete it. To organize your study of nervous tissue, glance over the Chapter 12 Framework now. Be sure to refer to the Framework frequently and note relationships among key terms in each section.

TOPIC OUTLINE AND OBJECTIVES

A. Overview of the nervous system

1. List the structures and describe the basic functions of the nervous system.
2. Describe the organization of the nervous system.

B. Histology of nervous tissue

3. Contrast the histological characteristics and the functions of neurons and neuroglia.
4. Distinguish between gray matter and white matter.

C. Electrical signals in neurons

5. Describe the cellular properties that permit communication among neurons and effectors.

6. Describe the factors that maintain a resting membrane potential.
7. List the sequence of events involved in the generation of a nerve action potential.

D. Signal transmission at synapses; neurotransmitters; circuits

8. Explain the events of signal transmission at a chemical synapse.
9. Distinguish between spatial and temporal summation.
10. Give examples of excitatory and inhibitory neurotransmitters and describe how they act.
11. Describe the classification and functions of neurotransmitters.
12. Identify the various types of neuronal circuits in the nervous system.

E. Regeneration and repair of nervous tissue; disorders

13. Describe plasticity and neurogenesis.
14. Describe the events in damage and repair of peripheral nerves.

WORDBYTES

Now become familiar with the language of this chapter by studying each wordbyte, its meaning, and an example of its use within a term. After you study the entire list, self-check your understanding by writing the meaning of each wordbyte on the line. As you continue through the *Learning Guide,* identify (and fill in) additional terms that contain the same wordbyte.

Wordbyte	Self-check	Meaning	Example(s)
af-	_____	toward	*af*ferent
astro-	_____	star	*astro*glia
dendr-	_____	tree	*dendr*ite
ef-	_____	away from	*ef*ferent
-ferent	_____	carried	af*ferent*
-glia	_____	glue	neuro*glia*
lemm-	_____	sheath	neuro*lemm*a
neuro-	_____	nerve	*neuro*n
olig-	_____	few	*olig*odendrocytes
saltat-	_____	leaping	*saltat*ory
-soma-	_____	body	axo*soma*tic
syn-	_____	together	*syn*apse

CHECKPOINTS

A. Overview of the nervous system (pages 378–381)

A1. Contrast functions of the nervous system and the endocrine system in maintaining homeostasis.

A2. Define *neurology.*

✓▲ **A3.** Match the components of the nervous system in the box with descriptions below.

CN.	**Cranial nerves**	**SC.**	**Spinal cord**
B.	**Brain**	**SN.**	**Spinal nerves**
EP.	**Enteric plexuses**	**SR.**	**Sensory receptors**
G.	**Ganglia**		

_____ a. Largest part of the nervous system; housed within the cranium

_____ b. Twelve pairs of nerves attached to the base of the brain

_____ c. Thirty-one pairs of nerves attached to the spinal cord

_____ d. Structure that extends inferiorly from foramen magnum within the vertebral canal

_____ e. Clusters of neuron cell bodies located outside the brain and spinal cord

_____ f. Networks of neurons within walls of GI organs

_____ g. Dendrites of neurons or complex clusters of cells that detect changes in environment

✓ **A4.** List the three basic functions of the nervous system.

✓ **A5.** Check your understanding of the organization of the nervous system by selecting answers that best fit descriptions below.

Aff.	**Afferent**	**ENS.**	**Enteric nervous system**
ANS.	**Autonomic nervous system**	**PNS.**	**Peripheral nervous system**
CNS.	**Central nervous system**	**SNS.**	**Somatic nervous system**
Eff.	**Efferent**		

_____ a. Brain and spinal cord

_____ b. Sensory nerves

_____ c. Carry information from CNS to skeletal muscles

_____ d. Consists of sympathetic and parasympathetic divisions

_____ e. Nerves that convey impulses to smooth muscle, cardiac muscle, glands, and adipose tissue; involuntary

_____ f. Cranial nerves, spinal nerves, ganglia, and sensory receptors

_____ g. "Brain of the gut"

B. Histology of nervous tissue (pages 381–386)

B1. On Figure LG 12.1, label all structures with leader lines. Next, draw arrows beside the figure to indicate direction of nerve impulses. Then color structures to match color code ovals.

✓ **B2.** Match the parts of a neuron listed in the box with the descriptions below.

A.	**Axon**	**M.**	**Mitochondria**
CB.	**Cell body (soma)**	**NB.**	**Nissl bodies**
D.	**Dendrite**	**NF.**	**Neurofibrils**
L.	**Lipofuscin**	**T.**	**Trigger zone**

_____ a. Contains nucleus; cannot regenerate

_____ b. Yellowish brown pigment that increases with age; appears to be product of lysosomes

_____ c. Provide energy for neurons

_____ d. Long, thin filaments that provide support and shape for the cell

_____ e. Orderly arrangement of rough ER; site of protein synthesis

_____ f. Conducts impulses toward cell body

_____ g. Conducts impulses away from cell body; has synaptic end bulbs that secrete neurotransmitter

_____ h. Located at junction of axon hillock and initial segment; nerve impulses arise here in many neurons

B3. *Clinical correlation.* Contrast the two types of axonal transport and tell which type is involved in the spread of the toxin that causes tetanus.

B4. Name and give a brief description of three types of neurons based on structural characteristics and three types of neurons classified according to functional differences.

Structural   **Description**

_____-polar _____

_____-polar _____

_____-polar _____

Functional   **Description**

_____-fferent _____

_____-fferent _____

_____-fferent _____

Name three types of interneurons.

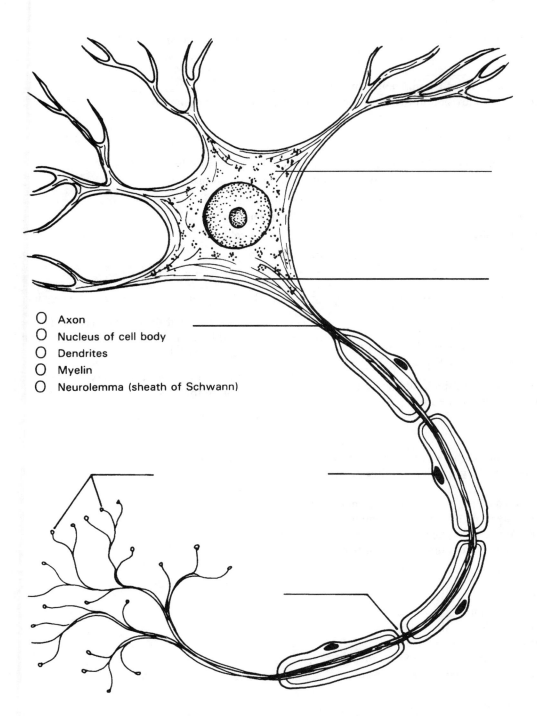

○ Axon
○ Nucleus of cell body
○ Dendrites
○ Myelin
○ Neurolemma (sheath of Schwann)

Figure LG 12.1 Structure of a typical neuron as exemplified by an efferent (motor) neuron. Complete the figure as directed in Checkpoint B1.

✓ **B5.** Refer to Checkpoint B4 above and place asterisks next to the most common:
    a. Structural type of neuron in the brain and spinal cord
    b. Functional type of neuron in the body

✓ **B6.** Write *neurons* or *neuroglia* after descriptions of these cells.
    a. Conduct impulses from one part of the nervous system to another: _____

    b. Provide support and protection for the nervous system: _____

c. Bind nervous tissue to blood vessels, form myelin, and serve phagocytic functions:

_____

d. Smaller in size, but much more abundant in number: _____

e. Can undergo cell division in the mature nervous system, possibly forming tumors:

_____

✓ **B7.** Check your understanding of neuroglia by selecting types of neuroglia in the box that fit descriptions below. Answers may be used once or more than once.

A.	Astrocytes	O.	Oligodendrocytes	
E.	Ependymal cells	SAT.	Satellite cells	
M.	Microglia	SCH.	Schwann	

_____ a. Neuroglia found in the PNS (two answers)

_____ b. Form myelin (two answers)

_____ c. Maintain proper balance of K^+ and Ca^{2+} needed for nerve impulses

_____ d. Help form the blood–brain barrier and provide nutrients to neurons

_____ e. Star-shaped cells

_____ f. Protective because they are phagocytic

_____ g. Line the fluid-filled spaces (ventricles) within the brain, and form cerebrospinal fluid (CSF) there

_____ h. Support neurons in clusters (ganglia) within the PNS

_____ i. Help regenerate PNS axons

✓ **B8.** Check your understanding of coverings over nerve fibers in this Checkpoint.

a. In the _(CNS? PNS?)_, myelination requires Schwann cells.

b. Myelin is formed of _____, whereas the neurolemma is formed of _____.
   A. Up to 100 layers of Schwann cell membrane
   B. The outer nucleated cytoplasmic layer of the Schwann cell

c. Arrange these layers from outermost to innermost: _____ _____ _____
   A. Axon
   B. Neurolemma
   C. Myelin sheath

d. A neurolemma is found _____.
   A. Only around axons in the CNS
   B. Only around axons in the PNS
   C. Around axons in both the CNS and the PNS

e. A neurolemma is involved in the process of _____ of injured axons.

f. Nodes of Ranvier are sections of axons that lack _____.
   Consequently, nerve impulses must "jump" from node to node; this process is called

   _____ conduction. The ultimate effect is that myelin arranged

   between nodes causes a(n) _____-crease in the rate of nerve impulse transmission.
   (Hint. See page 394 in your text.)

g. Do any CNS axons have a neurolemma? _(Yes? No?)_ State one consequence of this fact. (_Hint:_ See page 384 your text.)
   Do any CNS axons have a myelin sheath? _(Yes? No?)_ What cells produce this?

   _____

✓ **B9.** *The Big Picture: Looking Ahead.* Refer to figures in your text to complete this Checkpoint. Write G for gray matter or W for white matter next to related descriptions.

_____ a. Consists mostly of myelinated nerve fibers (axons or dendrites) (Figures 12.3 and 13.10b.)

_____ b. Consists mostly of cell bodies and unmyelinated nerve fibers, such as in ganglia (Figure 17.4)

_____ c. Forms the H-shaped inner portion of the spinal cord as seen in cross section (Figure 13.3a)

_____ d. Forms the outer part of the brain known as the cerebral cortex (Figure 14.11a)

_____ e. Forms nuclei (clusters of neuron cell bodies within the CNS) (Figure 14.9)

_____ f. Forms spinal nerves (Figure 13.16) and cranial nerves (Figure 14.5) in the PNS

C. Electrical signs in neurons (pages 386–396)

✓ **C1.** Complete this Checkpoint describing the properties of the plasma membrane of cells, such as those of muscles or glands, that may be stimulated (excited) by neurons.

a. Excitable cells exhibit a membrane _____ along which an electric

charge (or _____) may pass. In living cells, this current is usually carried by *(electrons? ions?)*.

b. Two types of potential may be developed: _____ potential and

_____ potential. These potentials can develop because of the

presence of _____ channels that open or close to control ion flow. These channels provide the main path for flow of current across the membrane because the phospholipid bilayer normally permits *(little? much?)* passage of ions.

✓ **C2.** Do this exercise about roles of ion channels in development of membrane potentials.
a. There are two kinds of ion channels: *(gated? leakage?)* channels that are always open

and _____ channels that require a stimulus to activate opening or closure. The plasma cell of a membrane is more permeable to *(K+? Na+?)* because there are more leakage channels for *(K+? Na+?)*.

b. Identify the type of gated channel that is responsive in each case described here. Choose from these answers: *(voltage? ligand? mechanically?)* gated channels.

1. Channels in plasma membranes of neurons and muscle cells: _____

2. Channels in auditory receptors stimulated by sound waves in ears: _____

3. Channels responsive to the neurotransmitter acetylcholine: _____

4. Channels in neurons responsive to touch (in skin): _____

✓ **C3.** Contrast direct vs. indirect change in membrane permeability to ions in this Checkpoint.
a. The neurotransmitter acetylcholine *(directly? indirectly?)* alters membrane permeability to ions such as Na^+, Ca^{2+}, or K^+.
b. Name chemicals involved in an *indirect* method to change membrane permeability.

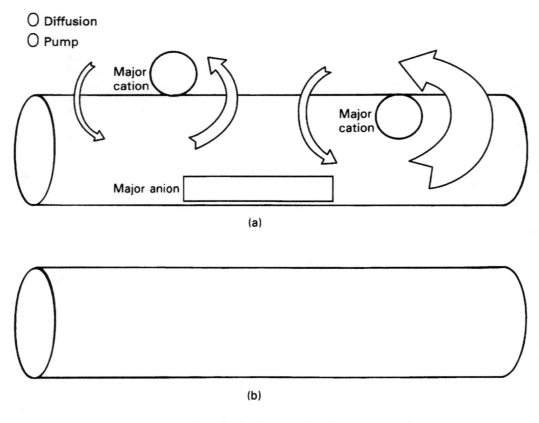

○ Diffusion
○ Pump

Major cation

Major cation

Major anion

(a)

(b)

Figure LG 12.2 Diagrams of a nerve cell. (a) Factors that contribute to membrane potential of a resting neuron. Label and color as directed in checkpoint C4. (b) Complete as directed in Checkpoint C8. ECF–Extracellular fluid; ICF–Intracellular fluid.

✓ **C4.** Refer to Figure LG 12.2a and describe the two principal factors that contribute to resting potential in this Checkpoint.

a. One factor is the unequal distribution of _____ across the membrane. In the circles on the figure write the names of the major cation outside the neuron (in ECF) and the major cation inside the neuron (in ICF). In the rectangle write the name of the major anion in ICF. (*Hint:* refer to text Figure 27.6 and also to Figure LG 3.2A, page 60 in the *Learning Guide.*)

b. Color the two arrows to indicate the second factor—the relative permeability of the

membrane to Na^+ and K^+. The membrane is about _____ times more permeable to K^+ than to Na^+, so K^+ tends to diffuse out much more than Na^+ diffuses in.

c. If K^+ did freely leak out of the membrane that is so permeable to it, the inside of the membrane would become increasingly *(negative? positive?)*, contributing to a *(positive? negative?)* membrane potential.

d. Three other factors that contribute to this potential are inability of large negative ions

such as _____ to exit from the neuron, a slight leakage of _____ into cells, and the Na^+/K^+ pump, which pumps more cations out than it pumps in. Specifically, *(three? two?)* Na+ are pumped out for every *(three? two?)* K^+ pumped in. Show this by coloring "pump" arrows on Figure LG 12.2a.

e. As a result of factors described in (c) and (d), the membrane potential of a resting

(nonactive) neuron is about _____ mV, or a range of _____ to _____. This means that the *(entire neuron? part of a neuron right next to the membrane?)* is much more *(positive? negative?)* than the fluid immediately outside the membrane.

✓ **C5.** Describe *graded potentials* in this exercise.

    a. Graded potentials are so named because they vary in amplitude (_____)

       depending on the strength of the _____. The size of the potentials varies *(directly? indirectly?)* with the number of gated ion channels open or closed as well as the amount of time they are open.

    b. These channels alter flow of _____ across the membrane. If the flow of ions makes the membrane more positive (such as –70 to –60 mV), it is said to be *(de? hyper?)*-polarized. If the potential becomes positive enough (such as –70 to –55 mV) to reach threshold (described below), an action potential *(is? is not?)* likely to occur. If the membrane potential becomes more negative (such as –70 to –80 mV),

       the membrane is said to be _____ polarized, a point farther away from starting an action potential (nerve impulse).

    c. Graded potentials have names based on the type of stimulus applied. For example, neurotransmitters affecting receptors of chemically gated ion channels at synapses lead to a post- _____. Sensory receptors respond to stimuli by pro-

       ducing graded potentials called _____ potentials, or _____ potentials.

    d. Graded potentials most often occur in *(dendrites or cell bodies? axons?)*. These potentials produce currents along *(just a few micrometers? the entire length?)* of a neuron, whereas action potentials produce currents along *(short distances? short or long distances?)* of a neuron such as along the entire length of axons.

✓ **C6.** Check your understanding of an action potential (nerve impulse) in this Checkpoint.

    a. A stimulus causes the nerve cell membrane to become *(more? less?)* permeable to Na^+.

       Na^+ can then enter the cell as voltage-gated Na^+ _____ become activated and open. Notice from Figure 12.12 in your text that Na^+ channels have two gates, much like a double door system. Which gate is located in the outer portion of the neuron cell membrane? *(Activation? Inactivation?)* gate. This gate opens in response to the stimulus.

    b. At rest, the membrane had a potential of _____ mV. As Na^+ enters the cell, the inside of the membrane becomes more *(positive? negative ?)*. The potential will tend to go toward *(–80? –55?)* mV. The process of *(polarization? depolarization?)* is occurring. This process causes changes in more Na+ channels so that even more Na^+ enters. This is an example of a *(positive? negative?)* feedback mechanism. The result is a nerve

       impulse or nerve _____.

    c. The membrane is completely depolarized at exactly *(–50? 0? +30?)* mV. Na^+ channels stay open until the inside of the membrane potential is *(reversed? repolarized?)* at +30 mV. Then inactivation gates close just milliseconds after the activation gates had opened, preventing more inward flow of Na^+.

    d. After a fraction of a second, K^+ voltage-gated channels at the site of the original stimulus open. K^+ is more concentrated *(outside? inside?)* the cell (as you showed on Figure 12.2a); therefore; K^+ diffuses *(in? out?)*. This causes the inside of the membrane to

       become more negative and return to its resting potential of _____ mV. The process is known as *(de? re ?)*-polarization. In fact, outflow of K^+ may be so great that

       _____-polarization occurs in which membrane potential becomes closer to *(–50? –90?)* mV.

    e. During depolarization, the nerve cannot be stimulated at all. This period is known as the *(absolute? relative?)* refractory period. *(Large? Small?)*-diameter axons have a longer absolute refractory period, with this about *(0.4? 4? 40?)* msec. In other words, slower impulses are likely to occur along *(large? small?)*-diameter neurons. Only a stronger-

than-normal stimulus will result in an action potential during the _____

refractory period, which corresponds roughly with *(de? re?)*-polarization.

f. The nerve impulse is propagated (or _____) along the nerve, as
adjacent areas are depolarized, causing more channels to be activated and more
(Na⁺? K⁺?) to enter.

g. What effect do anesthetics such as procaine (Novocaine) have on action potentials?

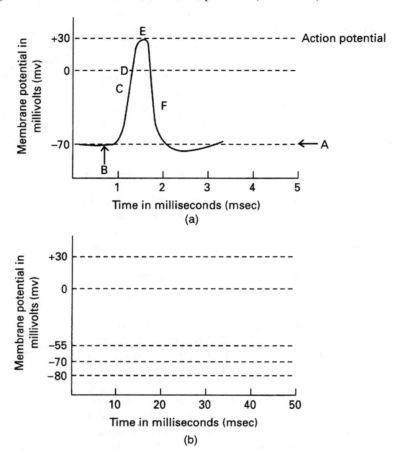

Figure LG 12.3 Diagrams for showing nerve action potentials. (a) Identify letter labels
in Checkpoint C7. (b) Complete as directed in Checkpoint D2(b) and (d).

✓ **C7.** Write the correct letter label from Figure LG 12.3a next to each description.

_____ a. The stimulus is applied at this point.

_____ b. Resting membrane potential is at this
level.

_____ c. Membrane becomes so permeable to
K⁺ that K⁺ diffuses rapidly out through
K⁺ channels.

_____ d. The membrane is becoming more positive
inside as Na⁺ enters; its potential is −30
mV. The process of depolarization is
occurring.

_____ e. The membrane is completely depolarized
at this point.

_____ f. The membrane is repolarizing at this
point.

_____ g. Reversed polarization occurs; enough Na⁺
has entered so that the region just inside
the membrane at this part of the cell is
more positive inside than outside.

C8. *For extra review.* Show the events that occur during initiation and propagation of a nerve impulse on Figure LG 12.2b. Compare your diagram to Figures 12.12 and 12.13 in the text.

C9. Explain how the all-or-none principle resembles a domino effect.

✓ **C10.** Describe how myelination, fiber thickness, and temperature affect speed of impulse propagation.

   a. Saltatory conduction occurs along *(myelinated? unmyelinated?)* nerve fibers. Saltatory transmission is *(faster? slower?)* and takes *(more? less?)* energy than continuous conduction. Explain why.

   b. Type A fibers have the *(largest? smallest?)* diameter and *(are? are not?)* myelinated. These fibers conduct impulses *(rapidly? slowly?)*. Give two examples of A fibers.

   c. Type C fibers have the *(largest? smallest?)* diameter and *(are? are not?)* myelinated. These fibers conduct impulses *(rapidly? slowly?)*. Give two examples of C fibers.

   d. *A clinical correlation. (Warm? Cool?)* nerve fibers conduct impulses faster. How can this information be applied clinically?

✓ **C11.** Contrast action potentials of nerve and muscle by completing this table.

Tissue	Typical Resting Membrane Potential	Duration of Nerve impulse
a. Neuron		
b. Muscle		

C12. In Checkpoint C5 we considered graded potentials. Contrast *graded potentials* with *action potentials* by completing this table.

Characteristic	Graded Potential	Action Potential
a. Amplitude		All-or-none potential, usually about 100 mV
b. Duration	Long; may last several minutes	
c. Types of ion channels		Only voltage-gated
d. Location where potential arises	Usually dendrites or cell bodies	
e. Propagation		Long distances
f. Refractory period	None, so summation may occur	

D. Signal transmission at synapses; neurotransmitters; circuits (pages 396–404)

✓ **D1.** Do this activity about types of synapses.

   a. In Chapter 10, you studied the point at which a neuron comes close to contacting a

      muscle. This is known as a _____ junction, shown in Figure LG 10.2.

      The minute space between two neurons is known as a _____.

   b. There are two types of synapses. A(n) *(chemical? electrical?)* synapse is designed to allow spread of an ionic current from one neuron to the next, in fact, in both direc-

      tions. These synapses consist of _____ junctions made of hundreds

      of proteinaceous tunnels called _____. Gap junctions are common in *(skeletal? smooth* and *cardiac?)* muscle where a group of muscle fibers works in synchrony.

   c. In chemical synapses membranes of the two cells *(do? do not?)* touch. Does a nerve impulse actually "jump" across the synaptic cleft? Explain.

   d. Now summarize the sequence of events at a chemical synapse by placing these

      events in order from first to last: _____ → _____ → _____

      A. Electrical signal: postsynaptic potential (graded potential; if threshold level, action potential results)

      B. Electrical signal: presynaptic potential (nerve impulse)

      C. Chemical signal: release of neurotransmitter into synaptic cleft

e. Write these structures in order to summarize the anatomical pathway of a chemical synapse.

EB.	End bulb of presynaptic neuron	SC.	Synaptic cleft
N.	Neurotransmitter	SV.	Synaptic vesicle
NR.	Neurotransmitter receptor		

_____ → _____ → _____ → _____ → _____
First   Last

f. Explain the role of calcium ions (Ca^{2+}) in nerve transmission at chemical synapses.

g. The rationale for *one-way information transfer* at chemical synapses is that only synaptic end bulbs of *(pre? post?)*-synaptic neurons release neurotransmitters, and only *(pre? post?)*-synaptic neurons have receptors.

✓ **D2.** Complete the following exercise about postsynaptic potentials.

a. If a neurotransmitter causes depolarization of a postsynaptic neuron, for example, from –70 to –65 mV, the postsynaptic potential (PSP) is called *(excitatory? inhibitory?)*; in other words, it is called an *(EPSP? IPSP?)*. EPSPs result primarily from inflow of *(K^+? Na^+? Ca^{2+}?)* through chemically gated chemicals.

b. Usually a single EPSP within a single neuron *(is? is not?)* sufficient to cause a threshold potential and initiate a nerve impulse. Instead, a single EPSP can cause *(partial? total?)* depolarization. Diagram a partial depolarization from –70 to –65 mV in green on Figure LG 12.3b.

c. If neurotransmitters are released from a number of presynaptic end bulbs at one

time, their combined effect may produce threshold EPSP of about _____ mV. This phenomenon is known as *(spatial? temporal?)* summation. Temporal summation is that due to accumulation of transmitters from *(one? many?)* presynaptic end bulb(s) over a period of time.

d. If a neurotransmitter causes inhibition of the postsynaptic neuron, the process is known as *(de? hyper?)*-polarization, and the PSP is *(excitatory? inhibitory?)*. IPSPs often result from influx of *(Cl^-? K^+? Na^+?)* and/or outflow of *(Cl^-? K^+? Na^+?)* through gated channels. An example of an IPSP would be change in the membrane PSP from –70 to *(–60? –80?)* my. Diagram such an IPSP in red on Figure 12.3b.

e. In other words, a neuron with an IPSP is *(closer? farther?)* from threshold than a neuron at resting membrane potential (RMP), so a cell with an IPSP is less likely to have an action potential. An example of use of an IPSP is inhibition of your triceps

brachii muscle as the _____ is stimulated (with an EPSP) to contract, or innervation of the heart by the vagus nerve (CN X) which *(in? de?)*-creases heart rate.

✓ **D3.** *A clinical correlation.* Once a neurotransmitter completes its job, it must be removed from the synaptic cleft. Describe two mechanisms for getting rid of these chemicals. Notice the consequences of alterations of these mechanisms.

a. Inactivation of a neurotransmitter may occur, such as breakdown of

_____ (ACh) by the enzyme named _____. Certain
drugs (such as physostigmine) destroy this enzyme, so postsynaptic neurons (or muscles) (*remain? are less?*) activated. Predict the effects of such a drug.

b. A neurotransmitter such as norepinephrine (NE) can be removed from the synapse
by a different mechanism. Describe this method.

c. Explain how the antidepressant Prozac alters transmission at synapses.

✓ **D4.** Answer these questions about different types of neurotransmitters (NTs).
  a. The number of known (or strongly suspected) NTs is about *(5? 30? 100?)*. A *(large?
tiny?)* amount of NT is released at a single synapse.
  b. Which type of NT acts more quickly at a synapse? One that (*opens or closes ion channels in the neuron membrane? acts by a second messenger system that influences reactions
inside the postsynaptic cell?*).
  c. (*All? Not all?*) NTs are made by neurons. Name one type of cell other than a neuron

that can synthesize NTs. _____ Name one type of cell within the

brain that can make hormones. _____

  d. We have already discussed acetylcholine (_____), an NT released at synapses and
neuromuscular junctions. ACh is (*excitatory? inhibitory.*) toward skeletal muscle, but
ACh released from the vagus nerve is (*excitatory? inhibitory?*) toward cardiac muscle.
So when the vagus nerve sends impulses to your heart, your pulse (heart rate)
becomes (*faster? slower?*).
  e. GABA and glycine are both (*excitatory? inhibitory?*) NTs. They act by opening (Cl^-?
Na^+?) channels, leading to IPSPs. The most common inhibitory transmitter in the brain

is (*GABA? glycine?*), whereas _____ is more commonly released by
neurons in the spinal cord.
  f. Strychnine is a chemical that blocks (*GABA? glycine?*) receptors so that muscles are
not properly inhibited (relaxed). Strychnine poisoning is likely to lead to death

because the muscles of the _____ cannot relax so that air that is
high in carbon dioxide cannot be exhaled.

g. List three NTs that are categorized as catecholamines. These NTs are removed from the synapse by (*enzymatic breakdown as by COMT or MAO? reuptake? both of these mechanisms?*).

h. Pharmaceuticals that mimic the body's NTs (or cause NTs to linger in the synapse) are called (*agonists? antagonists?*), whereas drugs that block the action of a NT are called (*agonists? antagonists?*).

✓ **D5.** Match the neurotransmitter with the description that best fits. The first one is done for you. Lines at the end of each description are for Checkpoint D6.

ACh.	Acetylcholine	GABA.	Gamma aminobutyric acid
DA.	Dopamine	Ser.	Serotonin
EED.	Enkephalins, endorphins, and dynorphins	SP.	Substance P

GABA a. The most common inhibitory neurotransmitter in the brain __AA__

_____ b. Transmits pain-related impulses _____

_____ c. Morphinelike chemicals that are the body's natural painkillers; suppress release of substance P _____

_____ d. Concentrated in brainstem neurons, this chemical helps regulate sleep, temperature, and mood _____

_____ e. Made by neurons involved with emotional responses and automatic movements of skeletal muscles _____

_____ f. Produced by neurons that degenerate in Parkinson's disease _____

✓ **D6.** Select answers in the box below to identify the correct category of each of the neurotransmitters and neuromodulators listed in Checkpoint D5. Write answers on lines after each of those descriptions. The first one is done for you.

ACh.	Acetylcholine	B.	Biogenic amines
AA.	Amino acids	N.	Neuropeptides

D7. Answer these questions about opioid peptides such as enkephalins.
a. Briefly explain the relationship between these chemicals and acupuncture.

b. Besides serving as painkillers, enkephalins have several other functions. List four or more of these functions.

✓ **D8.** Match the types of circuits in the box with related descriptions. Answers may be used more than once.

C.	**Converging**	**P.**	**Parallel after-discharge**
D.	**Diverging**	**R.**	**Reverberating**

_____ a. Impulse from a single presynaptic neuron causes stimulation of increasing numbers of cells along the circuit.

_____ b. An example is a single motor neuron in the brain that stimulates many motor neurons in the spinal cord, therefore activating many muscle fibers.

_____ c. Branches from a second and third neuron in a pathway may send impulses back to the first, so the signal may last for hours, as in coordinated muscle activities.

_____ d. One postsynaptic neuron receives impulses from several nerve fibers.

_____ e. A single presynaptic neuron stimulates intermediate neurons which synapse with a common postsynaptic neuron, allowing this neuron to send out a stream of impulses, as in precise mathematical calculations.

E. Regeneration and repair of nervous tissue; disorders (pages 404–406)

E1. Define *plasticity* and list several factors that account for this.

✓ **E2.** Which of the following can regenerate if destroyed? Briefly explain why in each case.
  a. Neuron cell body

  b. CNS nerve fiber

  c. PNS nerve fiber

E3. Discuss methods currently being investigated to promote regrowth of damaged neurons.

✓ **E4.** Answer these questions about nerve regeneration.
  a. In order for a damaged neuron to be repaired, it must have an intact cell body and

  also a _____.

b. Can axons in the CNS regenerate? Explain.

c. When a nerve fiber (axon or dendrite) is injured, the changes that follow in the cell
    body are called *(chromatolysis? Wallerian degeneration?)*. Those that occur in the

    portion of the fiber distal to the injury are known as _____.

✓ **E5.** Discuss multiple sclerosis (MS) in this Checkpoint.

a. This condition involves destruction of _____ surrounding axons.
    This condition most commonly manifests initial symptoms in *(women? men?)* in their
    (20s–30s? 50s–60s?).
b. The name multiple sclerosis refers to the many locations in which myelin is replaced

    by _____. MS *(is? is not?)* progressive and *(does? does not?)* involve
    remissions and relapses.

E7. Check your understanding of epilepsy in this Checkpoint.
a. Define the term epileptic seizure.

b. Most epileptic seizures are *(caused by changes in blood chemistry? idiopathic [of unknown
    cause]?)*. Epilepsy *(usually? almost never?)* affects intelligence.

ANSWERS TO SELECTED CHECKPOINTS: CHAPTER 12

A3. (a) B. (b) CN. (c) SN. (d) SC. (e) G. (f) EP.
    (g) SR.
A4. Sensation, integration, and motor or glandular
    response.
A5. (a) CNS. (b) Aff. (c) SNS or Eff (*Hint:* remember
    S A M E: Sensory = Afferent; *Motor* = *E*fferent).
    (d) ANS. (e) ANS. (f) PNS. (g) ENS.
B1. See Figure LG 12.1A to the right.
B2. (a) CB. (b) L. (c) M. (d) NF (e) NB (f) D. (g) A.
    (h) T.
B5. Multipolar. (b) Association.
B6. (a) Neurons. (b-e) Neuroglia.
B7. (a) SAT, SCH. (b) O, SCH. (c-e) A. (f) M. (g) E.
    (h) SAT. (i) SCH.
B8. (a) PNS. (b) A, B. (c) B C A. (d) B. (e) Regenera-
    tion or growth. (f) Myelin; saltatory; in. (g) No;
    injured CNS nerve fibers do not heal well; yes;
    oligodendrocytes.
B9. (a) W. (b–e) G. (f) W.
C1. (a) Potential, current; ions. (b) Graded, action; ion;
    little.
C2. (a) Leakage, gated; (b1) K$^+$, K$^+$. Voltage. (b2)
    Mechanically. (b3) Ligand. (b4) Mechanically.
C3. (a) Directly. (b) G protein and a second messenger
    system using molecules in cytosol.

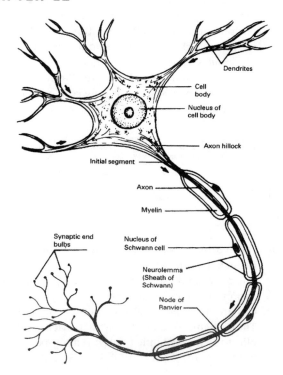

Figure LG 12.1A Structure of a typical neuron as exemplified
by an efferent (motor) neuron.

C4. (a) Ions; see Figure LG 12.2A. (b) 50 to 100. (c) Negative, negative. (d) Anions of proteins, C1; three, two; see Figure LG 12.2A (e) –70, –40 to –90; part of a neuron right next to the membrane, negative.

Figure 12.2aA Diagram of a nerve cell.

C5. (a) Size, stimulus; directly. (b) Ions; de; is; hyper. (c) Synaptic; receptor, generator. (d) Dendrites or cell bodies; just a few micrometers, short or long distances.

C6. (a) More; channels; activation. (b) –70; positive; –55; depolarization; positive; action potential. (c) 0; reversed. (d) Inside, out; –70; re; after-hyper, –90. (e) Absolute; small, 4; small; relative, re. (f) Transmitted, Na⁺. (g) Such anesthetics prevent opening of voltage-gated Na⁺ channels in pain fibers.

C7. (a) B. (b) A. (c) F. (d) C. (e) D. (f) F. (g) E.

C10. (a) Myelinated; faster because the nodes of Ranvier have a high density of voltage-gated Na⁺ channels; less energy because only small regions (the nodes) have inflow of Na⁺, which then must be pumped out by Na⁺/K⁺ pumps. (b) Largest, are; rapidly; large, sensory neurons, such as for touch, position of joints, and temperature, as well as nerves to skeletal muscles for quick reactions. (c) Smallest, are not; slowly; nerves that carry impulses to and from viscera. (d) Warm; ice or other cold applications can slow conduction of pain impulses.

C11.

Tissue	Typical Resting Membrane Potential	Duration of Nerve impulse
a. Neuron	Higher (–70 mV)	Shorter (0.5–2 msec:
b. Muscle	Lower (–90 mV)	Longer (1.0–5.0 msec: skeletal muscle; 10–300 msec: cardiac and smooth muscle)

D1. (a) Neuromuscular; synapse. (b) Electrical; gap, connexons; smooth and cardiac. (c) Do not; no; The nerve impulse travels along the presynaptic axon to the end of the axon, where synaptic vesicles are triggered to release neurotransmitters. These chemicals enter the synaptic cleft and then affect the postsynaptic neuron (either excite it or inhibit it). (d) B CA. (e) EB SV N SC NR. (f) The

nerve impulse in the presynaptic neuron opens Ca²⁺ channels, creating an influx of Ca²⁺ into the presynaptic end bulbs. The Ca²⁺ here triggers exocytosis of synaptic vesicles with release of neurotransmitters. (g) Pre, post.

D2. (a) Excitatory, EPSP; Na⁺. (b) Is not; partial; see Figure LG 12.3bA. (c) –55; spatial; one. (d) Hyper, inhibitory; C1, K⁺ –80; see Figure LG 12.3bA. (e) Farther; biceps, de.

Figure LG 12.3bA Diagram of a facilitation EPSP and an IPSP.

D3. (a) Acetylcholine, acetylcholinesterase (ACHase); remain. Such drugs may help activate muscles of persons with myasthenia gravis because many of their neurotransmitter receptors are unavailable; the extra activation of ACh can enhance response of remaining receptors. On the other hand, use of some powerful anticholinesterase drugs (as "nerve gases") cause profound and lethal effects. See page 296 of the text. (b) NE can be recycled into the presynaptic neuron. (c) Prozac inhibits recycling (or reuptake) of the neurotransmitter serotonin, which stays longer in the synapse, affecting postsynaptic action potentials that lower depression.

D4. (a) 100: tiny. (b) Opens or closes ion channels in the neuron membrane. (c) Not all; endocrine cells; neurosecretory. (d) ACh: excitatory, inhibitory; slower. (e) Inhibitory; C1⁻; GABA, glycine. (f) Glycine; diaphragm. (g), Norepinephrine (NE), epinephrine (epi), and dopamine (DA); both of these mechanisms. (h) Agonists; agonist.

D5, D6. (b) SP (N). (c) EED (N). (d) Ser (B). (e) DA (B). (f) DA (B).

D8. (a) D. (b) D. (c) R. (d) C. (e) P.

E2. (a) No. About six months after birth, the mitotic apparatus is lost. (b) No. CNS nerve fibers lack the neurolemma necessary for regeneration, and scar tissue builds up by proliferation of astroglia cells. (c) Yes. PNS fibers do have a neurolemma.

E4. (a) Neurolemma. (b) No. They lack a neurolemma. (c) Chromatolysis; Wallerian degeneration.

E5. (a) Myelin; women, 20's and 30's. (b) Hardened scars or plaques; is, does.

MORE CRITICAL THINKING: CHAPTER 12

1. Contrast axons and dendrites with regard to structure and function.
2. If you cut a nerve in your finger, it can regenerate. However, severed axons in the brain or spinal cord cannot heal. Explain why.
3. Contrast the process of myelination of neurons in the CNS and in the PNS.
4. Explain how the structure (thickness and presence or absence of myelination) affects the rate of conduction along "life-or-death" neurons (such as those that detect heat or pain in fingertips) versus the rate along neurons that innervate the intestine.
5. Explain how your sensory systems can detect the difference between a weak handshake and a firm one.
6. Contrast effects of the following neurotransmitters: glutamate, glycine, and catecholamines such as norepinephrine.
7. Describe the steps in repair of peripheral neurons.
8. Describe causes and symptoms of and methods for alleviating epileptic seizures.

MASTERY TEST: CHAPTER 12

Questions 1–2: Arrange the answers in correct sequence.

_____ _____ _____ 1. In order of transmission across synapse, from first structure to last:
   A. Presynaptic end bulb
   B. Postsynaptic neuron
   C. Synaptic cleft

_____ _____ _____ 2. Membrane potential values, from most negative to zero:
   A. Resting membrane potential
   B. Depolarized membrane potential
   C. Threshold potential

Questions 3–10: Circle the letter preceding the one best answer to each question.

3. Choose the one *false* statement.
   A. The membrane of a resting neuron has a membrane potential of –70 mV.
   B. In a resting membrane, permeability to K+ ions is about 100 times less than permeability to Na+ ions.
   C. C fibers are thin, unmyelinated fibers with a relatively slow rate of nerve transmission.
   D. C fibers are more likely to innervate the heart and bladder than structures (such as skeletal muscles) that must make instantaneous responses.
4. All of the following are listed with a correct function *except:*
   A. Schwann cell—myelination of neurons in the PNS
   B. Oligodendrocytes–myelination of neurons in the CNS
   C. Astrocytes—form an epithelial lining of the ventricles of the brain
   D. Microglia—phagocytic
5. Synaptic end bulbs are located:
   A. At ends of axon terminals
   B. On axon hillocks
   C. On neuron cell bodies
   D. At ends of dendrites
   E. At ends of both axons and dendrites

6. Which of these is equivalent to a nerve fiber?
   A. A neuron
   B. A ganglion
   C. An axon or dendrite
   D. A nerve, such as sciatic nerve
7. ACh is an abbreviation for:
   A. Acetylcholine   D. Serotonin
   B. Norepinephrine   E. Inhibitory post-
   C. Acetylcholinesterase   synaptic potential
8. A term that means the same thing as *afferent* is:
   A. Autonomic   D. Motor
   B. Somatic   E. Sensory
   C. Peripheral
9. An excitatory transmitter substance that changes the membrane potential from –70 to –65 mV causes:
   A. Impulse conduction   C. Inhibition
   B. Partial depolarization   D. Hyperpolarization
10. All of these statements are true EXCEPT:
   A. Neuroglia are more numerous than neurons in the human nervous system.
   B. Action potentials can spread across longer distances (such as along long axons) than graded potentials do.
   C. Graded potentials can vary in amplitude, but action potentials do not.
   D. Action potentials do exhibit temporal and spatial summation, whereas graded potentials do not.

Questions 11–20: Circle T (true) or F (false). If the statement is false, change the underlined word or phrase so that the statement is correct.

T F   11. Because CNS fibers contain no neurolemma and the neurolemma produces myelin, CNS <u>fibers are all unmyelinated</u>.

T F   12. Neurotransmitter substances are released at <u>synapses and also at neuromuscular junctions</u>.

T F   13. Generally, release of excitatory transmitter by <u>a single presynaptic end bulb</u> is sufficient to develop an action potential in the postsynaptic neuron.

T F   14. <u>Epilepsy</u> is a condition that involves abnormal electrical discharges within neurons of the brain.

T F   15. In the <u>converging</u> circuit, a single presynaptic neuron influences several postsynaptic neurons (or muscle or gland cells) at the same time.

T F   16. Action potentials are measured in <u>milliseconds, which are thousandths of a second</u>.

T F   17. Nerve fibers with a short absolute refractory period can respond to <u>more rapid</u> stimuli than nerve fibers with a long absolute refractory period.

T F   18. The concentration of potassium ions (K+) is considerably <u>greater</u> inside a resting cell than outside of it.

T F   19. A stimulus that is adequate will temporarily <u>increase permeability of the nerve membrane to Na+</u>.

T F   20. The <u>brain and spinal nerves</u> are parts of the peripheral nervous system (PNS).

Questions 21–25: Fill-ins. Complete each sentence with the word or phrase that best fits.

_____   21. The _____ nervous system consists of the sympathetic and parasympathetic divisions.

_____   22. Application of cold to a painful area can decrease pain in that area because _____.

_____   23. One-way nerve impulse transmission can be explained on the basis of release of transmitters only from the _____ of neurons.

_____   24. The neurolemma is found only around nerve fibers of the _____ nervous system so that these fibers can regenerate.

_____   25. In an axosomatic synapse, an axon's synaptic end bulb transmits nerve impulses to the _____ of a postsynaptic neuron.

ANSWERS TO MASTERY TEST: CHAPTER 12

Arrange

1. A C B
2. A C B

Multiple Choice

3. B
4. C
5. A
6. C
7. A
8. E
9. B
10. D

True–False

11. F. May be myelinated because oligodendrocytes myelinate CNS fibers
12. T
13. F. A number of presynaptic end bulbs
14. T
15. F. Diverging
16. T
17. T
18. T
19. T
20. F. Spinal nerves (as well as some other structures; but not the brain)

Fill-ins

21. Autonomic
22. Cooling of neurons slows down the speed of nerve transmission, for example, of pain impulses
23. End bulbs of axons
24. Peripheral
25. Cell body (soma)

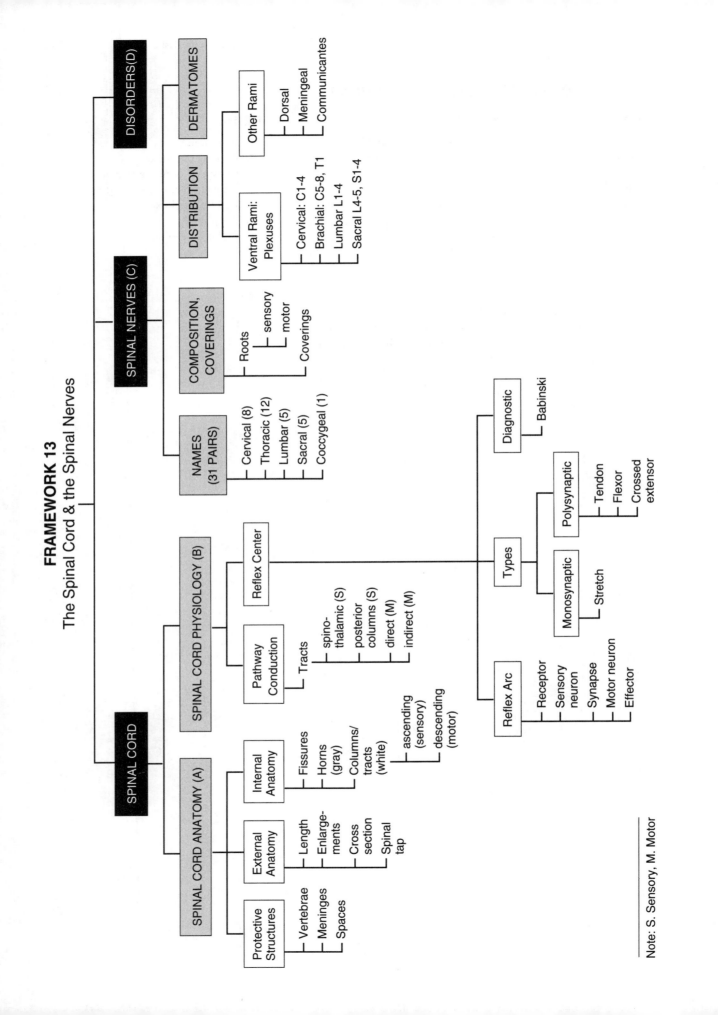

FRAMEWORK 13
The Spinal Cord & the Spinal Nerves

DISORDERS (D)

SPINAL NERVES (C)

- DERMATOMES

- DISTRIBUTION
  - Other Rami
    - Dorsal
    - Meningeal
    - Communicantes
  - Ventral Rami: Plexuses
    - Cervical: C1-4
    - Brachial: C5-8, T1
    - Lumbar L1-4
    - Sacral L4-5, S1-4

- COMPOSITION, COVERINGS
  - Roots
    - sensory
    - motor
  - Coverings

- NAMES (31 PAIRS)
  - Cervical (8)
  - Thoracic (12)
  - Lumbar (5)
  - Sacral (5)
  - Coccygeal (1)

SPINAL CORD

- SPINAL CORD PHYSIOLOGY (B)
  - Reflex Center
    - Diagnostic
      - Babinski
    - Types
      - Polysynaptic
        - Tendon
        - Flexor
        - Crossed extensor
      - Monosynaptic
        - Stretch
    - Reflex Arc
      - Receptor
      - Sensory neuron
      - Synapse
      - Motor neuron
      - Effector
  - Pathway Conduction
    - Tracts
      - spino-thalamic (S)
      - posterior columns (S)
      - direct (M)
      - indirect (M)

- SPINAL CORD ANATOMY (A)
  - Internal Anatomy
    - Fissures
    - Horns (gray)
    - Columns/ tracts (white)
      - ascending (sensory)
      - descending (motor)
  - External Anatomy
    - Length
    - Enlarge-ments
    - Cross section
    - Spinal tap
  - Protective Structures
    - Vertebrae
    - Meninges
    - Spaces

Note: S. Sensory, M. Motor

CHAPTER **13**

The Spinal Cord and Spinal Nerves

The spinal cord and spinal nerves serve as the major links in the communication pathways between the brain and all other parts of the body. Nerve impulses are conveyed along routes (or tracts) in the spinal cord, laid out much as train tracks: some head north to regions of the brain, and others carry nerve messages south from the brain toward specific body parts. Spinal nerves branch off from the spinal cord, perhaps like a series of bus lines that pick up passengers (nerve messages) to or from train depots (points along the spinal cord) en route to their final destinations. Organization is critical in this nerve impulse transportation network. Any structural breakdowns—by trauma, disease, or other disorders—can lead to interruption in service with resultant chaos (such as spasticity) or standstill (such as sensory loss or paralysis).

As you begin your study of the spinal cord and spinal nerves, carefully examine the Chapter 13 Topic Outline and Objectives; check off each one as you complete it. To organize your study of the spinal cord and spinal nerves, glance over the Chapter 13 Framework now. Be sure to refer to the Framework frequently and note relationships among key terms in each section.

TOPIC OUTLINE AND OBJECTIVES

A. Spinal cord anatomy

1. Describe the protective structures and gross anatomical features of the spinal cord.

B. Spinal cord physiology

2. Describe the functions of the principal sensory and motor tracts of the spinal cord.
3. Describe the functional components of a reflex arc and the ways reflexes maintain homeostasis.

C. Spinal nerves

4. Describe the components, connective tissue coverings, and branching of a spinal nerve.
5. Define plexus, and identify the distribution of nerves of the cervical, brachial, lumbar, and sacral plexuses.
6. Describe the clinical significance of dermatomes.
7. Describe the origin, distribution, and effects of damage to the cervical plexus.
8. Describe the origin, distribution, and effects of damage to the brachial plexus.
9. Describe the origin, distribution, and effects of damage to the lumbar plexus.
10. Describe the origin, distribution, and effects of damage to the sacral plexus.

D. Disorders

WORDBYTES

Now become familiar with the language of this chapter by studying each wordbyte, its meaning, and an example of its use within a term. After you study the entire list, self-check your understanding by writing the meaning of each wordbyte on the line. As you continue through the *Learning Guide*, identify (and fill in) additional terms that contain the same wordbyte.

Wordbyte	Self-check	Meaning	Example(s)
arachn-	_____	spider	*arachn*oid
cauda	_____	tail	*cauda* equina
conus	_____	cone	*conus* medullaris
dent-	_____	small tooth	*dent*iculated
dura	_____	hard	*dura* mater
equina	_____	horse	cauda *equina*
mater	_____	mother	dura *mater*
pia	_____	tender	*pia* mater
soma-	_____	body	*soma*tic
-tome	_____	thin section	derma*tome*

CHECKPOINTS

A. Spinal cord anatomy (pages 412–417)

✓ **A1.** List three general functions of the spinal cord.

✓ **A2.** The spinal cord is part of the *(central? peripheral?)* nervous system.

✓▲ **A3.** Refer to Figure LG 13.1 and do the following exercise.
   a. Color the meninges to match color code ovals.
   b. Label these spaces: *epidural space*, *subarachnoid space*, and *subdural* space.
   c. Write next to each label of a space the contents of that space. Choose from these answers: *cerebrospinal (CSF); interstitial fluid; fat and connective tissue.*

   d. Inflammation of meninges is a condition called _____.
   e. The denticulate ligaments are extensions of *(dura mater? arachnoid? pia mater?)* that

   attach laterally to _____ along the length of the cord. State functions of these ligaments.

Blood
vessel —

Spinal
cord
(or brain)

○ Arachnoid
○ Dura mater
○ Pia mater

Figure LG 13.1 Meninges. Color and label as indicated in Checkpoint A3.

✓ **A4.** Do this activity about your own spinal cord.
   a. Identify the location of your own spinal cord. It lies within the vertebral canal, extending from just inferior to the cranium to about the level of your *(waist? sacrum?)*. This level corresponds with about the level of *(L1–L2? L4–L5? S4–S5?)* vertebrae. In other words, the *(spinal cord? vertebral column?)* reaches a lower (more

   inferior) level in your body. The spinal cord is about 42 to 45 cm (_____ inches) in length.
   b. Circle the two regions of your spinal cord that have notable enlargements where nerves exit to your upper and lower limbs:

   cervical   thoracic   lumbar   sacral   coccygeal

✓▲ **A5.** Match the names of the structures listed in the box with descriptions given.

Ca.	**Cauda equina**	F.	**Filum terminale**
Co.	**Conus medullaris**	S.	**Spinal segment**

_____ a. Tapering inferior end of spinal cord

_____ b. Any region of spinal cord from which one pair of spinal nerves arises

_____ c. Nonnervous extension of pia mater; anchors cord in place

_____ d. "Horse's tail"; extension of spinal nerves in lumbar and sacral regions within sub-arachnoid space

✓▲ **A6.** *A clinical correlation.* Answer these questions about meninges and related clinical procedures.
   a. State two or more purposes of a spinal tap (or lumbar puncture).

Into which space does the needle enter? *(Epidural? Subdural? Subarachnoid?)*

b. At what level of the vertebral column (not spinal cord) is the needle for a spinal tap

(or lumbar puncture) inserted? Between _____ vertebrae. Explain
why this location is a relatively safe site for this procedure.

c. Identify level L4 of the vertebral column on yourself by listing two other landmarks

at this level. _____ _____

d. In some cases anesthetics are introduced into the epidural space, rather than into the
subarachnoid space (as in a spinal tap). Explain why an epidural is likely to exert its
effects on nerves with slower and more prolonged action than a spinal tap.

State one example of a procedure for which an epidural may be used. _____

Figure LG 13.2 Outline of the spinal cord, roots, and nerves.
Color and label according to Checkpoints A7 and C1g.

✓▲ **A7.** Identify the following structures on Figure LG 13.2.
a. Gray matter is found *(within? outside?)* the H-shaped outline, and white matter is

located _____ that outline. With a lead pencil, shade the gray matter
on the right side of the figure only.
b. Now label parts A–G on the left side of the figure.
c. On the right side of the figure, label H.

B. Spinal cord: physiology (Pages 417–426)

✓ **B1.** In Checkpoint Al, you listed main functions of the spinal cord. Expand that description in this Checkpoint.
a. One primary function of the spinal cord is to permit _____ between
nerves in the periphery, such as arms and legs, and the brain by means of the *(tracts?
nerves? ganglia?)* located in the white columns of the cord.

b. Another function of the cord is to serve as a _____ center by means of spinal nerves. These are attached by _____ roots. The *(anterior? posterior?)* root

contains sensory nerve fibers, and the _____ root contains motor fibers.

Fasciculus cuneatus
and gracilis

Lateral cortico-
spinal tract

Lateral spino-
thalamic tract

Anterior spino-
thalamic tract

Anterior cortico-
spinal tract

◯ Motor tracts

◯ Sensory tracts

Figure LG 13.3 Major tracts in the spinal cord.
Color and complete as directed in Checkpoints B2, B3, and C1.

✓▲**B2.** Refer to Figure LG 13.3 and do this activity on the conduction function of the
spinal cord.

a. Tracts conduct nerve impulses in the *(central? peripheral?)* nervous system. Function-

ally they are comparable to _____ in the peripheral nervous system.
Tracts are located in *(gray horns? white columns?)*. They appear white because they
consist of bundles of *(myelinated? unmyelinated?)* nerve fibers.

b. Ascending tracts are all *(sensory? motor?)*, conveying impulses between the spinal cord

and the _____. All motor tracts in the cord are *(ascending? descending?)*.

c. Color the five tracts on the figure, using color code ovals to demonstrate sensory or
motor function of tracts. (*Note:* Tracts are actually present on both sides of the cord
but are shown on only one side here.)

d. Write next to the name of each tract its correct functions. Use these answers:

Pain, temperature, crude touch, pressure

Precise voluntary movements

Proprioception, pressure, vibration, and two-point discriminative touch

e. The name lateral corticospinal indicates that the tract is located in the _____
white column, that it originates in the *(cerebral cortex? thalamus? spinal cord?)*, and

that it ends in the _____.

f. The lateral corticospinal tract is *(ascending, sensory? descending, motor?)* and it is a(n) *(extrapyramidal? pyramidal?)* tract.

g. The rubrospinal, tectospinal, and vestibulospinal tracts are *(extrapyramidal?*

pyramidal?) tracts. These are involved with control of: _____.
   A. Precise, voluntary movements
   B. Automatic movements such as maintenance of muscle tone, posture, and equilibrium

h. Now label all structures with leader lines on the right side of the figure.

B3. *For extra review.* Draw and label the following tracts on Figure LG 13.3. Then color them according to the color code ovals on the figure: *anterior spinocerebellar, posterior spinocerebellar, rubrospinal, tectospinal,* and *vestibulospinal.*

O   Motor
O   Sensory

Figure LG 13.4 Reflex arc: stretch reflex.
Color and label as directed in Checkpoints B4 and B5.

✓ **B4.** Checkpoint B1b describes the function of the spinal cord and spinal nerves in reflexes. Color lightly the sensory and motor roots of a spinal nerve on Figure LG 13.4. Select colors according to the color code ovals there.

✓ **B5.** Complete this activity on the function of the spinal cord as a reflex center. Label structures A–H on Figure LG 13.4, using the following terms: *effector, motor neuron axon, motor neuron cell body, receptor, sensory neuron axon, sensory neuron cell body, sensory neuron dendrite,* and *synapse* (integrating center). Note that these structures are lettered in alphabetical order along the conduction pathway of a reflex arc. Add arrows showing the direction of nerve transmission in the arc.

✓ **B6.** Answer these questions about the reflex arc in Figure LG 13.4.

a. How many neurons does this reflex contain? _____ The neuron that conveys the

impulse toward the spinal cord is a _____ neuron; the one that

carries the impulses toward the effector is a _____ neuron.

b. This is a *(monosynaptic? polysynaptic?)* reflex arc. The synapse, like all somatic synapses, is located in the *(CNS? PNS?)*.

c. Receptors, in this case located in skeletal muscle, are called _____.

   They are sensitive to changes in _____. This type of reflex might

   therefore be called a _____ reflex.

d. Because sensory impulses enter the cord on the same side as motor impulses leave, the reflex is called *(ipsilateral? contralateral? intersegmental?)*.

e. What structure is the effector? _____

f. One example of a stretch reflex is the _____, in which stretching of the patellar tendon initiates the reflex.

✓ **B7.** Explain how the brain may get the message that a stretch reflex (or other type of reflex) has occurred.

✓ **B8.** Do this exercise describing how tendon reflexes protect tendons.

a. Receptors located in tendons are named *(muscle spindles? tendon organs?)*. They are sensitive to changes in muscle *(length? tension?)*, as when the hamstring muscles are contracted excessively, pulling on tendons.

b. When this occurs, association neurons cause *(excitation? inhibition?)* of this same muscle, so that the hamstring fibers *(contract further? relax?)*.

c. Simultaneously, other association neurons fire impulses that stimulate *(synergistic? antagonistic?)* muscles (such as the quadriceps in this case). These muscles then *(contract? relax?)*.

d. The net effect of such a *(mono? poly?)*-synaptic tendon reflex is that the tendons are *(protected? injured?)*.

✓ **B9.** Contrast stretch, flexor, and crossed extensor reflexes in this learning activity.

a. A flexor reflex *(does? does not?)* involve association neurons, and so it is *(more? less?)* complex than a stretch reflex.

b. A flexor reflex sends impulses to *(one? several?)* muscle(s), whereas a stretch reflex, such as the knee jerk, activates *(one? several?)* muscle(s), such as the quadriceps. A

   flexor reflex is also known as a _____ reflex, and it is

   _____ -lateral.

c. If you simultaneously contract the flexor and extensor (biceps and triceps) muscles of

   your forearm with equal effort, what action occurs? (Try it.) _____
   In order for movement to occur, it is necessary for extensors (triceps) to be inhibited while flexors (biceps) are stimulated. The nervous system exhibits such control by a

   phenomenon known as _____ innervation.

d. Reciprocal innervation also occurs in the following instance. Suppose you step on a tack under your right foot. You quickly withdraw that foot by *(flexing? extending?)* your right leg using your hamstring muscles. (Stand up and try it.) What happens to your left leg? You *(flex? extend?)* it. This is an example of reciprocal innervation

   involving a _____ reflex.

    e. A crossed extensor reflex is *(ipsilateral? contralateral?)*. It *(may? may not?)* be interseg-mental. Therefore, many muscles may be contracted to extend your left thigh and leg to shift weight to your left side and provide balance during the tack episode.

✓ **B10.** Why are deep tendon reflexes (that involve stretching a tendon such as that of the quadriceps femoris muscle in the patellar reflex) particularly helpful diagnostically?

✓ **B11.** *Clinical correlation.* Discuss a commonly used method for assessing health of the nervous system in this Checkpoint.

    a. How is the Babinski sign assessed? Demonstrate the assessment technique on a friend.

    Note: For (b–d), choose from these answers:

    A. Extending the great toe with or without fanning of the other toes (the Babinski sign)

    B. Curling toes under (plantar flexion)

    b. In children over 18 months and in adults, what is the normal response? *(A? B?)*

    c. In babies under 18 months, what is the normal response? *(A? B?)*

    d. If corticospinal tracts lack myelination, which response is likely to occur? *(A? B?)* State two reasons why these tracts may have incomplete myelination.

C. Spinal nerves (pages 426–441)

✓▲ **C1.** Complete this exercise about spinal nerves.

    a. There are _____ pairs of spinal nerves. Write the number of pairs in each region.

        _____ Cervical _____ Thoracic _____ Lumbar _____ Sacral _____ Coccygeal

    b. Which of these spinal nerves form the cauda equina?

    c. Spinal nerves are attached by two roots. The posterior root is *(sensory? motor? mixed?)*,

    and the anterior root is _____, whereas the spinal nerve is

    _____.

    d. Individual nerve fibers are wrapped in a connective tissue covering known as *(endo-? epi-? peri-?)* neurium. Groups of nerve fibers are held in bundles (fascicles) by

    _____-neurium. The entire nerve is wrapped with _____-neurium.

    e. Spinal nerves branch when they leave the intervertebral foramen. These branches are

    called _____. Because they are extensions of spinal nerves, rami are *(sensory? motor? mixed?)*.

f. Which ramus is larger? *(Ventral? Dorsal?)* What areas does it supply?

g. What area does the dorsal ramus innervate? Label it on Figure LG 13.3.

h. Name two other branches (rami) and state their functions.

✓▲ **C2.** Match the plexus names in the box with descriptions. Refer to Figure 13.2 for help.

B.	**Brachial**	**L.**	**Lumbar**
C.	**Cervical**	**S.**	**Sacral**
I.	**Intercostal**		

_____ a. Provides the entire nerve supply for the arm

_____ b. Contains origin of phrenic nerve (nerve that supplies diaphragm)

_____ c. Forms median, radial, and axillary nerves

_____ d. Not a plexus at all, but rather segmentally arranged nerves

_____ e. Supplies nerves to scalp, neck, and part of shoulder and chest

_____ f. Supplies fibers to the femoral nerve, which innervates the quadriceps, so injury to this plexus would interfere with actions such as touching the toes

_____ g. Forms the largest nerve in the body (the sciatic), which supplies posterior of thigh and the leg

✓▲ **C3.** *For extra review.* Match names of nerves in the box with their descriptions. On the line following the description, write the site of origin of the nerve. The first one is done for you.

Axillary	Musculocutaneous	Radial
Inferior gluteal	Pudendal	Sciatic

____**Sciatic**____ a. Consists of two nerves, the tibial and common peroneal; supplies the hamstrings, abductors, and all muscles distal to the knee. __L4–S3__

_____ b. Supplies the deltoid muscle. _____

_____ c. Innervates the major flexors of the arm. _____

_____ d. Supplies most extensor muscles of the forearm, wrist, and fingers; may be damaged by extensive use of crutches. _____

_____ e. Supplies the gluteus maximus muscle. _____

_____ f. May be anesthetized in childbirth because it innervates external genitalia and lower part of the vagina. _____

▲ **C4.** *For additional review* of nerve supply to muscles, complete the table below. Write the name of the plexus with which the nerve is associated, as well as the distribution of that nerve. The first one is done for you. (*Hint:* one answer involves a nerve that is not a spinal nerve, so it is not a part of a plexus.)

Name of Nerve	Plexus	Distribution
a. Lateral pectoral	Brachial	
b. Medial pectoral		Pectoralis major and minor muscles
c. Lesser occipital		
d. Ulnar		
e.		Latissimus dorsi muscle
f. Long thoracic		
g.		Trapezius and sternocleidomastoid muscles (*Hint:* See Table 14.2, page 478 of text.)
h. Femoral		
i.	Lumbar	Cremaster muscle that pulls testes closer to pelvis
j. Obturator		
k.		Gastrocnemius, soleus, tibialis posterior
l.		Tibialis anterior

✓ **C5.** *Clinical thinking.* If the cord were completely transected (severed) just below the C7 spinal nerves, how would the functions listed below be affected? (Remember that nerves that originate below this point would not communicate with the brain and so would lose much of their function.) Explain your reasons in each case. (*Hint:* Refer to Figure 13.2 in the text.)

a.   Breathing via diaphragm

b.   Movement and sensation of thigh and leg

c. Movement and sensation of the arm

d. Use of muscles of facial expression and muscles that move jaw, tongue, eyeballs

✓▲ **C6.** *A clinical correlation.* Refer to diagrams in Figure 13.14 and descriptions in Exhibit 3.4 of the text and assume each of those positions. Then select names of each condition listed in the box that is associated with the nerve injury described below.

CH.	**Clawhand**	**SC.**	**Sciatica**
CV.	**Calcaneovalgus**	**WD.**	**Wristdrop**
ED.	**Erb-Duchenne palsy (waiter's tip)**	**WS.**	**Winged scapula**
FD.	**Footdrop (equinovarus)**		

_____ a. Radial nerve injury leading to inability to extend the wrist and fingers

_____ b. Ulnar nerve palsy with loss of sensation in the little finger

_____ c. Injury to superior roots of the brachial plexus (C5–C6)

_____ d. Injury of the long thoracic nerve supply to serratus anterior muscles

_____ e. Injury to the tibial portion of the sciatic nerve causing "knock-kneed" position

_____ f. Injury to the common peroneal nerve causing plantar flexion and eversion ("bowlegged") position

_____ g. Pain over buttocks and much of the posterior of the lower extremity often due to a herniated L5–S1 disc

C7. Describe the general pattern of dermatomes:
a. In the trunk

b. In the limbs

▲ **C8.** *A clinical correlation.* Explain how knowledge of dermatomes and related spinal cord segments can be used clinically.

D. Disorders (page 442)

✓ **D1.** Shingles is an infection of the *(central? peripheral?)* nervous system. The causative virus is also the agent of *(chickenpox? measles? cold sores?)*. Following recovery from chickenpox, the virus remains in the body in the *(spinal cord? dorsal root ganglia?)*. At times it is activated and travels along *(sensory? motor?)* neurons, causing *(pain? paralysis?)*.

✓▲ **D2.** Match names of disorders in the box with definitions below.

N.	**Neuritis**	Sc.	**Sciatica**
P.	**Poliomyelitis**	Sh.	**Shingles**

_____ a. Inflammation of a single nerve

_____ b. Also known as infantile paralysis; caused by a virus that may destroy motor cell bodies in the brainstem or in the anterior gray horn of the spinal cord

_____ c. Acute inflammation of the nervous system by *Herpes zoster* virus

_____ d. Neuritis of a nerve in the posterior of hip and thigh; often due to a slipped disc in the lower lumbar region

ANSWERS TO SELECTED CHECKPOINTS: CHAPTER 13

A1. Processing center for (reflexes), integration (summing) of afferent and efferent nerve impulses, and pathway for those impulses.

A2. Central (CNS).

A3.

Figure LG 13.1A Meninges.

(d) Meningitis. (e) Pia mater, dura mater; anchor the cord and protect it from displacement.

A4. (a) Waist; L1–L2; vertebral column; 16–18. (b) Cervical, lumbar.

A5. (a) Co. (b) S. (c) F. (d) Ca.

A6. (a) To insert anesthetics, antibiotics, chemotherapy or contrast media; to withdraw cerebrospinal fluid (CSF) for diagnostic purposes such as for analysis for blood or microorganisms; subarachnoid. (b) L3–L4 or L4–L5; the spinal cord ends at about L1–L2, so the cord is not likely to be injured at this lower level. (c) The iliac crest and umbilicus (navel) are both at about this level. (d) The anesthetic must penetrate epidural tissues and then all three layers of meninges before reaching nerve fibers, whereas an anesthetic in the subarachnoid space needs to penetrate only the pia mater to reach nerve tissue; childbirth (labor and delivery).

A7 (a) Within, outside; refer to Figure 13.3 of the text. (b) A, posterior columns; B, posterior gray horn; C, lateral white/gray columns; D, lateral white/gray horn; E, anterior gray/white horn; F, anterior gray/white columns; G, anterior median fissure.(c) Dorsal root ganglion (H).

B1. (a) Conduction (or a pathway), tracts. (b) Reflex; 2; posterior (dorsal), anterior (ventral). See Figure LG 13.3A.

B2. (a) Central; nerves; white columns; myelinated.
(b) Sensory, brain; descending. (c–d) See Figure
LG 13.3A. (e) Lateral, cerebral cortex, spinal cord
(in anterior gray horn). (f) Descending, motor,
pyramidal. (g) Extrapyramidal; B (h) See Figure
LG 13.3A.

Fasciculus cuneatus and gracilis
**(Proprioception, pressure,
vibration, two-point
discriminative touch) (S)**

Lateral cortico-spinal tract
**(Precise voluntary
movements) (M)**

Lateral spino-
thalamic tract

Anterior spino-
thalamic tract

**(Pain, temperature, crude
touch, pressure) (S)**

Anterior corticospinal tract
(Precise, voluntary movements) (M)

Lateral gray horn
Posterior gray horn
Sensory (posterior)
nerve root
Lateral white column
Dorsal
ramus

Motor (anterior)
nerve root
Anterior gray horn
Gray commissure
Anterior median
fissure

**M = motor
S = sensory**

Figure LG 13.3A Major tracts in the spinal cord.

B4. Refer to Figure LG 13.4A; B and D are sensory; G is
motor.

B5.

**A. Receptor
B. Sensory neuron dendrite
C. Sensory neuron cell body
D. Sensory neuron axon
E. Synapse (integrating center)
F. Motor neuron cell body
G. Motor neuron axon
H. Effector**

Figure LG 13.4A Reflex arc:
stretch reflex.

B6. (a) 2; sensory, motor. (b) Monosynaptic; CNS.
(c) Muscle spindles; length (or stretch); stretch.
(d) Ipsilateral. (e) Skeletal muscle. (f) Knee jerk
(patellar reflex).

B7. Branches from axons of sensory or association
neurons travel through tracts to the brain.
B8. (a) Tendon organs; tension. (b) Inhibition, relax.
(c) Antagonistic: contract. (d) Poly, protected.
B9. (a) Does, more. (b) Several, one; withdrawal, ipsi.
(c) No action; reciprocal. (d) Flexing; extend;
crossed extensor. (e) Contralateral; may.
B10. They can readily pinpoint a disorder of a specific
spinal nerve or plexus, or portion of the cord,
because they do not involve the brain.
B11. (a) Stroking the lateral margin of the sole of the
foot. (b) B. (c) A. (d) A; developmental stage
(under 18 months) or damage to these tracts in
older persons.
C1. (a) 31; 8, 12, 5, 5, 1. (b) Lumbar, sacral, and
coccygeal. (c) Sensory, motor, mixed. (d) Endo-;
peri-; epi-. (e) Rami; mixed. (f) Ventral: all of the
limbs and the ventral and lateral portions of the
trunk; see Figure LG 13.3A. (g) Muscles and skin
of the back; see Figure LG 13.3A. (h) Meningeal
branch supplies primarily vertebrae and meninges;
rami communicantes have autonomic functions.
C2. (a) B. (b) C. (c) B. (d) I. (e) C. (f) L. (g) S.
C3. (b) Axillary, C5–C6. (c) Musculocutaneous,
(d) Radial, C5–C8, T1. (e) Inferior gluteal, L5–S2.
(f) Pudendal, S2–S4.
C5. (a) Not affected because (phrenic) nerve to the
diaphragm originates from the cervical plexus (at
C3–C5), higher than the transection. So this nerve
continues to receive nerve impulses from the
brain. (b) Complete loss of sensation and paralysis
because lumbar and sacral plexuses originate
below the injury and therefore no longer communicate with the brain. (c) Most arm functions are
not affected. As shown on Figure 13.2 of the text,
the brachial plexus originates from C5 through T1,
so most nerves to the arm (those from C5 through
C7) still communicate with the brain. (d) Not
affected because all are supplied by cranial nerves
which originate from the brain.
C6. (a) WD. (b) CH. (c) ED. (d) WS. (e) CV. (f) FD.
(g) S.
D1. Peripheral; chickenpox; dorsal root ganglia sensory
pain
D2. (a) N. (b) P. (c) Sh. (d) Sc.

MORE CRITICAL THINKING: CHAPTER 13

1. Describe the anatomy and physiology of the meninges, as well as the spaces formed
between and surrounding meninges.
2. State several reasons why a lumbar puncture might be done. Give a brief description of
the procedure.
3. Describe a reflex arc, including the structure and function of its five components.
4. Contrast a stretch reflex with a flexor (withdrawal) reflex.
5. Describe how reflexes help you to maintain your balance when you quickly pick up your
foot in response to stepping on a piece of glass.
6. Define the term plexus and describe what is meant by ventral rami that form a plexus.
State examples of two plexuses and the major nerves they form.

MASTERY TEST: CHAPTER 13

Questions 1–4: Arrange the answers in correct sequence.

_____ _____ _____ 1. From superficial to deep:
        A. Subarachnoid space
        B. Epidural space
        C. Dura mater

_____ _____ _____ 2. From anterior to posterior in the spinal cord:
        A. Fasciculus gracilis and cuneatus
        B. Anterior spinothalamic tract
        C. Central canal of the spinal cord

_____ _____ _____ _____ 3. The plexuses, from superior to inferior:
        A. Lumbar   C. Cervical
        B. Brachial   D. Sacral

_____ _____ _____ _____ _____ 4. Order of structures in a conduction pathway, from origin to termination:
        A. Motor neuron   D. Receptor
        B. Sensory neuron   E. Effector
        C. Integrative center

Questions 5–11: Circle the letter preceding the one best answer to each question.

5. Herniation (or "slipping") of the disc between L4 and L5 vertebrae is most likely to result in damage to the _____ nerve.
    A. Femoral   C. Radial
    B. Sciatic   D. Musculocutaneous

6. All of these tracts are sensory *except:*
    A. Anterior spinothalamic
    B. Lateral spinothalamic
    C. Fasciculus cuneatus
    D. Lateral corticospinal
    E. Posterior spinocerebellar

7. Choose the *false* statement about the spinal cord.
    A. It has enlargements in the cervical and lumbar areas.
    B. It lies in the vertebral foramen.
    C. It extends from the medulla to the sacrum.
    D. It is surrounded by meninges.
    E. In cross section an H-shaped area of gray matter can be found.

8. All of these structures are composed of white matter *except:*
    A. Posterior root (spinal) ganglia
    B. Lumbar plexus
    C. Sciatic nerve
    D. Ventral ramus of a spinal nerve

9. Which is a *false* statement about the patellar reflex?
    A. It is also called the knee jerk.
    B. It involves a two-neuron, monosynaptic reflex arc.
    C. It results in extension of the leg by contraction of the quadriceps femoris.
    D. It is contralateral.

10. The cauda equina is:
    A. Another name for the cervical plexus
    B. The lumbar and sacral nerve roots extending below the end of the cord and resembling a horse's tail
    C. The inferior extension of the pia mater
    D. A denticulate ligament
    E. A canal running through the center of the spinal cord

11. Choose the one *false* statement.
    A. The Babinski sign *is normal* in a 10-month-old child but *not* in a 10-year-old child.
    B. Femoral nerve injury is more likely to cause loss of function of *hamstring* muscles than of *quadriceps* muscles.
    C. All spinal nerves except Cl emerge from below the vertebra of the same number.
    D. The axillary nerve supplies the deltoid muscle.

*Questions 12–20: Circle T (true) or F (false). If the statement is false,
change the underlined word or phrase so that the statement is correct.*

T F   12. The two main functions of the spinal cord are that it serves as a <u>reflex center</u> and it is <u>the site where sensations are felt</u>.

T F   13. Dorsal roots of spinal nerves are <u>sensory, ventral roots are motor, and spinal nerves are mixed</u>.

T F   14. A tract is a bundle of nerve fibers <u>inside the central nervous system (CNS)</u>.

T F   15. Synapses <u>are</u> present in posterior (dorsal) root ganglia.

T F   16. After a person reaches 18 months, the Babinski sign should be <u>negative, as indicated by plantar flexion (curling under of toes and foot)</u>.

T F   17. Visceral reflexes are used diagnostically <u>more often than somatic ones because it is easy to stimulate most visceral receptors</u>.

T F   18. Transection (cutting) of the spinal cord at level C6 will result in <u>greater</u> loss of function than transection at level T6.

T F   19. The ventral root of a spinal nerve contains <u>axons and dendrites</u> of <u>both motor and sensory</u> neurons.

T F   20. A lumbar puncture (spinal tap) is usually performed at about the level of vertebrae <u>L1 to L2</u> because the cord ends between about <u>L3 and L4</u>.

Questions 21–25: Fill-ins. Complete each sentence with the word or short answer that best fits.

_____   21. The phrenic nerve innervates the _____.

_____   22. An inflammation of the dura mater, arachnoid, and/or pia mater is known as.

_____   23. Tendon reflexes are _____-synaptic and _____-lateral.

_____   24. The layer of the meninges that gets its name from its delicate structure, which is much like a spider's web, is the _____.

_____   25. The filum terminale and denticulate ligaments are both composed of _____ mater.

ANSWERS TO MASTERY TEST: CHAPTER 13

Arrange

1. B C A
2. B C A
3. C B A D
4. D B C A E

Multiple Choice

5. B
6. D
7. C
8. A
9. D
10. B
11. B

True–False

12. F. Reflex center, a conduction site
13. T
14. T
15. F. Are not
16. T
17. F. Less often than somatic ones because it is difficult to stimulate most visceral receptors
18. T
19. F. Axons of motor
20. F. L3–L4, L1–L2

Fill-ins

21. Diaphragm
22. Meningitis
23. Poly, ipsi
24. Arachnoid
25. Pia

FRAMEWORK 14
Brain & Cranial Nerves

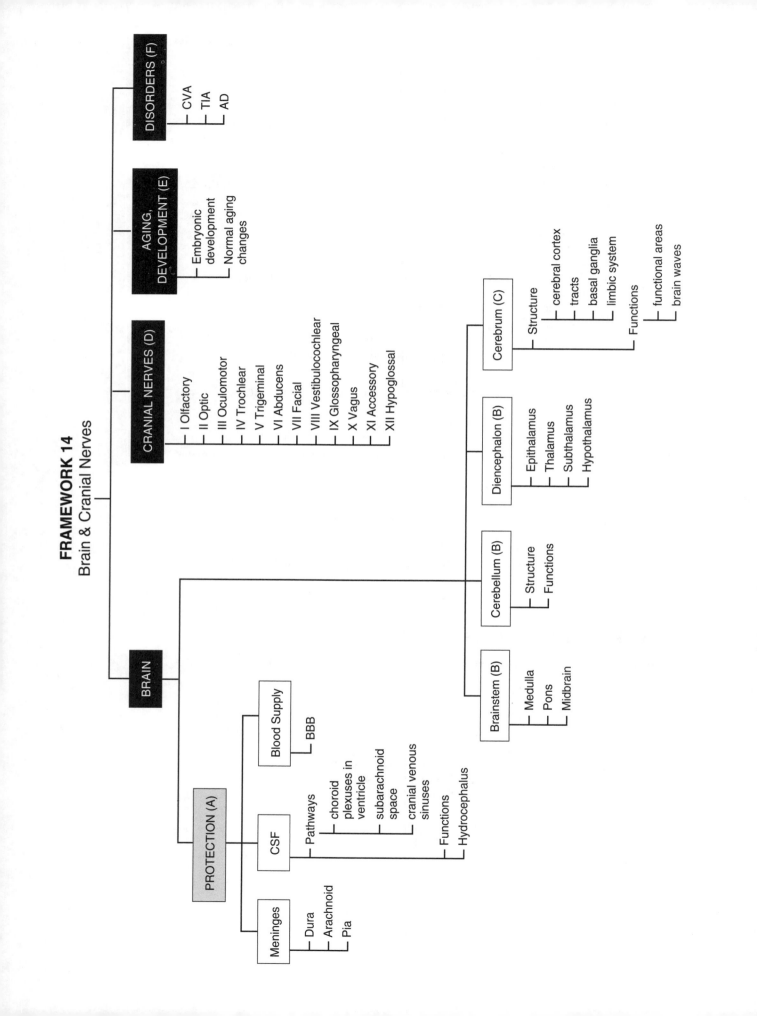

BRAIN

DISORDERS (F)
- CVA
- TIA
- AD

AGING, DEVELOPMENT (E)
- Embryonic development
- Normal aging changes

CRANIAL NERVES (D)
- I Olfactory
- II Optic
- III Oculomotor
- IV Trochlear
- V Trigeminal
- VI Abducens
- VII Facial
- VIII Vestibulocochlear
- IX Glossopharyngeal
- X Vagus
- XI Accessory
- XII Hypoglossal

PROTECTION (A)

Meninges
- Dura
- Arachnoid
- Pia

CSF
- Pathways
  - choroid plexuses in ventricle
  - subarachnoid space
  - cranial venous sinuses
- Functions
- Hydrocephalus

Blood Supply
- BBB

Brainstem (B)
- Medulla
- Pons
- Midbrain

Cerebellum (B)
- Structure
- Functions

Diencephalon (B)
- Epithalamus
- Thalamus
- Subthalamus
- Hypothalamus

Cerebrum (C)
- Structure
  - cerebral cortex
  - tracts
  - basal ganglia
  - limbic system
- Functions
  - functional areas
  - brain waves

C H A P T E R **14**

The Brain and Cranial Nerves

The brain is the major control center for the global communication network of the body: the nervous system. The brain requires round-the-clock protection and maintenance afforded by bones, meninges, cerebrospinal fluid, a special blood—brain barrier, along with a fail-safe blood supply. This vital control center consists of four major substructures: the brain stem, diencephalon, cerebrum, and cerebellum. Each brain part carries out specific functions and each releases specific chemical neurotransmitters. Twelve pairs of cranial nerves convey information to and from the brain.

Structural defects may occur in construction (brain development) and also with the normal wear and tear that accompanies aging. Just as a complex computer network may experience minor disruptions in service or major shutdowns, disorders within the brain or cranial nerves may lead to minor, temporary changes in nerve functions or profound and fatal outcomes.

As you begin your study of the brain, carefully examine the Chapter 14 Topic Outline and Objectives; check off each one as you complete it. To organize your study of the brain, glance over the Chapter 14 Framework now. Be sure to refer to the Framework frequently and note relationships among key terms in each section.

TOPIC OUTLINE AND OBJECTIVES

A. Overview of brain: organization, protection, blood supply

1. Identify the principal parts of the brain.
2. Describe how the brain is protected.
3. Explain the formation and circulation of cerebrospinal fluid.

B. The brain stem, cerebellum, and diencephalon

4. Describe the structure and functions of the brain stem.
5. Describe the structure and functions of the cerebellum.
6. Describe the components and functions of the diencephalon.

C. The cerebrum

7. Describe the structure and functions of the cerebrum.
8. Describe the locations and functions of the sensory, association, and motor areas of the cerebral cortex.

D. Cranial nerves

9. Identify the cranial nerves by name, number, and type, and give the function of each.

E. Developmental anatomy and aging of the nervous system

10. Describe how the parts of the brain develop.
11. Describe the effects of aging on the nervous system.

F. Disorders

WORDBYTES

Now become familiar with the language of this chapter by studying each wordbyte, its meaning, and an example of its use within a term. After you study the entire list, self-check your understanding by writing the meaning of each wordbyte on the line. As you continue through the *Learning Guide,* identify (and fill in) additional terms that contain the same wordbyte.

Wordbyte	Self-check	Meaning	Example(s)
-algia	_____	pain	neur*algia*, an*algesia*
cauda-	_____	tail	*cauda*te
cephalo-	_____	head	hydro*cephalic*
corpora-	_____	bodies	*corpora* quadrigemina
cortico-	_____	bark	cerebral *cortex*
di-	_____	two	*di*encephalon
enceph-	_____	brain	di*enceph*alon
falx	_____	sickle	*falx* cerebri
glossi-	_____	tongue	hypo*glossal*
hemi-	_____	half	*hemi*sphere
hypo-	_____	under	*hypo*thalamus
ophthalm-	_____	eye	*ophthalm*ic
pineal	_____	pineconelike	*pineal* gland
pons	_____	bridge	*pons*
quad-	_____	four	corpora *quad*rigemina

CHECKPOINTS

A. Brain: protection and blood supply (pages 446–452)

✓▲ **A1.** Identify numbered parts of the brain on Figure LG 14.1. Then complete this exercise.

a. Structures 1–3 are parts of the _____.

b. Structures 4 and 5 together form the _____.

c. Structure 6 is the largest part of the brain, the _____.

d. The second largest part of the brain is structure 7, the _____.

✓ **A2.** List three ways in which the brain is protected.

KEY:

1. Medulla oblongata
2. Pons
3. Midbrain
4. Hypothalamus
5. Thalamus
6. Cerebrum
7. Cerebellum

A. Arachnoid villus
B. Cranial venous sinus
C. Subarachnoid space of brain
D. Lateral venticle
E. Interventricular foramen
F. Third ventricle

G. Cerebral aqueduct
H. Fourth ventricle
I. Lateral aperture
J. Median aperture
K. Subarachnoid space of spinal cord

Figure LG 14.1 Brain and meninges seen in sagittal section. Parts of the brain are numbered; refer to Checkpoint A1. Letters indicate pathway of cerebrospinal fluid (CSF): refer to Checkpoint A8.

A3. Review the layers of the meninges covering the brain and spinal cord by listing them
here. *For extra review,* label the layers on Figure LG 14.1 and review LG Chapter 13,
Checkpoint A3.

✓▲ **A4.** Identify the extensions of dura mater in this Checkpoint. Use answers in the box.

Falx cerebelli	**Falx cerebri**	**Tentorium cerebelli**

   a. A tentlike structure that separates the cerebrum from the cerebellum:

     _____.

   b. Separates the two hemispheres of the cerebellum: _____.

   c. Separates the two hemispheres of the cerebrum: _____.

A5. Briefly state results of oxygen starvation of the brain for about 4 minutes.

     Explain the role of lysosomes in this process.

     Now list effects of glucose deprivation of the brain.

✓ **A6.** Describe the blood-brain barrier (BBB) in this exercise.
   a. Blood capillaries supplying the brain are *(more? less?)* leaky than most capillaries of
the body. This fact is related to a large number of *(gap? tight?)* junctions, as well as an

     abundance of neuroglia named _____ pressed against capillaries.
   b. What advantages are provided by this membrane?

   c. List two or more substances needed by the brain that do normally cross the BBB.

   d. What problems may result from the fact that some chemicals cannot cross this barrier?

✓ **A7.** Circle all correct answers about CSF.

a. The entire nervous system contains 80–150 mL of CSF. This amount is equal to approximately:

   A. 1 to 2 tablespoons
   B. ⅓ to ⅔ cup
   C. 1 to 2 cups
   D. 1 quart

b. The color of CSF is normally:

   A. Yellow
   B. Clear, colorless
   C. Red
   D. Green

c. Choose the function(s) of CSF.

   A. Serves as a shock absorber for brain and cord
   B. Contains red blood cells
   C. Contains some white blood cells
   D. Contains nutrients such as glucose and proteins, ions, and wastes such as urea

d. Which statement(s) describe its formation?

   A. It is formed by diffusion of substances from blood.
   B. It is formed by filtration and secretion.
   C. It is formed from blood in capillaries called choroid plexuses.
   D. It is formed by ependymal cells that line all four ventricles.

e. Which statement(s) describe its pathway?

   A. It circulates around the brain but not the cord.
   B. It flows inferior to the end of the spinal cord.
   C. It bathes the brain by flowing through the epidural space.
   D. It passes via projections (villi) of the arachnoid into blood vessels (venous sinuses) surrounding the brain.
   E. It is formed initially from blood and finally flows back to blood.

f. An excessive accumulation of CSF within the ventricles is a condition known as:

   A. Hydrarthrosis
   B. Hydrophobia
   C. Hydrocephalus
   D. Hydrocholecystitis

✓ **A8.** To check your understanding of the pathway of cerebrospinal fluid (CSF), list in order the structures through which it passes. Use key letters on Figure LG 14.1. Start at the site of formation of CSF.

B. Brain: brain stem, cerebellum, and diencephalon (pages 452–463)

✓▲ **B1.** Describe the principal functions of the medulla in this exercise.

a. The medulla serves as a _____ pathway for all ascending and descending tracts. Its white matter therefore transmits *(sensory? motor? both sensory and motor?)* impulses.

b. Included among these tracts are the triangular _____ tracts which are the principal *(sensory? motor?)* pathways. The main fibers that pass through the pyramids are the *(spinothalamic? corticospinal?)* tracts.

c. Crossing (or _____) of fibers occurs in the medulla. This explains why movements of your right hand are initiated by motor neurons that originate in the *(right? left?)* side of your cerebrum. The *(axons? dendrites?)* of these motor neurons decussate in the medulla to proceed down the right lateral

   _____-spinal tract.

    d.   The medulla contains gray areas as well as white matter. Several important nuclei lie in the medulla. Two of these are synapse points for sensory axons that have ascended in the posterior columns of the cord. These two nuclei, named _____

       and _____ , transmit impulses for sensations of _____ ,

       _____ , and _____ .

    e.   A hard blow to the base of the skull can be fatal because the medulla is the site of

       three vital centers: the _____ center, regulating the heart; the

       _____ center, adjusting the rhythm of breathing; and the

       _____ center, regulating blood pressure by altering the diameter of blood vessels.

    f.   Some input to the medulla arrives by means of cranial nerves; these nerves may serve motor functions also. Which cranial nerves are attached to the medulla?

       _____

       (Functions of these nerves will be discussed later in this chapter.)

✓▲ **B2.**   Summarize important aspects of the pons in this learning activity.

    a.   The name *pons* means _____ . It serves as a bridge in two ways.

       It contains longitudinally arranged fibers that connect the _____

       and _____ with the upper parts of the brain. It has transverse fibers

       that connect the two sides of the _____ .

    b.   Cell bodies associated with fibers in cranial nerves numbered _____ lie in nuclei in the pons.

    c.   *(Respiration? Heartbeat? Blood pressure?)* is controlled by the pneumotaxic and apneustic areas of the pons.

✓▲ **B3.**   Relate the midbrain to the pons and medulla in this exercise.

    a.   Like the pons and medulla, the midbrain is about 1 inch (_____ cm) long.

    b.   The midbrain is more *(anterior and superior? posterior and inferior?)* compared to the pons and medulla.

✓▲ **B4.**   You maintain a conscious state, or you wake up from sleep, thanks to the regions of

       the brain known as the RAS, or R_____ A_____

       S_____ . These areas are especially sensitive to sensations such as

       _____ from ears and _____ or _____ from skin. The RAS is part of the reticular formation. Where is the reticular formation located?

▲ **B5.** Describe the cerebellum
   a. Where is it located?

   b. Describe its structure. Include these terms in your description: *vermis, lobes, cortex,* and *arbor vitae.*

   c. Describe its functions. Use these key terms: *coordinated movements, posture, equilibrium,* and *emotions.*

✓ **B6.** Identify locations of each of the tracts known as *peduncles* that permit communication between the cerebellum and other brain parts. Fill in lines with the terms in the box.

Inferior	Middle	Superior

   a. Cerebellum → _____ cerebullar peduncle → Midbrain

   b. Cerebellum → _____ cerebullar peduncle → Pons

   c. Cerebellum → _____ cerebullar peduncle → Medulla

✓▲ **B7.** Describe the *diencephalon* in this Checkpoint.
   a. If you look for the diencephalon in most external views of the brain, you will not find it. It is located *(deep within? on the surface of?)* the cerebrum. The diencephalon

   consists of *(two? four? eight?)* main structures: the thalamus, as well as the _____

   thalamus, the _____ thalamus, and the _____ thalamus. Which part comprises

   80% of the diencephalon? _____
   b. The epithalamus forms the *(roof? lateral walls? floor?)* of the third ventricle. The epi-

   thalamus includes the _____ gland and the _____

   nuclei, as well as the _____ plexus of the third ventricle. The pineal

   gland secretes the hormone named _____ and serves as the body's

   biological _____. The habenular nucleus is involved in *(hearing? smell? vision?).*
   c. The subthalamus works in controlling *(sensations? movements?).*

d. The *(two? four?)* lobes of the thalamus form parts of the *(roof? lateral walls? floor?)* of the third ventricle. A "crossbar" between the two lobes passes through the center of

the _____ ventricle. (See Figure LG 14.1.)

e. The thalamus is the principal relay station for *(motor? sensory?)* impulses. For example, spinothalamic and lemniscal tracts convey general sensations such as pain,

_____, _____, _____, and temperature to the thalamus where they are relayed to the cerebral cortex. Special sense impulses (for vision and hearing) are relayed through the *(geniculate? reticular? ventral posterior?)* nuclei of the thalamus.

f. Although the thalamus plays a primary role in conduction of sensory impulses, it also has other functions. Name two or more.

g. The name hypothalamus indicates that this structure lies *(above? below?)* the thalamus,

forming the floor and part of the lateral walls of the _____ ventricle.

B8. Expanding on the key words listed below, write a sentence describing major hypothalamic functions.

a. Regulator of visceral activities via control of the ANS

b. Regulating factors to anterior pituitary

c. Feelings (rage)

d. Temperature

e. Thirst

f. Feeding and satiety center

g. Reticular formation: arousal and consciousness

✓▲ **B9.** *For extra review.* Check your understanding of these parts of the brain stem and diencephalon by matching them with the descriptions given below.

H.	**Hypothalamus**	**P.**	**Pons**
Med.	**Medulla**	**T.**	**Thalamus**
Mid.	**Midbrain**		

_____ a. It is the principal regulator of visceral
activities because it acts as a liaison
between the cerebral cortex and autonomic nerves that control viscera.

_____ b. It is the site of origin of rubrospinal tracts
concerned with muscle tone and posture.

_____ c. Cranial nerves III–IV attach to this brain
part.

_____ d. Cranial nerves V–VIII attach to this
brain part.

_____ e. Cranial nerves VIII–XII attach to this
brain part.

_____ f. Feelings of hunger, fullness, and thirst
stimulate centers here so that you can
respond accordingly.

_____ g. All sensations are relayed through here.

_____ h. Regulation of heart, blood pressure, and
respiration occurs by centers located here.

_____ i. It constitutes four-fifths of the diencephalon.

_____ j. It lies under the third ventricle, forming
its floor.

_____ k. It forms most of side walls of the third
ventricle.

_____ l. Tumor in this region could compress the
cerebral aqueduct and cause internal
hydrocephalus.

_____ m. It contains centers for coughing, sneezing,
hiccuping, swallowing, and vomiting.

_____ n. The olivary and vestibular nuclei associated with equilibrium and posture control
are located here.

✓▲ **B10.** *For extra review* of specific locations and functions of nuclei and other masses of gray matter, complete this table. Choose answers from the lists of brain parts and functions. Provide two answers in parts of the table indicated by (2).

Brain part		Functions
Cer.	Cerebellum	Emotions, memory
Epi.	Epithalamus	Hearing
Hypo.	Hypothalamus	Movements in general
Med.	Medulla	Movements of head or neck in response
Mid.	Midbrain	to sights or sounds
Sub.	Subthalamus	Regulation of anterior pituitary
Thal.	Thalamus	Regulation of posterior pituitary
		Smell
		Taste
		Touch, pressure, vibrations
		Vision

Name of Nucleus	Brain Part	Functions
a. Habenular nucleus	Epi	Smell
b. Subthalamic nucleus		
c. Red nucleus	(2)	
d. Colliculi		
e. Nuclei cuneatus and gracilis		
f. Anterior nucleus		
g. Lateral geniculate nucleus		
h. Medial geniculate nucleus		
i. Ventral posterior nucleus		(2)
j. Mammillary bodies		
k. Median eminence of tuber cinereum		
l. Supraoptic region		

✓ **B11.** Describe *circumventricular organs* (CVOs) in this exercise.

a. Name three organs that are CVOs.

b. CVOs are unique in the brain region in that they *(have? lack?)* the blood–brain barrier. State one advantage and one disadvantage of this structural feature.

C. Brain: cerebrum (pages 463–472)

✓▲ **C1.** Complete this exercise about cerebral structure.

a. The outer layer of the cerebrum is called _____. It is composed of *(white? gray?)* matter. This means that it contains mainly *(cell bodies? tracts?)*.

b. In the margin, draw a line the same length as the thickness of the cerebral cortex. Use a metric ruler. Note how thin the cortex.

c. The surface of the cerebrum looks much like a view of tightly packed mountains or

ridges, called _____. The parts where the cerebral cortex dips down

into valleys are called _____ (deep valleys) or _____ (shallow valleys).

d. The cerebrum is divided into halves called _____. Connecting them

is a band of *(white? gray?)* matter called the _____. Notice this structure in Figure LG 14.1 and in Figure 14.10 in your text.

e. The falx cerebri is composed of *(nerve fibers ? dura mater?)*. Where is it located?

_____ At its superior and inferior margins, the falx is dilated to form channels for venous blood flowing from the brain; these enclosures are called

_____.

✓▲ **C2.** Label the following structures using leader lines on Figure LG 14.2: *frontal lobe, occipital lobe, parietal lobe, temporal lobe, central sulcus, lateral cerebral sulcus, precentral gyrus, postcentral gyrus.*

Figure LG 14.2 Right lateral view of lobes and fissures of the cerebrum. Label and color as directed in Checkpoints C2 and 08.

○ Motor speech (Broca's) area   ○ Primary somatosensory (general sensory) area
○ Premotor area   ○ Primary visual area
○ Primary auditory area   ○ Auditory association area
○ Primary motor area

✓▲ **C3.** Match the three types of white matter fibers with these descriptions.

A. Association	**C. Commissural**	**P. Projection**

_____ a. The corpus callosum contains these fibers and connects the two cerebral hemispheres.

_____ b. Sensory and motor fibers passing between cerebrum and other parts of the CNS are this type of fiber; the internal capsule is an example.

_____ c. These fibers transmit impulses among different areas of the same hemisphere.

▲ **C4.** List names of structures that are considered parts of the basal ganglia.

Now list two functions of the basal ganglia.

C5. Describe the limbic system in this exercise.
a. This system consists of portions of the cerebrum, thalamus, and hypothalamus. List several component structures.

b. Explain why the limbic system is sometimes called the "visceral" or "emotional" brain.

c. One other function of the limbic system is _____. Forgetfulness. such as inability to recall recent events, results partly from impairment of this system.

✓ **C6.** Which condition can be described as visible bruising of the brain with extended loss of consciousness? *Concussion? Contusion?*

✓ **C7.** Draw two important generalizations about brain functions in this activity.
a. In general, the anterior of the cerebrum is more involved with *(motor? sensory?)* control, whereas the posterior of the cerebrum is more involved with *(motor? sensory?)* functions. (Take a moment to visualize those activities taking place in the front and back of your own brain.)
b. As a general rule, *(primary? association?)* sensory areas receive sensations and *(primary? association?)* sensory areas are involved with interpretation and memory of sensations. For example, your ability to see the outline of the cerebrum, as in Figure LG 14.2, depends on your *(primary? association?)* visual areas. The fact that you can distinguish this diagram as a cerebrum (not a hand or heart), along with your memory of its structure for your next test, is based on the health of the neurons in your *(primary? association?)* visual areas.

✓ **C8.** Color functional areas of the cerebral cortex listed with color code ovals on Figure 14.2.

✓▲ **C9.** Check your understanding of functional areas of the cerebral cortex by doing this matching exercise. Answers may be used more than once.

F.	**Frontal eye field**	PM.	**Primary motor area**
G.	**Gnostic area**	PG.	**Primary olfactory area**
M.	**Motor speech (Broca's) area**	PS.	**Primary somatosensory area**
PA.	**Primary auditory area**	PV.	**Primary visual area**

_____ a. In the occipital lobe

_____ b. In the postcentral gyrus

_____ c. Receives sensations of pain, touch, pressure, and temperature

_____ d. In the parietal lobe

_____ e. In the temporal lobe; permits hearing

_____ f. Integrates general and special sensations to form a common thought about them

_____ g. In precentral gyms of the frontal lobe; controls specific muscles or groups of muscles

_____ h. In the frontal lobe; translates thoughts into speech

_____ i. Controls smell

_____ j. Controls scanning movements of eyes, such as searching for a name in a telephone book

✓ **C10.** Do this exercise about control of speech.
   a. Language areas are usually located in the *(left? right?)* cerebral hemisphere. This is true for most *(persons who are right-handed? persons, regardless of handedness?)*.

   b. The term aphasia refers to inability to _____. *Fluent aphasia* means inability to *(articulate or form? understand?)* words. *Word blindness* refers to inability to understand *(spoken? written?)* words, whereas *word deafness* is inability to understand

   _____ words.

✓ **C11.** *Critical thinking.* You are helping to care for Laura, a hospital patient with a contusion of the right side of the brain. Which of the following would you be most likely to observe in Laura? Circle letters of correct answers.
   A. Her right leg is paralyzed.
   B. Her left arm is paralyzed.
   C. She cannot speak out loud to you nor write a note to you.
   D. She used to sing well but cannot seem to stay on tune now.
   E. She has difficulty with concepts involving numbers.

✓ **C12.** Identify the type of brain waves associated with each of the following situations.

A.	**Alpha**	D.	**Delta**
B.	**Beta**	T.	**Theta**

_____ a. Occur in persons experiencing stress and in certain brain disorders

_____ b. Lowest frequency brain waves, normally occurring when adult is in deep sleep; presence in awake adult indicates brain damage

_____ c. Highest frequency waves, noted during periods of mental activity

_____ d. Present when awake but resting, these waves are intermediate in frequency between beta and delta waves

D. Cranial nerves (pages 472; 475–479)

✓▲ **D1.** Answer these questions about cranial nerves.

    a.   There are _____ pairs of cranial nerves. They are all attached to the

        _____; they leave the _____ via foramina.

    b.   They are numbered by Roman numerals in the order that they leave the cranium. Which is most anterior? *(I? XII?)* Which is most posterior? *(I? XII?)*

    c.   All spinal nerves are *(purely sensory? purely motor? mixed?).* Are all cranial nerves mixed? *(Yes? No?)*

✓▲ **D2.** Complete the table about the 12 pairs of cranial nerves. (*For extra review* on locations of these nerves, identify their entrance or exit points through the cranial bones in LG Chapter 7, Checkpoint B7.)

Number	Name	Functions
a.	Olfactory	
b.		Vision (not pain or temperature of the eye)
c. III		
d.	Trochlear	
e. V		
f.		Stimulates lateral rectus muscle to abduct eye; proprioception of the lateral rectus
g.	Facial	
h.		Hearing; equilibrium
i. IX		
j.	Vagus	
k. XI		
l.		Supplies muscles of the tongue with motor and sensory fibers

✓▲ **D3.** *The Big Picture: Looking Ahead and Looking Back.* Refer to figures and exhibits in your text that emphasize structure and function of cranial nerves.

    a.   Refer to Figure 16.1 and identify the olfactory nerves, which form part of the pathway for *(smell? taste?).* These are located in mucosa of *(inferior? superior?)* portions of the nose. After passing through the cribriform plate of the *(ethmoid? frontal? sphenoid?)*

bone, axons of olfactory nerves synapse with neuron cell bodies located in the olfactory *(bulb? tract?)*. Axons of these neurons then pass through the olfactory *(bulb? tract?)*, which is located at the base of the *(frontal? parietal?)* lobes of the cerebrum.

b. Figure 16.5 shows the optic nerves, which are cranial nerves number *(I? II? VIII?)*. Optic nerves form part of the pathway for *(hearing? taste? vision?)*; each eye has *(one? many?)* optic nerve(s). Notice on Figure 16.8 that optic nerves consist of axons of *(rod or cone? bipolar? ganglion?)* cells within the retina of the eye. These axons may cross in the optic *(chiasma? tract?)* (Figure 16.15) or remain uncrossed before they end in the *(thalamus? hypothalamus?)*. There they synapse with neurons that terminate in the *(occipital? temporal?)* lobe of the cerebral cortex.

c. Cranial nerve VIII is associated with the *(ear? tongue?)* (Figure 16.16). The *(cochlear? vestibular?)* portion of this nerve conveys impulses associated with hearing from the snail-shaped *(cochlea? vestibule?)*. The three semicircular canals contain receptors that sense body position and movement; these neurons send axons to the thalamus via the *(cochlear? vestibular?)* branch of cranial nerve VIII.

d. Cranial nerve VII is the *(facial? trigeminal?)*; it can be seen on Figure 14.5, lying right next to cranial nerve *(II? VIII?)*. In fact, these two nerves both pass through the cranium via the same foramen, the *(internal? external?)* auditory meatus (Figure 16.16). The facial nerve is primarily involved with *(sensation? movement?)* of the face (Exhibit 11.1). It also carries sensations of taste from the *(anterior? posterior?)* two-thirds of the tongue (page 516 of the text).

e. Four cranial nerves transmit impulses as part of the autonomic nervous system

   (Figure 17.2). These are cranial nerves numbers _____, _____, _____, and _____.

   Which of these nerves stimulates production of tears from lacrimal glands? _____

   Which two pairs of nerves stimulate salivary glands? _____ _____

   Which pair of nerves innervates organs such as airways, heart, and intestine? _____

f. Which three pairs of cranial nerves control eye movements (Exhibit 11.2)? _____,

   _____; and _____. Which of these also stimulates the sphincter muscle of the

   iris to control the size of the pupil (Figure 17.2)? _____

✓▲ **D4.** Check your understanding of cranial nerves by completing this exercise. Write the name of the correct cranial nerve following the related description.

a. Differs from all other cranial nerves in that it originates from the brainstem and from

   the spinal cord: _____

b. Eighth cranial nerve (VIII): _____

c. Is widely distributed into neck, thorax, and abdomen: _____

d. Senses toothache, pain under a contact lens, wind on the face: _____

e. The largest cranial nerve; has three parts (ophthalmic, maxillary, and mandibular):

   _____

f. Two nerves that contain taste fibers and autonomic fibers to salivary glands:

   _____, _____

g. Three cranial nerves that are purely or mainly sensory: _____,

   _____, _____

✓ **D5.** Write the number of the cranial nerve related to each of the following disorders.

_____ a. Bell's palsy

_____ b. Inability to shrug shoulders or turn head

_____ c. Anosmia

_____ d. Strabismus and diplopia

_____ e. Blindness

_____ f. Trigeminal neuralgia (tic douloureux)

_____ g. Paralysis of vocal cords; loss of sensation
of many organs

_____ h. Vertigo and nystagmus

E. Developmental anatomy and aging of the nervous system (pages 472–474)

✓ **E1.** Check your understanding of the early development of the nervous system by completing this exercise.

a. The nervous system begins to develop during the third week of gestation when the

_____-derm forms a thickening called the _____
plate. Soon a longitudinal depression is found in the plate; this is the neural

_____. As neural folds on the sides of the groove grow and meet,

they form the neural _____.

b. Three types of cells form the walls of this tube. Two of these are the marginal layer
which forms _(gray? white?)_ matter and the mantle layer which becomes _(gray? white?)_
matter.

c. The neural crest forms _(peripheral? central?)_ nervous system structures such as ganglia

as well as spinal and cranial _____.

d. The anterior portion of the neural plate and tube develops into three enlarged

_____ by the fourth week. These fluid-filled cavities eventually

become the _____ which are filled with _____ fluid.

e. By week five the primary vesicles have flexed (bent) to form a total of _(five? ten?)_ secondary vesicles, described in Checkpoint E2.

✓ **E2.** Fill in the blanks in this table outlining brain development.

a. Primary Vesicles (3)	b. Secondary Vesicles (5)	c. Principal Parts of Brain Formed (7)
Prosencephalon (_____-brain) →	1. Diencephalon → 2. _____ →	1A. _____ 1B. _____ 1C. _____ 1D. _____ 2. Cerebrum
_____-encephalon → (Midbrain)	3. Mesencephalon →	3A. _____
_____-encephalon (_____) →	4. _____-encephalon → 5. Met-_____ →	4A. _____ 5A. Pons 5B. _____

✓ **E3.** Do this exercise describing effects of aging on the nervous system.

   a. The total number of nerve cells _____-creases with age. Because conduction veloc-

     ity *(increases? slows down?)*, the time required for a typical reflex is _____-creased.

   b. Parkinson's disease is the most common *(motor? sensory?)* disorder in the elderly.

   c. The sense of touch is likely to be *(impaired? heightened?)* in old age, placing the older person at *(higher? lower?)* risk for bums or other skin injury.

F. Disorders, medical terminology (pages 479–480)

  F1. CVA refers to _____

    CVAs are *(rare? common?)* brain disorders. Describe three causes of CVAs.

  F2. Contrast CVA with TIA.

  F3. Describe the following aspects of Alzheimer's disease (AD):

  a. Progressive changes in behavior

  b. Pathological findings of brain tissue

  c. Risk factors for AD

✓ **F4.** *For extra review.* Match the name of the disorder with the related description.

Agnosia	Neuralgia
Alzheimer's disease (AD)	Stupor
Cerebrovascular accident (CVA)	Transient ischemic attack (TIA)
Lethargy	

  a. Characterized by loss of neurons, amyloid plaques, and neurofibrillary tangles:

    _____

  b. Inability to recognize significance of sensory stimuli, such as tactile or gustatory:

    _____

  c. Temporary cerebral dysfunction caused by interference of blood supply to the brain:

    _____

  d. Most common brain disorder; also called stroke: _____

  e. Attack of pain along an entire nerve, for example, tic douloureux: _____

  f. Functional sluggishness: _____

g.   Unresponsiveness; can be aroused only briefly and by repeated, vigorous stimulation:

_____

ANSWERS TO SELECTED CHECKPOINTS: CHAPTER 14

A1. (a) Brain stem. (b) Diencephalon. (c) Cerebrum. (d) Cerebellum.

A2. Skull bones, meninges, cerebrospinal fluid (CSF).

A4. (a) Tentorium cerebelli. (b) Falx cerebelli. (c) Falx cerebri.

A6. (a) Less; tight, astrocytes. (b) The brain is protected from many substances that could harm the brain but are kept from the brain by this barrier. (c) Glucose, oxygen, carbon dioxide, and anesthetics. (d) Certain helpful medications, such as most antibiotics, and chemotherapeutic chemicals cannot cross this barrier.

A7. (a) B. (b) B. (c) A, C, D. (d) B, C, D. (e) B, D, E. (f) C.

A8. D E F G H I J K C A B.

B1. (a) Conduction; both sensory and motor. (b) Pyramidal, motor; corticospinal. (c) Decussation; left; axons, cortico. (d) Cuneatus, gracilis, two-point discriminatory touch, proprioception, vibrations. (e) Cardiac, respiratory, vasomotor (vasoconstrictor). (f) Part of VIII, as well as IX–XII.

B2. (a) Bridge; spinal cord, medulla; cerebellum. (b) V–VII and part of VIII. (c) Respiration.

B3. (a) 2.5. (b) Anterior and superior.

B4. Reticular activating system; sound, temperature, pain; medulla, pons, and midbrain, as well as portions of the diencephalon (thalamus and hypothalamus) and spinal cord.

B6. (a) Superior. (b) Middle. (c) Inferior. (Note that the names of these tracts are logical, based on locations of the midbrain, pons, and medulla.)

B7. (a) Deep within; four; epi-, sub-, hypo-; thalamus. (b) Roof; pineal, habenular, choroid; melatonin, clock; smell. (c) Movements. (d) Two, lateral walls; third. (e) Sensory; touch, pressure, proprioception; geniculate. (f) Motor control, emotions, and memory. (g) Below, third.

B9. (a) H. (b) Mid. (c) Mid. (d) P. (e) Med. (f) H. (g) T. (h) Med. (i) T. (j) H. (k) T. (1) Mid. (m) Med. (n) Mid.

B10. See Table at right.

B11. (a) Hypothalamus, pineal gland, and pituitary. (b) Lack; advantage: because their capillaries are permeable to most substances, they can monitor homeostatic activities such as fluid balance and hunger: disadvantage: they are permeable to harmful substances, for example, possibly the AIDS virus.

C1. (a) Cerebral cortex; gray; cell bodies. (b) 2 to 4 mm. (c) Gyri; fissures, sulci. (d) Hemispheres;

Name of Nucleus	Brain Part	Functions
a. Habenular nucleus	Epi	Smell
b. Subthalamic nucleus	**Sub**	**Movements in general**
c. Red nucleus	(2) **Mid, sub**	**Movements in general**
d. Colliculi	**Mid**	**Movement of head or neck in response to sight or sounds**
e. Nuclei cuneatus and gracilis	**Med**	**Touch, pressure, vibrations**
f. Anterior nucleus	**Thal**	**Emotions, memory**
g. Lateral geniculate nucleus	**Thal**	**Vision**
h. Medial geniculate nucleus	**Thal**	**Hearing**
i. Ventral posterior nucleus	**Thal**	(2) **Taste; touch, pressure, vibrations**
j. Mammillary bodies	**Hypo**	**Smell**
k. Median eminence of tuber cinereum	**Hypo**	**Regulation of anterior pituitary**
l. Supraoptic region	**Hypo**	**Regulation of posterior pituitary**

white, corpus callosum. (e) Dura mater; between cerebral hemispheres; superior and inferior sagittal sinuses (see Figure 14.2 in the text).

C2. See Figure LG 14.2A.

C3. (a) C. (b) P. (c) A.

C6. Contusion.

C7. (a) Motor, sensory. (b) Primary, association; primary; association.

C8. See Figure LG 14.2A.

C9. (a) PV. (b) PS. (c) PS. (d) PS. (e) PA. (f) G. (g) PM. (h) M. (i) PO. (j) F.

C10. (a) Left; persons, regardless of handedness. (b) Speak; understand; written, spoken.

Figure LG 14.2A Right lateral view of lobes and fissures of the cerebrum.

C11.B D

C12. (a) T. (b) D. (c) B. (d)A.

D1. (a) 12; brain, cranium. (b) I; XII. (c) Mixed; no (the first two are purely sensory).

D2. Refer to answers to LG Chapter 7, Activity B7, page 138.

D3 (a) Smell; superior; ethmoid, bulb; tract, frontal. (b) II; vision, one; ganglion; chiasma, thalamus; occipital. (c) Ear; cochlear, cochlea; vestibular. (d) Facial, VIII; internal; movement; anterior. (e) III, VII, IX, X; VII; VII, IX; X. (f) III, IV, VI; III.

D4. (a) Accessory. (b) Vestibulocochlear. (c) Vagus. (d) Trigeminal. (e) Trigeminal. (f) Facial, glossopharyngeal. (g) Olfactory, optic, vestibulocochlear.

D5. VII. (a) VII. (b) XI. (c) I. (d) III. (e) II. (f) V. (g) X. (h) VIII (vestibular branch).

E1. (a) Ecto, neural; groove; tube. (b) White, gray. (c) Peripheral, nerves. (d) Primary vesicles; ventricles of brain, cerebrospinal. (e) Five.

E2.

a. Primary Vesicles (3)	b. Secondary Vesicles (5)	c. Principal Parts of Brain Formed (7)
Prosencephalon (__Fore__-brain) →	1. Diencephalon →	1A. __Epithalamus__ 1B. __Thalamus__ 1C. __Subthalamus__ 1D. __Hypothalamus__
	2. __Telencephalon__ →	2. Cerebrum
__Mes__-encephalon → (Midbrain)	3. Mesencephalon →	3A. __Midbrain__
__Rhomb__-encephalon (__Hindbrain__) →	4. __Myel__-encephalon →	4A. Medulla oblongata
	5. Met-__encephalon__ →	5A. Pons 5B. __Cerebellum__

E3. (a) De; slows down, in. (b) motor. (c) Impaired, higher.

F4. (a) AD. (b) Agnosia. (c) TIA. (d) CVA. (e) Neuralgia. (f) Lethargy. (g) Stupor.

MORE CRITICAL THINKING: CHAPTER 14

1. Consider what effects are likely to result from the following: (a) blockage of the cerebral aqueduct by a tumor of the midbrain; (b) a severe blow to the medulla; (c) a CVA involving the occipital lobe.

2. Name the structures that form the circumventricular organs (CVOs) and explain why they are so named. Describe the structural characteristic that allows them to monitor chemical changes in the blood supply to the brain.

3. Describe the embryonic development of the brain.

4. Describe the structure and functions of the following: (a) limbic system, (b) diencephalon, (c) reticular formation.

5. Contrast motor and sensory functions of cranial nerves V, VII, and IX.

6. Identify structural changes of the brain that are associated with Alzheimer's disease. Describe roles of the following proteins in these changes: beta amyloid, APP, and A68.

7. Contrast a CVA with a TIA.

MASTERY TEST: CHAPTER 14

Questions 1–9: Circle the letter preceding the one best answer to each question.

1. All of these are located in the medulla *except:*
   A. Nucleus cuneatus and nucleus gracilis
   B. Cardiac center, which regulates heart
   C. Site of decussation (crossing) of pyramidal tracts
   D. The olive
   E. Origin of cranial nerves V to VIII

2. Which of these is a function of the postcentral gyms?
   A. Controls specific groups of muscles, causing their contraction
   B. Receives general sensations from skin, muscles, and viscera
   C. Receives olfactory impulses
   D. Primary visual reception area
   E. Somatosensory association area

3. All of these structures contain cerebrospinal fluid *except:*
    A. Subdural space
    B. Ventricles of the brain
    C. Central canal of the spinal cord
    D. Subarachnoid space
4. Damage to the accessory nerve would result in:
    A. Inability to turn the head or shrug shoulders
    B. Loss of normal speech function
    C. Hearing loss
    D. Changes in heart rate
    E. Anosmia
5. Which statement about the trigeminal nerve is *false?*
    A. It sends motor fibers to the muscles used for chewing.
    B. Pain in this nerve is called trigeminal neuralgia (tic douloureux).
    C. It carries sensory fibers for pain, temperature, and touch from the face, including eyes, lips, and teeth area.
    D. It is the smallest cranial nerve.
6. All of these are functions of the hypothalamus *except:*
    A. Control of body temperature
    B. Release of chemicals (regulating factors) that affect release or inhibition of hormones
    C. Principal relay station for sensory impulses
    D. Center for mind-over-body (psychosomatic) phenomena
    E. Involved in maintaining sleeping or waking state
7. All of these are located in the midbrain *except:*
    A. Apneustic area
    B. Cerebral peduncles
    C. Cerebral aqueduct
    D. Origin of cranial nerves Ill and IV
    E. Red nucleus
8. All of the following are mixed cranial nerves except cranial nerve number:
    A. II
    B. III
    C. V
    D. VII
    E. IX
9. All of the following contain parts of motor pathways except
    A. Red nucleus
    B. Substantia nigra
    C. Corpora quadrigemina
    D. Medial lemniscus

Questions 10–14: Arrange the answers in correct sequence.

_____ _____ _____ 10. From superior to inferior:
    A. Thalamus
    B. Hypothalamus
    C. Corpus callosum

_____ _____ _____ 11. Order in which impulses are relayed in conduction pathway for vision:
    A. Optic nerve
    B. Optic tract
    C. Optic chiasma

_____ _____ _____ _____ 12. Pathway of cerebrospinal fluid, from formation to final destination:
    A. Choroid plexus in ventricle
    B. Subarachnoid space
    C. Cranial venous sinus
    D. Arachnoid villi

_____ _____ _____ 13. From anterior to posterior:
    A. Fourth ventricle
    B. Pons and medulla
    C. Cerebellum

_____ _____ _____ 14. Location of parts of the cerebral cortex from anterior to posterior:
    A. Premotor cortex
    B. Somatosensory cortex
    C. Primary motor cortex
    D. Visual cortex

Questions 15–20: Circle T (true) or F (false). If the statement is false, change the underlined word or phrase so that the statement is correct.

T F   15. The thalamus, hypothalamus, and cerebrum are all developed from the <u>forebrain</u> (<u>prosencephalon</u>).

T F   16. The language areas are located in the <u>cerebellar cortex</u>.

T F   17. The limbic system functions in control of <u>emotional aspects of behavior</u>.

T F   18. The feeding, satiety, and thirst centers are all located in the <u>thalamus</u>.

T F   19. The corpus callosum contains <u>projection</u> fibers.

T F   20. A cerebrovascular accident (CVA) can be expected to produce <u>more lasting</u> effects than a transient ischemic attack (TIA).

Questions 21–25: Fill-ins. Complete each sentence with the word or phrase that best fits.

_____ 21. The _____ is the extension of dura mater that separates the cerebral hemispheres, whereas the _____ separates the cerebrum from the cerebellum.

_____ 22. The superior and inferior colliculi, associated with movements of eyeballs and head in response to visual stimuli, are located in the _____.

_____ 23. If the brain is deprived of oxygen for more than 4 minutes, _____ of brain cells release enzymes that cause these cells to self-destruct.

_____ 24. The _____ controls wakefulness.

_____ 25. The corpus striatum, including the caudate and lentiform nuclei, is part of the _____.

ANSWERS TO MASTERY TEST: CHAPTER 14

Multiple Choice
1. E
2. B
3. A
4. A
5. D
6. C
7. A
8. A
9. D

Arrange
10. C A B
11. A C B
12. A B D C
13. B A C
14. A C B D

True-False
15. T
16. F. Cerebral cortex
17. T
18. F. Hypothalamus
19. F. Commissural
20. T

Fill-ins
21. Falx cerebri; tentorium cerebelli
22. Midbrain
23. Lysosomes
24. Reticular formation or reticular activating system (RAS)
25. Basal ganglia

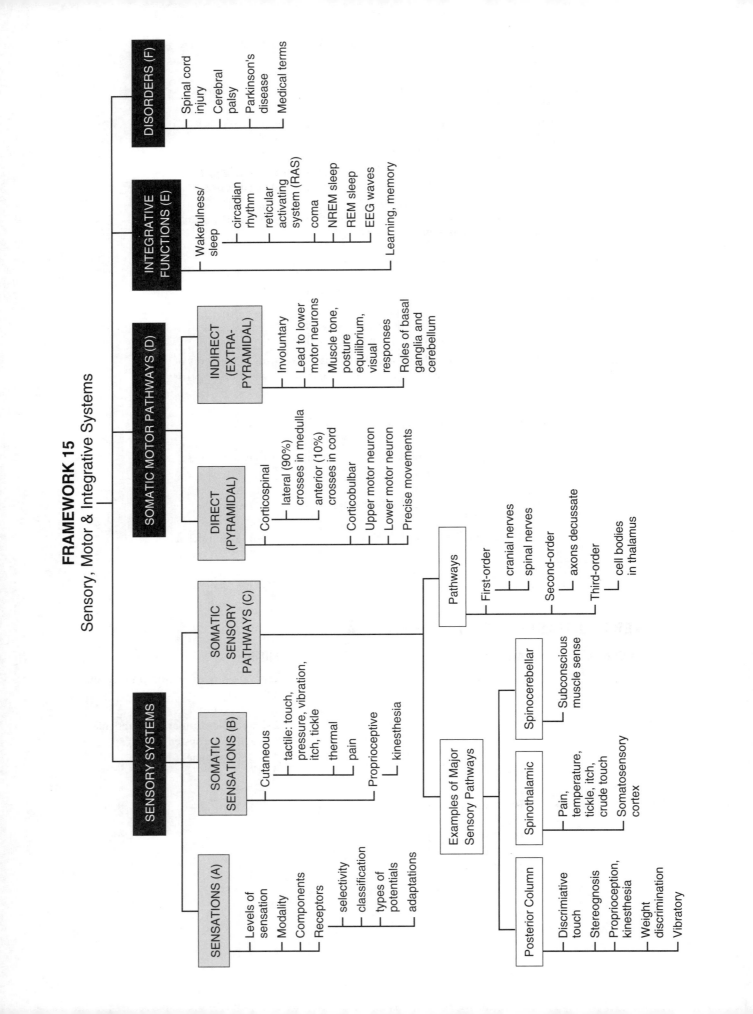

FRAMEWORK 15
Sensory, Motor & Integrative Systems

SENSORY SYSTEMS

SENSATIONS (A)
- Levels of sensation
- Modality
- Components
- Receptors
  - selectivity
  - classification
  - types of potentials
  - adaptations

SOMATIC SENSATIONS (B)
- Cutaneous
  - tactile: touch, pressure, vibration, itch, tickle
  - thermal
  - pain
- Proprioceptive
  - kinesthesia

SOMATIC SENSORY PATHWAYS (C)

Examples of Major Sensory Pathways

- Posterior Column
  - Discrimiative touch
  - Stereognosis
  - Proprioception, kinesthesia
  - Weight discrimination
  - Vibratory
- Spinothalamic
  - Pain, temperature, tickle, itch, crude touch
  - Somatosensory cortex
- Spinocerebellar
  - Subconscious muscle sense

Pathways
- First-order
  - cranial nerves
  - spinal nerves
- Second-order
  - axons decussate
- Third-order
  - cell bodies in thalamus

SOMATIC MOTOR PATHWAYS (D)

DIRECT (PYRAMIDAL)
- Corticospinal
  - lateral (90%) crosses in medulla
  - anterior (10%) crosses in cord
- Corticobulbar
- Upper motor neuron
- Lower motor neuron
- Precise movements

INDIRECT (EXTRA-PYRAMIDAL)
- Involuntary
- Lead to lower motor neurons
- Muscle tone, posture equilibrium, visual responses
- Roles of basal ganglia and cerebellum

INTEGRATIVE FUNCTIONS (E)
- Wakefulness/sleep
  - circadian rhythm
  - reticular activating system (RAS)
  - coma
  - NREM sleep
  - REM sleep
  - EEG waves
- Learning, memory

DISORDERS (F)
- Spinal cord injury
- Cerebral palsy
- Parkinson's disease
- Medical terms

C H A P T E R **15**

Sensory, Motor, and Integrative Systems

The last three chapters have introduced the components of the nervous system and demonstrated the arrangement of neurons in the spinal cord, spinal nerves, brain, and cranial nerves. Chapter 15 integrates this information in the study of major nerve pathways including those for (a) general senses (such as touch, pain, and temperature), (b) motor control routes for precise, conscious movements and for unconscious movements, and (c) complex integrative functions such as memory and sleep.

As you begin your study of sensory, motor, and integrative systems, carefully examine the Chapter 15 Topic Outline and Objectives; check off each one as you complete it. To organize your study of this chapter glance over the Chapter 15 Framework now. Be sure to refer to the Framework frequently and note relationships among key terms in each section.

TOPIC OUTLINE AND OBJECTIVES

A. Sensations

1. Define a sensation and discuss its components.
2. Describe the different ways to classify sensory receptors.

B. Somatic sensations

3. Describe the location and functions of the receptors for tactile, thermal, and pain sensations.
4. Identify the receptors for proprioception, and describe their functions.

C. Somatic sensory pathways

5. Describe the neuronal components and functions of the posterior column-medial lemniscus, the anterolateral, and the spinocerebellar pathways.

D. Somatic motor pathways

6. Compare the locations and functions of the direct and indirect motor pathways.
7. Explain how the basal ganglia and cerebellum contribute to motor responses.

E. Integrative functions

8. Compare the integrative cerebral functions of wakefulness and sleep, and learning and memory.

F. Disorders

WORDBYTES

Now become familiar with the language of this chapter by studying each wordbyte, its meaning, and an example of its use within a term. After you study the entire list, self-check your understanding by writing the meaning of each wordbyte on the line. As you continue through the *Learning Guide,* identify (and fill in) additional terms that contain the same wordbyte.

Wordbyte	Self-check	Meaning	Example(s)
a-, an-	_____	without	*an*esthesia
-ceptor	_____	receiver	proprio*ceptor*
circa-	_____	about	*circa*dian
dia-	_____	a day	circa*dia*n
-esthesia	_____	sensation	an*esthesia*, par*esthesia*
hemi-	_____	half	*hemi*plegia, *hemi*sphere
kin-	_____	motion	*kin*esthesia
noci-	_____	hurt, pain	*noci*ceptor
para-	_____	abnormal, beyond	*para*esthesia
-plegia	_____	paralysis or stroke	hemi*plegia*
propio-	_____	one's own	*proprio*ception
quad-	_____	four	*quad*riplegia
rhiz-	_____	root	*rhiz*otomy
tact-	_____	touch	*tact*ile

CHECKPOINTS

A. Sensations (pages 485–489)

✓ **A1.** Review the three basic functions of the nervous system by listing them in sequence here:

a. Receiving _____ input

b. _____; _____; _____ information

c. Transmitting _____ impulses to _____ or _____

A2. For a moment, visualize what your life would be like if you were unable to experience any sensations. List the three types of sensations that you believe you would miss most.

_____ _____ _____ Now write a sentence describing how your health and safety might be endangered by lack of ability to perceive sensations.

✓ **A3.** Do this activity about levels of sensation.

a. *(Perception? Sensation?)* is the awareness (either conscious or unconscious) of stimuli,

such as the act of seeing a bird, whereas _____ is the conscious awareness and interpretation of sensation, such as the recognition of that bird as a bluebird and not a robin.

b. Identify levels of sensation by matching parts of the nervous system listed in the box with descriptions below.

B. Brain stem	**S. Spinal cord**
C. Cerebral cortex	**T. Thalamus**

_____ 1. Immediate reflex action is possible without involvement of the brain.

_____ 2. Involves subconscious motor reactions, such as those that facilitate balance and equilibrium.

_____ 3. Offers a general (or "crude") awareness of locale and type of sensations, such as

awareness of pressure within the leg or that something is being heard; offers no specifics

_____ 4. Gives precise information about sensations, such as distinguishing Bach from Baez from the B-52s, as well as memories of hearing a particular musical work

✓ **A4.** The quality that makes the sensation of pain different from the sense of touch, or

which distinguishes hearing from vision is called the _____ of sensation. Any single sensory neuron carries sensations of *(one modality? many modalities?)*. Different modalities of sensations are distinguished precisely at the level of the *(spinal cord? thalamus? cerebral cortex?)*.

✓ **A5.** Select from answers in the box to correctly categorize different modalities of sensations. Write one answer on each line provided. One answer will not be used; other answers will each be used once.

Awareness of joint positions	**Pain**	**Vibration**
Awareness of movements	**Pressure**	**Vision**
Cold	**Smell**	**Warm**
Equilibrium	**Taste**	
Hearing	**Touch**	

a. Thermal sensations: _____ _____

b. Proprioceptive sensations: _____ _____

c. Tactile sensations: _____ _____ _____

d. Special senses: _____ _____ _____

_____ _____

✓ **A6.** Arrange in correct order the components in the pathway of sensation.

IGP.	**Impulse generation and propagation**	**S.**	**Stimulation**
Int.	**Integration**	**T.**	**Transduction**

1. _____ → 2. _____ → 3. _____ → 4. _____

Now match each of these components of sensation with the correct description below.

_____ a. Activation of a sensory neuron

_____ b. Stimulus → graded potential

_____ c. Graded potential → nerve impulse; impulse is conveyed along a first-order

neuron into the spinal cord or brainstem up to a higher level of the brain.

_____ d. Nerve impulse → sensation; usually occurs in the cerebral cortex

✓ **A7.** Complete Table 15.1, categorizing the nature of different sensory receptors.

Table LG 15.1 Classification of sensory receptors

Receptor	Structure	Type of Graded Potential	Location (-ceptor)	Type of Stimulus Detected
	Free Encapsulated Separate cells	Generator Receptor	Extero- Intero- Proprio-	Mechanoreceptor Thermoreceptor Nociceptor Photoreceptor Chemoreceptor
1. Pressure		Generator		
2. Pain (skin)	Free		Extero-	
3. Pain (stomach)				Nociceptor
4. Vision		Receptor		Photoreceptor
5. Equilibrium			Proprio-	
6. Smell	Separate cells			
7. Cold on skin	Free			Thermoreceptor
8. Changes of O₂ level in blood	(Omit)	(Omit)	Intero-	

✓ **A8.** Do this exercise on the characteristics of sensations.
   a. Arrange these receptors from simplest to most complex structurally:

   _____ _____ _____

   A. Receptors for special senses such as vision or hearing
   B. Free nerve endings, such as receptors for pain
   C. Receptors for touch or pressure
   b. Describe what is meant by the *selectivity* of receptors.

c. Recall getting dressed this morning. The fact that you felt the shirt touching your back just after you put on that shirt, but that the awareness of that touch has

dissipated over time, is known as the characteristic of _____. Circle the type of receptors that adapt most slowly *(Pain? Pressure? Smell? Touch?)*. Explain how such slow adaptation is advantageous to you.

B. Somatic senses (pages 489–494)

B1. Contrast these terms: *somatic sensations/cutaneous sensations*

✓ **B2.** Do the following exercise about tactile sensations.
a. Name the tactile sensations.

These sensations are all sensed by _____-receptors. Receptors for these sensations are *(evenly? unevenly?)* distributed throughout the skin of the body.
b. *(Discriminative? Crude?)* touch refers to ability to determine that something has touched the skin, but its precise location, shape, size, or texture cannot be distinguished.
c. In general, pressure receptors are more *(superficial? deep?)* in location than touch receptors.
d. Sensations of _____ result from rapid repetition of sensory impulses from tactile receptors.

✓ **B3.** Match names of receptors with their descriptions. Answers may be used more than once.

C. Corpuscles of touch (Meissner's corpuscles)	**Type I. Type I cutaneous mechanoreceptors** (tactile or Merkel discs)
H. Hair root plexuses	**Type II. Type II cutaneous mechanoreceptors** (end organs of Ruffini)
L. Lamellated (Pacinian) corpuscles	
N. Nociceptors	

_____ a. Egg-shaped masses located in dermal papillae, especially in fingertips, palms of hands, and soles of feet

_____ b. Onion-shaped structures sensitive to pressure and high-frequency vibration

_____ c. Free nerve endings that sense pain

_____ d. Slowly adapting touch receptors (three answers)

_____ e. may respond to any type of stimulus if stimulus is strong enough to cause tissue damage

_____ f. Surround hair follicles; can detect breeze blowing against hairs on skin.

B4. *For extra review.* Refer to Figure LG 5.1. Identify and label types of receptors on that figure. Add your own drawings of three other types of receptors in appropriate layers of skin. Label those receptors also.

✓ **B5.** For additional review of cutaneous sensations, do this Checkpoint.

    a. Explain how the sensation of itch develops.

    b. Name three chemicals that may be released from injured tissues that may stimulate

        pain receptors. _____, _____, _____

    c. Fast pain is felt in *(deep? superficial?)* tissues of the body. Circle the qualities that are associated with fast pain.

        Aching   **Acute**   **Burning**   **Chronic**   **Pricking**   **Sharp**   **Throbbing**

    d. Which type of pain impulses travel along unmyelinated C nerve fibers? *(Fast? Slow?)*

    e. Circle the type of pain that arises from receptors in skin.

        Deep somatic pain   **Superficial somatic pain**   **Visceral pain**

✓▲ **B6.** *The Big Picture: Looking Ahead and a clinical challenge.* Refer to figures in upcoming chapters of your text to help you determine why patients experiencing visceral pain (as during a heart attack or gallbladder attack) may feel pain in locations quite distant from these two organs.

    a. Pain impulses that originate from the heart, for example, during a "heart attack" involving the left ventricle (Figure 20.4), enter the spinal cord at the same level as do

        sensory fibers from skin covering the _____. This level of the cord is

        about _____ to _____.

    b. Refer to Figure 15.3. Notice that some liver and gallbladder pain is felt in a region

        quite distant from these organs, that is, in the _____ region. These

        organs lie just inferior to the dome-shaped _____, which is supplied by the phrenic nerve from the *(cervical? brachial? lumbar?)* plexus. A painful gallbladder can send impulses to the cord via the phrenic nerve, which enters the cord at the

        same level (about C _____ or C _____) as nerves from the neck and shoulder. So

        gallbladder pain is said to be _____ to the neck and shoulder area.

✓ **B7.** *Critical thinking.* Mr. Connors was in a motorcycle accident that resulted in his left leg being severed above the knee. Months later he is referred to the university pain clinic for an assessment. He states, "I feel this pain in my left foot, and sometimes my toes itch like crazy, but they're not there to scratch!" How might you interpret his experience?

✓ **B8.** *A clinical correlation.* Match methods of pain relief listed in the box with the following descriptions.

A.	Anesthetic (Novocaine)	O.	Opioids such as morphine,
AAI.	Aspirin, acetaminophen (Tylenol), ibuprofen (Motrin, Advil)		codeine, or meperidine (Demerol)

_____ a. Medication that alters the quality of pain perception so that pain feels less noxious

_____ b. Medication that inhibits formation of chemicals (prostaglandins) that stimulate pain receptors.

_____ c. Medication that blocks conduction of nerve impulses in first-order neurons

✓ **B9.** Close your eyes for a moment. Reflect on the position of your arms and legs as well as the degree of tension in your muscles and tendons. This awareness of body

position and movements is known as _____. Now match names of proprioceptors listed in the box with their descriptions below. Answers may be used more than once.

H.	Hair cells of the inner ear	M.	Muscle spindles
J.	Joint kinesthetic receptors	T.	Tendon organs

_____ a. Located within and around articular capsules of synovial joints; help protect joints

_____ b. Type Ia and type II sensory fibers sensitive to stretch help prevent overstretching that could injure muscles

_____ c. Consists of a thin capsule penetrated by type Ib fibers

_____ d. Monitor the force of contraction of muscles; help to protect tendons (and their muscles) from excessive tension

C. Somatic sensory pathways (pages 494–498)

✓ **C1.** Write terms listed in the box after correct definitions below to identify types of sensations you experience.

Crude touch	Stereognosis
Kinesthesia	Two-point discriminative touch
Proprioception	

a. Ability to sense touch on the back of your neck but inability to specify precise

location of the touch: _____

b. Ability to recognize exact location of a sharp pencil touched to your fingertip and ability to distinguish whether one or two sharp pencils are touching nearby points on

that fingertip: _____

c. Ability to tell—with your eyes closed—the exact position of your arms and legs:

_____

d. Awareness of precise movements of your fingers as you type a paper: _____

e. Ability to distinguish a dime from a button based on size, shape, and texture:

_____

✓ **C2.** Contrast functions of major sensory pathways by selecting answers from the box that match the types of sensations they transmit.

APSC.	**Anterior and posterior spinocerebellar tract**
AST.	**Anterior spinothalamic tract**
LST.	**Lateral spinothalamic tract**
PCML.	**Posterior column and medical lemniscus tract**

_____ a. Pain and temperature

_____ b. Crude touch or pressure

_____ c. Discriminative, two-point discrimination

_____ d. Proprioception, kinesthesia, stereognosis

_____ e. Weight discrimination and vibration

_____ f. Tickle and itch

_____ g. Subconscious proprioceptive impulses to the cerebellum needed for posture, balance, and coordination

✓ **C3.** Sensory pathways involve *(one? two? three?)*-neuron sets that allow nerve impulses from receptors, for example, in skin or muscles, to reach the primary somatosensory area of the cerebral cortex. Identify first-order (I), second-order (II), and third-order (III) neurons by writing I, II, or III next to the related neuron.

_____ a. Crosses to the opposite side of the spinal cord or brainstem and ascends to the thalamus

_____ b. Reaches the spinal cord or brainstem

_____ c. Passes to the postcentral gyrus of the cerebral cortex

✓ **C4.** Write in locations of cell bodies of the three neurons in these two pathways.

Pathway	First-Order	Second-Order	Third-Order
a. Posterior column–medial lemniscus			
b. Anterolateral (spinothalamic)			

✓▲ **C5.** Check your understanding of pathways by drawing them according to the directions below. Try to draw them from memory; do not refer to the text figures or the table until you finish. Label the *first-*, *second-*, and *third-order neurons* as *I*, *II*, and *III*.
  a. Draw in *blue* on Figure LG 15.1 the pathway that allows you to accomplish stereognosis so that you can recognize the shape of a key in your left hand. (*Hint:* The correct recognition of the key involves discriminative touch, proprioception, and weight discrimination.)
  b. Now draw in *red* the pathway for sensing pain and temperature as the index finger of the left hand comes too close to a flame. For simplification, draw only the lateral (not anterior) component.

✓▲ **C6.** Do this learning activity about sensory reception in the cerebral cortex.
  a. Picture the brain as roughly similar in shape to a small, rounded loaf of bread. The

  crust would be comparable to the cerebral _____ .
  b. Suppose you cut a slice of bread in midloaf that represents the postcentral gyrus. It looks somewhat like the "slice" or section of brain shown at the top of Figure LG 15.1.

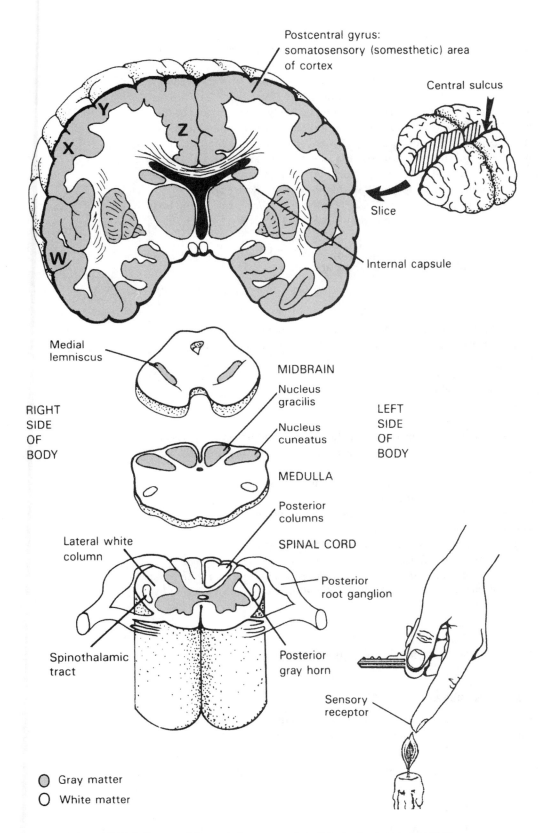

Postcentral gyrus: somatosensory (somesthetic) area of cortex

Central sulcus

Slice

Internal capsule

Medial lemniscus

MIDBRAIN

Nucleus gracilis

RIGHT SIDE OF BODY

LEFT SIDE OF BODY

Nucleus cuneatus

MEDULLA

Posterior columns

Lateral white column

SPINAL CORD

Posterior root ganglion

Spinothalamic tract

Posterior gray horn

Sensory receptor

○ Gray matter
○ White matter

Figure LG 15.1 Diagram of the central nervous system. Refer to Checkpoints C5 and C6.

Locate the Z on that figure. This portion (next to the _____ fissure) receives sensations from the *(face? hand? hip? foot?)*. (Refer to Figure 15.6a, of your text for help.)

c. Suppose a person suffered a loss of blood supply (CVA or "stroke") to the area of the brain marked X in Figure LG 15.1. Loss of sensation in the *(face? hand? leg?)* area would result. Damage to the area marked W would be likely to lead to loss of the sense of *(vision? hearing? taste?)* because the W is located on the temporal lobe.

d. The letters W, X, Y, and Z are marked on the *(right? left?)* side of the brain. (Note that this is an anterior view of this brain section. If a friend stands and faces you, the *right* side of your friend's brain will be in your *left* field of view.) What would be the most profound effects of brain damage to area Y of the postcentral gyrus? Loss of *(sensation? movement?)* to the *(face? hand? foot?)* on the *(right? left?)* side or the body.

e. Certain regions of the body are represented by large areas of the

_____ _____ cortex, indicating that these areas *(are? are not?)* very sensitive. Name two of these areas.

✓ **C7.** Discuss sensory pathways to the cerebellum in this Checkpoint.
   a. Circle the two pathways that are the major routes for proprioceptive impulses to the cerebellum.
      A. Anterior spinocerebellar tract
      B. Posterior spinocerebellar tract
      C. Cuneocerebellar
      D. Rostrocerebellar
   b. Proprioception in the left leg is transmitted by the anterior and posterior spinocerebellar tracts to the *(right? left?)* side of the cerebellum. These sensory impulses *(are? are not?)* consciously perceived.
   c. Circle all functions of the spinocerebellar pathways in this list:
      A. Provide awareness of touch and tickle
      B. Needed for balance and coordination, for example, in walking
      C. Needed to help maintain posture
      D. Provide motor impulses that initiate fine motor movements
   d. Syphilis is a sexually transmitted disease (STD) caused by a *(bacterium? fungus? virus?)*. Which stage occurs latest in the life of a syphilis patient? *(Primary? Secondary? Tertiary?)* syphilis. Which one of the tracts listed above (in C7a) is most affected by

      tertiary syphilis? _____

   C8. Review locations of tracts in the spinal cord by referring to LG Chapter 13, Checkpoint B3.

D. Somatic motor pathways (pages 498–502)

✓ **D1.** The linking of sensory to motor response occurs at different levels in the central nervous system. Indicate which is the highest level required for each of these responses.

BG. Basal ganglia	C. Cerebellum	S. Spinal cord

_____ a. Quick withdrawal of a hand when it touches a hot object

_____ b. Unconscious responses to proprioceptive impulses to permit smooth, coordinated movements (two answers)

_____ c. Indirect pathways may synapse here (two answers)

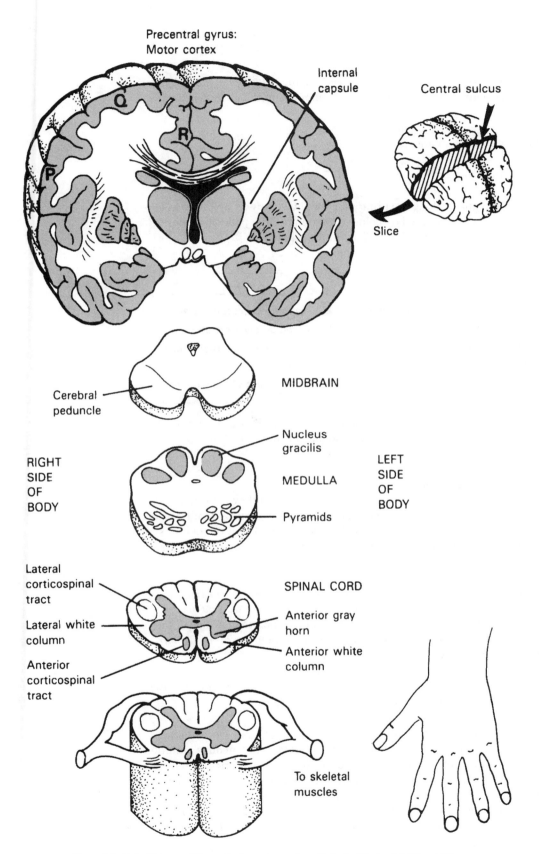

Figure LG 15.2 Diagram of the central nervous system. Refer to Checkpoint D2 and D4.

✓▲ **D2.** Refer to Figure LG 15.2 and do the following exercise about motor control.
   a. This "slice" or section of the brain containing the primary motor cortex is located in the *(frontal? parietal?)* lobe.
   b. Area P in this figure controls *(sensation? movement?)* to the *(left? right?)* *(arm? leg? side of the face?)*. Motor neurons in area P would send impulses to the pons to activate

   cranial nerve _____ to facial muscles.
   c. Describe effects of damage to area R.

✓ **D3.** Show the route of impulses along the principal pyramidal pathway by listing in correct sequence the structures in the box below that comprise the pathway. Refer to Figure 15.7 in the text if you have difficulty with this activity.

   _____ _____ _____ _____ _____ _____ _____ _____

A.	Anterior gray horn (lower motor neuron)	E.	Lateral corticospinal tract
B.	Midbrain and pons	F.	Medulla, decussation site
C.	Effector (skeletal muscle)	G.	Precentral gyms (upper motor
D.	Internal capsule		neuron)
		H.	Ventral root of spinal nerve

✓▲ **D4.** Now draw this pathway on Figure LG 15.2. Show the control of muscles in the left hand by means of neurons in the lateral corticospinal pathway. Label the *upper* and *lower motor neurons*.

✓ **D5.** Complete this exercise on motor pathways.
   a. Corticospinal tracts are classified as *(direct? indirect?)* pathways; they control *(precise, voluntary? involuntary?)* movements. Based on their structure, corticospinal tracts are also known as *(pyramidal? extrapyramidal?)* pathways. About 60% of the neurons in these tracts originate in the *(pre? post?)*-central gyrus, whereas about 40% originate

   in the _____ gyrus.
   b. Axons of about 90% of upper motor neurons (UMNs) pass through *(anterior? lateral?)* corticospinal tracts; these are axons that *(decussated or crossed? did not decussate or cross?)* in the medulla. Lateral corticospinal tracts on the left side of the spinal cord consist of axons that originated in upper motor neurons on the *(right? left?)* side of the motor cortex.
   c. These tracts (left lateral corticospinal) then convey impulses to LMNs located in the *(right? left?)* *(anterior? lateral? posterior?)* gray horns of the spinal cord. In most cases there *(are? are not?)* interneurons between UMNs and LMNs.

   d. About _____% of UMN axons pass through _____ corticospinal tracts. Because axons of these UMNs *(do? do not?)* decussate in the medulla, anterior corticospinal tracts on the left side of the cord consist of axons that originated on the *(right? left?)* side of the motor cortex. These tracts are therefore *(contra? ipsi?)*-lateral. Most of these *(do? do not?)* eventually decussate at lower levels.
   e. Corticobulbar tracts consist of UMNs that terminate in the *(brainstem? anterior gray horn of the cord?)*. Lower motor neurons (LMNs) there pass out through *(cranial? spinal?)* nerves.
   f. Destruction of *(upper? lower?)* motor neurons results in spastic paralysis because the brain is not controlling movements. Flaccid paralysis results from damage of *(upper? lower?)* motor neurons.

g. Why are lower motor neurons called the final *common pathway?*

✓▲ **D6.** *Critical thinking.* Refer to Figure 21.20c as you consider possible effects of cerebrovascular accidents (CVAs or "strokes") involving specific arteries to the brain.

   a. Locate the anterior cerebral arteries on the figure. These proceed anteriorly and pass between the two cerebral hemispheres, supplying area R of Figure LG 15.2. Paralysis and/or sensory loss of *(arm and hand? leg and foot? face?)* is most likely to result from interruption of blood flow through this vessel.

   b. Locate the middle cerebral arteries. These proceed laterally and then pass medial to the temporal lobe to supply pre- and postcentral gyri. A stroke involving this artery is likely to cause effects on *(sensation only? movement only? sensation and movement?).*

   c. Identify the vertebral and basilar arteries and their branches. Strokes involving these vessels *(are? are not?)* likely to cause effects on extrapyramidal pathways. State your rationale.

✓ **D7.** Describe locations of indirect (extrapyramidal) tracts in this exercise:

   a. All of these tracts lie *(within? outside of?)* corticospinal tracts. Most of them originate

      in areas *(within? outside of?)* the cerebral cortex and end in the _____.

   b. Select names of tracts in the box that fit descriptions below. Write one answer on each line.

LRS.	**Lateral reticulospinal**	**TS.**	**Tectospinal**
MRS.	**Medial reticulospinal**	**VS.**	**Vestibulospinal**
RS.	**Rubrospinal**		

_____ 1. You tighten your back muscles as you sit up straight in front of your computer.

_____ 2. Neurons in the left side of the midbrain convey messages to cranial nerves to direct your head and eyes to the right side of the computer monitor.

_____ 3. As you turn your head to the right, muscle tone on your right side is adjusted slightly to help you maintain your balance.

_____ 4. Neurons in the left side of the midbrain send messages to the spinal cord and spinal nerves to move the fingers of your right hand on the keyboard.

D8. Summarize how the basal ganglia and cerebellum, together with associated neurons, help you to carry out coordinated, precise movements, such as those involved in a game of tennis.

✓ **D9.** *A clinical challenge.* Describe two conditions that involve damage to basal ganglia in this exercise.

    a. *(Huntington's? Parkinson's?)* disease involves changes in basal ganglia associated with decrease in the neurotransmitter *(acetylcholine? dopamine? GABA?)*. Signs of this condition are *(jerky movements and facial twitches? muscle rigidity and tremors?)*.

    b. A predominant sign of Huntington's disease (HD) is jerky movements (or "dance")

      known as _____ related to loss of neurons that make the inhibitory

      transmitter named _____. Signs of this progressive condition are most likely to be first noted at about age *(4 years? 40 years? 70 years?)*. A child of a parent with HD has *(almost no? a 50–50?)* chance of having this condition.

E. Integrative functions (pages 502–506)

✓ **E1.** List three types of activities that require complex integration processes by the brain.

✓ **E2.** Complete this exercise about sleep and wakefulness.

    a. _____ rhythm is a term given to the usual daily pattern of sleep and wakefulness. This rhythm appears to be established by a nucleus within the *(basal ganglia? hypothalamus? thalamus?)*.

    b. The RAS (or R_____ A_____ S_____) is a portion of the reticular formation. List three types of sensory input that can stimulate the RAS.

    c. Stimulation of the reticular formation leads to *(increased? decreased?)* activity of the cerebral cortex, as indicated by recordings of brain waves. If RAS and cerebral activation are continued, the state of consciousness is maintained. Inactivation or damage

      of these areas leads to _____ or _____. In the state of *(coma? sleep?)*, the person cannot be roused.

    d. Name one chemical that can induce sleep: _____. This chemical

      binds to _____ receptors so neurons of the RAS are *(activated? inhibited?)*. Explain how coffee or tea can keep you awake.

✓ **E3.** Do the following learning activity about sleep.
    a. Name two kinds of sleep.

    b. In NREM sleep, a person gradually progresses from stage *(1 to 4? 4 to 1?)* into deep sleep. Alpha waves are present in stage(s) *(1? 3 and 4?)*, whereas slow delta waves characterize stage(s) *(1? 3 and 4?)* sleep. Sleep spindles characterize stages

      _____ and _____ of NREM sleep.

c. After moving from stage 1 to stage 4 sleep, a person *(moves directly into? ascends to stages 3 and 2 and then moves into?)* REM sleep.

d. REM is much *(more active? slower?)* sleep than NREM sleep. Respirations and pulse

are _____-creased compared to their levels in stage 4. Alpha waves are present as

in stage *(1? 4?)* sleep. The name REM indicates that _____ movements can be observed during REM sleep. Dreaming occurs during *(stages 3–4 of NREM? REM?)* sleep.

e. Periods of REM and NREM sleep alternate throughout the night in about _____- minute cycles. During the early part of an 8-hour sleep period, REM periods last about *(5–10? 30? 50?)* minutes; they gradually increase in length until the final REM

period lasts about _____ minutes.

f. With aging, the percentage of REM sleep normally _____-creases.

E4. Define these terms: *learning, memory, plasticity.*

E5. List several parts of the brain that are considered to assist with memory.

✓ **E6.** Do this exercise on memory.

a. Looking up a phone number, dialing it, and then quickly forgetting it is an example of *(short-term? long-term?)* memory. One theory of such memory is that the phone num-

ber is remembered only as long as a _____ neuronal circuit is active.

b. Repeated use of a telephone number can commit it to long-term memory. Such rein-

forcement is known as memory _____. *(Most? Very little?)* of the information that comes to conscious attention goes into long-term memory because the brain is selective about what it retains.

c. Some evidence indicates that *(short-term? long-term?)* memory involves electrical and chemical events rather than anatomical changes. Supporting this idea is the fact that chemicals used for anesthesia and shock treatments interfere with *(recent? long-term?)* memory.

E7. Discuss changes in the following anatomical structures or chemicals that may be associated with enhanced long-term memory:

a. Presynaptic terminals, synaptic end bulbs, dendritic branches

b. Glutamate, NMDA glutamate receptors, nitric oxide (NO)

c. Acetylcholine (ACh)

F. Disorders (pages 507–508)

F1. List several causes of spinal cord injury.

✓ **F2.** Match the terms in the box with descriptions below.

H. Hemiplegia	**P. Paraplegia**
M. Monoplegia	**Q. Quadriplegia**

_____ a. Paralysis of one extremity only

_____ b. Paralysis of both legs

_____ c. Paralysis of both arms and both legs

_____ d. Paralysis of the arm, leg, and trunk on one side of the body

✓ **F3.** *A clinical correlation.* Describe changes associated with transection of the spinal cord in this activity.
   a. Hemisection of the cord refers to transection of *(half? all?)* of the spinal cord.
   b. If the posterior columns (cuneatus and gracilis) and lateral corticospinal tracts on the right side of the cord are severed at level T11–T12, symptoms of loss of awareness of muscle sensations (proprioception), loss of touch sensations, and paralysis are likely to occur in the *(right arm? left arm? right leg? left leg?).* (Circle all that apply.)
   c. The period of spinal shock is likely to last for several *(days to weeks? years?).* During this time, the person is likely to experience *(areflexia? hyperreflexia?),* meaning *(exaggerated? no?)* reflexes.
   d. When reflexes gradually return to the injured person, they are likely to return in a specific sequence. Arrange these reflexes in order from first to last, according to their

   return. _____ _____ _____
   A. Stretch reflex
   B. Flexion reflex
   C. Crossed extensor reflex

✓ **F4.** Cerebral palsy (CP) is a group of disorders affecting *(movements? sensations?).* This condition results from brain damage *(in fetal life, at birth, or in infancy? in early adulthood? in later years?).* CP *(is? is not?)* a progressive disease; it *(is? is not?)* reversible.

✓ **F5.** Complete this Checkpoint about Parkinson's disease (PD).

  a. Parkinson's typically affects persons *(in fetal life, at birth, or in infancy? in early adult-hood? in later years?)*. PD *(is? is not?)* a progressive disease; it *(is? is not?)* usually hereditary.

  b. This condition involves a decrease in the release of the neurotransmitter

_____. This chemical is normally produced by cell bodies located in

the substantia nigra, which is part of the _____. Axons lead from

here to the _____, where dopamine (DA) is released. DA is an *(excitatory? inhibitory?)* transmitter.

  c. The basal ganglia neurons produce a neurotransmitter _____ (ACh), which is *(excitatory? inhibitory?)* to skeletal muscles. The result is a relative increase in the ratio of ACh to DA in Parkinson's patients. Explain how this imbalance leads to muscle tremor and rigidity.

  d. Voluntary movements may be slower than normal; this is the condition of *(brady? tachy?)*-kinesia.

  e. Because persons with parkinsonism have a deficit in dopamine, it seems logical to treat the condition by administration of this neurotransmitter. Does this work? *(Yes? No?)* Explain.

  f. Why do levodopa and/or anticholinergic medications provide some relief to these patients?

  g. How do MAO-inhibitor drugs help PD patients?

ANSWERS TO SELECTED CHECKPOINTS: CHAPTER 15

A1. (a) Sensory. (b) Integrating, associating, and storing. (c) Motor (efferent), muscles or glands.

A3. (a) Sensation, perception. (b) 1: S. 2: B. 3: T. 4: C.

A4. Modality; one modality; cerebral cortex.

A5. (a) Cold, warm. (b) Awareness of joint positions, awareness of movements. (c) Pressure, touch, vibration. (d) Equilibrium, hearing, smell, taste, vision.

A6. 1. S → 2. T → 3. IGP → 4. Int. (a) S. (b) T. (c) IGP. (d) Int.

A7. See Table LG 15.1 at right.

A8. (a) B C A. (b) Vigorous response of each type of receptor to a specific type of stimulus (such as retina of eye to light) with little or no response to other types of stimuli (such as lack of retinal response to sound waves). (c) Adaptation; pain; protects you by continuing to provide warning signals of painful stimuli.

B2. (a) Touch, pressure, vibration, as well as itch and tickle; mechano; unevenly. (See Mastery Test Question 6.) (b) Crude. (c) Deep. (d) Vibration.

B3 (a) C. (b) L. (c) N. (d) N, Type I, Type II. (e) N. (f) H.

B5. (a) A local inflammation, for example, from a mosquito bite, causes release of chemicals such as bradykinin, which stimulates free nerve endings and lamellated corpuscles. (b) Prostaglandins, kinins, K⁺. (c) Superficial; acute, sharp, pricking. (d) Slow. (e) Superficial somatic pain.

B6: (a) Medial aspects of left arm; T1, T4. (b) Shoulder and neck (right side); diaphragm, cervical; 3, 4; referred.

B7. Phantom pain (phantom limb sensation). This results from irritation of nerves in the stump. Pathways from the stump to the cord and brain are still intact and may be stimulated even though the original sensory receptors in his foot are missing. Remember that he feels pain or itching in his brain, not in his leg or foot.

B8. (a) 0. (b) AA1. (c) A.

B9. Proprioception. (a) J. (b) M. (c) T. (d) T.

C1. (a) Crude touch. (b) Two-point discriminative touch. (c) Proprioception. (d) Kinesthesia. (e) Stereognosis.

C2. (a) LST. (b) AST. (c–e) PCML. (f) AST. (g) APSC.

C3. Three. (a) II. (b) I. (c) III.

C4.

Pathway	First-Order	Second-Order	Third-Order
a. Posterior column–medial lemniscus	Posterior root ganglion	Medulla: nucleus cuneatus and nucleus gracilis	Thalamus
b. Anterolateral (spinothalamic)	Posterior root ganglion	Posterior gray horn of spinal cord	Thalamus

Table LG 15.1 Classification of sensory receptors

Receptor	Structure	Type of Graded Potential	Location (-ceptor)	Type of Stimulus Detected
	Free Encapsulated Separate cells	Generator Receptor	Extero- Intero- Proprio-	Mechanoreceptor Thermoreceptor Nociceptor Photoreceptor Chemoreceptor
1. Pressure	**Encapsulated**	Generator	**Extero-**	**Mechanoreceptor**
2. Pain (skin)	Free	**Generator**	Extero-	**Nociceptor**
3. Pain (stomach)	**Free**	**Generator**	**Intero-**	Nociceptor
4. Vision	**Seperate cells**	Receptor	**Extero-**	Photoreceptor
5. Equilibrium	**Seperate cells**	**Receptor**	Proprio-	**Mechanoreceptor**
6. Smell	Separate cells	**Receptor**	**Extero-**	**Chemoreceptor**
7. Cold on skin	Free	Generator	Extero-	Thermoreceptor
8. Changes of O₂ level in blood	(Omit)	(Omit)	Intero-	Chemoreceptor

C5. See Figure LG 15.1A.

Figure LG 15.1A Diagram of the central nervous system.

C6. (a) Cortex. (b) Longitudinal, foot. (c) Face; hearing. (d) Right; sensation, hand, left. (Note: some left-handed paralysis also, since 40% of motor control pathways are in the postcentral gyrus; see page 499 of the text.) (e) General sensory or somatosensory, are; lips, face, thumb, and other fingers. (See Figure 15.6a in the text.)

C7. (a) A, B. (b) Left; are not. (c) B, C. (d) Bacterium; tertiary; B.

D1. (a) S. (b–c) BG and C.

D2. (a) Frontal. (b) Movement, left side of the face; VII. (c) Movement of left foot affected (spasticity).

D3. G D B F E A H C.

D4. See Figure 15.2A.

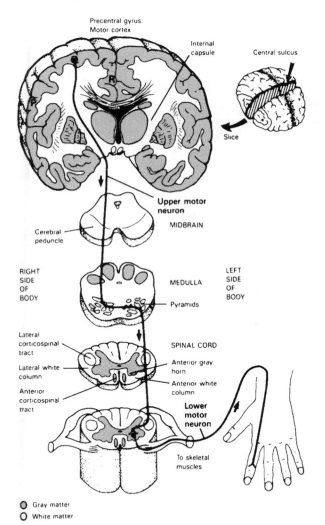

○ Gray matter
○ White matter

Figure LG 15.2A Diagram of the central nervous system.

D5. (a) Direct; precise, voluntary; pyramidal; pre, postcentral. (b) Lateral, decussated or crossed; right. (c) Left, anterior; are. (d) 10, anterior; do not, left; ipsi; do. (e) Brainstem; cranial. (f) Upper; lower. (g) Because these neurons, which are activated by input from a variety of sources (within the brain or spinal cord), must function for any movement to occur.

D6. (a) Leg and foot. (b) Sensation and movement. (c) Are, because the brainstem (including reticular formation) and cerebellum are sites of origin of these pathways.

D7. (a) Outside of; outside of, spinal cord. (b1) MRS. (b2) TS. (b3). VS. (b4) RS.

D9. (a) Parkinson's, dopamine; muscle rigidity and tremors. (b) Chorea, GABA; 40 years; 50–50.

E1. Wakefulness and sleep, learning and memory, and emotional responses.

E2. (a) Circadian; hypothalamus. (b) Reticular activating system; pain, proprioception, visual or auditory stimuli. (c) Increased, EEG sleep or coma; coma. (d) Adenosine; Al, inhibited; chemicals (caffeine or theophylline) block Al receptors; the RAS remains activated.

E3. (a) Nonrapid eye movement (NREM) and rapid eye movement (REM). (b) 1 to 4; 1; 3 and 4; 2, 3. (c) Ascends to stages 3 and 2 and then moves into. (d) More active; in; 1; rapid eye; REM. (e) 90; 5–10; 50. (f) De.

E6. Short-term; reverberating. (b) Consolidation; very little. (c) Short-term, recent.

F2. (a)M. (b)P. (c)Q. (d)H.

F3. (a) Half. (b) Right leg (same side as cord is severed), and in areas served by nerves inferior to the T11–T12 transection such as the right leg. (c) Days to weeks; areflexia, no. (d) A B C.

F4. (a) Movements; in fetal life, at birth, or in infancy; is not; is not.

F5. (a) In later years; is; is not. (b) Dopamine (DA); midbrain; basal ganglia; inhibitory. (c) Acetylcholine, excitatory; lack of inhibitory DA causes a double-negative effect so that extra, unnecessary movements occur (tremor) that may then interfere with normal movements (such as extension at the same time as flexion), resulting in rigidity and masklike facial appearance. (d) Brady. (e) No, DA cannot cross the blood—brain barrier. (f) Levodopa (L-dopa) is a precursor (leads to formation) of DA and anticholinergic medications counteract effects of ACh. (g) MAO (monoamine oxidase) is an enzyme made in the body that normally destroys catecholamines such as norepinephrine or dopamine. Use of MAO inhibitors permits the dopamine to continue acting for longer periods.

MORE CRITICAL THINKING: CHAPTER 15

1. Describe structures and mechanisms that help prevent overstretching of your muscles, tendons, and joints. Be sure to include roles of receptors and describe nerve pathways.
2. Contrast the sensations of proprioception, kinesthesia, and stereognosis. Tell what information these senses provide to you.
3. Contrast each of the following sensory receptors according to the following four characteristics: receptor structure and location, type of graded potential generated, and type of stimulus detected: (a) taste, (b) hearing, (c) light touch, and (d) awareness of muscle tension.
4. Contrast upper motor neurons (UMNs) with lower motor neurons (LMNs) with regard to locations of cell bodies and axons, functions, and effects if damaged.
5. Contrast pyramidal and extrapyramidal pathways according to their specific functions and also locations of the upper and lower motor neurons in each pathway.
6. Discuss whether a CVA affecting a 0.5 cm^2 area of the cerebral cortex or a 0.5 cm^2 area of the internal capsule would be likely to cause greater loss of motor function.
7. Describe the pathological changes and treatment associated with Parkinson's disease.

MASTERY TEST: CHAPTER 15

Questions 1–5: Circle the letter preceding the one best answer to each question.

1. All of these sensations are conveyed by the posterior column—medial lemniscus pathway *except:*
   A. Pain and temperature
   B. Proprioception
   C. Fine touch, two-point discrimination
   D. Vibration
   E. Stereognosis
2. Choose the *false* statement about REM sleep.
   A. Infant sleep consists of a higher percentage of REM sleep than does adult sleep.
   B. Dreaming occurs during this type of sleep.
   C. The eyes move rapidly behind closed lids during REM sleep.
   D. EEG readings are similar to those of stage 4 of NREM sleep.
   E. REM sleep occurs periodically throughout a typical 8-hour sleep period.
3. All of the following are correct statements about the left fasciculus gracilis *except:*
   A. It is part of the posterior columns.
   B. It consists of axons whose cell bodies lie in the left dorsal root ganglia.
   C. It transmits impulses along axons that terminate in the medulla.
   D. It transmits nerve impulses that originated in the left upper limb.
4. Choose the *false* statement about pain receptors.
   A. They may be stimulated by any type of stimulus, such as heat, cold, or pressure.
   B. They have a simple structure with no capsule.
   C. They are characterized by a high level of adaptation.
   D. They are found in almost every tissue of the body.
   E. They are important in helping to maintain homeostasis.
5. "Change in sensitivity to a long-lasting stimulus" is a description of the characteristic of sensation called:
   A. Adaptation
   B. Modality
   C. Selectivity
   D. Proprioception

Questions 6–9: Arrange the answers in correct sequence.

_____ _____ _____   6. Density of sense receptors, from most dense to least:

    A. In tip of tongue

    B. In tip of finger

    C. In back of neck

_____ _____ _____ _____   7. In order of occurrence in a sensation:

    A. Transduction of a stimulus to a generator potential

    B. Integration of the impulse into a sensation

    C. Impulse generation and propagation along a nervous pathway

    D. A stimulus

_____ _____ _____ _____   8. Levels of sensation, from those causing simplest, least precise reflexes to those causing most complex and precise responses:

    A. Thalamus

    B. Brain stem

    C. Cerebral cortex

    D. Spinal cord

_____ _____ _____ _____ _____   9. Pathway for conduction of most of the impulses for voluntary movement of muscles:

    A. Anterior gray horn of the spinal cord

    B. Precentral gyms

    C. Internal capsule

    D. Location where decussation occurs

    E. Lateral corticospinal tract

Questions 10–20: Circle T (true) or F (false). If the statement is false, change the underlined word or phrase so that the statement is correct.

T F   10.   In general, the <u>left</u> side of the brain controls the right side of the body.

T F   11.   The final common pathway consists of <u>upper motor</u> neurons.

T F   12.   Damage to the final common pathway is likely to result in <u>flaccid paralysis</u>.

T F   13.   The fasciculus cuneatus and the nucleus cuneatus form parts of the pathway for <u>proprioception and most tactile sensations from neck, upper chest, and upper arms.</u>

T F   14.   The fasciculus cuneatus and the nucleus cuneatus within posterior columns contain <u>second-order neurons</u> in a <u>sensory</u> pathway.

T F   15.   Perception is the <u>conscious or unconscious awareness</u> of sensations.

T F   16.   The neuron that crosses to the opposite side in sensory pathways is typically the <u>second-order</u> neuron.

T F   17.   Damage to the left lateral spinothalamic tract would be most likely to result in loss of awareness of <u>pain and vibration sensations in the left</u> side of the body.

T F   18.   Conscious sensations, such as those of sight and touch, can occur only in the <u>cerebral cortex</u>.

T F   19.   Pain experienced by an amputee, as if the amputated limb were still there, is an example of <u>phantom pain</u>.

T F   20.   Circadian rhythm pertains to the <u>complex feedback circuits involved in producing coordinated movements</u>.

Questions 21–25: Fill-ins. Complete each sentence with the word or phrase that best fits.

_____ 21. _____ refers to the awareness of muscles, tendons, joints, balance, and equilibrium.

_____ 22. Inactivation of the _____ results in the state of sleep.

_____ 23. Rubrospinal, tectospinal, and vestibulospinal tracts are all

classified as _____ tracts.

_____ 24. A CVA affecting the medial portion of the postcentral gyms of the right hemisphere (area Z in Figure LG 15.1, page 291) is most likely

to result in symptoms such as _____.

_____ 25. Voluntary motor impulses are conveyed from motor cortex to

neurons in the spinal cord by _____ pathways.

ANSWERS TO MASTERY TEST: CHAPTER 15

Multiple Choice

1. A
2. D
3. D
4. C
5. A

Arrange

6. A B C
7. D A C B
8. D B A C
9. B C D E A

True-False

10. T
11. F. Lower motor
12. T
13. T
14. F. First-order neurons in a sensory.
15. F. Conscious awareness and interpretation of meaning
16. T
17. F. Pain and temperature sensations in the right
18. T
19. T
20. F. Events that occur at approximately 24-hour intervals, such as sleeping and waking

Fill-ins

21. Proprioception
22. Reticular activating system (RAS)
23. Extrapyramidal
24. Loss of sensation of left foot
25. Direct (pyramidal or corticospinal) pathways

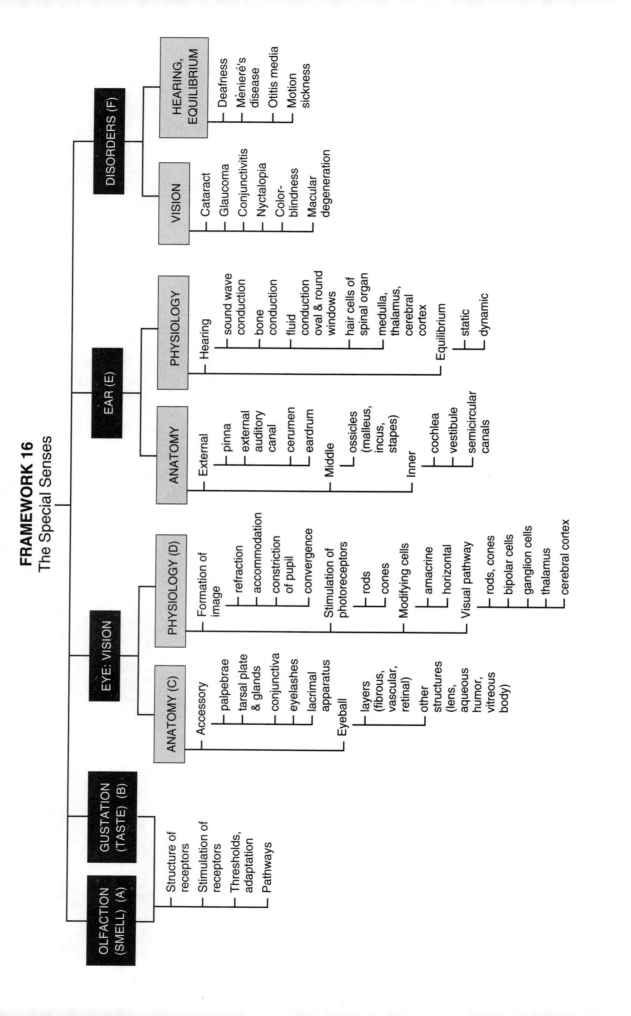

FRAMEWORK 16
The Special Senses

The Special Senses

All sensory organs contain receptors for increasing sensitivity to the environment. In Chapter 15 you considered the relatively simple receptors and pathways for the general senses of touch, pressure, temperature, pain, and vibration. The special senses include smell, taste, vision, hearing, and equilibrium. Special afferent pathways in many ways resemble general afferent pathways. However, a major point of differentiation is the arrangement of special sense receptors in complex sensory organs, specifically the nose, tongue, eyes, and ears.

As you begin your study of the special senses, carefully examine the Chapter 16 Topic Outline and Objectives; check off each one as you complete it. To organize your study of special senses, glance over the Chapter 16 Framework now. Be sure to refer to the Framework frequently and note relationships among key terms in each section.

TOPIC OUTLINE AND OBJECTIVES

A. Olfaction: sense of smell

    1. Describe the olfactory receptors and the neural pathway for olfaction.

B. Gustation: sense of taste

    2. Describe the gustatory receptors and the neural pathway for gustation.

C. Vision: anatomy

    3. List and describe the accessory structures of the eye and the structural components of the eyeball.

    4. Discuss image formation by describing refraction, accommodation, and constriction of the pupil.

D. Vision: physiology

    5. Describe how photoreceptors and photopigments function in vision.

    6. Describe the processing of visual signals in the retina and the normal pathway for vision.

E. Hearing and equilibrium

    7. Describe the anatomy of the structures in the three principal regions of the ear.

    8. List the principal events in the physiology of hearing.

    9. Identify the receptor organs for equilibrium, and describe how they function.

F. Disorders and medical terminology

WORDBYTES

Now become familiar with the language of this chapter by studying each wordbyte, its meaning, and an example of its use within a term. After you study the entire list, self-check your understanding by writing the meaning of each wordbyte on the line. As you continue through the *Learning Guide*, identify (and fill in) additional terms that contain the same wordbyte.

Wordbyte	Self-check	Meaning	Example(s)
aqua-	_____	water	*aqua*eous humor
blephar-	_____	eyelid	*blephar*oplasty
kerat-	_____	cornea	radial *kerat*otomy
macula	_____	spot	*macula* lutea
med-	_____	middle	otitis *med*ia
ocul-	_____	eye	*ocul*omotor nerve
olfact-	_____	smell	*olfact*ory
ophthalm-	_____	eye	*ophthalm*ologist
opt-	_____	eye	*opt*ic nerve, optician
orbit-	_____	eye socket	supra*orbit*al
ossi-	_____	bone	*ossi*cle
ot(o)	_____	ear	*ot*olith
presby-	_____	old	*presby*opia
sclera	_____	hard	*sclera*, oto*scler*osis
tympano-	_____	drum	*tympan*ic membrane
vitr-	_____	glassy	*vitr*eous humor

CHECKPOINTS

A. Olfaction: sense of smell (pages 512–514)

✓ **A1.** Describe olfactory receptors in this exercise.

  a. Receptors for smell are located in the *(superior? inferior?)* portion of the nasal cavity.

   Receptors consist of _____-polar neurons which are located between

   _____ cells. The distal end of each olfactory receptor cell consists of

   dendrites that have cilia known as olfactory _____.

  b. What is unique about the function of basal cells?

c. What is the function of olfactory (Bowman's) glands in the nose?

d. Because smell is a _____ sense, receptors in olfactory hair mem-
    branes respond to different chemical molecules, leading to a

    _____ and _____.

e. Adaptation to smell occurs *(slowly? rapidly?)* at first and then happens at a much
    slower rate.

✓▲ **A2.** Complete this exercise about the nerves associated with the nose.

a. Upon stimulation of olfactory hairs, impulses pass to cell bodies and axons; the
    axons of these olfactory cells form cranial nerves *(I? II? III?)*, the olfactory nerves.
    These pass from the nasal cavity to the cranium to terminate in the olfactory

    _____ located just inferior to the _____ lobes of the
    cerebrum.

b. Cell bodies in the olfactory bulb then send axons through the olfactory *(nerves?
    tracts?)* along two pathways. One route extends to the limbic system. What is the
    effect of projection of sensations of smell to this area?

c. The other olfactory pathway passes to the thalamus and then to the frontal lobes of

    the _____ cortex.

B. Gustation: sense of taste (pages 514–516)

✓ **B1.** Describe receptors for taste in this exercise.

a. Receptors for taste, or _____ sensation, are located in taste buds.

    These receptors are protected by a cluster of _____ cells, as well as

    the _____ cells that produce the supporting cells. The latter cells
    ultimately become receptor cells, which have a life span of *(10 days? 10 years?)*.

b. Within each taste bud are about *(5? 50?)* gustatory receptor cells, each with a gusta-
    tory hair that projects from the taste bud pore. Stimulation of this hair appears to

    initiate a _____ potential, which results in release of

    _____ from receptor cells. This chemical then activates sensory
    neurons in cranial nerves (see Checkpoint B2d).

c. Taste buds are located on elevated projections of the tongue called

    _____. The largest of these are located in a V-formation at the back

    of the tongue. These are _____ papillae. _____
    papillae are mushroom-shaped and are located over the entire surface of the tongue.

    Pointed _____ papillae also cover the entire surface of the tongue,
    and these *(do? do not?)* contain many taste buds.

✓ **B2.** Do the following exercise about taste.

a. There are only four primary taste sensations: _____,

_____, _____, and _____. The

taste buds at the tip of the tongue are most sensitive to _____ and

_____ tastes, whereas the posterior portion is most sensitive to

_____ taste.

b. The four primary tastes are modified by _____ sensations to produce the wide variety of "tastes" experienced. If you have a cold (with a "stuffy nose"), you may not be able to discriminate the usual variety of tastes, largely because of loss

of the sense of _____.

c. The threshold for smell is quite *(high? low?)*, meaning that *(a large? only a small?)* amount of odor must be present in order for smell to occur. Threshold varies for the four primary tastes; threshold for *(sour? sweet and salt? bitter?)* is lowest,

while those for _____ are highest.

d. Taste impulses are conveyed to the brainstem by cranial nerves _____, _____,

and _____; there, nerve messages travel to lower parts of the brain, including the

_____, _____, and _____, and finally stimulate the cerebral cortex.

C. Vision: anatomy (pages 516-525)

C1. Examine your own eye structure in a mirror. With the help of Figure 16.3 in your text), identify each of the accessory structures of the eye listed below.

Conjunctiva	**Lacrimal caruncle**	**Medial commissure**
Eyebrow	**Lacrimal glands: identify site**	**Palpebrae**
Eyelashes		

✓ **C2.** *For extra review.* Match the structures listed above with the following descriptions. Write the name of the structure on the line provided.

a. Eyelid: _____

b. Covers the orbicularis oculis muscle; provides protection: _____

c. Short hairs; infection of glands at the base of these hairs is *sty:* _____

d. Located in superolateral region of orbit; secrete tears: _____

e. Mucous membrane lining the eyelids and covering anterior surface of the eye:

_____

f. Point at which the two palpebrae meet laterally: _____

g. Reddish elevation at site of medial commissure; produces whitish secretions:

_____

✓▲ **C3.** Do this exercise on accessory eye structures.

   a. Lacrimal glands produce lacrimal fluid, also known as _____, at the

   rate of about _____ mL per day. Identify the location of these glands on yourself.

   Arrange these structures in correct order in the pathway of tears:

   _____ _____ _____ _____

A. **Lacrimal canals and nasolacrimal duct**	C. **Excretory lacrimal ducts**
B. **Lacrimal glands**	D. **Lacrimal punctae**

   State two or more functions of tears.

   b. Each eye is moved by *(3? 4? 6?)* extrinsic eye muscles; these are innervated by

   cranial nerves _____, _____, and _____. *(For extra review,* see Chapter 11,
   Checkpoint C3.

   c. The eye sits in the bony socket known as the _____ with only the
   anterior *(one-sixth? one-half?)* of the eyeball exposed.

✓▲ **C4.** On Figure LG 16.1, color the three layers (or tunics) of the eye using color code ovals.
   Next to the ovals write the letters of the labeled structures that form each layer. One
   is done for you. Then label all structures in the eye and check your answers against
   the key.

✓ **C5.** *Critical thinking.* A cornea that is surgically transplanted *(is? is not?)* likely to be
   rejected. State your rationale.

✓ **C6.** Complete this exercise about blood vessels of the eye.
   a. Retinal blood vessels *(can? cannot?)* be viewed through an ophthalmoscope. Of what
   advantage is this?

   b. The central retinal artery enters the eye at the *(optic disc? central fovea?).*

   C7. Contrast functions of the following eye structures:
   a. Circular and radial muscles of the iris

   b. Visual and nonvisual portions of the retina

KEY

A. Pupil
B. Scleral venous sinus
   (Canal of Schlemm)
C. Suspensory ligament
D. Ciliary body
E. Retina
F. Choroid
G. Sclera
H. Optic disk (blind spot)
I. Optic nerve

J. Conjunctiva
K. Cornea
L. Anterior chamber of anterior cavity
M. Iris
N. Posterior chamber of anterior cavity
O. Lens
P. Ora serrata
Q. Extrinsic eye muscle
R. Posterior cavity with vitreous body
S. Central fovea (in macula lutea)

○ Fibrous tunic   __**G**__ ____

○ Vascular tunic   ____ ____ ____

○ Nervous tunic   ____ ____

Figure LG 16.1 Structure of the eyeball in horizontal section. Color as directed in Checkpoint C4.

✓ **C8.** Do this exercise that describes the retina.
   a. After light strikes the back of the retina, nerve impulses pass anteriorly through three

   zones of neurons. First is the region of rods and cones, or _____

   zone; next is the _____ layer; most anterior is the

   _____ layer.
   b. Axons of the ganglion layer converge to form the optic _____. This

   nerve exits from the eye at the point known as the _____ disc, or

_____. The name indicates that no image formation can occur here because no rods or cones are present.

c. What function(s) do *amacrine cells* and *horizontal cells* serve?

d. The point of sharpest vision is known as the area of highest visual _____. This site is the *(central fovea ? optic disc?)*. It contains only the photoreceptor cells named *(rods? cones?)* and *(no? many?)* bipolar and ganglion cells.

✓ **C9.** Contrast the two types of photoreceptor cells by writing R for rods or C for cones before the related descriptions.

_____ a. About 6 million in each eye; most concentrated in the central fovea of the macula lutea.

_____ b. Over 120 million in each eye; located mainly in peripheral regions of the eye.

_____ c. Sense color and acute (sharp) vision.

_____ d. Used for night vision since they have a low threshold.

✓ **C10.** Describe the lens and the cavities of the eye in this exercise.

a. The lens is composed of *(lipid? protein?)* arranged in layers, much like an onion. Normally the lens is *(clear? cloudy?)*. A loss of transparency of the lens occurs in the

condition called _____.

b. The lens divides the eye into anterior and posterior *(chambers? cavities?)*. The posterior

cavity is filled with _____ body, which has a *(watery? jellylike?)*

consistency. This body helps to hold the _____ in place. Vitreous body is formed during embryonic life and *(is? is not?)* replaced in later life. Bits of

debris in the vitreous body are known as vitreal _____.

c. The anterior cavity is subdivided by the _____ into two chambers. Which is larger? *(Anterior chamber? Posterior chamber?)* A watery fluid called

_____ humor is present in the anterior cavity. This fluid is formed by

the _____ processes, and it is normally replaced about every 90 *(minutes? days?)*.

d. *Critical thinking.* How is the formation and final destination of aqueous fluid similar to that of cerebrospinal fluid (CSF)?

e. State two functions of aqueous humor.

    f. Aqueous fluid creates pressure within the eye; in glaucoma this pressure *(increases? decreases?)*. What effects may occur?

✓▲ **C11.** *For extra review.* Check your understanding by matching eye structures in the box with descriptions below. Use each answer once.

CF.	**Central fovea**	**OD.**	**Optic disc**
Cho.	**Choroid**	**P.**	**Pupil**
CM.	**Ciliary muscle**	**R.**	**Retina**
Cor.	**Cornea**	**S.**	**Sclera**
I.	**Iris**	**SVS.**	**Scleral venous sinus**
L.	**Lens**		**(canal of Schlemm)**

_____ a. "Whites of the eyes"

_____ b. Normally clear; if cloudy, it is a cataract

_____ c. Blind spot; area in which there are no cones or rods

_____ d. Area of sharpest vision; area of densest concentration of cones

_____ e. Anteriorly located, avascular coat responsible for about 75% of all focusing done by the eye

_____ f. Layer containing neurons

_____ g. Brown-black layers prevent reflection of light rays; also nourishes retina because it is vascular

_____ h. A hole; appears black, like a circular doorway leading into a dark room

_____ i. Muscular tissue that regulates the amount of light entering the eye; colored part of the eye

_____ j. Attaches to the lens by means of radially arranged fibers called the suspensory ligaments

_____ k. Located at the junction of iris and cornea; drains aqueous humor

    C12. Define each of these processes involved in *image formation* on the retina.

    a. *Refraction* of light

    b. *Accommodation* of the lens

    c. *Constriction* of the pupil

    d. *Convergence* of the eyes

✓ **C13.** Answer these questions about formation of an image on the retina.

a. Bending of light rays so that images focus exactly upon the retina is a process known

as _____. The *(cornea? lens?)* accomplishes most (75%) of the refraction within the eye. Explain why you think this is so.

b. Draw how a letter "e" would look as it is focused on the retina. _____
c. Explain why you see things right side up even though refraction produces inverted images on your retina.

✓ **C14.** Explain how accommodation enables your eyes to focus clearly.

a. A *(concave? convex?)* lens will bend light rays so that they converge and finally intersect. The lens in each of your eyes is *(biconcave? biconvex?)*.
b. When you bring your finger toward your eye, light rays from your finger must converge *(more? less?)* so that they will focus on your retina. Thus for near vision, the anterior surface of your lens must become *(more? less?)* convex. This occurs by a thickening and bulging forward of the lens caused by *(contraction? relaxation?)* of the

_____ muscle. In fact, after long periods of close work (such as reading), you may experience eye strain.

c. What is the *near point of vision* for a young adult? _____ cm (_____ inches) For

a 60-year-old? _____ cm (_____ inches) What factor explains this difference?

✓ **C15.** *A clinical correlation.* Describe errors in refraction and accommodation in this exercise.

a. The normal (or _____ eye) can refract rays from a distance of _____

meters (_____ feet) and form a clear image on the retina.

b. In the nearsighted (or _____) eye, images focus *(in front of? behind?)* the retina. This may be due to an eyeball that is too *(long? short?)*. Corrective lenses should be slightly *(concave? convex?)* so that they refract rays less and allow them to focus farther back on the retina.

c. The farsighted person can see *(near? far?)* well but has difficulty seeing *(near? far?)*

without the aid of corrective lenses. The farsighted (or _____) person requires *(concave? convex?)* lenses. After about age 40, most persons lose the ability

to see near objects clearly. This condition is known as _____.

✓ **C16.** Choose the correct answers to complete each statement.

a. Both accommodation of the lens and constriction of the pupil involve *(extrinsic? intrinsic?)* muscles, whereas convergence of the eyes involves *(extrinsic? intrinsic?)* muscles.

b. The pupil *(dilates? constricts?)* in the presence of bright light; the pupil *(dilates? constricts?)* in stressful situations when sympathetic nerves stimulate the iris.

D. Vision: physiology (page 525–529)

✓ **Dl.** List three principal processes that are necessary for vision to occur.

a.

b.

c.

✓ **D2.** Do this exercise on photoreceptors and photopigments.

a. The basis for the names *rods* and *cones* rests in the shape of the *(inner? outer?)* segments of these cells, that is, the segment closest to the *(choroid? vitreous body?)*. In which segment are photopigments such as rhodopsin located? *(Inner? Outer?)* In which segment can the nucleus, Golgi complex, and mitochondria be found? *(Inner? Outer?)*

b. How many types of photopigment are found in the human retina? _____ Rhodopsin is the photopigment found in *(rods? cones?)*. How many types of photopigments are found in cones?

c. Photopigments are also known as _____ pigments and consist of colored *(proteins? lipids?)* in outer segment plasma membranes. In rods, these chemicals pinch off to form stacks of *(bubbles? discs?)*. A new disc is formed every 20–60 *(minutes? days? months?)*; old discs are destroyed by the process of _____.

d. Any photopigment consists of two parts: *retinal* and *opsin*. What is the function of retinal?

e. Retinal is a vitamin _____ derivative and is formed from carotenoids. List several carotenoid-rich foods.

Tell how such foods may help to prevent *nyctalopia* (night blindness)?

✓ **D3.** Refer to Figure 16.13 in your text, and complete this Checkpoint about response of photopigments to light and darkness.

a. *(Retinal? Opsin?)* appears as a colored, dumbbell-shaped protein embedded in a disc in a rod or cone. Retinal fits snugly into opsin when retinal is in a *(bent? straightened?)* shape. This form of retinal is known as the *(cis? trans?)* form, and it is found when eyes are in *(darkness? light?)*.

b. Conversion of *cis-* to *trans*-retinal is known as _____. This occurs when retinal *(absorbs light? sits in darkness?)*. Within a minute of exposure to light, the *(cis? trans ?)*-retinal slips out of place in opsin. The separate products (opsin and *trans*retinal) are *(colored? colorless or bleached?)*. This is the first step in the process of

conversion of a _____ (light) into a generator potential; this process is known as *(transduction? integration?)*. As a result, bipolar and then ganglion cells will be activated to begin nerve transmission along the visual pathway (see Checkpoints D6–D10).

c. To make further visual transductions possible, *trans*-retinal must be converted back

to *cis*-retinal, a process known as _____. This process requires

(light? darkness?) and action of an enzyme known as retinal _____. Which photopigment regenerates more quickly? *(Rhodopsin? Cone photopigments?)*

✓ **D4.** Explain what happens in your retinas after you walk from a bright lobby into a darkened movie theater.

a. In the lobby the rhodopsin in your rods had been bleached (broken into

_____ retinal and opsin) as the initial step in vision involving rods.

b. Only a period of darkness will permit adequate regeneration of retinal in the *cis* form

that can bind to opsin and form _____. After complete bleaching of rhodopsin, it takes *(5? 1.5?)* minutes to regenerate 50% of rhodopsin and about

_____ minutes for full regeneration of this photopigment. This process is called *(light? dark?)* adaptation.

c. Also necessary for rhodopsin formation is adequate vitamin A. Explain why a detached retina can result in reduced rhodopsin formation and nyctalopia.

✓ **D5.** Explain how cones help you to see in "living color" by completing this paragraph. Cones contain pigments which *(do? do not?)* require bright light for breakdown (and therefore for a generator potential). Three different pigments are present, sensitive

to three colors: _____, _____, and _____. A person who is red-green color-blind lacks some of the cones receptive to two colors

(_____ and _____) and so cannot distinguish between these colors. This condition occurs more often in *(males? females?)*.

✓ **D6.** In Checkpoint D3 you reviewed the transduction in photoreceptor cells in response to a stimulus (dim light). Refer to Figure LG 16.2 and do this activity describing what happens next in the pathway to vision.

a. First color the large arrows indicating pathways of light and of nerve impulses (visual data processing) and all cells on the figure according to the color code ovals.

b. At point * in the figure, a *(rod? cone?)* synapses with a *(bipolar? ganglion?)* cell. In darkness, an inhibitory neurotransmitter named *(cyclic GMP? glutamate?)* is continually released at this synapse. The effect is to *(de? hyper?)*-polarize bipolar cells so that they *(become activated? stay inactive?)* and no visual information is transmitted to the brain.

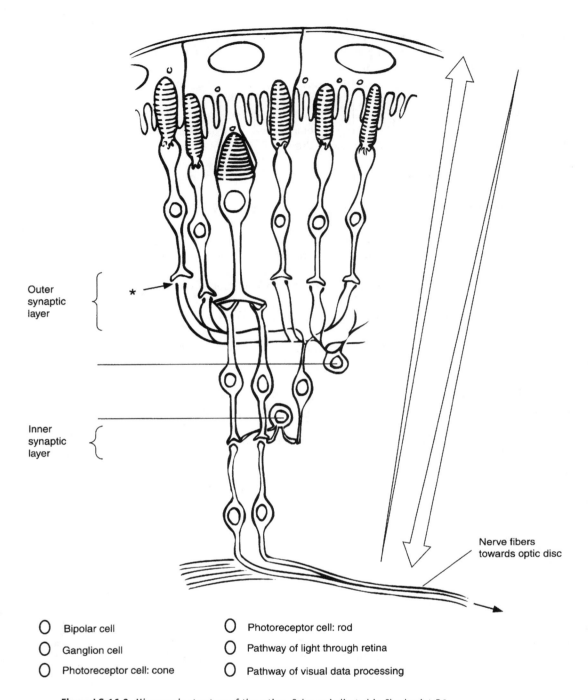

Outer
synaptic
layer

*

Inner
synaptic
layer

Nerve fibers
towards optic disc

○ Bipolar cell

○ Ganglion cell

○ Photoreceptor cell: cone

○ Photoreceptor cell: rod

○ Pathway of light through retina

○ Pathway of visual data processing

Figure LG 16.2 Microscopic structure of the retina. Color as indicated in Checkpoint D6.

c. What causes glutamate to be continually released at this synapse in darkness? Inflow

of *(Ca²⁺? Na⁺? K⁺?)*, known as the "_____ current," causes release of
glutamate from synaptic terminals of rods. But what keeps these Na⁺ channels open
during darkness is a chemical called cyclic *(AMP? GMP?)*. Review these steps by referring to the left side of Figure 16.14 in your text.

d. Now review events (right side of Figure 16.14) that follow a stimulus of dim light
exposure to rods. Effects will be roughly *(the same as? opposite to?)* those just described
in (b) and (c). An enzyme is activated to break down GMP. Na⁺ channels are therefore *(opened? closed?)*, and the inhibitory glutamate is *(still? no longer?)* released so that
bipolar cells (and the rest of the visual pathway) can become activated.

✓ **D7.** Refer again to Figure LG 16.2 and complete this Checkpoint about retinal cells.

a. The figure shows that the retina contains a larger number of *(photoreceptor? bipolar and ganglion?)* cells. In fact, there are between _____ and _____ rods for every one bipolar cell. This arrangement demonstrates *(convergence? divergence?)*.

b. Do cones also exhibit convergence? *(Yes? No?)* State the significance of this fact.

c. What functions do horizontal cells and amacrine cells perform?

✓ **D8.** Describe the conduction pathway for vision by arranging these structures in sequence. Write the letters in correct order on the lines provided. Also, indicate the four points where synapsing occurs by placing an asterisk (*) between the letters.

B.	**Bipolar cells**	**ON.**	**Optic nerve**
C.	**Cerebral cortex (visual areas)**	**OT.**	**Optic tract**
G.	**Ganglion cells**	**P.**	**Photoreceptor cells (rods, cones)**
OC.	**Optic chiasma**	**T.**	**Thalamus**

_____ _____ _____ _____ _____ _____

✓ **D9.** Answer these questions about the visual pathway to the brain. It may be helpful to refer to Figure 16.15 in the text.

a. Hold your left hand up high and to the left so you can still see it. Your hand is in the *(temporal? nasal?)* visual field of your left eye and in the _____ visual field of your right eye.

b. Due to refraction, the image of your hand will be projected onto the *(left? right?)* *(upper? lower?)* portion of the retinas of your eyes.

c. All nerve fibers from these areas of your retinas reach the *(left? right?)* side of your thalamus and cerebral cortex.

d. Damage to the right optic tract, right side of thalamus, or right visual cortex would result in loss of sight of the *(left? right?)* visual fields of each eye.

✓ **D10.** *The Big Picture: Looking Back.* Refer to Figure 14.5, page 398 of your text. The optic chiasm is located just anterior to the _____ gland. Visual disturbances are likely to signal a tumor in this gland. Pressure of the tumor is likely to be exerted against the more medial fibers within the chiasm. These are fibers that have passed from *(nasal? temporal?)* regions of the retinas and that enter "crossed fiber" regions of optic tracts to reach opposite sides of the brain. (Refer to Figure 16.14 in your text.) As a result of loss of function of nasal retinas, the patient would experience blindness in *(nasal? temporal?)* visual fields, a condition known as "tunnel vision."

E. Hearing and equilibrium (pages 529–542)

✓▲ **E1.** Refer to Figure LG 16.3 and do the following exercise.

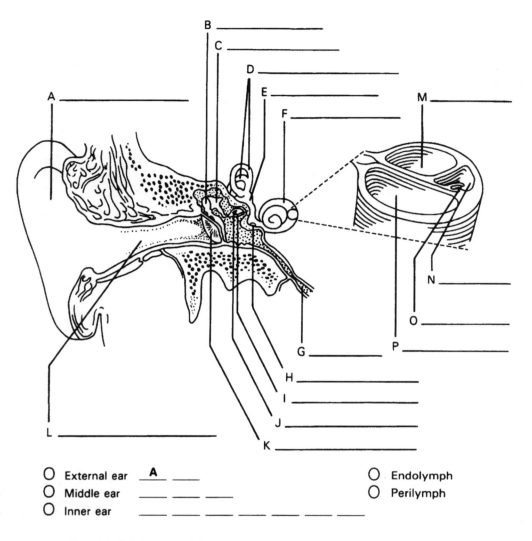

B _____

C _____

D _____

A _____

E _____

F _____

M _____

N _____

O _____

P _____

G _____

H _____

I _____

J _____

K _____

L _____

O External ear ___**A**___ _____

O Middle ear _____ _____ _____

O Inner ear _____ _____ _____ _____ _____ _____ _____

O Endolymph

O Perilymph

Figure LG 16.3 Structure of the ear. Labels as directed in Checkpoints E1, and E7.

a. Color the three parts of the ear using color code ovals. Then write on lines next to the ovals the letters of structures that are located in each part of the ear. One is done for you.

b. Label each lettered structure using leader lines on the figure.

✓▲ **E2.** *For extra review.* Select the ear structures in the box that fit the descriptions below. Not all answers will be used.

A.	**Auricle (pinna)**	**OW.**	**Oval window**
AT.	**Auditory (Eustachian) tube**	**RW.**	**Round window**
I.	**Incus**	**S.**	**Stapes**
M.	**Malleus**	**TM.**	**Tympanic membrane**

_____ a. Tube used to equalize pressure on either side of tympanic membrane

_____ b. Eardrum

_____ c. Structure on which stapes exerts piston-like action

_____ d. Ossicle adjacent to eardrum

_____ e. Anvil-shaped ear bone

_____ f. Portion of the external ear shaped like the flared end of a trumpet

▲ **E3.** Contrast locations, anatomy, and physiology of the following structures:

   a. *External auditory canal—auditory (Eustachian) tube*

   b. *Tensor tympani muscle—stapedius muscle*

   c. *Oval window—round window*

   d. *Bony labyrinth—membranous labyrinth*

   e. *Utricle—saccule*

   f. *Modiolus—helicotrema*

   g. *Vestibular membrane—basilar membrane*

   h. *Ossicles—otoliths*

E4. Describe the structure of the spiral organ (organ of Corti).

Identify specific roles of the following parts of the spiral organ:
a.   Inner and outer hair cells

b.   Stereocilia and tip link proteins

c.   Tectorial membrane

✓ **E5.** Fill in the blanks and circle correct answers about sound waves.
a.   Sound waves heard by humans range from frequencies of 20 to 20,000 cycles/sec

(Hz). Humans can best hear sounds in the range of _____ Hz.
b.   A musical high note has a *(higher? lower?)* frequency than a low note. So frequency is *(directly? indirectly?)* related to pitch.
c.   Sound intensity (loudness) is measured in units called _____.

Normal conversation is at a level of about _____ dB, whereas sounds at _____ dB can cause pain.

✓▲ **E6.** Summarize events in the process of hearing in this activity. It may help to refer to Figure 16.20 in the text.
a.   Sound waves travel through the _____ and strike the _____ membrane. Sound waves are magnified by the action of the three

_____ in the middle ear.
b.   The ear bone named *(malleus? incus? stapes?)* strikes the *(round? oval?)* window, setting up waves in *(endo-? peri-?)* lymph. This pushes on the floor of the upper scala *(vestibuli? tympani?)*. As a result the cochlear duct is moved, and so is the perilymph in the lower canal, the scala *(vestibuli? tympani?)*. The pressure of the perilymph is finally expended by bulging out the *(round? oval?)* window.
c.   As the cochlear duct moves, tiny hair cells embedded in the floor of the duct are

stimulated. These hair cells are part of the _____ organ; its name is

based on its spiral arrangement on the _____ membrane all the way around the 2³⁄4 coils of the cochlear duct. *(High? Low?)*-pitched sounds are sensed best toward the apex (helicotrema region) of the cochlea.
d.   As spiral organ hair cells are moved by waves in endolymph, hairs move against the

_____ gelatinous membrane. This movement generates receptor potentials in hair cells that excite nearby sensory neurons of the

_____ branch of cranial nerve _____. The pathway continues to the brainstem, _____ (relay center), and finally to the _____ lobe of the cerebral cortex.

✓ **E7.** Color endolymph and perilymph and related color code ovals in Figure LG 16.3 on page LG 320. Endolymph contains an unusually *(high? low?)* K⁺ level for an extracellular fluid (ECF). State the significance of this fact.

✓▲ **E8.** Check your understanding of the pathway of fluid conduction by placing the following structures in correct sequence. Write the letters on the lines provided.

_____ _____ _____ _____ _____ _____

B.	**Basilar membrane**	**ST.**	**Scala tympani (perilymph)**
C.	**Cochlear duct (endolymph)**	**SV.**	**Scala vestibuli (perilymph)**
O.	**Oval window**	**VM.**	**Vestibular membrane**
RW.	**Round window**		

E9. Explain how hair cells of the spiral organ may be damaged by loud noises.

✓ **E10.** Contrast receptors for hearing and for equilibrium in this summary of the inner ear.
a. Receptors for hearing and equilibrium are all located in the *(middle? inner?)* ear.

All consist of supporting cells and _____ cells that are covered by a

_____ membrane.
b. In the spiral organ, which senses _____, the gelatinous membrane is

called the _____ membrane. Hair cells move against this membrane as a result of *(sound waves? change in body position?)*.
c. In the macula, located in the *(semicircular canals? vestibule?)*, the gelatinous mem-

brane is embedded with calcium carbonate crystals called _____.
These respond to gravity in such a way that the macula is the main receptor for *(static? dynamic?)* equilibrium. An example of such equilibrium occurs as you are aware of your *(position while lying down? change in position on a careening roller coaster?)*.

d. In the semicircular canals the gelatinous membrane is called the _____.
Its shape is *(flat? like an inverted cup?)*. The cupula is part of the *(crista? saccule?)* located in the ampulla. Change in direction (as in a roller coaster) causes

_____ to bend hairs in the cupula. Cristae in semicircular canals are therefore receptors primarily for *(static? dynamic?)* equilibrium.

E11. Describe vestibular pathways, including the role of the cerebellum in maintaining equilibrium.

F. Disorders, medical terminology (pages 542–543)

F1. Match the name of the disorder with the description.

Cat.	Cataract	MD.	Macular degeneration
Con.	Conjunctivitis	Myo.	Myopia
G.	Glaucoma	OM.	Otitis media
H.	Hyperacusia	Pr.	Presbyopia
K.	Keratitis	Str.	Strabismus
Men.	Ménière's syndrome	T.	Trachoma
		V.	Vertigo

_____ a. Condition requiring corrective lenses to focus distant objects

_____ b. Excessive intraocular pressure resulting in blindness; second most common cause of blindness

_____ c. Earache

_____ d. Pinkeye

_____ e. Abnormally sensitive hearing

_____ f. The leading cause of blindness in persons over age 75

_____ g. The greatest single cause of blindness in the world; caused by the sexually transmitted microorganism, *Chlamydia*

_____ h. Disturbance of the inner ear with excessive endolymph

_____ i. Loss of transparency of the lens

_____ j. Farsightedness due to loss of elasticity of lens, especially after age 40

_____ k. Inflammation of the cornea

_____ l. Sense of spinning or whirling

F2. Discuss causes, signs and symptoms, and treatments of:

a. Macular degeneration

b. Otitis media

F3. Contrast sensorineural deafness with conduction deafness.

ANSWERS TO SELECTED CHECKPOINTS: CHAPTER 16

A1. Superior; bi, supporting; hairs. (b) They continually produce new olfactory receptors that live for only about a month. Because receptor cells are neurons, this "breaks the rule" that mature neurons are never replaced. (c) Produce mucus that acts as a solvent for odoriferous substances. (d) Chemical, generator potential, nerve impulse. (e) Rapidly.

A2. (a) I; bulb, frontal. (b) Tracts; awareness of certain smells, such as putrid odors or fragrance of roses, may lead to emotional responses. (c) Cerebral.

B1. (a) Gustatory; supporting, basal; 10 days. (b) 50; receptor, neurotransmitter. (c) Papillae; circumvallate; fungiform; filiform, do not.

B2. (a) Sweet, sour, salt, bitter; sweet and salty, bitter. (b) Olfactory; smell. (c) Low, only a small; bitter; sweet and salt. (d) VII, IX, X, thalamus, hypothalamus, limbic system.

C2. (a) Palpebra. (b) Eyebrow. (c) Eyelashes. (d) Lacrimal glands. (e) Conjunctiva. (f) Lateral commissure (g) Lacrimal caruncle.

C3. (a) Tears, 1; superolateral to each eyeball: see Figure 16.4b in your text; B C D A; tears protect against infection via the bactericidal enzyme lysozyme and by flushing away irritating substances; tears are signs of emotions. (b) 6; III, IV, VI. (c) Orbit, one-sixth.

C4. Fibrous tunic: G, K; vascular tunic or uvea: D, F, M; nervous tunic: E, S.

C5. Is not. Because the cornea is avascular, antibodies that cause rejection do not circulate there.

C6. (a) Can; health of blood vessels can be readily assessed, for example, in persons with diabetes. (b) Optic disc.

C8. (a) Photoreceptor, bipolar, ganglion. (b) Nerve; optic, blind spot. (c) They modify signals transmitted along optic pathways. (d) Acuity; central fovea; cones, no.

C9. (a) C. (b) R. (c) C. (d) R.

C10. (a) Protein; clear; cataract. (b) Cavities; vitreous, jellylike; retina; is not; floaters. (c) Iris; anterior chamber; aqueous; ciliary, minutes. (d) Both are formed from blood vessels (choroid plexuses) and the fluid finally returns to venous blood. (e) Provides nutrients and oxygen and removes wastes from the cornea and lens; also provides pressure to separate the cornea from lens. (t) Increases; damage to the retina and optic nerve with possible blindness.

C11. (a) S (b) L. (c) OD. (d) CF. (e) Cor. (f) R. (g) Cho. (h) P. (i) I. (j) CM. (k) SVS.

C13. (a) Refraction; cornea; the density of the cornea differs considerably from the air anterior to it. (b) "ə" ("e" inverted 180° and much smaller). (c) Your brain learned early in your life how to "turn" images so that what you see corresponds with locations of objects you touch.

C14. (a) Convex; biconvex. (b) More; more; contraction, ciliary. (c) 10, 4, 80, 31. The lens loses elasticity (presbyopia) with aging.

C15. (a) Emmetropic, 6, 20. (b) Myopic, in front of; 4, long; concave. (c) Far, near; hypermetropic, convex; presbyopia.

C16. (a) Intrinsic, extrinsic. (b) Constricts; dilates.

D1. (a) Formation of an image on the retina. (b) Stimulation of photoreceptors so that a light stimulus is converted into an electrical stimulus (receptor potential and nerve impulse). (c) Transmission of the impulse along neural pathways to the thalamus and visual cortex.

D2. (a) Outer, choroid; outer; inner. (b) 4; rods; one in each of three types of cones that are sensitive to different colors. (c) Visual, proteins; discs; minutes (one to three per hour), phagocytosis. (d) Retinal is the light-absorbing portion of all photopigments (rhodopsin and the three opsins in cones), so retinal begins the process of transduction of a stimulus (light) into a receptor potential. (e) A; carrots, yellow squash, broccoli, spinach, and liver; these foods lead to production of retinal in rods or cones.

D3. (a) Opsin; bent, *cis*, darkness. (b) Isomerization; absorbs light; *trans*; colorless or bleached; stimulus, transduction. (c) Regeneration; darkness, isomerase; cone photopigments.

D4. (a) *Trans*. (b) Rhodopsin; 5; 30–40; dark. (c) Because the retina has detached from the adjacent pigment epithelium, which is the storage area for vitamin A needed for synthesis of retinal for rhodopsin; difficulty seeing in dim light (nyctalopia) may result.

D5. Do; yellow to red, green, blue; red, green; males.

Figure LG 16.2A Microscopic structure of the retina.

Figure LG 16.3A Diagram of the ear in frontal section

D6. (a) See Figure LG 16.2A. (b) Rod, bipolar; glutamate; hyper, stay inactive. (c) Na^+, dark; GMP (d) Opposite to; closed, no longer.

D7. Photoreceptor; 6, 600; convergence. (b) No; one-on-one (cone to bipolar cell) synapsing allows for greater acuity (sharpness) of vision. (c) Enhancement of contrast and differentiation of colors.

D8. P * B * G ON OC OT * T * C. (Note that some crossing but no synapsing occurs in the optic chiasma.)

D9. (a) Temporal, nasal. (b) Right, lower. (c) Right. (d) Left.

D10. Pituitary; nasal; temporal.

El. See Figure LG 16.3A.

E2. (a) AT. (b) TM. (c) OW. (d) M. (e) I. (f) A.

E5. (a) 1,000–4,000. (b) Higher; directly. (c) Decibels

(dB); 45, 115–120.

E6. (a) External auditory canal, tympanic; ossicles. (b) Stapes, oval, peri-; vestibuli; tympani; round. (c) Spiral; basilar; low. (d) Tectorial; cochlear, VIII; thalamus, temporal.

E7. See Figure LG 16.3A. High. Entrance of K^+ into hair cells produces depolarization, thereby initiating nerve impulses in sensory neurons of cochlear nerves.

E8. O SV VM C B ST RW.

E10. (a) Inner; hair, gelatinous. (b) Hearing, tectorial; sound waves. (c) Vestibule, otoliths; static; position while lying down. (d) Cupula; like an inverted cup; crista; endolymph; dynamic.

Fl. (a) Myo. (b) G. (c) OM. (d) Con. (e) H. (f) MD. (g) T. (h) Men. (i) Cat. (j) Pr. (k) K. (1) V.

MORE CRITICAL THINKING: CHAPTER 16

1. Contrast smell and taste according to the following criteria: (a) rate of adaptation, (b) threshold, (c) cranial nerve pathways for each. Also describe how you could demonstrate that much of what most people think of as taste is actually smell.

2. Compare formation, locations, and functions of lacrimal fluid and aqueous fluid.

3. Discuss causes and potential damage to the eye associated with the following conditions: (a) detachment of the retina, (b) glaucoma, (c) cataracts, (d) trachoma.

4. Contrast roles of ions (Na^+, K^+, and Ca^{2+}) in activation of photoreceptor cells of the retina versus hair cells of the cochlea.

5. Explain how cochlear implants function to take the place of cells of the spiral organ (organ of Corti).

6. Explain roles of the outer hair cells in producing otoacoustic emission, and describe how these sounds are used for assessment of hearing.

7. Contrast the anatomy of receptors for static equilibrium with those for dynamic equilibrium.

MASTERY TEST: CHAPTER 16

Questions 1–5: Circle T (true) or F (false). If the statement is false, change the underlined word or phrase so that the statement is correct.

T F 1. <u>Parasympathetic</u> nerves stimulate the circular iris muscle to constrict the pupil, whereas <u>sympathetic</u> nerves cause dilation of the pupil.

T F 2. <u>Convergence and accommodation are both results</u> of contraction of smooth muscle of the eye.

T F 3. <u>Crista, macula, otolith, and spiral organ</u> are all structures located in the inner ear.

T F 4. <u>Both the aqueous body and vitreous humor are</u> replaced constantly throughout your life.

T F 5. The receptor organs for special senses are <u>less</u> complex structurally than those for general senses.

Questions 6–12: Arrange the answers in correct sequence.

_____ _____ _____ 6. Layers of the eye, from superficial to deep:
  A. Sclera
  B. Retina
  C. Choroid

_____ _____ _____ _____ 7. From anterior to posterior:
  A. Vitreous body
  B. Optic nerve
  C. Cornea
  D. Lens

_____ _____ _____ _____ 8. Pathway of aqueous humor, from site of formation to destination:
  A. Anterior chamber
  B. Scleral venous sinus
  C. Ciliary body
  D. Posterior chamber

_____ _____ _____ _____ _____ 9. From anterior to posterior:
  A. Anterior chamber
  B. Iris
  C. Lens
  D. Posterior cavity
  E. Posterior chamber

_____ _____ _____ _____ _____ 10. Pathway of sound waves and resulting mechanical action:
  A. External auditory canal
  B. Stapes
  C. Malleus and incus
  D. Oval window
  E. Eardrum

_____ _____ _____ _____ _____ 11. Pathway of tears, from site of formation to entrance to nose:
  A. Lacrimal gland
  B. Excretory lacrimal duct
  C. Nasolacrimal duct
  D. Surface of conjunctiva
  E. Lacrimal puncta and lacrimal canals

_____ _____ _____ _____ _____ 12. Order of impulses along conduction pathway for smell:
  A. Olfactory bulb
  B. Olfactory hairs
  C. Olfactory nerves
  D. Olfactory tract
  E. Primary olfactory area of cortex

Questions 13–20: Circle the letter preceding the one best answer to each question.

13. Which chemical within rods and cones has the function of absorbing light?
    A. Glutamate
    B. Opsin
    C. Retinal
    D. Magnesium

14. Infections in the throat (pharynx) are most likely to lead to ear infections in the following manner. Bacteria spread through the:
    A. External auditory meatus to the external ear
    B. Auditory (eustachian) tube to the middle ear
    C. Oval window to the inner ear
    D. Round window to the inner ear

15. Choose the *false* statement about rods.
    A. There are more rods than cones in the eye.
    B. Rods are concentrated in the fovea and are less dense around the periphery.
    C. Rods enable you to see in dim (not bright) light.
    D. Rods contain rhodopsin.
    E. No rods are present at the optic disc.

16. Choose the *false* statement about the lens of the eye.
    A. It is biconvex.
    B. It is avascular.
    C. It becomes more rounded (convex) as you look at distant objects.
    D. Its shape is changed by contraction of the ciliary muscle.
    E. Change in the curvature of the lens is called accommodation.

17. Choose the *false* statement about the middle ear.
    A. It contains three ear bones called ossicles.
    B. Infection in the middle ear is called otitis media.
    C. It functions in conduction of sound from the external ear to the inner ear.
    D. The cochlea is located here.

18. Choose the *false* statement about the semicircular canals.
    A. They are located in the inner ear.
    B. They sense acceleration or changes in position.
    C. Nerve impulses begun here are conveyed to the brain by the vestibular branch of cranial nerve VIII.
    D. There are four semicircular canals in each ear.
    E. Each canal has an enlarged portion called an ampulla.

19. Destruction of the left optic tract would result in:
    A. Blindness in the left eye
    B. Loss of left visual field of each eye
    C. Loss of right visual field of each eye
    D. Loss of lateral field of view of each eye ("tunnel vision")
    E. No effects on the eye

20. Mr. Frederick has a detached retina of the lower right portion of one eye. As a result he is unable to see objects in which area in the visual field of that eye?
    A. High in the left
    B. High in the right
    C. Low in the left
    D. Low in the right

Questions 21–25: Fill-ins. Complete each sentence with the word or phrase that best fits.

_____ 21. In darkness, Na⁺ channels are held open by a molecule called

_____.

_____ 22. Name the four processes necessary for formation of an image on

the retina: _____ of light rays, _____ of the lens, _____

of the pupil, and _____ of the eyes.

_____ 23. _____ is the study of the structure, functions, and diseases of
the eye.

_____ 24. The names of three ossicles are _____, _____, and _____.

_____ 25. Two functions of the ciliary body are _____ and _____.

ANSWERS TO MASTERY TEST: CHAPTER 16

True–False

1. T
2. F. Accommodation, but not convergence, is a result.
3. T
4. F. Aqueous humor, but not vitreous humor, is
5. F. More

Arrange

6. A C B
7. C D A B
8. C D A B
9. A B E C D
10. A E C B D
11. A B D E C
12. B C A D E

Multiple Choice

13. C
14. B
15. B
16. C
17. D
18. D
19. C
20. A

Fill-ins

21. Cyclic GMP
22. Refraction, accommodation, constriction, convergence
23. Ophthalmology
24. Malleus, incus, and stapes
25. Production of aqueous humor and alteration of lens shape for accommodation

FRAMEWORK 17
The Autonomic Nervous System (ANS)

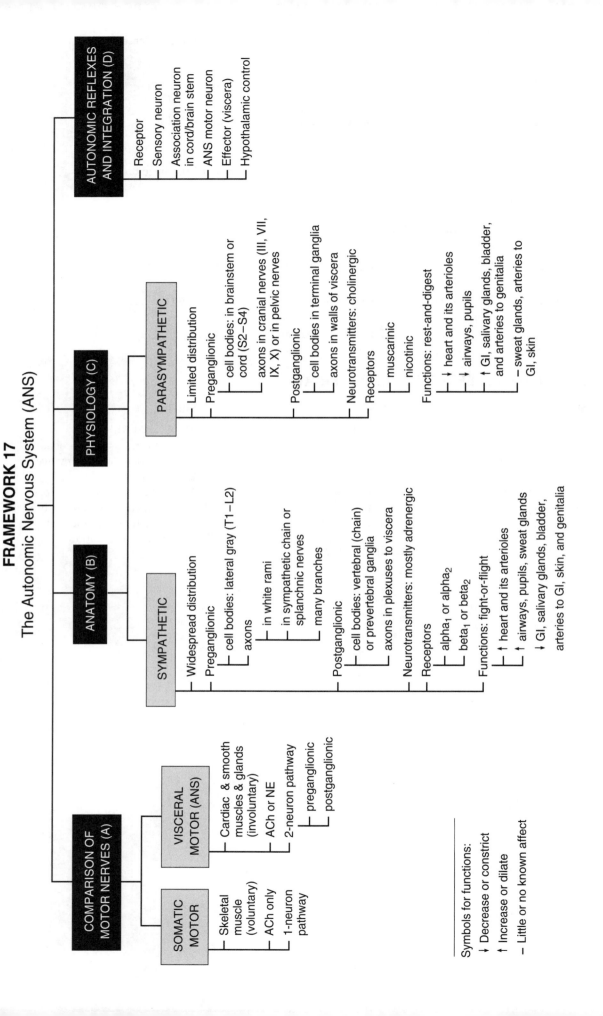

COMPARISON OF MOTOR NERVES (A)

SOMATIC MOTOR
- Skeletal muscle (voluntary)
- ACh only
- 1-neuron pathway

VISCERAL MOTOR (ANS)
- Cardiac & smooth muscles & glands (involuntary)
- ACh or NE
- 2-neuron pathway
  - preganglionic
  - postganglionic

ANATOMY (B)

SYMPATHETIC
- Widespread distribution
- Preganglionic
  - cell bodies: lateral gray (T1–L2)
  - axons
    - in white rami
    - in sympathetic chain or splanchnic nerves
      - many branches
- Postganglionic
  - cell bodies: vertebral (chain) or prevertebral ganglia
  - axons in plexuses to viscera
- Neurotransmitters: mostly adrenergic
- Receptors
  - alpha$_1$ or alpha$_2$
  - beta$_1$ or beta$_2$
- Functions: fight-or-flight
  - ↑ heart and its arterioles
  - ↑ airways, pupils, sweat glands
  - ↓ GI, salivary glands, bladder, arteries to GI, skin, and genitalia

PHYSIOLOGY (C)

PARASYMPATHETIC
- Limited distribution
- Preganglionic
  - cell bodies: in brainstem or cord (S2–S4)
  - axons in cranial nerves (III, VII, IX, X) or in pelvic nerves
- Postganglionic
  - cell bodies in terminal ganglia
  - axons in walls of viscera
- Neurotransmitters: cholinergic
- Receptors
  - muscarinic
  - nicotinic
- Functions: rest-and-digest
  - ↓ heart and its arterioles
  - ↓ airways, pupils
  - ↑ GI, salivary glands, bladder, and arteries to genitalia
  - – sweat glands, arteries to GI, skin

AUTONOMIC REFLEXES AND INTEGRATION (D)
- Receptor
- Sensory neuron
- Association neuron in cord/brain stem
- ANS motor neuron
- Effector (viscera)
- Hypothalamic control

Symbols for functions:
↓ Decrease or constrict
↑ Increase or dilate
– Little or no known affect

The Autonomic Nervous System

The autonomic nervous system (ANS) consists of the special branch of the nervous system that exerts unconscious control over viscera. The ANS regulates activities such as heart rate and blood pressure, glandular secretion, and digestion. The two divisions of the ANS—the sympathetic and parasympathetic—carry out a continual balancing act, readying the body for response to stress or facilitating rest and relaxation, according to momentary demands. ANS neurotransmitters hold particular significance clinically because they are often mimicked or inhibited by medications.

As you begin your study of the autonomic nervous system, carefully examine the Chapter 17 Topic Outline and Objectives; check off each one as you complete it. To organize your study of the autonomic nervous system, glance over the Chapter 17 Framework now. Be sure to refer to the Framework frequently and note relationships among key terms in each section.

TOPIC OUTLINE AND OBJECTIVES

A. Comparison of somatic and autonomic nervous systems

1. Compare the structural and functional differences of the somatic and autonomic portions of the nervous system.

B. Anatomy of autonomic motor pathways

2. Describe preganglionic and postganglionic neurons of the autonomic nervous system.
3. Compare the anatomical structure of the sympathetic and parasympathetic divisions of the autonomic nervous system.

C. ANS neurotransmitters and receptors

4. Describe the neurotransmitters and receptors in the autonomic responses.

D. Integration and control of autonomic functions

5. Describe the components of an autonomic reflex.
6. Explain the relationship of the hypothalamus to the ANS.

WORDBYTES

Now become familiar with the language of this chapter by studying each wordbyte, its meaning, and an example of its use within a term. After you study the entire list, self-check your understanding by writing the meaning of each wordbyte on the line. As you continue through the *Learning Guide*, identify (and fill in) additional terms that contain the same wordbyte.

Wordbyte	Self-check	Meaning	Example(s)
auto-	_____	self	*auto*nomic
nomos	_____	law, governing	auto*nomic*
para-	_____	near, beside	*para*vertebral, *para*thyroid
post-	_____	after	*post*ganglionic
pre-	_____	before	*pre*ganglionic
ram(i)-	_____	branch	white *ramus*
splanch-	_____	viscera	*splanch*nic

CHECKPOINTS

A. Comparison of somatic and autonomic nervous systems (pages 548–550)

✓ **A1.** Why is the autonomic nervous system (ANS) so named? Is the ANS entirely independent of higher control centers? Explain.

✓ **A2.** Contrast the somatic and autonomic nervous systems in this table.

	Somatic	Autonomic
a. Sensations and movements are mostly *(conscious? unconscious/automatic?)*		
b. Types of tissue innervated by efferent nerves	Skeletal muscle	
c. Efferent neurons are excitatory (E), inhibitory (I), or both (E + 1)		

	Somatic	Autonomic
d. Examples of sensations (input)		Mostly visceral sensory (such as changes in CO_2 level of blood or stretching or visceral wall)
e. Number of neurons in efferent pathway		
f. Neurotransmitter(s) released by motor neurons	Acetylcholine (ACh)	

✓▲ **A3.** Name the two divisions of the ANS. _____ _____

   A4. Many visceral organs have dual innervation by the autonomic system.
   a. What does dual innervation mean?

✓▲ b. How do the sympathetic and parasympathetic divisions work in harmony to control viscera?

B. Anatomy of autonomic motor pathways (pages 550–556)

✓ **B1.** Complete this exercise describing structural differences between visceral efferent and somatic efferent pathways.
   a. Between the spinal cord and effector, somatic pathways include _____ neuron(s),

   whereas visceral pathways require neuron(s), known as the pre- _____

   and _____ neurons.
   b. Somatic pathways begin at *(all? only certain?)* levels of the cord, whereas visceral

   routes begin at _____ levels of the cord.

✓▲ **B2.** Contrast preganglionic with postganglionic fibers in this exercise.
   a. Consider the sympathetic division first. Its preganglionic cell bodies lie in the lateral

   gray of segments _____ to _____ of the cord. For this reason, the sympathetic

   division is also known as the _____ outflow.
   b. Preganglionic axons *(are? are not?)* myelinated and therefore they appear white.

   These axons may branch to reach a number of different _____ where they synapse. These ganglia contain *(pre-? post-?)* ganglionic neuron cell bodies whose axons proceed to the appropriate organ. Postganglionic axons are *(gray? white?)* because they *(do? do not?)* have a myelin covering.

c. Parasympathetic preganglionic cell bodies are located in two areas. One is the

_____, from which axons pass out in cranial nerves _____

_____ _____ and _____. A second location is the *(anterior? lateral?)* gray

horns of segments _____ through _____ of the cord.

d. Based on location of *(pre? post?)*-ganglionic neuron cell bodies, the parasympathetic

division is also known as the _____ outflow. The axons of these neurons extend to *(the same? different?)* autonomic ganglia from those in sympathetic
pathways (see Checkpoint B4). There, postganglionic cell bodies send out short *(gray? white?)* axons to innervate viscera.

e. Most viscera *(do? do not?)* receive fibers from both the sympathetic and the parasympathetic divisions. However, the origin of these divisions in the CNS and the
pathways taken to reach viscera *(are the same? differ?)*.

✓ **B3.** Contrast these two types of ganglia.

	Posterior Root Ganglia	**Autonomic Ganglia**
a. Contains neurons that are *(sensory? motor?)*.		
b. Contains neurons in *(somatic? visceral? both?)* pathways.		
c. Synapsing *(does? does not?)* occur here.		

✓▲ **B4.** Complete the table contrasting types of autonomic ganglia.

	Sympathetic Trunk	**Prevertebral**	**Terminal**
a. Sympathetic or parasympathetic		Sympathetic	
b. Alternate name	Paravertebral ganglia or vertebral chain		
c. General location			Close to or in walls of effectors

✓ **B5.** Refer to Figure 17.4 in your text. Trace with your finger along the route of a preganglionic neuron. Now describe the pathway common to *all* sympathetic preganglionic neurons

by listing the structures in correct sequence. _____ _____ _____ _____

> **A.** **Ventral root of spinal nerve**
> **B.** **Sympathetic trunk ganglion**
> **C.** **Lateral gray of spinal cord (between Ti and L2)**
> **D.** **White ramus communicans**

✓▲ **B6.** Suppose you could walk along the route of nerve impulses in sympathetic pathways. Start at the sympathetic trunk ganglion. It may be helpful to refer to the "map" provided by Figure 17.4 in your text.

a. What is the shortest possible path you could take to reach a synapse with a postganglionic neuron cell body?

b. You could then ascend and/or descend the sympathetic chain to reach trunk ganglia at other levels. This is important because sympathetic preganglionic cell bodies are

limited in location to _____ to _____ levels of the cord, and yet the

sympathetic trunk extends from _____ to _____ levels of the vertebral column. Sympathetic fibers must be able to reach these distant areas to provide the entire body with sympathetic innervation.

c. Now "walk" the sympathetic nerve pathway to sweat glands, blood vessels, or hair muscles in skin or extremities. Again trace this route on Figure 17.4 in the text. Preganglionic neurons synapse with postganglionic cell bodies in sympathetic trunk ganglia (at the level of entry or after ascending or descending). Postganglionic fibers then

pass through _____, which connect to _____ and then convey impulses to skin or extremities.

d. Some preganglionic fibers do not synapse as described in (b) or (c) but pass on through trunk ganglia without synapsing there. They course through

_____ nerves to _____ ganglia. Follow their route as

they synapse with postganglionic neurons whose axons form _____ en route to viscera.

e. Prevertebral ganglia are located only in the _____. In the neck, thorax, and pelvis the only sympathetic ganglia are those of the trunk (see Figure 17.2 in the text). As a result, cardiac nerves, for example, contain only *(preganglionic? postganglionic?)* fibers, as indicated by broken lines in the figure.

f. A given sympathetic preganglionic neuron is likely to have *(few? many?)* branches; these may take any of the paths you have just "walked." Once a branch synapses, it *(does? does not?)* synapse again, because any autonomic pathway consists of just

_____ neurons (preganglionic and postganglionic).

✓ **B7.** Test your understanding of these routes by completing the sympathetic pathway shown in Figure LG 17.1. A preganglionic neuron located at level T5 of the cord has its axon drawn as far as the white ramus. Finish this pathway by showing how the axon may branch and synapse to eventually innervate these three organs: the heart, a sweat gland in skin of the thoracic region, and the stomach. Draw the preganglionic fibers in solid lines and the postganglionic fibers in broken lines.

Heart

Cardiac nerve

T3

Dorsal ramus of spinal nerve

Ventral ramus of spinal nerve

Posterior gray horn

Posterior root

Posterior root ganglion

Lateral gray horn

Gray ramus communicans

White ramus communicans

T4

Sympathetic chain

Anterior gray horn

Spinal nerve

Spinal cord T5

Anterior root

Sympathetic trunk ganglion

T5

Splanchnic nerve

Prevertebral ganglion (Celiac ganglion)

T6

Stomach

T7

Sweat gland

○————< Preganglionic ANS

○---------< Postganglionic ANS

○·············< Afferent

Figure LG 17.1 Typical sympathetic pathway beginning at level T5 of the cord. Complete pathways according to directions in Checkpoints B7 and D2.

✓▲ **B8.** In Checkpoint B4 you contrasted two types of sympathetic trunk ganglia, namely sympathetic trunk and prevertebral ganglia. For more details on these ganglia, complete this Checkpoint and also B9.

   a. The sympathetic trunk ganglia lie in two chains close to vertebrae, ribs, and sacrum. Each chain has *(12–14? 22–25? 31–33?)* ganglia. *(Three? Seven? Eight?)* ganglia lie in the cervical region. The *(superior? middle? inferior?)* cervical ganglion lies next to the second cervical vertebra; its postganglionic fibers innervate *(muscles and glands of the head? the heart?)*.

   b. The middle and inferior cervical ganglia lie near the *(third and fourth? sixth and seventh?)* vertebrae. Postganglionic fibers from these ganglia innervate the

     _____.

   c. There are *(8–9? 11–12?)* trunk ganglia in the thoracic area; they lie against the *(anterior? posterior?)* of ribs. List six types of viscera innervated by postganglionic fibers that arise from these ganglia.

   d. There are _____ sympathetic trunk ganglia in the lumbar region, and _____ in the sacral region.

✓▲ **B9.** Do this Checkpoint on autonomic pathways involving splanchnic nerves. These nerves contain *(sympathetic? parasympathetic?)* fibers. Splanchnic nerves are composed of *(pre? post?)*-ganglion fibers that pass through trunk ganglia and *(do? do not?)* synapse there; instead they extend to *(prevertebral? terminal?)* ganglia. Postganglionic

cells there pass axons on to viscera located in the _____. Complete this table about these nerves.

Splanchnic Nerve	Prevertebral Ganglion: Site of Postganglionic Cell Bodies	Viscera Supplied
a. Greater splanchnic n.		
b. Lesser splanchnic n.		
c. Lumbar splanchnic n.		

✓ **B10.** In the past several Checkpoints, you have been focusing on the *(sympathetic? parasympathetic?)* division of the ANS. Now that you are somewhat familiar with the rather complex sympathetic pathways, you will notice that parasympathetic pathways are much simpler. This is due to the fact that there is no parasympathetic

chain of ganglia; the only parasympathetic ganglia are _____

_____, which are in or close to the organs innervated. Also, parasympathetic preganglionic fibers synapse with *(more? fewer?)* postganglionic neurons. What is the significance of the structural simplicity of the parasympathetic system?

✓▲ **B11.** Complete the table about the cranial portion of the parasympathetic system.

| Cranial Nerve | | Name of Terminal | Structures |
Number	Name	Ganglion	Innervated
a.			Iris and ciliary muscle of eye
b.		1. Pterygopalatine 2. Submandibular	1. 2.
c.	Glossopharyngeal		
d.		Ganglia in cardiac and pulmonary plexuses and in abdominal plexuses	

✓▲ **B12.** *For extra review.* Write S if the description applies to the sympathetic division of the ANS, P if it applies to the parasympathetic division, and P, S if it applies to both.

_____ a. Also called thoracolumbar outflow

_____ b. Has long preganglionic fibers leading to terminal ganglia and very short postganglionic fibers

_____ c. Celiac and superior mesenteric ganglia are sites of postganglionic neuron cell bodies

_____ d. Sends some preganglionic fibers through cranial nerves

_____ e. Has some preganglionic fibers synapsing in vertebral chain (trunk)

_____ f. Has more widespread effect in the body, affecting more organs

_____ g. Has fibers running in gray rami communicantes to supply sweat glands, hair muscles, and blood vessels

_____ h. Has fibers in white rami (connecting spinal nerve with vertebral chain)

_____ i. Contains fibers that supply viscera with motor impulses

_____ j. The greater and lesser splanchnic nerves contain axons of this division

_____ k. The pelvic splanchnic nerves contain axons of this division

✓ **B13.** In Activity B1, we emphasized that autonomic pathways require two neurons. One

exception to this rule is the pathway to the _____ glands. Because this pair of glands function like modified *(sympathetic? parasympathetic?)* ganglia, no postganglionic neurons lead out to any other organ.

C. ANS neurotransmitters and receptors (pages 556–562)

✓ **C1.** *(All? Most? No?)* viscera receive both sympathetic and parasympathetic innervation,

known as _____ innervation of the ANS. List examples of exceptions here:

a. Sympathetic only: _____

_____

b. Parasympathetic only: _____

✓ **C2.** Fill in the blanks to indicate which neurotransmitters are released by sympathetic (S) or parasympathetic (P) neurons or by somatic neurons. (It may help to refer to Figure LG 17.2.) Use these answers:

ACh.   **Acetyl<u>choline</u> (<u>cholin</u>ergic)**

NE.   **Norepinephrine = nor<u>adren</u>alin, or possibly epinephrine = <u>adren</u>aline (<u>adren</u>ergic)**

_____ a. All preganglionic neurons (both S and P) release this neurotransmitter

_____ b. Most S postganglionic neurons

_____ c. A few S postganglionic neurons, namely those to sweat glands

_____ d. All P postganglionic neurons

_____ e. All somatic neurons

For extra review, color the neurotransmitters on Figure LG 17.2, matching colors you use in the key with those you use in the figure.

✓ **C3.** Do this activity summarizing the effects of autonomic neurotransmitters.

a. Axons that release the transmitter acetylcholine (ACh) are known as *(adrenergic? cholinergic?)*. Those that release norepinephrine (NE, also called noradrenalin) are

called _____. Most sympathetic postganglionic nerves are said to be

(cholinergic? adrenergic?), whereas all parasympathetic nerves are _____.

b. During stress, the *(sympathetic? parasympathetic?)* division of the ANS prevails, so stress responses are primarily *(adrenergic? cholinergic?)* responses.

c. Another source of NE, as well as epinephrine, is the gland known as the

_____ _____. Therefore, chemicals released by this gland (as in stress) will mimic the action of the *(sympathetic? parasympathetic?)* division of the ANS.

✓ **C4.** Again, refer to Figure LG 17.2 and do this exercise on neurotransmitters and related receptors. Be sure to match the color you use for receptor symbols in the key to the color you use for receptors in the figure. (The symbol S denotes *sympathetic* and P indicates *parasympathetic* in the following statements.)

a. First identify all nicotinic receptors in the figure. Notice that these receptors are found on cell bodies of *(P postganglionic neurons? S postganglionic neurons? autonomic effector cells? somatic effector cells [skeletal muscle]?)*. (Circle all that apply.) In other words, nicotinic receptors are all sites for action of the neurotransmitter *(ACh? NE? either ACh or NE?)*. Activation of nicotinic receptors is *(always? sometimes? never?)* excitatory.

b. On what other types of receptors can ACh act? _____. Identify these receptors on the figure. As you do, notice that these receptors are found on all effec-

Key: Neurotransmitters   Receptors

ⓜ ACh. Acetylcholine (cholinergic)   Alpha   ⋀ Muscarinic
                                       Beta₁   ⌂ Nicotinic
◻ NE. Norepinephrine   Beta₂
   (adrenergic)

Figure LG 17.2 Diagram of autonomic and somatic neurons, neurotransmitters, and receptors.
Note that neurotransmitters and receptors are relatively enlarged. Refer to Checkpoints C2 and C4 to C6.

tors stimulated by *(P? S?)* postganglionic neurons. In addition, muscarinic receptors

are found on a few S effectors, such as _____. Activation of
muscarinic receptors is *(always? sometimes? never?)* excitatory. For example, as P
nerves (vagus) release ACh to a muscarinic receptor in cardiac muscle, heart activity
(increases? decreases?), but when P nerves (vagus) release ACh to a muscarinic receptor in the wall of the stomach or intestine gastrointestinal activity *(increases?
decreases?)*.

c. Explain what accounts for the names nicotinic and muscarinic for ACh receptors. The

   names are based on the fact that the action of _____ on these receptors is mimicked by action of _____ on nicotinic receptors and by a

poison named _____ from mushrooms on muscarinic receptors.

d. *Alpha* and *beta* receptors are found only on effectors innervated by *(P? S?)* nerves; these effectors are stimulated by *(ACh? NE and epinephrine?).* NE and epinephrine are

categorized as _____ neurotransmitters.

e. *Critical thinking.* Alex receives word that his brother has been seriously injured in an automobile accident. Among other immediate physical responses, his skin "goes pale." Explain this response. (Refer to Table 17.3 in your text, for help.) Which type of receptor is found in smooth muscle of blood vessels of skin? *(Alpha$_1$? Beta$_1$? Beta$_2$?)* Alpha$_1$ receptors are usually *(excitatory? inhibitory?).* As sympathetic (the "stress system") nerves release NE, smooth muscle of blood vessels are excited, causing them to *(constrict/narrow? dilate/widen?).* As a result, blood flows *(into? out of?)* skin, and the skin pales. How may this response help Alex?

C5. *A clinical correlation.* Certain medications can mimic the effects of the body's own transmitters. Again refer to Figure LG 17.2 and do this exercise.

a. A beta$_1$-stimulator (or beta-exciter) mimics *(ACh? NE?),* causing the heart rate and

strength of contraction to _____-crease. Name one such drug. (For help, refer to a

pharmacology text.) _____ Such a drug has a structure that is close enough to the shape of NE that it can "sit on" the beta$_1$ receptor as NE would and activate it. Such a drug is a beta$_1$ *(agonist? antagonist?).*

b. A beta-blocker (or beta-inhibitor) also has a shape similar to that of NE and so can sit

on _____ receptors. Name one such drug. _____. However, this drug cannot activate the receptor but simply prevents NE from doing

so. So a beta-blocker _____-creases heart rate and force of contraction. A beta-blocker is a beta *(agonist? antagonist?).*

c. Other drugs known as anticholinergics (or ACh-blockers) can take up residence on ACh receptors so that ACh being released from vagal nerve stimulation cannot exert its normal effects. A person taking such a medication may exhibit a(n) *(increased? decreased?)* heart rate.

C6. *Critical thinking.* Do this additional exercise on effects of medications on ANS nerves.

a. What types of receptors are located on smooth muscle of bronchi and bronchioles (airways)? *(Alpha? Beta$_1$? Beta$_2$?)* For help refer to Table 17.3 in your text. The effect of the sympathetic stimulation of lungs is to cause *(dilation? constriction?)* of airways, making breathing *(easier? more difficult?)* during stressful times.

b. The effect of NE on the smooth muscle of the stomach, intestines, bladder, and uterus

is also *(contraction? relaxation?),* so that during stress the body _____-creases activity of these organs and can focus on more vital activities such as heart contractions.

c. From Activity C4e, you may recognize that the sympathetic response during stress causes many blood vessels, such as those in skin, to *(constrict? dilate?).* Prolonged stress may therefore lead to *(high? low?)* blood pressure. One type of medication used to lower blood pressure is an alpha-*(blocker? stimulator?)* because it tends to dilate these vessels and "pool" blood in nonessential areas like skin, thereby avoiding overloading the heart with blood flow.

✓ **C7.** Do this exercise relating the fate of ANS neurotransmitters and overall effects of the two divisions of the ANS.

   a. On the arrows below, write the initials of the enzymes that destroy each transmitter:

   ACh ⟶ ~~ACh~~   NE ⟶ ~~NE~~

   b. The ANS neurotransmitter that "hangs around" longer (has more lasting effect) is *(ACh? NE?)*. This fact provides one explanation for the longer-lasting effects of *(sympathetic/adrenergic? parasympathetic/cholinergic?)* neurons.

   c. State three other rationales for the more lasting and widespread effects of the sympathetic system:

   1.

   2.

   3.

✓ **C8.** Use arrows to show whether parasympathetic (P) or sympathetic (S) fibers stimulate (↑) or inhibit (↓) each of the following activities. Use a dash (—) to indicate that there is no (or virtually no) parasympathetic innervation. The first one is done for you.

   a. P __↓__   S __↑__   Dilation of pupil

   b. P _____   S _____   Heart rate and blood flow to coronary (heart muscle) blood vessels

   c. P _____   S _____   Constriction of blood vessels of skin and abdominal viscera

   d. P _____   S _____   Salivation and digestive organ contractions

   e. P _____   S _____   Erection or engorgement of genitalia

   f. P _____   S _____   Dilation of bronchioles for easier breathing

   g. P _____   S _____   Contraction of bladder and relaxation of internal urethral sphincter causing urination

   h. P _____   S _____   Contraction of arrector pili of hair follicles causing "goose bumps"

   i. P _____   S _____   Contraction of spleen which transfers some of its blood to general circulation, causing increase in blood pressure

   j. P _____   S _____   Release of epinephrine and norepinephrine from adrenal medulla

   k. P _____   S _____   Secretion of insulin and digestive enzymes from the pancreas

    l. P _____   S _____   Increase in blood glucose level by pancreatic secretion of the hormone glucagon and by formation of glucose in the liver

    m. P _____   S _____   Coping with stress, fight-or-flight response

C9. Write a paragraph outlining the changes effected by the ANS during a fight-or-flight response, such as running in fear.

C10. List functions of parasympathetic nerves; use the following hints:
a. S L U D D

b. The three "decreases"

D. Integration and control of autonomic functions (page 562–563)

✓ **D1.** Autonomic motor neurons can be stimulated by reflexes initiated by *(somatic sensory neurons? visceral sensory neurons? either somatic or visceral sensory neurons?)*. Visceral sensory receptors are mostly *(extero? intero? proprio?)*-receptors.

✓ **D2.** On Figure LG 17.1 draw an afferent neuron (in contrasting color or with a dotted line) to show how the sympathetic neuron could be stimulated by stomach pain.

✓ **D3.** Arrange in correct sequence structures in the pathway for a painful stimulus at your fingertip to cause an autonomic (visceral) reflex, such as sweating. Write the letters of the structures in order on the lines provided.

_____ _____ _____ _____ _____ _____

A. Association neuron in spinal cord	**F.** Nerve fiber in anterior root of spinal nerve
B. Postganglionic fiber in gray ramus	
C. Pain receptor in skin	**G.** Nerve fiber in white ramus
D. Cell body of postganglionic neuron in trunk ganglion	**H.** Postganglionic fiber in spinal nerve in brachial plexus
E. Cell body of preganglionic neuron in lateral gray of cord	**I.** Sweat gland
	J. Somatic sensory neuron

✓ **D4.** Most visceral sensations *(do? do not?)* reach the cerebral cortex, so most visceral sensations are at *(conscious? subconscious?)* levels.

D5. Explain roles of the hypothalamus in control of autonomic function by answering the following questions.
a. What types of sensory input reach the hypothalamus to evoke autonomic responses?

b. What types of output go to autonomic centers in the brainstem?

c. What types of output go to autonomic centers in the spinal cord?

d. Which parts of the hypothalamus control sympathetic nerves? Which parts control parasympathetic nerves?

D6. Describe how sympathetic nerves are affected by the following conditions. List specific signs and symptoms of each condition.
a. Horner's syndrome

b. Raynaud's disease

c. Autonomic dysreflexia

ANSWERS TO SELECTED CHECKPOINTS: CHAPTER 17

A1. It was thought to be autonomous (self-governing). However, it is regulated by brain centers such as the hypothalamus and brainstem with input from the limbic system and other parts of the cerebrum.

A2.

	Somatic	Autonomic
a. Sensations and movements are mostly *(conscious? unconscious/automatic?)*	**Conscious**	**Unconscious/automatic**
b. Types of tissue innervated by efferent nerves	Skeletal muscle	**Cardiac muscle, smooth muscle, and most glands**
c. Efferent neurons are excitatory (E), inhibitory (I), or both (E + I)	**E**	**E + I**
d. Examples of sensations (input)	**Special senses, general somatic senses, or proprioceptors, and possibly visceral sensory, as when pain of appendicitis or menstrual cramping causes a person to curl up (flexion of thighs, legs)**	Mostly visceral sensory (such as changes in CO_2 level of blood or stretching or visceral wall)
e. Number of neurons in efferent pathway	**One (from spinal cord to effector)**	**Two (pre- and postganglionic**
f. Neurotransmitter(s) released by motor neurons	Acetylcholine (ACh)	**ACh or norepinephrine (NE)**

A3. Sympathetic and parasympathetic.

A4. (a) Most viscera are innervated by both sympathetic and parasympathetic divisions. (b) One division excites and the other division inhibits the organ's activity.

B1. (a) 1, 2, ganglionic, postganglionic. (b) All, only certain (see Checkpoint B2).

B2. (a) T1, L2; thoracolumbar. (b) Are; ganglia; post-; gray, do not. (c) Brainstem, III, VII, IX, X; lateral, S2, S4. (d) Pre-, craniosacral; different; gray. (e) Do; differ.

B3. (a) Sensory, motor. (b) Both, visceral. (c) Does not, does.

B4.

	Sympathetic Trunk	Prevertebral	Terminal
a. Sympathetic or para-sympathetic	**Sympathetic**	Sympathetic	**Parasympathetic**
b. Alternate name	Paravertebral ganglia or vertebral chain	**Collateral or prevertebral (celiac, superior and inferior mesenteric)**	**Intramural (in walls of viscera)**
c. General location	**In vertical chain along both sides of vertebral bodies from base of skull to coccyx**	**In three sites anterior to spinal cord and close to major abdominal arteries**	Close to or in walls of effectors

B5. C A D B.

B6. (a) Immediately synapse in trunk ganglion. (b) T1, L2, C3, sacral. (c) Gray rami, spinal nerves. (d) Splanchnic, prevertebral; plexuses. (e) Abdomen; postganglionic. (f) Many; does not, 2.

B7.

Figure LG 17.1A connections and labels:
- Heart
- Cardiac nerve
- T3
- Dorsal ramus of spinal nerve
- Ventral ramus of spinal nerve
- Posterior gray horn
- Posterior root
- Lateral gray horn
- Posterior root ganglion
- Gray ramus communicans
- White ramus communicans
- T4
- Sympathetic chain
- Anterior gray horn
- Spinal nerve
- Spinal cord T5
- Anterior root
- Sympathetic trunk ganglion
- Splanchnic nerve
- T5
- Prevertebral ganglion (Celiac ganglion)
- T6
- Visceral afferent receptor
- Stomach
- T7
- Sweat gland

○———< Preganglionic ANS
○- - - - -< Postganglionic ANS
○·········< Afferent

Figure LG 17.1A Typical sympathetic pathway beginning at level T5 of the cord.

B8. (a) 22–25; three; superior, muscles and glands of the head (b) Sixth and seventh; heart. (c) 11–12, posterior; heart, lungs, bronchi, sweat glands, smooth muscle of blood vessels, and hair (arrector pili) muscles. (d) 4–5, 4–5.

B9. Sympathetic; pre, do not; prevertebral; abdomen or pelvis.

Splanchnic Nerve	Prevertebral Ganglion: Site of Postganglionic Cell Bodies	Viscera Supplied
a. Greater splanchnic n.	**Celiac**	**Stomach, spleen, liver, kidneys, small intestine**
b. Lesser splanchnic n.	**Superior mesenteric**	**Small intestine, colon**
c. Lumbar splanchnic n.	**Inferior mesenteric**	**Distal colon, rectum; urinary bladder, genitalia**

B10. Sympathetic; terminal ganglia; fewer. Parasympathetic effects are much less widespread than sympathetic, distributed to fewer areas of the body, and characterized by a response of perhaps just one organ, rather than an integrated response in which many organs respond in *sympathy* with one another.

B11.

Cranial Nerve Number	Cranial Nerve Name	Name of Terminal Ganglion	Structures Innervated
a. III	Oculomotor	Ciliary	Iris and ciliary muscle of eye
b. VII	Facial	1. Pterygopalatine 2. Submandibular	1. **Nasal mucosa, palate, pharynx, lacrimal glands** 2. **Submandibular and sublingual salivary glands**
c. IX	Glossopharyngeal	Otic	Parotid salivary gland
d. X	Vagus	Ganglia in cardiac and pulmonary plexuses and in abdominal plexuses	**Thoracic and abdominal viscera**

B12.(a) S. (b) P. (c) S. (d) P. (e) S. (t) S. (g) S. (h) S. (i) P, S. (j) S. (k) P.

B13. Adrenal (medulla); sympathetic.

Cl. Most, dual. (a) Sweat glands, arrector pili (hair) muscles, fat cells, kidneys, adrenal glands, spleen, and most blood vessels. (b) Lacrimal (tear) glands. (Refer to Table 17.4 and Figure 17.2 in your text.)

C2. (a) ACh. (b) NE. (c–e) ACh. (Refer to Figure LG 17.2.)

C3. (a) Cholinergic; adrenergic; adrenergic, cholinergic. (b) Sympathetic, adrenergic. (c) Adrenal medulla; sympathetic.

C4. (a) Refer to key for Figure LG 17.2; both P and S postganglionic neurons and somatic effector cells (skeletal muscle); ACh; always. (b) Muscarinic (*Hint:* Take time to notice that in this figure and the muscarinic receptors are ones shaped so that ACh "sits down" in an "M" [for muscarinic]

position, whereas nicotinic receptors have ACh attach in the "W" position); P (notice this at top of Table 17.3 in the text; many sweat glands; sometimes; decreases, increases. (c) ACh, nicotine, muscarine. (d) NE and epinephrine; catecholamine. (e) $Alpha_1$; excitatory; constrict/narrow; out of; blood flows into major vessels to increase blood pressure, which may assist in a stress response.

C5. (a) NE, in; isoproterinol (Isoprel) or epinephrine; agonist. (b) Beta; propranolol (Inderal); de; antagonist. (c) Increased (a double negative: the inhibiting vagus is itself inhibited).

C6. (a) $Beta_2$ dilation, easier. (b) Relaxation, de. (c) Constrict; high; blocker.

C7. (a)

$$ACh \xrightarrow{\text{AChE}} \cancel{ACh} \qquad NE \xrightarrow{\text{COMT or MAO}} \cancel{NE}$$

(b) NE; sympathetic/adrenergic. (c1) Greater divergence of sympathetic neurons, for example, through the extensive sympathetic trunk chain and through gray rami to all spinal nerves. (c2) Mimicking of sympathetic responses via release of catecholamines (epinephrine and NE) from the adrenal medulla. (c3) Adrenergic neurotransmitters (NE and epinephrine) linger longer in the synaptic cleft so stimulate the effector cells longer.

C8. (b) P↓ S↑. (c) P—,S↑. (d) P↑, S↓. (e) P↑, S↓. (f) P↓, S↑. (g) P↑, S↓. (h) P—, S↑. (i) P—, S↑. (j) P—, S↑. (k) P↑, S↓. (l) P—, S↑. (m) P↓, S↑.

Dl. Either somatic or visceral afferents; intero.

D2. See Figure LG 17.1A above.

D3. C J A E F G D B H I.

D4. Do not, subconscious.

MORE CRITICAL THINKING: CHAPTER 17

1. Does the human ANS include more white or gray rami communicantes? Explain.
2. Identify 12 specific organs in your own body that are supplied by autonomic neurons.
3. Explain why the names craniosacral and thoracolumbar are given to the two divisions of the ANS.
4. Contrast locations of autonomic ganglia in sympathetic and parasympathetic divisions of the ANS.
5. Describe a "sympathetic response" that you might exhibit when faced with a threatening situation. Include specific changes likely to occur in your viscera.

MASTERY TEST: CHAPTER 17

Questions 1–6: Circle letters preceding all correct answers to each question.

1. Choose all *true* statements about the vagus nerve.
   A. Its autonomic fibers are sympathetic.
   B. It supplies ANS fibers to viscera in the thorax and abdomen but not in the pelvis.
   C. It causes digestive glands to increase secretions.
   D. Its ANS fibers are mainly preganglionic.
   E. It is a cranial nerve originating from the medulla.

2. Which activities are characteristic of the stress response, or fight-or-flight reaction?
   A. The liver breaks down glycogen to glucose.
   B. The heart rate decreases.
   C. Kidneys increase urine production because blood is shunted to kidneys.
   D. There is increased blood flow to genitalia, causing erect or engorged state.
   E. Hairs stand on end ("goose pimples") due to contraction of arrector pili muscles.
   F. In general, the sympathetic system is active.

3. Which fibers are classified as autonomic?
   A. Any visceral efferent nerve fiber
   B. Any visceral motor sensory nerve fiber
   C. Nerves to salivary glands and sweat glands
   D. Pain fibers from ulcer in stomach wall
   E. Sympathetic fibers carrying impulses to blood vessels
   F. All nerve fibers within cranial nerves
   G. All parasympathetic nerve fibers within cranial nerves

4. Choose all *true* statements about gray rami.
   A. They contain only sympathetic nerve fibers.
   B. They contain only postganglionic nerve fibers.
   C. They carry impulses from trunk ganglia to spinal nerves.
   D. They are located at all levels of the vertebral column (from cervical to coccyx).
   E. They carry impulses between lateral ganglia and collateral ganglia.
   F. They carry preganglionic neurons from anterior ramus of spinal nerve to trunk ganglion.

5. Which of the following structures contain some sympathetic preganglionic nerve fibers?
   A. Splanchnic nerves
   B. White rami
   C. Sciatic nerve
   D. Cardiac nerves
   E. Ventral roots of spinal nerves
   F. The sympathetic chains

6. Which are structural features of the parasympathetic system?
   A. Ganglia lie close to the CNS and distant from the effector
   B. Forms the craniosacral outflow
   C. Distributed throughout the body, including extremities
   D. Supplies nerves to blood vessels, sweat glands, and adrenal glands
   E. Has some of its nerve fibers passing through lateral (paravertebral) ganglia

Questions 7–12; Circle the letter preceding the one best answer to each question.

7. All of the following axons are cholinergic *except:*
   A. Parasympathetic preganglionic
   B. Parasympathetic postganglionic
   C. Sympathetic preganglionic
   D. Sympathetic postganglionic to sweat glands
   E. Sympathetic postganglionic to heart muscle

8. All of the following are collateral ganglia *except:*
   A. Superior cervical
   B. Prevertebral ganglia
   C. Celiac ganglion
   D. Superior mesenteric ganglion
   E. Inferior mesenteric ganglion

9. Which region of the cord contains no preganglionic cell bodies at all?
   A. Sacral
   B. Lumbar
   C. Cervical
   D. Thoracic

10. Which statement about postganglionic neurons is *false*?
    A. They all lie entirely outside of the CNS.
    B. Their axons are unmyelinated.
    C. They terminate in visceral effectors.
    D. Their cell bodies lie in the lateral gray matter of the cord.
    E. They are very short in the parasympathetic.

11. All of the following answers match types of receptors with correct locations *except:*
    A. Alpha$_1$: blood vessels in skin and in walls of abdominal viscera
    B. Alpha$_1$: salivary gland cells
    C. Beta$_1$: smooth muscle fibers in walls of airways
    D. Nicotinic: motor end plates of skeletal muscle fibers
    E. Muscarinic: most sweat glands

12. In which condition are fingers and toes cool and pale due to vasoconstriction of blood vessels there?
    A. Autonomic dysreflexia
    B. Homer's syndrome
    C. Raynaud's disease

Questions 13–20: Circle T (true) or F (false). If the statement is false, change the underlined word or phrase so that the statement is correct.

T F   13.   Synapsing occurs in both <u>sympathetic and parasympathetic</u> ganglia.

T F   14.   The sciatic, brachial, and femoral nerves all contain <u>sympathetic postganglionic</u> nerve fibers.

T F   15.   Under stress conditions the <u>sympathetic system</u> dominates over the parasympathetic.

T F   16.   When one side of the body is deprived of its sympathetic nerve supply, as in Horner's syndrome, the following symptoms can be expected: <u>constricted pupil (miosis) and lack of sweating (anhidrosis)</u>.

T F   17.   Sympathetic cardiac nerves <u>stimulate</u> heart rate, and the vagus <u>slows down</u> heart rate.

T F   18.   Viscera <u>do have sensory nerve fibers, and they are included in the autonomic nervous system</u>.

T F   19.   <u>All</u> viscera have dual innervation by sympathetic and parasympathetic divisions of the ANS.

T F   20.   The sympathetic system has a <u>more</u> widespread effect in the body than the parasympathetic does.

Questions 21–25: Fill-ins. Complete each sentence with the word or phrase that best fits.

_____ 21. The three major types of tissue (effectors) innervated by the ANS

nerves are _____.

_____ 22. _____ nerves convey impulses from the sympathetic trunk
ganglia to collateral ganglia.

_____ 23. About 80% of the craniosacral outflow (parasympathetic nerves)

is located in the _____ nerves.

_____ 24. Alpha (α) and beta (β) receptors are stimulated by the transmitter

_____.

_____ 25. Drugs that block (inhibit) beta receptors in the heart will cause

_____-crease in heart rate and blood pressure.

ANSWERS TO MASTERY TEST: CHAPTER 17

Multiple Answers

1. B C D E
2. A E F
3. A B C D E G
4. A B C D
5. A B E F
6. B

Multiple Choice

7. E
8. A
9. C
10. D
11. C
12. C

True–False

13. T
14. T
15. T
16. T
17. T
18. T
19. F Most (or some)
20. T

Fill-ins

21. Cardiac muscle, smooth muscle, and glandular epithelium
22. Splanchnic
23. Vagus
24. NE (and also epinephrine)
25. De

FRAMEWORK 18
The Endocrine System

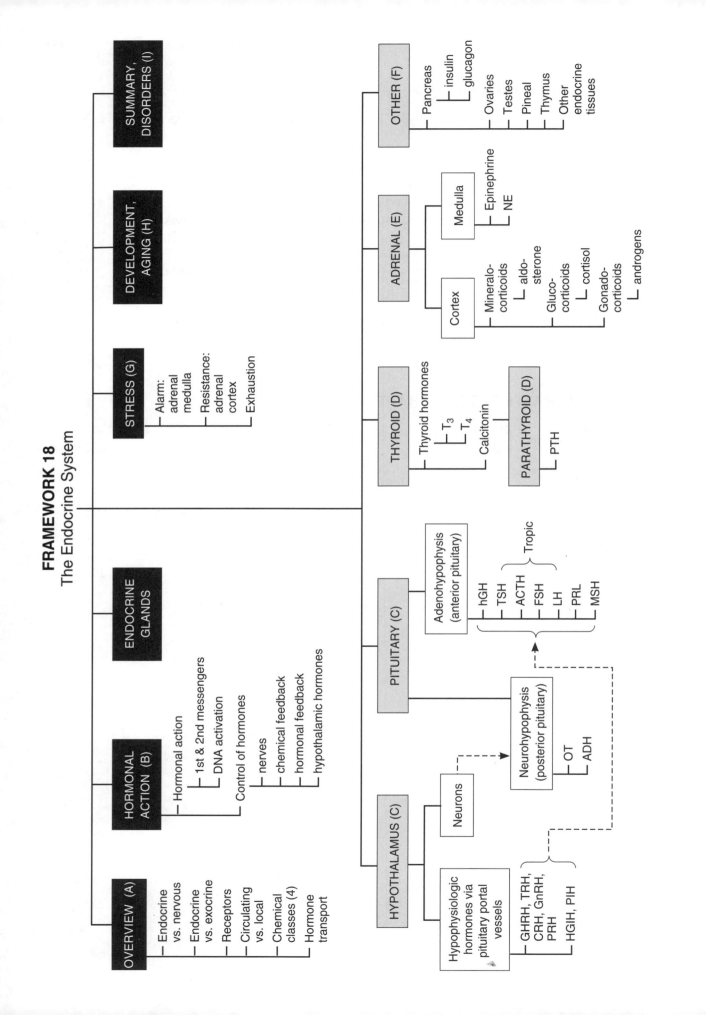

OVERVIEW (A)
- Endocrine vs. nervous
- Endocrine vs. exocrine
- Receptors
- Circulating vs. local
- Chemical classes (4)
- Hormone transport

HORMONAL ACTION (B)
- Hormonal action
  - 1st & 2nd messengers
  - DNA activation
- Control of hormones
  - nerves
  - chemical feedback
  - hormonal feedback
  - hypothalamic hormones

ENDOCRINE GLANDS

STRESS (G)
- Alarm: adrenal medulla
- Resistance: adrenal cortex
- Exhaustion

DEVELOPMENT, AGING (H)

SUMMARY, DISORDERS (I)

HYPOTHALAMUS (C)
- Neurons
- Hypophysiologic hormones via pituitary portal vessels
  - GHRH, TRH, CRH, GnRH, PRH
  - HGIH, PIH

PITUITARY (C)
- Adenohypophysis (anterior pituitary)
  - hGH
  - TSH
  - ACTH ⎫
  - FSH ⎬ Tropic
  - LH   ⎭
  - PRL
  - MSH
- Neurohypophysis (posterior pituitary)
  - OT
  - ADH

THYROID (D)
- Thyroid hormones
  - T_3
  - T_4
- Calcitonin

PARATHYROID (D)
- PTH

ADRENAL (E)
- Cortex
  - Mineralo-corticoids
    - aldosterone
  - Gluco-corticoids
    - cortisol
  - Gonado-corticoids
    - androgens
- Medulla
  - Epinephrine
  - NE

OTHER (F)
- Pancreas
  - insulin
  - glucagon
- Ovaries
- Testes
- Pineal
- Thymus
- Other endocrine tissues

C H A P T E R   **18**

The Endocrine System

Hormones produced by at least 18 organs exert widespread effects on just about every body tissue. Hormones are released into the bloodstream, which serves as the vehicle for distribution throughout the body. In fact, analysis of blood levels of hormones can provide information about the function of specific endocrine glands. How do hormones "know" which cells to affect? How does the body "know" when to release more or less of a hormone? Mechanisms of action and regulation of hormone levels are discussed in this chapter. The roles of all major hormones and effects of excesses or deficiencies in those hormones are also included. The human body is continually exposed to stressors, such as rapid environmental temperature changes, a piercing sound, or serious viral infection. Adaptations to stressors are vital to survival: however, certain adaptations lead to negative consequences. A discussion of stress and adaptation concludes Chapter 18.

As you begin your study of the endocrine system, carefully examine the Chapter 18 Topic Outline and Objectives: check off each one as you complete it. To organize your study of the endocrine system, glance over the Chapter 18 Framework now. Be sure to refer to the Framework frequently and note relationships among key terms in each section.

TOPIC OUTLINE AND OBJECTIVES

A. Overview: definitions, hormone activity

1. Distinguish between exocrine and endocrine glands.
2. Describe how hormones interact with target-cell receptors.
3. Compare the two chemical classes of hormones based on their solubility.

B. Mechanisms of hormonal action and control of hormone secretion

4. Describe the two general mechanisms of hormonal action.
5. Describe the three types of signals that can control hormone secretion.

C. Hypothalamus and pituitary gland

6. Explain why the hypothalamus is an endocrine gland.
7. Describe the location, histology, hormones, and functions of the anterior and posterior pituitary glands.

D. Thyroid and parathyroids

8. Describe the location, histology, hormones, and functions of the thyroid gland.
9. Describe the location, histology, hormones, and functions of the parathyroid glands.

E. Adrenal glands

10. Describe the location, histology, hormones, and functions of the adrenal glands.

F. Pancreas, gonads, pineal gland, thymus gland, and other endocrine cells

11. Describe the location, histology, hormones, and functions of the pancreas.
12. Describe the location, histology, hormones, and functions of the male and female gonads.
13. Describe the location, histology, hormones, and functions of the pineal gland.
14. Describe the location, histology, hormones, and functions of the thymus gland.
15. List the hormones secreted by cells in tissues and organs other than endocrine glands, and describe their functions.
16. Explain the action of eicosanoids.
17. List six important growth factors.

G. Stress and the general adaptation syndrome

18. Describe the general adaptation syndrome.

H. Developmental anatomy and aging of the endocrine system

19. Describe the development of the endocrine glands.
20. Describe the effects of aging on the endocrine system.

I. Summary of hormones and related disorders

WORDBYTES

Now become familiar with the language of this chapter by studying each wordbyte, its meaning, and an example of its use within a term. After you study the entire list, self-check your understanding by writing the meaning of each wordbyte on the line. As you continue through the *Learning Guide,* identify (and fill in) additional terms that contain the same wordbyte.

Wordbyte	Self-check	Meaning	Example(s)
adeno-	_____	gland	*adeno*hypophysis
andro-	_____	man	*andro*gens
crin-	_____	to secrete	endo*crine*
endo-	_____	within	*endo*crine
exo-	_____	outside	*exo*crine
gen-	_____	to create	diabeto*genic*
gonado-	_____	seed	*gonado*tropins
horm-	_____	excite, urge on	*horm*one
insipid-	_____	without taste	diabetes *insipid*us
mellit-	_____	sweet	diabetes *mellit*us
oxy(s)-	_____	swift	*oxy*tocin
para-	_____	around	*para*thyroid, *para*crine
somato-	_____	body	*somato*tropin
thyro-	_____	shield	*thyro*tropin
-tocin	_____	childbirth	oxy*tocin*
-tropin	_____	change, turn	somato*tropin*

A. Overview: definitions, hormone activity (pages 566–570)

✓ **A1.** Compare the ways in which the nervous and endocrine systems exert control over the body. Write *N* (nervous) or *E* (endocrine) next to descriptions that fit each system.

_____ a. Hormones are mediator molecules of this system.

_____ b. Neurotransmitters are mediator molecules of this system.

_____ c. Sends messages to muscles, glands, and neurons only.

_____ d. Sends messages to virtually any part of the body.

_____ e. Effects are generally faster and shorter lived.

✓ **A2.** Complete the table contrasting exocrine and endocrine glands.

	Secretions Transported In	Examples
a. Exocrine		
b. Endocrine		

✓▲ **A3.** Refer to Figure LG 18.1 and do this exercise.
   a. Color and label each endocrine gland next to letters A–G.
   b. Next to numbers 1–13, label each organ that contains endocrine tissue but is not an endocrine gland exclusively.

✓ **A4.** Write a sentence explaining how a specific hormone "knows" to affect a specific target organ, for example, how thyroid-stimulating hormone (TSH) "knows" to affect the thyroid gland rather than cells of pancreas, ovaries, or muscles.

How does this same principle explain the effectiveness of RU486 (mifepristone)?

✓ **A5.** Describe regulation of hormones in this exercise.
   a. When excessive amounts of hormones are present, the number of related receptors is likely to *(decrease? increase?)* so that the overabundant hormone is less effective. This effect is called *(down? up?)*-regulation.
   b. Up-regulation makes a target organ *(less? more?)* sensitive to hormones by increase of receptors, for example, when hormone level is *(deficient? excessive?)*.

✓ **A6.** Contrast types of hormones by filling in blanks. Use these answers: *auto, endo, para.*
   a. Circulating hormones that exert effects on distant target cells are called

      _____-crines.

   b. Local hormones include _____-crines that act upon the same cell

      that secreted the hormone, as well as _____-crines that act on
      nearby cells.
   c. Which of these types of hormones have longest lasting effects? _____

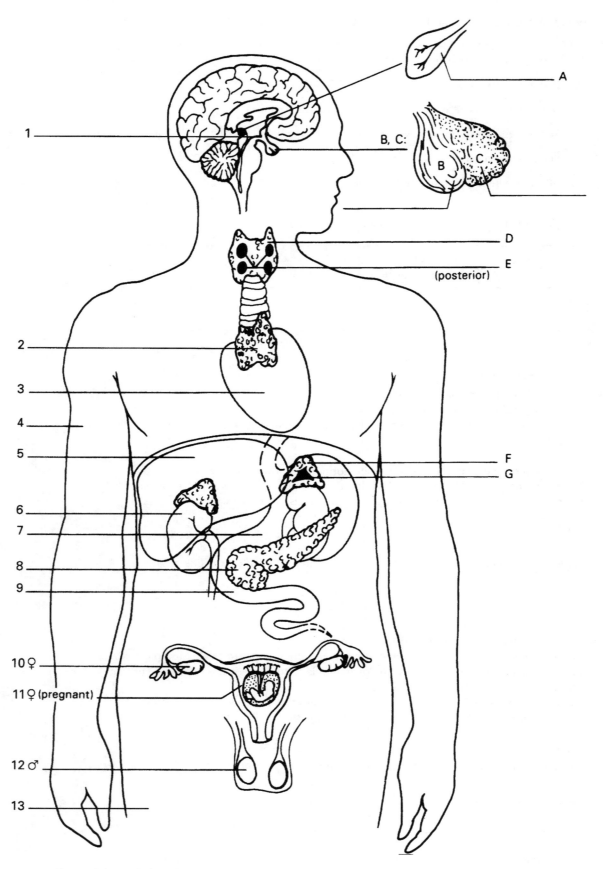

Figure LG 18.1 Diagram of endocrine glands labeled A–G and organs containing endocrine tissue labeled 1–13. Label and color as directed in Checkpoints A3 and C1.

✓ **A7.** Carefully examine Exhibit 18.2 in your text. Match the following chemical classes with descriptions of hormones listed below. The two short lines following each description will be used in Checkpoints A8 and B1.

A.	Amines	PP.	Peptide, protein
E.	Eicosanoid	S.	Steroid

_____ a. Chains of 3 to about 200 amino acids that may form glycoproteins; most

　　　　　hormones are in this category. _____ _____

_____ b. Catecholamines, including epinephrine, norepinephrine, and dopamine.

　　　　　_____ _____

_____ c. Thyroid hormones (T_3 and T_4). _____ _____

_____ d. Derived from 20-carbon fatty acid; examples are prostaglandins and leukotrienes.

　　　　　_____ _____

_____ e. Derived from cholesterol; includes hormones from adrenal cortex, ovary,

　　　　　and testis. _____ _____

✓ **A8.** Look back at Checkpoint A7 and write *L* for lipid-soluble or *W* for water-soluble on the first short line following each hormone description in Checkpoint A7.

✓ **A9.** Which category of hormones tend to circulate in the blood attached to transport proteins? *(Lipid-soluble? Water-soluble?)* Are any of these hormones ever not bound to a transport protein?

B. Mechanisms of hormonal action and control of hormone secretion (pages 570–573)

✓ **B1.** Look back at Activity A7. On the second lines following descriptions of hormones (a–e), write letters indicating the expected mechanism of hormonal action. (*Hint:* Remember that each action depends on the chemistry of the hormone.) Use these answers:

> **I.** This lipid-soluble hormone can pass through the lipid bilayer of the plasma membrane and enter the cell to use *intracellular receptors* that activate DNA to direct synthesis of new proteins.
>
> **P.** This water-soluble hormone cannot pass through the lipid bilayer of the plasma membrane and enter the cell, so it uses *plasma membrane receptors* to activate second messengers such as cyclic AMP.

✓ **B2.** Study the cyclic AMP mechanism in Figure 18.4 in your text. Then complete this exercise.

a. A hormone such as antidiuretic hormone (ADH) acts as the _____

messenger as it carries a message from the _____ where it is secreted

(in this case the hypothalamus) to the outside of the _____ (in this case kidney cells).

b. The hormone then binds to a receptor on the *(inner? outer?)* surface of the plasma

membrane, causing activation of _____-proteins.

c. Such activation increases activity of the enzyme _____, located *(outside? inside?)* the plasma membrane. This enzyme catalyzes conversion of ATP to

_____, which is known as the _____ messenger.

d. Cyclic AMP then activates one or more enzymes known as _____ to help transfer *(calcium? phosphate?)* from ATP to a protein, usually an enzyme. The resulting chemical can then set off the target cell's response. In the case of ADH, this is an increase in permeability of kidney cells, with a resulting decrease in

_____ production.

e. Effects of cyclic AMP are *(short? long?)* lived. This chemical is rapidly degraded by an

enzyme named _____.

f. A number of hormones are known to act by means of the cyclic AMP mechanism.

Included are most of the _____-soluble hormones. Name several.

g. Name two or more chemicals besides cyclic AMP that may act as second messengers.

B3. Explain what is meant by amplification of hormone effects.

✓ **B4.** Describe hormonal interactions in this activity.
   a. The interaction by which effects of progesterone on the uterus in preparation for pregnancy are enhanced by earlier exposure to estrogen is known as a(n) *(antagonistic? permissive? synergistic?)* effect.

   b. Insulin and glucagon exert _____ effects on blood glucose levels.

✓ **B5.** Precise regulation of hormone levels is critical to homeostasis. Briefly describe control mechanisms for each hormone listed below. Use these answers.

C. **Blood level of a chemical that is controlled by the hormone**	
H. **Blood level of another hormone**	**N.** **Nerve impulses**

_____ a. Epinephrine

_____ b. Parathyroid hormone

_____ c. Cortisol

✓ **B6.** *(Negative? Positive?)* feedback mechanisms are utilized for most hormonal regulation. For example, a low level of Ca^{2+} circulating through the parathyroid gland causes a(n) *(increase? decrease?)* in release of parathyroid hormone (PTH), which will increase blood level of Ca^{2+}. State one or more example(s) of hormone control by *positive* feedback.

C. Hypothalamus and pituitary gland (pages 573–581)

✓▲ **C1.** Complete this exercise about the pituitary gland.

 a. The pituitary is also known as the _____.
 b. Defend or dispute this statement: "The pituitary is the master gland."

 c. Where is it located?

 d. Seventy-five percent of the gland consists of the *(anterior? posterior?)* lobe, called

the _____-hypophysis. The posterior lobe (or _____-hypophysis) is somewhat smaller.
 e. The anterior pituitary is known to secrete *(2? 5? 7?)* different hormones. Write abbreviations for each of these hormones next to the diagram of the anterior pituitary on Figure LG 18.1.
 f. Arrange in correct sequence from first to last the vessels that form the blood pathway connecting the hypothalamus to the anterior pituitary. (*Hint:* it may help to refer to Figure 18.5 in your text.)

AHV.	**Anterior hypophyseal veins**
HPV.	**Hypophyseal portal veins**
PPC.	**Primary plexus of capillaries**
SPC.	**Secondary plexus of capillaries**
SHA.	**Superior hypophyseal arteries**

_____ _____ _____ _____ _____

Hypothalamic releasing and inhibiting factors are known as _____ hormones. Now place an asterisk (*) next to the one answer listed above that consists of vessels located within the anterior pituitary (adenohypophysis) that serve as an exchange site for both hypophysiotropic hormones and anterior pituitary hormones between blood and the anterior pituitary.
 g. What advantage is afforded by this special arrangement of blood vessels that link the hypothalamus with the anterior pituitary?

h. Figure 18.5 in your text, shows that the inferior hypophyseal artery, plexus of the infundibular process, and posterior hypophyseal veins form the blood supply to the *(anterior? posterior?)* lobe of the pituitary.

i. List functions of these five types of anterior pituitary cells.

1. Corticotrophs: _____

2. Gonadotrophs: _____

3. Lactotrophs: _____

4. Somatotrophs: _____

5. Thyrotrophs: _____

j. Define *tropic* hormones.

Which four of the anterior pituitary hormones are tropic hormones?

✓ **C2.** Tropic hormones are involved in feedback mechanisms summarized in steps 1 to 4 in Figure LG 18.2. Trace the pathway of those four steps as you do this exercise.

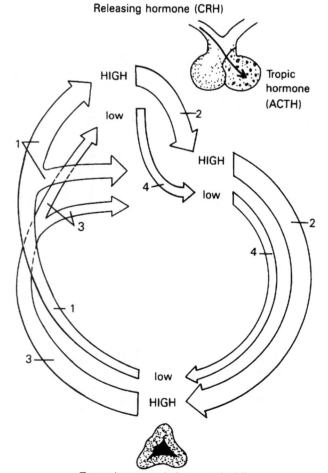

Releasing hormone (CRH)

Tropic hormone (ACTH)

HIGH
low
HIGH
low

Target hormone (adrenocorticoid)

Figure LG 18.2 Control of hormone secretion: example of releasing, tropic, and target hormones. Thickness of arrows indicates amounts of hormones. Color arrows as directed in Checkpoint C2. The branching of arrows 1 and 3 indicates that target hormones affect both the hypothalamus (releasing hormones) and the anterior pituitary (tropic hormones).

O 1   O 3
O 2   O 4

a. Color the arrow for step 1, in which a low blood level of target hormone, in this
    example, adrenocorticoid such as cortisone produced by the target gland

    _____, triggers the hypothalamus to secrete a *(high? low?)* level of
    releasing hormone (CRH). Simultaneously, the low level of target hormones directly
    stimulates the anterior pituitary to produce a *(high? low?)* level of ACTH.

b. Color the two arrows for step 2. First the high level of releasing hormone (CRH)

    passes through blood vessels to the _____ and stimulates more release
    of ACTH. A high level of this *(tropic? target?)* hormone in blood passing through the
    adrenal cortex then stimulates secretion of a *(high? low?)* level of target hormone.

c. Color steps 3 and 4, and note thickness of arrows. Both steps 1 and 3 are *(positive?
    negative?)* feedback mechanisms, whereas steps 2 and 4 are *(positive? negative?)*
    feedback mechanisms.

✓ **C3.** Complete this table listing releasing-tropic-target hormone relationships. Fill in the
   name of a hormone next to each letter and number.

Hypothalamic hormone →	Anterior Pituitary Hormone →	Target Hormone
a. GHRH (growth hormone releasing hormone or somatocrinin)		None
b. CRH (corticotropin releasing hormone)	1. 2.	1. 2. None
c.	TSH (thyrotropin)	1. 2. None
d. GnRH (gonadotropic releasing hormone	1. FSH (follicle stimulating hormone) 2. LH (luteinizing hormone)	1. 2. or testosterone
e.	Prolactin	None

✓ **C4.** Fill in this table naming two hypothalamic inhibiting hormones and their related
   anterior pituitary hormones. Hypothalamic inhibiting hormones provide additional
   regulation for the anterior pituitary hormones that *(are? are not?)* tropic hormones.
   (Notice this in the right column in the table below.)

Hypothalamic hormone →	Anterior Pituitary Hormone →	Target Hormone
a. GHRH (growth hormone releasing hormone or somatostatin)		None
b.	Prolactin	

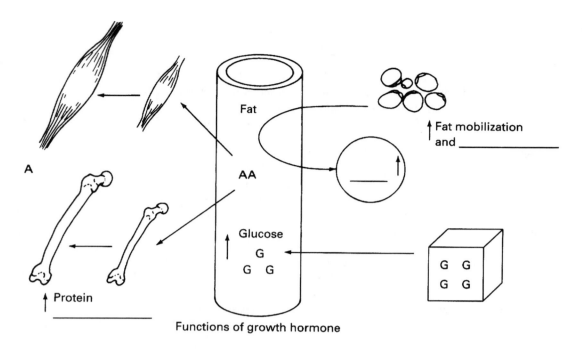

Functions of growth hormone

Figure LG 18.3 Functions of growth hormone in regulating metabolism.
AA, amino acid; G, glucose. Complete as directed in Checkpoint C5.

✓ **C5.** Refer to Figure LG 18.3 as you do this Checkpoint about human growth hormone hGH

   a. Effects of hGH are likely to be *(direct? indirect?)* because hGH may work via

       _____-like growth factors (IGFs). List four sites in the body where
       IGFs are made in response to hGH.

   b. A main function of hGH and IGFs is to stimulate growth and maintain size of

       _____ and _____. The alternative name for hGH,

       _____, indicates its function: to "change" or "turn" *(trop-)* the body
       (soma-) to growth. Note that somatotropic hormone *(is? is not?)* classified as a true
       tropic hormone because it *(does? does not?)* stimulate another endocrine gland to pro-
       duce a hormone.

   c. hGH stimulates growth by *(accelerating? inhibiting?)* entrance of amino acids into

       cells where they can be used for _____ synthesis. Indicate this by fill-
       ing in the blank on the left side of Figure LG 18.3.

   d. hGH promotes release of stored fats and *(synthesis? breakdown?)* of fats, leading to

       production of _____. Show this by filling in the blank on the right
       side of the figure.

   e. Because cells now turn to fats as a source of energy, the level of glucose in blood

       _____-creases and may lead to a condition known as *(hyper? hypo ?)*-glycemia. In
       this capacity, action of hGH is *(similar? opposite?)* to action of insulin. In fact, because
       hGH-induced hyperglycemia may overstimulate and "burn out" *(alpha? beta?)* cells

       of the pancreas, hGH may be considered _____-genic.

f. hGH secretion is controlled by two regulating factors: _____ and

_____. One stimulus that promotes release of hGH is low blood sugar (for example, during growth when cells require much energy). This condition, *(hvper-? hypo-?)* glycemia, stimulates secretion of *(GHRH? GHIH?)*, which then causes release of *(high? low?)* levels of hGH.

C6. *For extra review.* As you proceed through this chapter, complete the table describing factors that affect production of each of the following hormones. The first one is done for you.

Hormone	How Affected	Factors
a. hGH	Decreased by	HGIH, hGH (by negative feedback), REM sleep, obesity, increased fatty acids and decreased amino acids in blood; emotional deprivation; low thyroid
b. hGH	Increased by	
c. ACTH	Increased by	
d. ADH	Increased by	

✓ **C7.** Do this exercise about hormonal control of the mammary glands.
  a. Milk is produced in mammary glands following stimulation by the hormone

  _____ (_____) which is secreted by the *(anterior? posterior?)* pituitary. A different hormone causes ejection of milk from the glands at the time of

  the baby's suckling. This hormone, named _____, is released by the *(anterior? posterior?)* pituitary.
  b. Name several hormones that must "prime" breasts for milk production before prolactin and oxytocin can exert their effects.

  c. During pregnancy, prolactin levels *(increase? decrease?)* due to increased levels of *(PIH? PRH?)*. During most of a woman's lifetime prolactin levels are low as a result of high levels of *(PIH? PRH?)*. An exception is the time just prior to each menstrual flow when PIH levels *(increase? decrease?)*. Due to this lack of inhibition, prolactin secretion increases, activating breast tissues (and causing tenderness). State two reasons why breasts do not normally produce milk during this premenstrual phase.

✓ **C8.** Complete this Checkpoint on posterior pituitary hormones.

a. Name the two hormones released by the posterior pituitary: _____

(OT) and _____ (ADH).

b. Although these are released by the posterior pituitary, they are actually synthesized by *(endocrine cells? neurons?)* in the *(anterior pituitary? hypothalamus?)*. These chemicals are then transported by *(fast axonal transport? pituitary portal blood vessels?)* to the posterior pituitary where they are stored.

c. Oxytocin and ADH *(are? are not?)* regulated by tropic-target hormone relationships, as are the four anterior pituitary hormones TSH, _____ _____, and _____. OT and ADH *(are? are not?)* regulated by releasing or inhibiting hormones from the hypothalamus.

d. Functions of oxytocin are better known in human *(females? males?)*. The two target tissues of oxytocin are the uterus, which is stimulated to *(contract? relax?)*, and the breasts, in which milk *(production? letdown?)* is stimulated.

e. On a day when your body becomes dehydrated, for example, by loss of much sweat, your ADH production is likely to _____-crease. ADH causes kidneys to *(retain? eliminate?)* more water, _____-creasing urine output; ADH also _____-creases sweat production. ADH *(dilates? constricts?)* small blood vessels, which allows more blood to be retained in large arteries. Because of this latter function, ADH is also known as _____. As a result of all of these mechanisms, ADH _____-crease blood pressure.

f. Effects of diuretic medications are *(similar to? opposite of?)* ADH. In other words, diuretics _____-crease urine production and _____-crease blood volume and blood pressure.

g. Inadequate ADH production, as occurs in diabetes *(mellitus? insipidus?)* results in *(enormous? minute?)* daily volumes of urine. Unlike the urine produced in diabetes mellitus, this urine *(does? does not?)* contain sugar. It is bland or insipid. (Refer to page 603 of your text.)

h. Alcohol *(stimulates? inhibits?)* ADH secretion, which contributes to _____-creased urine production after alcohol intake. For extra review of other factors affecting ADH production, refer to Checkpoint C6.

C9. Complete parts (1–9) of Table LG 18.1, summarizing names and major functions of the seven hormones made by the anterior pituitary and the two hormones released by the posterior pituitary.

D. Thyroid and parathyroids (pages 581–585)

✓ **D1.** Describe the histology and hormone secretion of the thyroid gland in this exercise.

a. The thyroid is composed of two types of glandular cells. *(Follicular? Parafollicular?)* cells produce the two hormones _____ and _____.

Parafollicular cells manufacture the hormone _____.

b. The ion *(CL⁻? I⁻? Br⁻?)* is highly concentrated in the thyroid because it is an essential component of T_3 and T_4. Each molecule of thyroxine (T_4) contains four of the I^- ions, whereas T_3 contains _____ ions of I^-. *(T_3? T_4?)* is found in greater abundance in thyroid hormone, but *(T_3? T_4?)* is more potent.

c. Once formed, thyroid hormones *(are released immediately? may be stored for months?)*. T_3 and T_4 travel in the bloodstream bound to plasma *(lipids? proteins?)*.

Table 18.1 Summary of major endocrine glands and hormones.

Abbreviation	Name	Source	Principal Functions
1. GH		Anterior lobe of pituitary	
2.	Thyroid-stimulating hormone		
3. ACTH			
4. FSH			
5. LH			
6. PRL	Prolactin		
7. MSH			Increases skin pigmentation
8.	Oxytocin		
9. ADH			
10. —	Thyroxin		
11.			Decreases blood calcium
12.		Parathyroids	
13. —	Aldosterone		
14. —	Cortisol, cortisone		
15. —	Sex hormones (2)	(4)	
16.	(2)	Adrenal medulla	
17. —	Insulin		
18.		Alpha cells of pancreas	

D2. Briefly describe steps in the formation, storage, and release of thyroid hormones. Be sure to include these terms in your description: *trapping, tyrosines, colloid, peroxidase, T_1 and T_2, pinocytosis, secretion, TBG.*

✓ **D3.** Fill in blanks and circle answers about functions of thyroid hormones contrasted to those of growth hormone (hGH).

a. Like hGH, thyroid hormone stimulates protein *(synthesis? breakdown?).*

b. Unlike hGH, thyroid hormone _____-creases use of glucose for ATP production. Therefore, thyroid hormone *(increases? decreases?)* basal metabolic rate (BMR), releasing heat and *(increasing? decreasing?)* body temperature. This is known as the

_____ effect of thyroid hormones.

c. Thyroid hormone increases *(synthesis? breakdown?)* of fats and excretion of cholesterol

in bile, therefore _____-creasing blood cholesterol level.

d. Thyroid hormone *(increases? decreases?)* heart rate and blood pressure and *(increases? decreases?)* nervousness.

e. This hormone is important for development of normal bones and muscles, as well as brain tissue. Therefore, deficiency in thyroid hormone during childhood (a condition

called _____) *(does? does not?)* lead to retardation as well as small stature. Note that in childhood hGH deficiency, retardation of mental development *(occurs also? does not occur?).*

D4. Release of thyroid hormone is under influence of releasing factor

_____ and tropic hormone _____. Review the table in Checkpoint C3. List factors that stimulate TRF and TSH production.

✓ **D5.** Color arrows on Figure LG 18.4 to show hormonal control of calcium. This figure emphasizes that CT and PTH are *(antagonists? synergists?)* with regard to effects on calcium. One way that PTH increases blood calcium level is by stimulating kidney

production of the active form of vitamin _____ (also known as _____),

which facilitates _____ absorption in the intestine. Both CT and PTH act synergistically to *(lower? raise?)* blood phosphate levels.

Figure LG 18.4 Hormonal control of calcium ion (Ca²⁺). (a) Increased calcium storage in bones lowers blood calcium (hypocalcemic effect). (b) Calcium from three sources increases blood calcium level (hypercalcemic effect). Arrows show direction of calcium flow. Color arrows as directed in Checkpoint D5.

D6. Complete parts (10–12) of Table LG 18.1, summarizing names and major functions of hormones made by the thyroid and parathyroid glands.

E. Adrenals (pages 585–592)

✓ **E1.** Contrast the two regions of the adrenal glands in this Checkpoint. Write *C* (cortex) or

_____ a. More external in location   _____ d. Develops from ectoderm

_____ b. Makes up 80–90% of the adrenal glands   _____ e. Essential for life

_____ c. Develops from mesoderm

M (medulla) next to related descriptions.

E2. In the space below, draw an outline of the adrenal cortex (actual size). Label the *adrenal medulla* and the *adrenal cortex*. Then name the three zones (*zona*) of the adrenal cortex from outermost to innermost. Finally, label your figure with the primary hormones(s) secreted by each zone.

Figure LG 18.5 Hormonal control of fluid and electrolyte balance and blood pressure (BP) by renin, angiotensin, aldosterone, and ADH. Fill in blanks 1–10 as directed in Checkpoint E3.

✓ **E3.** Fill in blank lines on Figure LG 18.5 to show mechanisms that control fluid balance and blood pressure.

✓ **E4.** *A clinical correlation.* Determine whether each of the following conditions is likely to *increase* or *decrease* blood pressure.

   a. Excessive renin production: _____.

   b. Use of an angiotensin converting enzyme (A.C.E.)-inhibitor medication:

      _____.

   c. Aldosteronism (for example, due to a tumor of the zona glomerulosa):

      _____.

✓ **E5.** Aldosterone is a hormone made by the adrenal *(cortex? medulla?)*. Aldosterone release *(is? is not?)* regulated by adrenocorticotropic hormone (ACTH). State two factors that stimulate aldosterone production.

✓ **E6.** Write the name of the adrenal cortical hormone that is most abundant in each category:

   a. Mineralocorticoids: _____

   b. Glucocorticoids: _____

   c. Androgens: _____

✓ **E7.** In general, glucocorticoids are involved with metabolism and response to stress. Describe their effects in this exercise.

   a. These hormones promote *(synthesis? breakdown?)* of muscle proteins with release of amino acids then available for other body functions. In this respect their effects are *(similar? opposite?)* to those of both growth hormone and thyroid hormone.

   b. However, glucocorticoids may stimulate conversion of amino acids and fats to

      glucose, a process known as _____. In this way glucocorticoids (like hGH) *(increase? decrease?)* blood glucose.

   c. Refer to Figure 18.21 in your text. Note that glucocorticoid (and also hGH) production *(increases? decreases?)* during times of stress. By raising blood glucose levels, these hormones make energy available to cells most vital for a response to stress. During stress glucocorticoids also act to *(increase? decrease?)* blood pressure and to *(enhance? limit?)* the inflammatory process. (They are *anti-inflammatory.*)

   d. Fill in the blanks or circle answers to describe how glucocorticoids exert their specific anti-inflammatory effects.

      (1) _____-crease the number of mast cells (that release the chemical

      _____, which dilates and _____-creases permeability of blood

      vessels). The result is _____-crease of swelling at the inflamed site.

      (2) _____-crease activity of phagocytes and *(activate? stabilize?)* lysosomal enzymes, thereby *(enhancing? reducing?)* tissue destruction at the inflamed site.

   e. *For extra review* of regulation of glucocorticoids, refer to Checkpoints C2, C3, and C6.

✓ **E8.** Do this exercise on the adrenal medulla.

a. Name the two principal hormones secreted by the adrenal medulla.

_____   _____

Circle the one that accounts for 80% of adrenal medulla secretions.
These hormones are made by *(chromaffin? zona glomerulosa? zona fasciculata?)* cells.

b. In general, effects of these hormones mimic the *(sympathetic? parasympathetic?)* nervous system. List three or more effects.

c. Describe how the adrenal medulla is stimulated to release its hormones.

E9. Complete parts (13–16) of Table LG 18.1, summarizing names and major functions of hormones made by the adrenal glands.

F. Pancreas, gonads, pineal gland, thymus, and other endocrine cells (page 592–599)

✓ **Fl.** The *islets of Langerhans* are located in the _____. Do this exercise describing hormones produced there.

a. The name islets of Langerhans suggests that these clusters of _____-crine cells lie amidst a "sea" of exocrine cells within the pancreas. Name the four types of islet cells.

b. Refer to Figure LG 18.6a. Glucagon, produced by *(alpha? beta? delta?)* cells, *(increases? decreases?)* blood sugar in two ways. First, it stimulates the breakdown of

_____ to glucose, a process known as *(glycogenesis? glycogenolysis? gluconeogenesis?)*. Second, it stimulates the conversion of amino acids and other

compounds to glucose. This process is called _____.

c. Is glucagon controlled directly by an anterior pituitary tropic hormone? *(Yes? No?)*

In fact, control is by effect of blood _____ level directly upon the pancreas. When blood glucose is low, then a *(high? low?)* level of glucagon will be produced. This will raise blood sugar.

d. *For extra review.* Define the terms *glycogen* and *glucagon.*

e. Now look at Figure LG 18.6b. The action of insulin is *(the same as? opposite?)* that of glucagon. In other words, *insulin decreases blood sugar.* (Repeat that statement three times; it is important!) Insulin acts by several mechanisms. Two are shown in the figure. First, insulin *(helps? hinders?)* transport of glucose from blood into cells. Second,

it accelerates conversion of glucose to _____, the process called

_____.

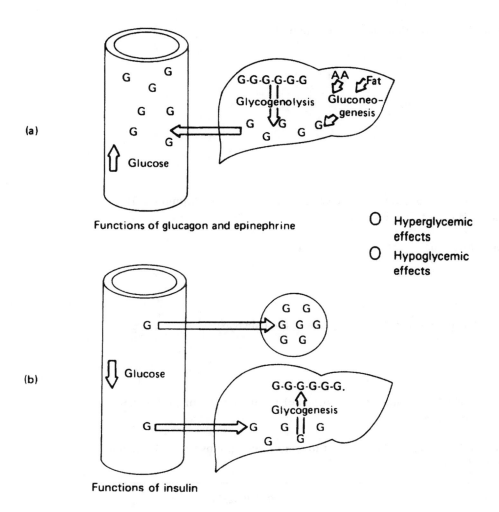

(a)

Functions of glucagon and epinephrine

○ Hyperglycemic
  effects

○ Hypoglycemic
  effects

(b)

Functions of insulin

Figure LG 18.6 Hormones that regulate blood glucose level. AA, amino acid; G, glucose; G-G-G-G-G-G, glycogen. Color according to directions in Checkpoint F1.

f. Hormones that raise blood glucose levels are called *(hypo? hyper?)*-glycemic hormones. Insulin, the one main hormone that lowers blood glucose level is said to

   be _____-glycemic. To show this, color arrows in Figure 18.6.

✓ **F2.** Complete the table contrasting effects of insulin and glucagon by drawing arrows up or down. The first one is done for you.

Metabolic Process	Insulin	Glucagon
a. Glycogenesis (formation of glycogen)	↑	Omit
b. Glycogenolysis (breakdown of glycogen to glucose)		
c. Gluconeogenesis (formation of glucose from sources other than glycogen)		
d. Protein synthesis (from amino acids)		Omit
e. Lipogenesis (formation of fatty acids)		Omit
f. Entrance of glucose into cells, especially skeletal muscle fibers		Omit
g. Blood glucose level		

F3. List three or more factors that increase secretion of:

a. Insulin

b. Glucagon

F4. Identify types of pancreatic cells that make each of the following hormones. Then write three functions of each hormone.

a. Somatostatin (made by _____ cells of the pancreas):

b. Pancreatic polypeptide (made by _____ cells of the pancreas):

F5. Complete parts (17–18) of Table LG 18.1, summarizing names and major functions of hormones made by the pancreas.

✓ **F6.** Write names of hormones in the box next to related descriptions below. Write one answer on each line.

E.	**Estrogens**	**R.**	**Relaxin**
I.	**Inhibin**	**T.**	**Testosterone**
P.	**Progesterone**		

a. Produced by pregnant women; dilates cervix and increases flexibility of the

symphysis pubic during labor and delivery: _____

b. The primary androgen in males: _____

c. The principal female sex hormones: _____ _____
d. Made in ovaries and testes; inhibits FSH in both genders: _____ _____ _____ _____

✓▲ **F7.** Sex hormones are produced in several locations in the body. List four locations:

▲ **F8.** Where is the pineal gland (body) located?

Secretion of one pineal hormone, melatonin, increases in *(light? darkness?)*. How may this hormone be related to the form of depression known as seasonal affective disorder (SAD)?

✓ **F9.** What is the general function of the thymus? _____ Name four
hormones made by this gland.

✓▲ **F10.** *The Big Picture: Looking Ahead.* Refer to upcoming chapters and answer questions
about hormones produced by tissues that not only contain endocrine cells, but that
also have other functions. Select answers from the box.

Atrial natriuretic peptide	**Erythropoietin**
CCK, gastrin, and secretin	**Human chorionic gonadotropin (HCG)**
Eicosanoids: thromboxane A2 and PG	**Vitamin D**

a. Chemical that promotes development of red blood cells (erythropoiesis) (Figure 19.5):

_____.

b. Chemical that decreases blood pressure by promoting loss of Na^+ and water, and by

dilating blood vessels (Table 21.1, page 686) _____.

c. Chemicals that constrict blood vessels and therefore increase blood pressure (page 686):

_____.

d. Hormones that alter secretion of gastric or intestinal secretions and movements

(Table 24.4): _____.

e. Hormone made in skin and kidney; aids in absorption of calcium (Exhibit 25.6, page

906): _____.

f. Hormone made by the chorion layers of the placenta that stimulates production of
estrogens and progesterone to maintain pregnancy (Figure 29.8):

_____.

F11. Write a brief description of functions of each of the following categories of chemicals
produced by the body.

a. Leukotrienes (LTs)

b. Prostaglandins (PGs)

c. Growth factors

✓ **F12.** *A clinical correlation.* Aspirin and nonsteroidal anti-inflammatories (NSAIDs) inhibit synthesis of one of the categories of chemicals listed in Checkpoint F11. Which one?

_____ Write three therapeutic effects of NSAIDs.

✓ **F13.** Match growth factors listed in the box with descriptions below.

EGF.	**Epidermal growth factor**
PDGF.	**Platelet-derived growth factor**
TA.	**Tumor angiogenesis factor**

_____ a. Produced by normal and tumor cells;   _____ c. Produced in platelets; appears to play a
stimulates growth of new blood vessels   role in wound healing

_____ b. Produced in salivary glands; stimulates
growth of epithelial cells, fibroblasts, and
neurons

G. Stress and the general adaptation syndrome (pages 599–601)

G1. Define two categories of stressors and give two examples of *stressors.*

✓ **G2.** Summarize the body's response to stressors in this exercise.
a. The part of the brain that senses the stress and initiates response is the

_____. It responds by two main mechanisms.
b. First is the *(alarm? resistance?)* reaction involving the adrenal *(medulla? cortex?)*

and the _____ division of the autonomic nervous system.

c. Second is the _____ reaction. Playing key roles in this response are

three anterior pituitary hormones: _____, _____

and _____ and their target hormones. Therefore, the adrenal
(medulla? cortex?) hormones are activated.

✓ **G3.** Now describe events that characterize the first (or alarm) stage of response to stress.
Complete this exercise.
a. The alarm reaction is sometimes called the _____ response.
b. During this stage blood glucose *(increases? decreases?)* by a number of hormonal

mechanisms, including those of adrenal medulla hormones _____

and _____. Glucose must be available for cells to have energy for
the stress response.
c. Oxygen must also be available to tissues; the respiratory system *(increases? decreases?)*
its activity. Heart rate and blood pressure *(increase? decrease?)*. Blood is

shunted to vital tissues such as the _____, _____,

and _____ and is directed away from reservoir organs such as the

_____, _____, and _____.

  d. Nonessential activities such as digestion *(increase? decrease?)*. Sweating *(increases? decreases?)* in order to control body temperature and eliminate wastes.

G4. Describe the effects of each of these hormones in the resistance stage.
  a. Mineralocorticoids

  b. Glucocorticoids

  c. Thyroxin

  d. hGH

✓ **G5.** Identify activities of the three stages of response stress in this Checkpoint. Use answers in the box; they may be used once, more than once, or not at all.

A. Alarm stage	**E.** Exhaustion stage	**R.** Resistance stage

_____ a. Makes large amounts of glucose and oxy-gen available for a fight-or-flight response.

_____ b. This stage is short lived.

_____ c. Initiated by sympathetic nerve impulses and adrenal medulla hormones (epinephrine and norepinephrine).

_____ d. Mediated by hypophysiotropic hormones CRH, GHRH, and TRH, which then stimu-late anterior pituitary hormones ACTH, hGH, and TSH

_____ e. Sodium and water retention is promoted by effects of aldosterone in this stage.

_____ f. During this stage, blood chemistry returns to nearly normal.

_____ g. Depletion of body reserves and failure of resistance mechanisms occur.

H. Development and aging of the endocrine system (pages 601–603)

✓ **H1.** Match the endocrine gland with the correct description of its embryological origin.

AC.	Adrenal cortex		P.	Pancreas
AM.	Adrenal medulla		PT.	Parathyroid
AP.	Anterior pituitary		T.	Thymus

_____ a. Develops from the foregut area that later becomes part of small intestine

_____ b. Derived from the roof of the mouth; called the hypophyseal (Rathke's) pouch

_____ c. Originates from the neural crest which also produces sympathetic ganglia

_____ d. Derived from tissue from the same region that forms gonads

_____ e. Arise from pharyngeal pouches (two answers)

✓ **H2.** Write arrows next to hormone names to indicate whether production of each hormone increases (↑) or decreases (↓) with aging. If the hormone level remains stable in elderly persons, write a dash (—). Note that in most cases tropic hormones (*) tend to increase when their target hormones have decreased.

a. Estrogen (women): _____

b. Testosterone (men): _____

c. FSH and LH (both genders)*: _____

d. T_3/T_4: _____

e. TSH*: _____

f. hGH: _____

g. Adrenocorticoids: _____

h. Insulin: _____

i. Thymic hormones: _____

j. Epinephrine and norepinephrine: _____

I. Summary of hormones; disorders and medical terminology (pages 603–605)

✓ **I1.** *For extra review.* Test your understanding of these hormones by writing the name (or abbreviation where possible) of the correct hormone after each of the following descriptions.

a. Stimulates release of growth hormone: _____

b. Stimulates testes to produce testosterone: _____

c. Promotes protein synthesis of bones and muscles: _____

d. Tropic hormone for thyroxin: _____

e. Stimulates development of ova in females and sperm in males: _____

f. Present in bloodstream of nonpregnant women to prevent lactation:

_____

g. Induces sleep: _____

h. Stimulates uterine contractions and also breast milk let-down: _____

i. Mimics many effects of sympathetic nerves: _____

j. Stimulates kidney tubules to produce small volume of concentrated urine:

_____

k. Target hormone of ACTH: _____

l. Increases blood calcium and decreases blood phosphate: _____

m. Decreases blood calcium and phosphate: _____

n. Contains iodine as an important component: _____

o. Raises blood glucose (three or more answers): _____,

_____, _____

p. Lowers blood glucose: _____

q. Serves anti-inflammatory functions: _____

✓ **I2.** *For extra review.* Match the disorder with the hormonal imbalance.

Ac.	**Acromegaly**		**GD.**	**Graves' disease**
AD.	**Addison's disease**		**M.**	**Myxedema**
C.	**Cretinism**		**PC.**	**Pheochromocytoma**
DI.	**Diabetes insipidus**		**PD.**	**Pituitary dwarfism**
DM.	**Diabetes mellitus**		**T.**	**Tetany**
Go.	**Goiter**			

_____ a. Deficiency of hGH in child; slow bone growth

_____ b. Excess of hGH in adult; enlargement of hands, feet, and jawbones

_____ c. Deficiency of ADH; production of enormous quantities of "insipid" (nonsugary) urine

_____ d. Deficiency of effective insulin; hyperglycemia and glycosuria (sugary urine)

_____ e. Deficiency of thyroxin in child; short stature and mental retardation

_____ f. Deficiency of thyroxin in adult; edematous facial tissues, lethargy

_____ g. Excess of thyroxin; protruding eyes, "nervousness," weight loss

_____ h. Deficiency of thyroxin due to lack of iodine; most common reason for enlarged thyroid gland

_____ i. Result of deficiency of PTH; decreased calcium in blood and fluids around muscles resulting in abnormal muscle contraction

_____ j. Deficiency of adrenocorticoids; increased K^+ and decreased Na^+ resulting in low blood pressure and dehydration

_____ k. Tumor of adrenal medulla causing sympathetic-type responses, such as increased pulse and blood pressure, hyperglycemia, and sweating

✓ **I3.** Hypocalcemia can result from *(hypo? hyper?)*-parathyroid conditions. Decreased

blood calcium results in abnormal _____-crease in nerve impulses to muscles, a

condition known as _____. In other words, there is a(n) *(direct? inverse?)* relationship between blood level of calcium and nerve and muscle activity.

✓ **I4.** Summarize effects of hormones on electrolytes by completing this table. Write *In* for increase and *De* for decrease to indicate the effect of a hormone on blood levels of each ion. Omit portions of the table with—.

Hormone	Ca^{2+}	Mg^{2+}	$HPO_4{}^{2-}$	K^+	Na^+
a. PTH	In			—	—
b. Calcitonin		—		—	—
c. Aldosterone	—	—	—		

✓ **I5.** Summarize effects of different hormones on metabolism by completing this table.

	Stimulates Protein Synthesis? Breakdown?	Stimulates Lipolysis?	Hyperglycemic Effect?
a. hGH		Yes? No?	Yes? No?
b. Thyroid hormone		Yes? No?	Yes? No?
c. Cortisol		Yes? No?	Yes? No?
d. Insulin		Yes? No?	Yes? No?

✓ **I6.** Write *A* (for Addison's disease) or *C* (for Cushing's syndrome) next to the manifestations of those adrenal cortex disorders listed below.

_____ a. Abdominal striae

_____ b. Hyperglycemia

_____ c. "Moon face" and "buffalo hump"

_____ d. Decreased resistance to stress and infection

_____ e. Low blood pressure

_____ f. High level of ACTH

✓ **I7.** Contrast type I and type II diabetes mellitus (DM) by writing I or II next to the related descriptions.

_____ a. Also known as maturity-onset DM

_____ b. The more common type of DM

_____ c. Related to insensitivity of body cells to insulin, rather than to absolute insulin deficiency

_____ d. Also known as insulin-dependent diabetes mellitus (IDDM)

_____ e. More likely to lead to serious complications such as ketoacidosis

✓ **I8.** *A clinical correlation.* Mrs. Jefferson has diabetes mellitus. In the absence of sufficient insulin, her cells are deprived of glucose. Explain why she experiences the following symptoms.

a. Hyperglycemia and glucosuria

b. Increased urine production (poly-_____) and increased water intake

(poly-_____)

c. Ketoacidosis, a form of _____-osis

d. Atherosclerosis

ANSWERS TO SELECTED CHECKPOINTS: CHAPTER 18

A1. (a) E. (b) N. (c) N. (d) E. (e) N.

A2.

	Secretions Transported In	Examples
a. Exocrine	Ducts	Sweat, oil, mucus, digestive juices
b. Endocrine	Bloodstream	Hormones

A3. (a) A, pineal; B, posterior pituitary; C, anterior pituitary; D, thyroid; E, parathyroid; F, adrenal cortex; G, adrenal medulla. (b) 1, hypothalamus; 2, thymus; 3, heart; 4, skin; 5, liver; 6, kidney; 7, stomach; 8, pancreas; 9, intestine; 10, ovary: 11, placenta; 12, testis; 13, adipose tissue.

A4. Hormones affect only cells with specific receptors that "fit" the hormone. RU486 blocks receptors for the hormone progesterone and therefore prevents pregnancy because this hormone is needed to establish and maintain a luxuriant uterine lining required for embryonic development.

A5 (a) Decrease; down. (b) More, deficient.

A6. (a) Endo. (b) Auto, para. (c) Endo.

A7 (a) PP. (b) A. (c) A. (d) E. (e) S.

A8. (a) W. (b) W. (c) L (since derived from the non-polar amino acid, tyrosine). (d) W. (e) L.

A9. Lipid-soluble. Yes, about 0.1–10% of lipid-soluble hormones; these are diffusing out of blood so they can bind to receptors and exert their actions.

B1. (a–b) P. (c) I. (d) P. (e) I.

B2. (a) First, endocrine gland, target cell. (b) Outer, G. (c) Adenylate cyclase, inside; cyclic AMP, second. (d) Protein kinases, phosphate; urine. (e) Short; phosphodiesterase. (f) Water; ADH, TSH, ACTH, glucagon, epinephrine, the hypothalamic releasing hormones. (g) Ca^{2+}, cGMP, IP_3, or DAG.

B4. (a) Permissive. (b) Antagonistic.

B5. (a) N. (b) C. (c) H.

B6. Negative; increase. Examples of positive feedback: uterine contractions stimulate more oxytocin, or high estrogen level stimulates high LH level and ovulation.

C1. (a) Hypophysis. (b) The pituitary regulates many body activities; yet this gland is "ruled" by releasing and inhibiting hormones from the hypothalamus and also by negative feedback mechanisms. (c) In the sella turcica just inferior to the hypothalamus. (d) Anterior, adeno; neuro. (e) 7; hGH, PRL, MSH, ACTH, TSH, FSH, LH. (f) SHA PPC HPV SPC* AHV; hypophysiotropic. (g) Hypothalamic hormones reach the pituitary quickly and without dilution because they do not have to traverse systemic circulation. (h) Posterior. (i) l, Corticotrophs synthesize ACTH, and some make MSH; 2, gonadotrophs make FSH and LH; 3, lactotrophs produce PRL; 4, somatotrophs make hGH; 5, thyrotrophs produce TSH. (j) Hormones that influence other endocrine glands; ACTH, FSH, LH, and TSH.

C2. (a) Adrenal cortex, high; high. (b) Anterior pituitary; tropic, high. (c) Negative, positive.

C3.

Hypothalamic hormone	→ Anterior Pituitary Hormone	→ Target Hormone
a. GHRH (growth hormone releasing hormone or somatocrinin)	hGH (human growth hormone)	None
b. CRH (corticotropin releasing hormone)	1. ACTH (corticotropin or adrenocorticotropic hormone) 2. MSH (melanocyte stimulating hormone)	1. Cortisol 2. None
c. TRH (thyrotropin releasing hormone)	TSH (thyrotropin) or thyroid stimulating hormone	1. Thyroxine 2. None
d. GnRH (gonadotropic releasing hormone	1. FSH (follicle stimulating hormone) 2. LH (luteinizing hormone)	1. Estrogen 2. Estrogen, progesterone or testosterone
e. PRH (prolactin releasing hormone)	Prolactin	None

C4. Are not. (a) hGH (human growth hormone) and also TSH (thyroid-stimulating hormone). (b) PIH (prolactin inhibiting hormone), which is dopamine; none.

C5. (a) Indirect, insulin; liver, muscle, cartilage, and bone. (b) Bones, skeletal muscles; somatotropin; is not, does not. (c) Accelerating, protein; write *synthesis* on the line at left. (d) Breakdown, ATP; write *breakdown* on the line and *ATP* in the circle. (e) In, hyper; opposite; beta, diabeto. (f) GHRH (somatocrinin), GHRH (somatostatin); hypo, GHRH, high.

C7. (a) Prolactin (PRL), anterior; oxytocin (OT), posterior. (b) Estrogens, progesterone, glucocorticoids, hGH, thyroxine, and insulin. (c) Increase, PRH; PIH; decrease; the high PRL level does not last long enough, and mammary glands are not "primed" for milk production as they are during pregnancy.

C8. (a) Oxytocin, antidiuretic hormone. (b) Neurons, hypothalamus; fast axonal transport. (c) Are not, FSH, LH, and ACTH; are not. (d) Females; contract, letdown. (e) In; retain, de, de; constricts; vasopressin; in. (f) Opposite of; in, de. (g) Insipidus, enormous; does not. (h) Inhibits; in.

Dl. (a) Follicular, thyroxine, triiodothyronine: calcitonin. (b) I^-; 3; T_4, T_3. (c) May be stored for months; proteins.

D3. (a) Synthesis. (b) In, increases; increasing; calorigenic. (c) Breakdown, de. (d) Increases, increases. (e) Cretinism, does; does not occur.

D5. Labels on Figure LG 18.4: (a) Calcitonin (CT). (b) Parathyroid hormone (PT). Antagonists; D, calcitriol or 1, 25-dihydroxy vitamin D_3, calcium, phosphate, and magnesium; lower.

El. (a–c) C. (d) M. (e) C.

E3. 1, The protein angiotensinogen; 2, renin; 3, angiotensin 1; 4, angiotensin II; 5, constrictor; 6, aldosterone; 7, Na^+ (and H_2O by osmosis, and Cl^- and HCO_3^- to achieve electrochemical balance); 8, K^+ and H^+; 9, H_2O; 10, In.

E4. (a) Increase (known as *high renin hypertension*). (b) Decrease, such as the antihypertensive captopril (Capoten). (c) Increase.

E5. Cortex; is not; angiotensin II (resulting from the renin-angiotensin-aldosterone pathway) and increased blood level of K^+ (hyperkalemia) since aldosterone helps to lower blood K^+.

E6. (a) Aldosterone. (b) Cortisol. (c) DHEA (dehydroepiandrosterone).

E7. (a) Breakdown; opposite. (b) Gluconeogenesis; increase. (c) Increases; increase, limit. (d1) De, histamine, in; de. (d2) De, stabilize, reducing.

E8. (a) Epinephrine or adrenaline (80%) and norepinephrine or noradrenaline; chromaffin. (b) Sympathetic: fight-or-flight response such as increase of heart rate, force of heart contraction, and blood pressure; constrict blood vessels and dilate airways; increase blood glucose level and decrease digestion (c) Direct stimulation by preganglionic sympathetic nerves during stressful periods.

Fl. Pancreas. (a) Endo; alpha, beta, delta, and F. (b) Alpha, increases, glycogen, glycogenolysis; gluconeogenesis. (c) No; glucose; high. (d) *Glycogen* is a storage form of glucose; *glucagon* is a hormone that stimulates breakdown of glycogen to glucose. (e) Opposite; helps; glycogen, glycogenesis. (f) Hyper; hypo; glucagon and epinephrine are hyperglycemic; insulin is hypoglycemic.

F2. (b–c) ↓, ↑. (d–f) ↑, omit. (g) ↓, ↑.

F6. (a) R. (b) T. (c) E, P. (d) E, P, T, I.

F7. Ovaries, testes, adrenal cortex, and placenta.

F9. Immunity; thymosin, thymic humoral factor, thymic factor, and thymopoietin.

F10. (a) Erythropoietin. (b) Atrial natriuretic peptide. (c) Eicosanoids: thromboxane A and PG F. (d) CCK, gastrin, GIP, and secretin. (e) Vitamin D. (f) Human chorionic gonadotropin (HCG).

F12. Prostaglandins (PGs); reduce fever, pain, and inflammation.

F13. (a) TAF. (b) EGF. (c) PDGF.

G2. (a) Hypothalamus. (b) Alarm, medulla, sympathetic. (c) Resistance; ACTH, hGH, TSH; cortex.

G3. (a) Fight-or-flight. (b) Increases. norepinephrine, epinephrine. (c) Increases; increase; skeletal muscles, heart, brain; spleen, GI tract, skin. (d) Decrease; increases.

G5. (a–c) A. (d–f) R. (g) E.

H1. (a) P. (b) AP. (c) AM. (d) AC. (e) PT, T.

H2. (a–b) ↓. (c) ↑. (d) ↓. (e) ↑. (f–i) ↓. (j) —.

I1. (a) GHRH. (b) LH. (c) hGH, thyroid hormone (T_3 and T_4), and testosterone. (d) TSH. (e) FSH. (f) PIH. (g) Melatonin (h) OT. (i) Epinephrine and norepinephrine. (j) ADH. (k) Adrenocorticoid or glucocorticoid, such as cortisol. (1) PTH. (m) CT. (n) Thyroid hormone (T_3 and T_4). (o) Epinephrine. norepinephrine, glucagon, hGH (and indirectly CRH, ACTH, GHRH). (p) Insulin. (q) Adrenocorticoid or glucocorticoid, such as cortisol (and indirectly CRH and ACTH).

I2. (a) PD. (b) Ac. (c) DI. (d) DM. (e) C. (f) M. (g) GD. (h) Go. (i) T. (j) AD. (k) PC.

I3. Hypo, in, tetany; inverse

I4.

Hormone	Ca^2	Mg^{2+}	HPO_4^{2-}	K^+	Na^+
a. PTH	In	In	De	—	—
b. Calcitonin	De	—	De	—	—
c. Aldosterone	—	—	—	De	In

I5.

	Stimulates Protein Synthesis? Breakdown?	Stimulates Lipolysis?	Hyperglycemic Effect?
a. hGH	Synthesis	Yes	Yes
b. Thyroid hormone	Synthesis	Yes	No (increases use of glucose for ATP production)
c. Cortisol	Breakdown	Yes	Yes
d. Insulin	Synthesis	No	No

I6. (a–d) C. (e-f) A.

I7. (a–c) II. (d-e) I.

I8. (a) Lack of insulin prevents glucose from entering cells, so more is left in blood and spills into urine. (b) Uria, dipsia: glucose in urine draws water (osmotically), and as result of water loss, the patient is thirsty. (c) Acid: because cells have little glucose to use, they turn to excessive fat catabolism which leads to keto-acid formation (discussed in Chapter 25). (d) As fats are transported in extra amounts (for catabolism), some fats are deposited in walls of vessels; fatty (= *athero*) thickening (= *sclerosis*) results.

MORE CRITICAL THINKING: CHAPTER 18

1. Contrast blood supply and hormone production of the anterior and posterior lobes of the pituitary gland.
2. Contrast sources and functions of releasing hormones and tropic hormones.
3. Contrast functions of prolactin and oxytocin in lactation process.
4. Describe hormonal regulation of blood levels of the following ions: (a) calcium: (b) potassium.
5. Describe the renin-angiotensin-aldosterone mechanism for controlling blood pressure.
6. Contrast functions of the following hormones in control of blood pressure: ADH (vasopressin), aldosterone, epinephrine.
7. Discuss advantages and disadvantages of anti-inflammatory effects of glucocorticoids.
8. Discuss the hormone imbalance (whether excessive or deficient) and symptoms of each of the following conditions: acromegaly, Addison's disease, Cushing's syndrome, diabetes insipidus, diabetes mellitus, and myxedema.
9. Describe the roles of feedback systems in the following disorders: (a) a condition in which dietary iodine is decreased, leading to enlargement of the thyroid gland (simple goiter); (b) a condition in which cortisol is decreased (congenital adrenal hyperplasia), leading to enlarged adrenal glands.

MASTERY TEST: CHAPTER 18

Questions 1–2: Arrange the answers in correct sequence.

_____ _____ _____ _____ _____ 1. Steps in action of TSH upon a target cell:
A. Adenylate cyclase breaks down ATP.
B. TSH attaches to receptor site.
C. Cyclic AMP is produced.
D. Protein kinase is activated to cause the specific effect mediated by TSH, namely, stimulation of thyroid hormone production.
E. G-protein is activated.

_____ _____ _____ _____ 2. Steps in renin-angiotensin mechanism to increase blood pressure:
A. Angiotensin I is converted to angiotensin II.
B. Renin converts angiotensinogen, a plasma protein, to angiotensin I.
C. Angiotensin II causes vasoconstriction and stimulates aldosterone production which causes water conservation and raises blood pressure.
D. Low blood pressure or low blood Na+ level causes secretion of enzyme renin from cells in kidneys.

Questions 3–10: Circle the letter preceding the one best answer to each question.

3. All of these compounds are synthesized in the hypothalamus *except:*
A. ADH
B. GHRH
C. CT
D. PIH
E. Oxytocin

4. Choose the *false* statement about endocrine glands.
A. They secrete chemicals called hormones.
B. Their secretions enter extracellular spaces and then pass into blood.
C. Sweat and sebaceous glands are endocrine glands.
D. Endocrine glands are ductless.

5. All of the following correctly match hormonal imbalance with related signs or symptoms *except:*
   A. Deficiency of ADH—excessive urinary output
   B. Excessive aldosterone—high blood pressure
   C. Deficiency of aldosterone—low blood level of potassium and muscle weakness
   D. Excess of parathyroid production—demineralization of bones; high blood level of calcium

6. All of the following hormones are secreted by the anterior pituitary *except:*
   A. ACTH   D. Oxytocin
   B. FSH   E. hGH
   C. Prolactin

7. Which one of the following is a function of adrenocorticoid hormones?
   A. Lower blood pressure
   B. Raise blood level of calcium
   C. Convert glucose to amino acids
   D. Lower blood level of sodium
   E. Anti-inflammatory

8. All of these hormones lead to increased blood glucose *except:*
   A. ACTH   D. Growth hormone
   B. Insulin   E. Epinephrine
   C. Glucagon

9. Choose the false statement about the anterior pituitary.
   A. It secretes at least seven hormones.
   B. It develops from mesoderm.
   C. It secretes tropic hormones.
   D. It is stimulated by releasing factors from the hypothalamus.
   E. It is also known as the adenohypophysis.

10. Choose the false statement.
   A. Both ADH and aldosterone tend to lead to retention of water and increase in blood pressure.
   B. PTH activates vitamin D, which enhances calcium absorption.
   C. PTH activates osteoclasts, leading to bone destruction.
   D. Tetany is a sign of hypocalcemia.
   E. ADH and OT pass from hypothalamus to posterior pituitary via pituitary portal veins.

Questions 11–20: Circle T (true) or F (false). If the statement is false, change the underlined word or phrase so that the statement is correct.

T F   11. Polyphagia, polyuria, and hypoglycemia are all symptoms associated with insulin deficiency.

T F   12. Hypophysiotropic hormones are all secreted by the anterior pituitary and affect the hypothalamus.

T F   13. The hormones of the adrenal medulla mimic the action of parasympathetic nerves.

T F   14. Prostaglandins are classified as eicosanoids and have a broad range of effects.

T F   15. Epinephrine and cortisol are both adrenocorticoids.

T F   16. Secretion of growth hormone, insulin, and glucagon is controlled (directly or indirectly) by blood glucose level.

T F   17. Most of the feedback mechanisms that regulate hormones are positive (rather than negative) feedback mechanisms.

T F   18. The alarm reaction occurs before the resistance reaction to stress.

T F   19. Most hormones studied so far appear to act by a mechanism involving cyclic AMP rather than by the gene activation mechanism.

T F   20. A high level of thyroxine circulating in the blood will tend to lead to a low level of TRH and TSH.

Questions 21–25: Fill-ins. Complete each sentence with the word or phrase that best fits.

_____ 21. _____ is a hormone used to induce labor because it stimulates contractions of smooth muscle of the uterus.

_____ 22. _____ and _____ are two hormones that are class chemically as amines.

_____ 23. _____ is a term that refers to the control mechanism which the number of receptors for a hormone will be decreased when that hormone is present in excessive amounts.

_____ 24. Type _____ diabetes mellitus is a non-insulin-dependent condition in which insulin level is adequate but cells have decreased sensitivity to insulin.

_____ 25. Three factors that stimulate release of ADH are _____.

ANSWERS TO MASTERY TEST: CHAPTER 18

Arrange

1. B E A C D
2. D B A C

Multiple Choice

3. C		7. E	
4. C		8. B	
5. C		9. B	
6. D		10. E	

True–False

11. F. Polyphagia and polyuria are both or Polyphagia, polyuria, and hyperglycemia are all
12. F. Hypothalamus and affect the anterior pituitary
13. F. Sympathetic
14. T
15. F. Cortisol, aldosterone, and androgens are all
16. T
17. F. Negative (rather than positive)
18. T
19. T
20. T

Fill-ins

21. Oxytocin (OT or pitocin)
22. Thyroid hormones (T_3 and T_4), epinephrine or norepinephrine
23. Down-regulation
24. II (maturity onset)
25. Decreased extracellular water, pain, stress, trauma, anxiety, nicotine, morphine, tranquilizers

FRAMEWORK 19
Blood

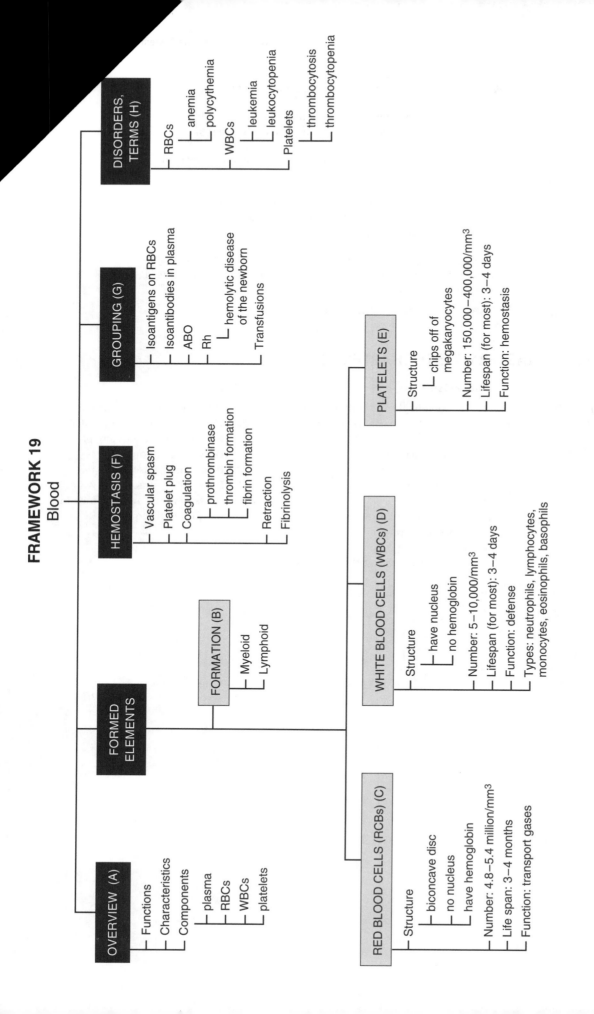

OVERVIEW (A)
- Functions
- Characteristics
- Components
  - plasma
  - RBCs
  - WBCs
  - platelets

FORMED ELEMENTS

FORMATION (B)
- Myeloid
- Lymphoid

RED BLOOD CELLS (RCBs) (C)
- Structure
  - biconcave disc
  - no nucleus
  - have hemoglobin
- Number: 4.8–5.4 million/mm³
- Life span: 3–4 months
- Function: transport gases

WHITE BLOOD CELLS (WBCs) (D)
- Structure
  - have nucleus
  - no hemoglobin
- Number: 5–10,000/mm³
- Lifespan (for most): 3–4 days
- Function: defense
- Types: neutrophils, lymphocytes, monocytes, eosinophils, basophils

PLATELETS (E)
- Structure
  - chips off of megakaryocytes
- Number: 150,000–400,000/mm³
- Lifespan (for most): 3–4 days
- Function: hemostasis

HEMOSTASIS (F)
- Vascular spasm
- Platelet plug
- Coagulation
  - prothrombinase
  - thrombin formation
  - fibrin formation
- Retraction
- Fibrinolysis

GROUPING (G)
- Isoantigens on RBCs
- Isoantibodies in plasma
- ABO
- Rh
  - hemolytic disease of the newborn
- Transfusions

DISORDERS, TERMS (H)
- RBCs
  - anemia
  - polycythemia
- WBCs
  - leukemia
  - leukocytopenia
- Platelets
  - thrombocytosis
  - thrombocytopenia

C H A P T E R   **19**

The Cardiovascular System: The Blood

The major role of the blood is transportation through an intricate system of channels—blood vessels—that reach virtually every part of the body. Blood constantly courses through the vessels, carrying gases, fluids, nutrients, electrolytes, hormones, and wastes to and from body cells. Most of these chemicals "ride" within the fluid portion of blood (plasma). Some, such as oxygen, piggyback on red blood cells. White blood cell "passengers" take short trips in the blood and depart from the vessels at sites calling for defense. Platelets and other clotting factors congregate to plug holes in the network of vessels.

As you begin your study of the blood, carefully examine the Chapter 19 Topic Outline and Objectives; check off each one as you complete it. To organize your study of blood, glance over the Chapter 19 Framework now. Be sure to refer to the Framework frequently and note relationships among key terms in each section.

TOPIC OUTLINE AND OBJECTIVES

A. Overview: functions, characteristics, and components of blood

1. List and describe the functions of blood.
2. List the principal physical characteristics of blood.
3. Describe the principal components of blood.

B. Formation of blood cells

4. Explain the origin of blood cells.

C. Red blood cells

5. Describe the structure, functions, life cycle, and production of red blood cells.

D. White blood cells

6. Describe the structure, functions, and production of white blood cells.

E. Platelets

7. Describe the structure, function, and origin of platelets.

F. Hemostasis

8. Describe the mechanisms that contribute to hemostasis.
9. Describe the stages involved in blood clotting, and explain the various factors that promote and inhibit blood clotting.

G. Blood groups and blood types

10. Explain the ABO and Rh blood groups.

H. Disorders, medical terminology

WORDBYTES

Now become familiar with the language of this chapter by studying each wordbyte, its meaning, and an example of its use within a term. After you study the entire list, self-check your understanding by writing the meaning of each wordbyte on the line. As you continue through the *Learning Guide*, identify (and fill in) additional terms that contain the same wordbyte.

Wordbyte	Self-check	Meaning	Example(s)
a-, an-	_____	not, without	*an*emia
-crit	_____	to separate	hemato*crit*
-cyte	_____	cell	leuko*cyte*
dia-	_____	through	*dia*phragm
-emia	_____	blood	leuk*emia*
erythro-	_____	red	*erythro*poiesis
ferr-	_____	iron	trans*ferr*in
heme-, hemo-	_____	blood	*hemo*stasis
leuko-	_____	white	*leuko*cytosis
mega-	_____	large	*mega*karyocyte
-osis	_____	condition	leukocyt*osis*
-pedesis	_____	moving	dia*pedesis*
-penia	_____	want, lack	leukocyto*penia*
-phil-	_____	love	hemo*phil*iac
-poiesis	_____	formation	hemo*poiesis*
poly-	_____	many	*poly*cythemia
-rhag-	_____	burst forth	hemor*rhag*e
-stasis	_____	standing still	hemo*stasis*
thromb-	_____	clot	*thromb*us, pro*thromb*in
trans-	_____	across	*trans*ferrin

CHECKPOINTS

A. Overview: functions, characteristics, and components of blood (pages 610–613)

A1. Review relationships among these three body fluids: blood, interstitial fluid, and lymph. (Refer to Figure LG 1.1 of the *Learning Guide*.)

✓ **A2.** Describe functions of blood in this Checkpoint.
  a. Blood transports many substances. List six or more.

  _____   _____   _____

  _____   _____   _____

b. Name the four body systems to which blood delivers wastes or chemicals that lead to wastes.

c. List four aspects of homeostasis regulated by blood. One is done for you.

_____pH_____   _____   _____

_____   _____   _____

d. Explain how your blood protects you.

✓ **A3.** Circle the answers that correctly describe characteristics of blood.
   a. Average temperature:
      36°C (96.8°C)   37°C (98.6°C)   38°C (100.4°C)
   b. pH:
      6.8   7.0–7.1   7.35–7.45
   c. Volume of blood in average adult:
      1.5–3 liters   4–6 liters   8–10 liters
   d. Three hormones that regulate volume and osmotic pressure of blood:
      ADH   aldosterone   ANP   epinephrine   PTH

A4. Contrast the following procedures: *venipuncture/finger stick.*

✓ **A5.** Refer to Figure 19.1a in your text, and complete this checkpoint. Use these answers.

E. Erythrocytes	LP. Leukocytes and platelets	P. Plasma

_____ a. In centrifuged blood, layers of blood from top to bottom of the tube are:

_____ _____ _____

_____ b. In centrifuged blood, volume of layers of blood from greatest to least are:

_____ _____ _____

_____ c. Which layer contains the red blood cells? _____

_____ d. Which layer forms the buffy coat? _____

✓ **A6.** Match the names of components of plasma with their descriptions.

A.	Albumins	GAF.	Glucose, amino acids, and fats
E.	Electrolytes	HE.	Hormones and enzymes
F.	Fibrinogen	UC.	Urea, creatinine
G.	Globulins	W.	Water

_____ a. Makes up about 92% of plasma

_____ b. Regulatory substances carried in blood

_____ c. Cations and anions carried in plasma

_____ d. Constitutes about 54% of plasma protein

_____ e. Made by liver; a protein used in clotting

_____ f. Antibody protein

_____ g. Wastes carried to kidneys or sweat glands

_____ h. Food substances carried in blood

_____ i. Of albumin, fibrinogen, and globulin, found in least abundance in plasma

✓ **A7.** Contrast antigens (Ag) and antibodies (Ab) in this exercise by writing *Ag* or *Ab* next to related descriptions below.

_____ a. Viruses or bacteria that enter the body as "invaders"

_____ b. Chemicals made by the body to defend against invaders

_____ c. Immunoglobulins (Ig's) produced by plasma cells and found in plasma

✓ **A8.** Check your understanding of formed elements of blood in this Checkpoint. Use the answers in the box.

E.	Erythrocytes	L.	Leukocytes	P.	Platelets

_____ a. Not truly cells, but fragments of cells

_____ b. Also known as thrombocytes

_____ c. Includes neutrophils, lymphocytes, monocytes, eosinophils, and basophils

_____ d. Also known as white blood cells

_____ e. Makes up over 99% of all formed elements

✓ **A9.** *A clinical correlation.* Answer these questions about hematocrits.
a. A hematocrit consists of the percentage of centrifuged blood that consists of

_____ blood cells.
b. A normal range for hematocrit for adult women is _____%; adult

males should have hematocrits in the range of _____%. State two reasons for this gender difference in hematocrits.

c. Polycythemia occurs when red blood cell count is *(higher? lower?)* than normal. For example, a hematocrit of *(22? 40? 62?)* would indicate polycythemia. Explain why dehydration may lead to polycythemia.

d. In centrifuged flood, which components of blood are found in the *buffy coat*?

A10. Describe blood doping in this Checkpoint.
a. State one reason why it is performed.

b. Describe the procedure.

c. Explain why this practice is dangerous.

B. Formation of blood cells (pages 614–615)

✓▲ **B1.** Check your knowledge about blood formation in this activity.
a. Regulation of *(erythrocyte? leukocyte?)* number occurs by a negative feedback mechanism, whereas alteration of *(erythrocyte? leukocyte?)* number occurs in response to invading microorganisms and antigens.
b. Blood formation is a process known as _____. All blood cells arise

from _____ cells.
c. Fill in the blanks in the table with names of cells that are stages in the formation of each type of blood cell derived from pluripotent stem cells.

Stem cell →	Progenitor cell →	Precursor cell →	Mature cell
1. Lymphoid stem cell	Prothymocyte		
2.		B lymphoblast	
3.			Neutrophil
4.		Monoblast	
5.	CFU-E		
6.	CFU-Meg		Platelet

d. Which of the mature cells above (in Checkpoint B1c) passes through a stage as a

reticulocyte immediately before the mature stage? _____ (see page 619 of the text). A reticulocyte *(has? lacks?)* a nucleus. Which of the mature cells

above (in Checkpoint B1c) later becomes a macrophage? _____.
e. Name several structures where blood formation takes place before birth.

   f. After birth, most hemopoiesis takes place in the *(red bone marrow? liver? lymph nodes?)*. Name six or more bones in which red blood cells are formed after birth.

   g. Which type of white blood cells complete their development in lymphoid tissues?

     _____ All other types of blood cells develop totally within myeloid

     tissue, which is another name for _____.

   h. Erythropoietin is a hormone made mostly in the _____. This hormone stimulates production of *(RBCs? WBCs? platelets?)*.

   i. Two types of cytokines that stimulate WBC formation are _____-

     _____ factors (CSFs) and _____ (ILs).

✓ **B2.** *A clinical correlation*. Explain how the following chemicals produced via recombinant DNA may be helpful clinically:

   a. *EPO* for patients with kidney failure

   b. *GM-CSF* and *G-CSF* for patients with immune deficiency

   c. *TPO*

C. Red blood cells (pages 615–619)

✓ **C1.** Refer to Figure LG 19.1. Which diagram represents a mature erythrocyte? _____ Note that it *(does? does not?)* contain a nucleus. The chemical named

     _____ accounts for the color of red blood cells (RBCs). Label and color the RBC.

✓ **C2.** Explain the significance of the structure and shape of red blood cells (RBCs) in this Checkpoint.

   a. RBCs *(do? do not?)* have a nucleus and mitochondria. State one advantage of this fact.

     State one disadvantage.

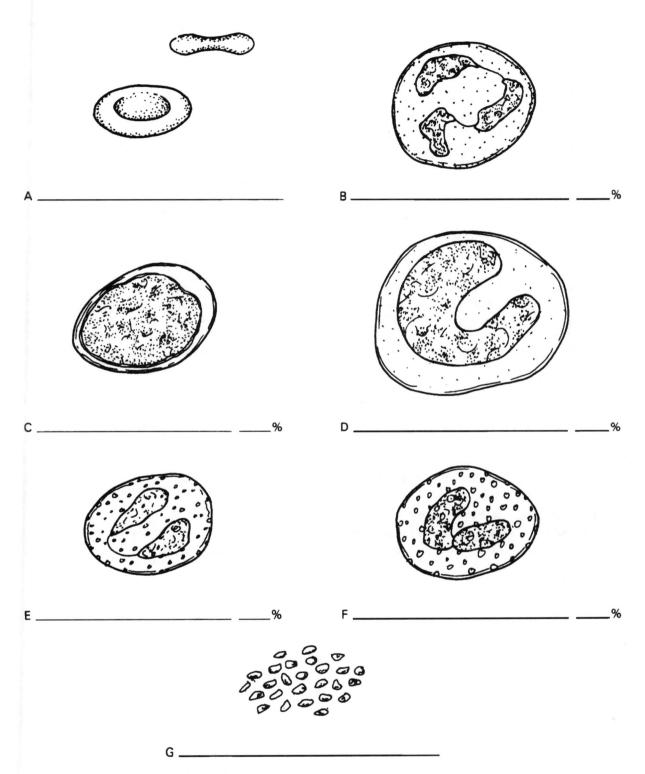

A _____

B _____ ____%

C _____ ____%

D _____ ____%

E _____ ____%

F _____ ____%

G _____

Figure LG 19.1 Diagrams of blood cells. Label, color, and complete as directed in Checkpoints C1, D1, and E2.

b.   RBCs are normally shaped like _____ disks. What advantage does this shape offer?

✓ **C3.** Describe hemoglobin in this exercise.

a.   The hemoglobin molecule consists of a central portion, which is the protein *(heme? globin?)* with four polypeptide chains called *(hemes? globins?)*.

b.   Each heme contains one _____ atom on which a molecule of *(oxygen? carbon dioxide?)* can be transported so that each hemoglobin molecule can carry *(1? 4?)* oxygen molecule(s). Almost all oxygen is transported in this manner.

c.   Carbon dioxide has one combining site on hemoglobin; it is the amino acid portion of

the *(heme? globin?)*; therefore, it contributes to formation of _____- hemoglobin. About *(23? 70? 97?)* % of carbon dioxide is carried in this state.

d.   Besides O_2 and CO_2, hemoglobin also carries two other gases. The gas *(NO? SNO?)* is picked up by hemoglobin as blood passes through lungs. This gas is released in tissues, where it causes blood vessels to *(dilate? constrict?)*. Then hemoglobin picks up the gas *(NO? SNO?)* and transports it to be exhaled in lungs along with CO_2. NO (nitrous oxide) is a vaso-*(constrictor? dilator?)*, which increases blood pressure, whereas SNO's

effects on blood vessels _____-creases blood pressure.

✓ **C4.** Fill in blanks in the following description of the RBC life cycle. Select answers from the box. Each answer will be used once except one which will be used three times. It may help to refer to Figure 19.5 in your text.

Amino acids	**Ferritin and hemosiderin**	**Stercobilin**
Bile	**Iron**	**Transferrin**
Bilirubin	**Red bone marrow**	**Urobilin**
Biliverdin	**Spleen, liver, or red bone marrow**	**Urobilinogen**

a.   Old RBCs are phagocytosed by macrophages in the _____.

b.   Hemoglobin is then recycled: globin is broken into _____ and hemes

are degraded into the metal _____ and a noniron portion; see (e) below.

c.   Iron is carried through blood attached to the protein _____; iron

may be stored in muscles attached to proteins _____ and

_____. Once released from muscles, iron is again combined with

_____. Iron absorbed from foods in the intestine is also attached to

_____ for transport through blood.

d.   Ultimately iron reaches _____, where it is used for formation of new

_____ molecules.

e.   What happens to the noniron portion of heme? It is converted to the green pigment named _____ and the orange pigment _____.

f. Bilirubin has two possible fates. It may pass in blood from bone marrow to liver and

be used to form _____, or bilirubin may be transformed by intestinal

bacteria into _____. Urobilinogen may be used to form

_____ that gives feces its brown color or _____ that
contributes to the yellow color of urine.

✓ **C5.** After you study Figure 19.6 in your text, explain the roles of the terms in the box in
red blood cell production.

B₁₂. Vitamin B₁₂	**H.** Hypoxia
E. Erythropoietin	**IF.** Intrinsic factor

_____ a. Decrease of oxygen in cells; serves as a signal that erythropoiesis is needed (to help provide more oxygen to tissues)

_____ b. Hormone produced by kidneys when they are hypoxic; stimulates erythropoiesis in red bone marrow

_____ c. Vitamin necessary for normal hemoglobin formation

_____ d. Substance produced by the stomach lining and necessary for normal vitamin B₁₂ absorption

✓ **C6.** Circle the most normal blood values. (Note that values vary slightly according to age
and gender.)
a. Average life of a red blood cell: 4 hours   4 days   4 months   4 years
b. RBC count in cube this size: ■ (mm³)   500   5,000   250,000   5 million
c. Number of RBCs produced each second by a healthy human adult:
20   200   2,000   20,000   2 million
d. Number of hemoglobin molecules in one red blood cell:
280   2,800   280,000   280 million
e. Reticulocyte count of person undergoing rapid erythropoiesis, for example, in
response to anti-anemic medications:
0.3%   1.0%   2.0%
f. Hemoglobin in adults (g/100 mL blood. See page 622 of the text):
1   8   15   27   41

D. White blood cells (pages 619–621)

✓ **D1.** Refer to Figure LG 19.1 and do this exercise about leukocytes.
a. Leukocytes *(have? lack?)* hemoglobin, and so these cells are known as *(red blood cells or RBCs? white blood cells or WBCs?)*.
b. Label each leukocyte (WBC) on Figure 19.1, and indicate what percentage of the total WBC count is accounted for by each type of WBC. Such a breakdown of white blood

cells is known as a _____ WBC count.
c. Each white blood cell (WBC) *(has? lacks?)* a nucleus. Which type of WBC has a large

kidney-shaped nucleus? _____ Which WBC has a nucleus that

occupies most of the cell? _____. Which is polymorphonuclear?

_____
d. Which three WBCs are granulocytes? _____
e. *For extra review.* Color nucleus, cytoplasm, and granules of all WBCs.

D2. Contrast terms in each pair.

a. *Granular leukocytes/agranular leukocytes*

b. *Polymorphonuclear leukocytes (PMNs)/bands*

c. *Fixed macrophages/wandering macrophages*

✓ **D3.** What are MHC antigens and how do they affect the success rate of transplants?

✓ **D4.** Contrast WBCs with RBCs in this activity.

a. A normal RBC count is about *(700? 5,000–10,000? 250,000? 5 million?)* cells/mm^3,

whereas a typical WBC count is _____ cells/mm^3. In other words, the

ratio of RBCs to WBCs is about _____:1.

b. Which cells normally live longer? *(RBCs? WBCs?)* The typical lifespan of a WBC is a few *(days? months? years?)*.

✓ **D5.** Match the answers in the box with descriptions of processes involved in inflammation.

A.	**Adhesion molecules**	**E.**	**Emigration**
C.	**Chemotaxis**	**P.**	**Phagocytosis**

_____ a. The process by which WBCs stick to capillary wall cells, slow down, stop, and then squeeze out through the vessel wall

_____ b. Chemicals known as selectins made by injured or inflamed capillary cells stick to chemicals on neutrophils, slowing them down

_____ c. Attraction of phagocytes to chemicals such as kinins or CSFs in microbes or in inflamed tissue

_____ d. Ingestion and disposal of microbes or damaged tissue by neutrophils or macrophages

✓ **D6.** Check your understanding of types of WBCs by matching names of WBCs with descriptions. Answers may be used more than once.

B. **Basophils**	**L.** **Lymphocytes**	**N.** **Neutrophils**
E. **Eosinophils**	**M.** **Monocytes**	

_____ a. Constitute the largest percentage of WBCs

_____ b. Produce defensins, lysozymes, and oxidants with antibiotic activity

_____ c. Types of these cells include B cells, T cells, and natural killer (NK) cells

_____ d. They leave the blood, develop into mast cells, and release heparin, histamine, and serotonin, which are involved in allergic and inflammatory responses

_____ e. Involved in allergic reactions, combat histamines, and provide protection against parasitic worms

_____ f. Form fixed and wandering macrophages that clean up sites of infection

_____ g. Important in phagocytosis (two answers)

_____ h. Classified as agranular leukocytes (two answers)

_____ i. Known as bands in immature state

_____ j. In stained cells, lilac-colored granules are visible

_____ k. In stained cells, red-orange granules are visible

✓ **D7.** Contrast types of lymphocytes in this activity.
  a. B lymphocytes (or B cells) are especially effective against *(bacteria and their toxins? cancer or transplanted cells, viruses, fungi, and some bacteria?)*.
  b. T cells are particularly effective against *(bacteria and their toxins? cancer or transplanted cells, viruses, fungi, and some bacteria?)*.
  c. Natural killer (NK) cells combat a *(few selected? wide variety?)* of microbes plus some cancer cells.

✓ **D8.** *A clinical correlation.* Mrs. Doud arrives at a health clinic with a suspected acute infection. Complete this Checkpoint about her.
  a. During infection, it is likely that Mrs. Doud's leukocyte count will *(in? de?)*-crease. A count of *(4000? 8000? 12,000?)* leukocytes/mm3 blood is most likely. This condition is known as *(leukocytosis? leukopenia?)*.
  b. A differential increase in the number of WBCs named _____ is most indicative of enhanced phagocytic activity during infection. Neutrophils are most likely to account for *(48? 62? 76?)*% of the total white count in Mrs. Doud's blood. A sign of a chronic infection (such as tuberculosis) is increase in the percentage of cells that can become macrophages; these are _____.

✓ **D9.** *A clinical correlation.* Do this exercise on diagnostic clues offered in a differential WBC count. Select the conditions that might he indicated for each specific WBC change listed below. Select answers from the box.

A. **Allergic reactions**
B. **Prolonged, severe illness and/or immunosuppression**
C. **Lupus, vitamin B$_{12}$ deficiency, radiation**
D. **Pregnancy, ovulation, stress, and hyperthyroidism**

_____ a. Decrease in lymphocyte count

_____ b. Decrease in basophil count

_____ c. Decrease in neutrophil count

_____ d. Increase in basophil or eosinophil count

D10. Briefly describe the procedure of *bone marrow transplant*. Describe the process as well as conditions for which this procedure is used.

E. Platelets (pages 621–622)

✓ **E1.** Circle correct answers related to platelets.

a. Platelets are also known as:

   antibodies   thrombocytes   red blood cells   white blood cells

b. Platelets are formed in:

   bone marrow   tonsils and lymph nodes   spleen

c. Platelets are:

   entire cells   chips off of megakaryocytes

d. A normal range for platelet count is _____ /mm^3

   5000–1000   4.5–5.5 million   150,000–400,000

e. The primary function of platelets is related to:

   O_2 and CO_2 transport   blood clotting   defense   blood typing

✓ **E2.** Label platelets on Figure LG 19.1.

✓ **E3.** List the components of a complete blood count (CBC).

F. Hemostasis (pages 622–627)

✓ **F1.** Hemostasis literally means _____ (*hemo*)-_____ (*stasis*).
List the three basic mechanisms of hemostasis.

a. Vascular _____

b. _____ plug formation

c. _____ (clotting)

F2. List chemicals present within granules of platelets.

a. In alpha granules:

b. In dense granules:

✓ **F3.** After reviewing Figure 19.9 in your text, check your understanding of the first two steps of hemostasis in this Checkpoint.

a. When platelets snag on the inner lining of a damaged blood vessel, their shape is altered. They become *(small, disc-shaped? large, irregular-shaped?)*. Additional platelets aggregate on the uneven projections. This phase is known as platelet *(adhesion? release reaction? plug?)*.

b. Platelets then release chemicals. Some of these chemicals *(dilate? constrict?)* vessels. How does this help hemostasis?

   Name two of these chemicals. _____, _____

c. State two roles of the ADP that is released by platelets.

d. The resulting platelet plug is useful in preventing blood loss in *(large? small?)* vessels.

✓ **F4.** The third step in hemostasis, coagulation, requires clotting factors derived from several sources described in this Checkpoint.

a. Most clotting factors are synthesized by *(bone marrow? kidney? liver?)* cells. (Notice this in Table 19.4 of your text.)

b. Others are released by *(red blood cells? platelets?)*.

c. One important factor, called _____ factor (TF) or thromboplastin, is released from damaged cells. If these cells are within blood or in the lining of blood vessels, then the *(extrinsic? intrinsic?)* pathway to clotting is initiated. If other tissues are traumatized and release tissue factor, then the *(extrinsic? intrinsic?)* pathway is initiated.

✓ **F5.** Summarize the three major stages in the cascade of events in coagulation by filling in blanks A–E in Figure LG 19.2. Choose from the list of terms on the figure.

✓ **F6.** *For extra review.* of events in the process of blood clotting, refer to Figure LG 19.2 and Table 19.4 in your text and do this exercise.

a. Which protein is an insoluble protein? *(Fibrin? Fibrinogen?)* Which of these proteins is always found in blood plasma? *(Fibrin? Fibrinogen?)* What would happen if fibrin were always present in plasma?

b. Fibrinogen is not converted to fibrin unless the enzyme _____ (D in Figure LG 19.2) is present. What would happen if thrombin were always present in blood?

c. The enzyme required to convert prothrombin to thrombin is _____. Fill in the two short lines on Figure LG 19.2 with Roman numerals to indicate which two factors combine to form prothrombinase.

d. The figure indicates that thrombin exerts a *(negative? positive?)* feedback effect upon prothrombinase, and thrombin also activates _____. As a result, the clot *(enlarges? shrinks?)* in size. Which chemical finally absorbs most of the thrombin to prevent snowballing of the clot? _____

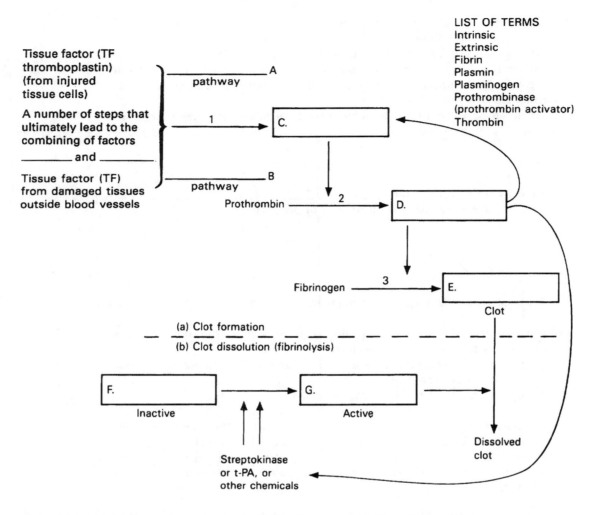

Figure LG 19.2 (a) Summary of three main steps in clot formation. Complete as directed in Checkpoints F5 and F6. (b) Summary of steps in clot dissolution (fibrinolysis). Complete as directed in Checkpoint F7.

e. Notice on Table 19.4 in the text that clotting factors are numbered from 1 to _____;

however there is no Factor _____. Use that table to help you identify the Roman numerals of other clotting chemicals shown in Figure LG 19.2. Write those numerals on the figure.

f. The ion *(Ca²⁺? K⁺? Na⁺?)*, also known as factor _____, plays a critical role in many aspects of coagulation. Write this ion on the figure next to each of the three major steps.

g. How does the cascade of events in blood clotting help to prevent excessive clot formation?

✓ **F7.** Following clot formation, clots undergo two changes. Summarize these.

a. First, clot _____ occurs.

b. The lower portion of Figure LG 19.2 shows the next step: clot dissolution or

_____. An inactive enzyme (F on the figure) is activated to (G), which dissolves clots. Label F and G on the figure. F is activated by chemicals produced within the body, for example, tissue plasminogen activator (t-PA), and also by (D).

✓ **F8.** *A clinical correlation.* Match each chemical with its role in clot formation or dissolution.

H.	**Heparin**	**S, t-PA.**	**Streptokinase and t-PA**
K.	**Vitamin K**	**W.**	**Warfarin (Coumadin)**
PR.	**Prostacyclin**		

_____ a. Required for synthesis of factors II, VII, IX, and X in clot formation. Commercial preparations may be required for newborns who lack the intestinal bacteria that synthesize this chemical and also for persons with malabsorption of fat because this is a fat soluble chemical.

_____ b. Slow-acting anticoagulant that is antagonistic to vitamin K, so decreases synthesis of prothrombin in liver; used to prevent clots in persons with history of clot formation; also the active ingredient in rat poison.

_____ c. Fast-acting anticoagulant that blocks action of thrombin; produced in the body by mast cells and basophils, but also available in pharmacologic preparations.

_____ d. Commercial preparations introduced into coronary arteries can limit clot size and prevent severe heart attacks; classified as fibrinolytic agents because they activate plasminogen (shown on Figure LG 19.2) to dissolve clot.

_____ e. A chemical made by white blood cells and cells lining blood vessels; opposes actions of thromboxane A2.

✓ **F9.** *A clinical correlation.* Match the correct term with the description.

E.	**Embolus**	**T.**	**Thrombus**

_____ a. A blood clot

_____ b. A "clot-on-the-run" dislodged from the site at which it formed (usually a deep vein of the leg); also fat from broken bone, bubble of air, or amniotic fluid traveling through blood, possibly to lung (pulmonary) vessels

✓ **F10.** *Critical thinking.* Address factors related to clotting in this Checkpoint.
   a. Mr. B, a chronic alcoholic, needs to have a surgical procedure. Explain why he might be at risk for excessive bleeding.

   b. Aspirin is used to *(promote? prevent?)* hemostasis, so it can lower risk of thrombi in cerebral, coronary, and peripheral arteries. Circle the step(s) of hemostasis that it affects. *(Vasoconstriction? Platelet plug formation? Coagulation?)*

G. Blood groups and blood types (pages 627–630)

✓ **G1.** Contrast *agglutination* and *coagulation.*

   a. Coagulation (or _____) involves coagulation factors found in platelets, plasma, or other tissue fluids. The process *(requires? can occur in absence of?)* red blood cells.
   b. Agglutination (or "clumping") of erythrocytes is an antigen-_____ process that *(does? does not?)* require red blood cells because these are sites of

   _____ used in the agglutination reaction.

✓ **G2.** Do this exercise about factors responsible for blood groups.
   a. The surface of RBCs contains chemicals that act as *(antigens? antibodies?)*, also known

as iso-_____. Based on these chemicals, blood can be categorized into at least *(2? 4? 24? 100?)* different blood groups. One of these groups, the ABO group, consists of four possible blood types. Name them:

type _____, type _____, type _____, and type _____.

b. Type B blood has *(A? B?)* antigens on RBCs, and *(anti-A? anti-B)* antibodies in blood plasma. In other words, people have antibodies in plasma that *(would? would not?)* react with their own RBC antigens.

G3. Complete the first four columns of the table contrasting blood types within the ABO group.

Blood Group	Percentage of White Population	Percentage of Black Population	Sketch of Blood Showing Correct Antigens and Antibodies	Can Donate Safely to	Can Receive Blood Safely from
a. Type A					A, O
b.	13				
c. Type AB		4	A B B ◯ A A ◡ B neither anti-A B A nor anti-B		
d.				A, B, AB, O	

✓ **G4.** Complete this exercise about the *Rh system.*

a. The Rh *(+? –?)* group is more common. Rh *(+? –?)* blood has Rh antigens on the surfaces of RBCs.

b. Under normal circumstances plasma of *(Rh+ blood? Rh– blood? both Rh groups? neither Rh group?)* contains anti-Rh antibodies.

c. Rh *(+? –?)* persons can develop these antibodies when they are exposed to Rh *(+? –?)* blood, for example, by transfusion.

d. An example of this occurs in fetal-maternal incompatibility when a mother who is Rh *(+? –?)* has a baby who is Rh *(+? –?)* and some of the baby's blood enters the mother's bloodstream. The mother develops anti-Rh antibodies that may cross the placenta in future pregnancies and hemolyze the RBCs of Rh *(+? –?)* babies. Such a condition is known as _____.

e. *A clinical correlation.* Discuss precautions taken with Rh- mothers soon after delivery, miscarriage, or abortion of an Rh+ baby to prevent future problems with Rh incompatibility.

✓ **G5.** Complete this exercise about blood transfusions.
   a. Type A blood has *(A? B? both A and B? neither A nor B?)* antigen(s) on RBCs.
   b. Type A blood most likely has the *(anti-A? anti-B? both anti-A and anti-B? neither anti-A nor anti-B?)* antibodies in blood plasma. If a person with type A blood receives type B blood, the recipient's *(anti-A? anti-B?)* antibodies will attack the *(A? B?)* antigens on the donor's cells. Are the donor's anti-A antibodies likely to attack the recipient's type A cells and cause a major incompatibility reaction?

   c. Type O blood has often been called the universal *(donor? recipient?)* because blood

   lacks _____ of the ABO group. Conversely, Type _____ has been called the universal recipient. Explain why these terms (universal donor and universal recipient) are misleading.

   G6. Now complete the last two columns in the table in Checkpoint G2.

✓ **G7.** Discuss blood typing in this Checkpoint.
   a. Blood typing involves checking for the presence of *(antigens? antibodies?)* in blood.
   b. A drop of serum from a bottle containing anti-A antibodies is mixed with a sample

   drop of blood. If the blood clumps (or _____), then the A antigen *(is? is not?)* present and the blood does contain A antigens. The serum in the bottle used here is comparable to serum from blood that is type *(A? B?)*. Why is serum (rather than plasma) used in this procedure?

   c. If similar procedures are performed using anti-B and anti-Rh antisera, and neither of those results demonstrates agglutination, what blood type is in the sample drop of

   blood? _____
   d. If all three drops of blood agglutinate using the above procedure, what blood type is

   in the sample drop of blood? _____

H. Disorders, medical terminology (pages 630–632)

✓ **H1.** Match names of types of anemia with descriptions below.

A.	Aplastic	Hr.	Hemorrhagic	P.	Pernicious
Hl.	Hemolytic	N.	Nutritional	S.	Sickle cell

_____ a. Condition resulting from inadequate diet such as deficiency of iron or amino acids

_____ b. Condition in which intrinsic factor is not produced, so absorption of vitamin B_{12} is inadequate

_____ c. Inherited condition in which hemoglobin forms stiff rodlike structures causing erythrocytes to assume sickle shape and rupture, reducing oxygen supply to tissues

_____ d. Rupture of red blood cell membranes due to variety of causes, such as parasites, toxins, or antibodies

_____ e. Condition due to excessive bleeding, as from wounds, gastric ulcers, heavy menstrual flow

_____ f. Inadequate erythropoiesis as a result of destruction or inhibition of red bone marrow

✓ **H2.** Check your understanding of sickle cell anemia in this Checkpoint.

    a. Describe the normal shape of red blood cells: _____

    b. What shape may RBCs assume in the disorder sickle-cell anemia (SCA)?

        _____ Under what conditions are the RBCs likely to do this?

        _____ What are the consequences?

    c. Explain why persons with sickle cell anemia are likely to have greater resistance to malaria.

✓ **H3.** Check your understanding of hemophilia in this Checkpoint.
    a. Although there are several different forms of hemophilia, all types involve deficiency

        of _____. Types A and B occur primarily in *(males? females?)*.
    b. In the most common form of hemophilia, known as hemophilia *(A? B? C?)*, factor

        _____ is missing. State the primary function of this factor.

    c. Write three or more signs or symptoms of hemophilia.

    H4. Explain why patients with disseminated intravascular clotting (DIC) have both excessive clotting and hemorrhaging occurring simultaneously.

    H5. Define and explain the clinical significance of each of these terms.
    a. Jaundice

    b. Thrombocytopenia

c. Septicemia

d. Cyanosis

ANSWERS TO SELECTED CHECKPOINTS: CHAPTER 19

A2. (a) Oxygen, carbon dioxide and other wastes, nutrients, hormones, enzymes, heat. (b) Respiratory, urinary, integumentary, and digestive. (c) pH, temperature, water (fluids) and dissolved chemicals (such as electrolytes). (d) Contains phagocytic cells, blood clotting proteins, and defense proteins such as antibodies, complement, and interferon.

A3. (a) 38°C (100.4°C). (b) 7.35–7.45. (c) 4–6 liters. (d) ADH, aldosterone, and ANP.

A5. (a) P LP E. (b) P E LP. (c) E. (d) LP.

A6. (a) W. (b) HE. (c) E. (d) A. (e) F. (f) G. (g) UC. (h) GAF. (i) F.

A7. (a) Ag. (b) Ab. (c) Ab.

A8. (a-b) P. (c-d) L. (e) E.

A9. (a) Red. (b) 38-46; 40-54; women lose blood during menstruation, and men produce more red blood cells due to testosterone levels. (c) Higher; 62; loss of water decreases blood plasma and leads to a relative increase in red blood cells. (d) White blood cells and platelets.

B1. (a) Erythrocyte, leukocyte. (b) Hemopoiesis (or hematopoiesis); pluripotent stem. (c)

Stem cell →	Progenitor cell →	Precursor cell →	Mature cell
1. Lymphoid stem cell	Prothymocyte	**T lymphoblast**	**T lymphocyte**
2. **Lymphoid stem cell**	Pre-B cell	B lymphoblast	**B lymphocyte**
3. **Myeloid stem cell**	**CFU-GM**	**Myeloblast**	Neutrophil
4. **Myeloid stem cell**	**CFU-GM**	Monoblast	**Monocyte**
5. **Myeloid stem cell**	CFU-E	**Proerythroblast**	**Erythrocyte**
6. **Myeloid stem cell**	CFU-Meg	**Megakaryocyte**	Platelet

(d) Red blood cell (or erythrocyte); lacks; monocyte. (e) Yolk sac, liver, spleen, thymus, lymph glands, bone marrow. (f) Red bone marrow; proximal epiphyses of femurs and humeri, flat bones of the skull, sternum, ribs, vertebrae, and hip bones. (g) B and T lymphocytes; red bone marrow. (h) Kidneys, RBCs. (i) Colony-stimulating factors (CSFs) and interleukins (ILs).

B2. EPO stimulates RBC production in persons whose failing kidneys no longer produce erythropoietin. (b) GM-CSF and G-CSF stimulate white blood cell formation in persons whose bone marrow function is inadequate, such as transplant, chemotherapy, or AIDS patients. (c) TPO helps to prevent depletion of platelets during chemotherapy.

C1. A; does not; hemoglobin; see Figure 19.4 in your text.

C2. (a) Do not; absence of these organelles leaves more room for O_2 transport; RBCs have a limited life span, partly because the anuclear cell cannot synthesize new components. (b) Biconcave; greater plasma membrane surface area for better diffusion of gas molecules and flexibility for squeezing through small capillaries.

C3. (a) Globin, hemes. (b) Iron (Fe), oxygen, 4. (c) Globin, carbamino; 23. (d) SNO; dilate; NO; constrictor, de.

C4. (a) Spleen, liver, or red bone marrow. (b) Amino acids, iron. (c) Transferrin, ferritin and hemosiderin; transferrin; transferrin. (d) Red bone marrow, heme. (e) Biliverdin, bilirubin. (f) Bile; urobilinogen; stercobilin, urobilin.

C5. (a) H. (b) E. (c) B_{12}. (d) IF.

C6. (a) 4 months. (b) 5 million. (c) 2 million. (d) 280 million. (e) 2. (f) 15.

D1. (a) Lack, white blood cells or WBCs. (b) B, neutrophil (60–70); C, lymphocyte (20–25); D, monocyte (3–8); E, eosinophil (2–4); F, basophil (0.5–1.0); differential. (c) Has; monocyte (D); lymphocyte (C); neutrophil (B). (d) B E F. (e) See Figure 19.7 and Table 19.3 in your text.

D3. MHC antigens are proteins on cell surfaces; they are unique for each person. The greater the similarity between major histocompatibility (MHC) antigens of donor and recipient, the less likely is rejection of the transplant.

D4. (a) 5 million, 5,000–10,000; 700. (b) RBCs; days.

D5. (a) E. (b) A. (c) C. (d) P.

D6. (a) N. (b) N. (c) L. (d) B. (e) E. (f) M. (g) M, N. (h) L, M. (i) N. (j) N. (k) E.

D7. (a) Bacteria and their toxins. (b) Cancer or transplanted cells, viruses, fungi, and some bacteria. (c) Wide variety of.

D8. (a) In; 12,000; leukocytosis. (b) Neutrophils (PMNs or polys); 76; monocytes.

D9. (a) B. (b) D. (c) C. (d) A.

E1. (a) Thrombocytes. (b) Bone marrow. (c) Chips off of megakaryocytes. (d) 150,000–400,000. (e) Blood clotting.

E2. G.

E3. RBC, WBC, and platelet counts; differential WBC count; hemoglobin and hematocrit (H&H).

F1. Blood standing still (or stopping bleeding).
(a) Spasm. (b) Platelet. (C) Coagulation.

F3. (a) Large. irregular-shaped; adhesion. (b) Constrict; reduces blood flow into injured area; serotonin. thromboxane A2. (c) Activates platelets and makes them stickier, thereby inviting more platelets to the scene. This phenomenon is known as platelet aggregation. (d) Small.

F4. (a) Liver. (b) Platelets. (c) Tissue; intrinsic; extrinsic.

F5. A, extrinsic. B, intrinsic. C, Prothrombinase (pro-thrombin activator). D, Thrombin. E. Fibrin.

F6. (a) Fibrin; fibrinogen; fibrin clots would form and block vessels inappropriately. (b) Thrombin; fib-rinogen would be converted to fibrin excessively, so excessive numbers of clots would form. (c) Pro-thrombinase; V and X. (d) Positive, platelets; enlarges; fibrin. (e) XIII; VI; I, fibrinogen; II, prothrombin; III, tissue factor (thromboplastin). (f) Ca^{2+}; IV. (g) Active clotting factors (prothrom-binase, thrombin, and fibrin) are not normally formed until circumstances call for clot formation.

F7. (a) Retraction. (b) Fibrinolysis; F, plasminogen; G, plasmin; thrombin.

F8. (a) K. (b) W. (c) H. (d) S, t-PA. (e) PR.

F9. (a) T. (b) E.

F10. (a) Most coagulation factors are normally synthe-sized in the liver, which is severely damaged by cirrhosis, a liver disease that can be caused by excessive consumption of alcohol. (b) Prevent; all three: vasoconstriction. platelet plug formation, and coagulation.

G1. (a) Clotting; can occur in absence of. (b) Antibody. does, antigens.

G2. (a) Antigens; antigens; 24; A, B, AB, and 0. (b) B, anti-A; would not.

G4. (a) +; +, (b) Neither Rh group. (c) –, +. (d) –, +; +, hemolytic disease of the newborn (HDN) (e) The chemical RhoGAM (anti-Rh antibodies), given soon after delivery, binds to any fetal Rh antigens within the mother, destroying them. Therefore the Rh-mother will not produce anti-Rh antibodies that would attack Rh antigens of future Rh^+ fetuses.

G5. (a) A. (b) Anti-B; anti-B, B. A major reaction is not likely to occur because the donor's antibodies are rapidly diluted in the recipient's plasma, but a mild (minor) reaction may occur because of this incompatibility. (c) Donor, antigens; AB. Blood contains antigens and antibodies of blood groups other than the ABO blood group.

G7. (a) Antigens. (b) Agglutinates, is; B; blood would clot if clotting factors (in plasma) were added. (c) A-negative. (d) AB-positive.

H1. (a) N. (b) P. (c) S. (d) H1. (e) Hr. (f) A.

H2. (a) Biconcave discs. (b) Sickle; upon exposure to low oxygen: sickled cells are more rigid so they may lodge in and block small vessels, (c) The gene that causes RBCs to sickle also increases perme-ability of RBCs to potassium, which then leaves the RBC. The malaria parasite cannot live in RBCs with such low potassium levels.

H3. (a) Some clotting factor; males. (b) A, VIII; it is known as fibrin stabilizing factor and it does just that, forming fibrin threads into a sturdy clot. (c) Hemorrhaging, including nosebleeds; blood in urine (hematuria); and joint damage and pain.

MORE CRITICAL THINKING: CHAPTER 19

1. Describe mechanisms for assuring homeostasis of red blood cell count.
2. Describe the recycling of red blood cells. Be sure to use all of these terms in your essay: hemoglobin, iron, transferrin, ferritin, and hemosiderin; bilirubin, bile, urobilinogen, uro-bilin, and stercobilin.
3. Describe roles of the following types of cells in inflammation (including phagocytosis) and immune responses: neutrophils, monocytes, and lymphocytes.
4. Contrast advantages and disadvantages of blood clotting in the human body. Then describe control mechanisms that provide checks and balances so that clotting does not get out of hand.
5. Explain why athletes might be likely to train in a high-altitude city such as Denver for several weeks immediately before competing in that city.
6. Contrast effects of blood type incidence as you address these questions: (a) Is hemolytic of the newborn more likely to occur among Asians and Native Americans or among white persons? Explain why. (b) Which cultural group has the greatest incidence of Type O blood, and the lowest incidence of the other three types?

MASTERY TEST: CHAPTER 19

Questions 1–3: Arrange the answers in correct sequence.

_____ _____ _____ 1. Events in the coagulation process:
   A. Retraction or tightening of fibrin clot
   B. Fibrinolysis or clot dissolution by plasma
   C. Clot formation

_____ _____ _____ 2. Stages in the clotting process:
   A. Formation of prothrombin activator (prothrombinase)
   B. Conversion of prothrombin to thrombin
   C. Conversion of fibrinogen to fibrin

_____ _____ _____ 3. Arrange from greatest to least in number per mm^3
   A. WBCs
   B. RBCs
   C. Platelets

Questions 4–17: Circle the letter preceding the one best answer to each question.

4. All of the following types of formed elements develop entirely within bone marrow *except:*
   A. Neutrophils   D. Erythrocytes
   B. Basophils   E. Lymphocytes
   C. Platelets

5. Which chemicals are forms in which bilirubin is excreted in urine or feces?
   A. Ferritin and hemosiderin
   B. Biliverdin and transferrin
   C. Urobilinogen and stercobilin
   D. Interleukin 5 and 7

6. Megakaryocytes are involved in formation of:
   A. Red blood cells   D. Platelets
   B. Basophils   E. Neutrophils
   C. Lymphocytes

7. Choose the *false* statement about Norine, who has type O blood.
   A. Norine has neither A nor B antigens on her red blood cells.
   B. She has both anti-A and anti-B antibodies in her plasma.
   C. She is called the universal donor.
   D. She can receive blood safely from both type O and type AB persons.

8. Choose the *false* statement about blood.
   A. Blood is thicker than water.
   B. It normally has a pH of 7.0.
   C. The human body normally contains about 4 to 6 liters of it.
   D. It normally consists of more plasma than cells.

9. Choose the *false* statement about factors related to erythropoiesis.
   A. Erythropoietin stimulates RBC formation.
   B. Oxygen deficiency in tissues serves as a stimulus for erythropoiesis.
   C. Intrinsic factor, which is necessary for RBC formation, is produced in the kidneys.
   D. A reticulocyte count of over 1.5% of circulating RBCs indicates that erythropoiesis is occurring rapidly.

10. Choose the *false* statement about plasma.
   A. It is red in color.
   B. It is composed mainly of water.
   C. Its concentration of protein is greater than that of interstitial fluid.
   D. It contains plasma proteins, primarily albumins.

11. Choose the *false* statement about neutrophils.
   A. They are actively phagocytic.
   B. They are the most abundant type of leukocyte.
   C. Neutrophil count decreases during most infections.
   D. An increase in their number would be a form of leukocytosis.

12. All of the following correctly match parts of blood with principal functions *except:*
   A. RBC: carry oxygen and CO_2
   B. Plasma: carries nutrients, wastes, hormones, enzymes
   C. WBCs: defense
   D. Platelets: determine blood type

13. Which of the following chemicals is an enzyme that converts fibrinogen to fibrin?
   A. Heparin   D. Coagulation factor VI
   B. Thrombin   E. Tissue factor
   C. Prothrombin

14. A type of anemia that is due to a deficiency of intrinsic factor production, caused, for example, by alterations in the lining of the stomach, is:
    A. Aplastic   C. Hemolytic
    B. Sickle cell   D. Pernicious

15. A type of anemia in which hemoglobin molecules assume rod shapes that alter the shape of RBCs.
    A. Iron-deficiency   C. Hemorrhagic
    B. Sickle cell   D. Pernicious

16. A person with a hematocrit of 66 and hemoglobin of 22 is most likely to have the condition named:
    A. Anemia
    B. Polycythemia
    C. Infectious mononucleosis
    D. Leukemia

17. A hematocrit value is normally about _____ the value of the hemoglobin.
    A. The same   C. Three times
    B. One third   D. 700 times

Questions 18–20: Circle T (true) or F (false). If the statement is false, change the underlined word or phrase so that the statement is correct.

T F   18. Plasmin is an enzyme that facilitates clot <u>formation</u>.

T F   19. Hemolytic disease of the newborn is most likely to occur with an Rh <u>positive mother and her Rh negative babies</u>.

T F   20. In both black and white populations in the U.S., type O is <u>most</u> common and type AB is <u>least</u> common of the ABO blood groups.

Questions 21–25: Fill-ins. Answer questions or complete sentences with the word or phrase that best fits.

_____ 21. Name five types of substances transported by blood.

_____ 22. Name three types of plasma proteins.

_____ 23. Name the type of leukocyte that is commonly known as a "PMN" or a "poly."

_____ 24. Write a value for a normal leukocyte count: _____/mm^3.

_____ 25. All blood cells and platelets are derived from ancestor cells

             called _____.

ANSWERS TO MASTERY TEST: CHAPTER 19

Arrange

1. C A B
2. A B C
3. B C A

Multiple Choice

4. E	11. C
5. C	12. D
6. D	13. B
7. D	14. D
8. B	15. B
9. C	16. B
10. A	17. C

True–False

18. F. Dissolution or fibrinolysis
19. F. Negative mother and her Rh positive babies
20. T

Fill-ins

21. Oxyge
    waste:
22. Albun
    fibrinc
23. Neutr
24. 5,000- ..,...
25. Pluripotent stem cells

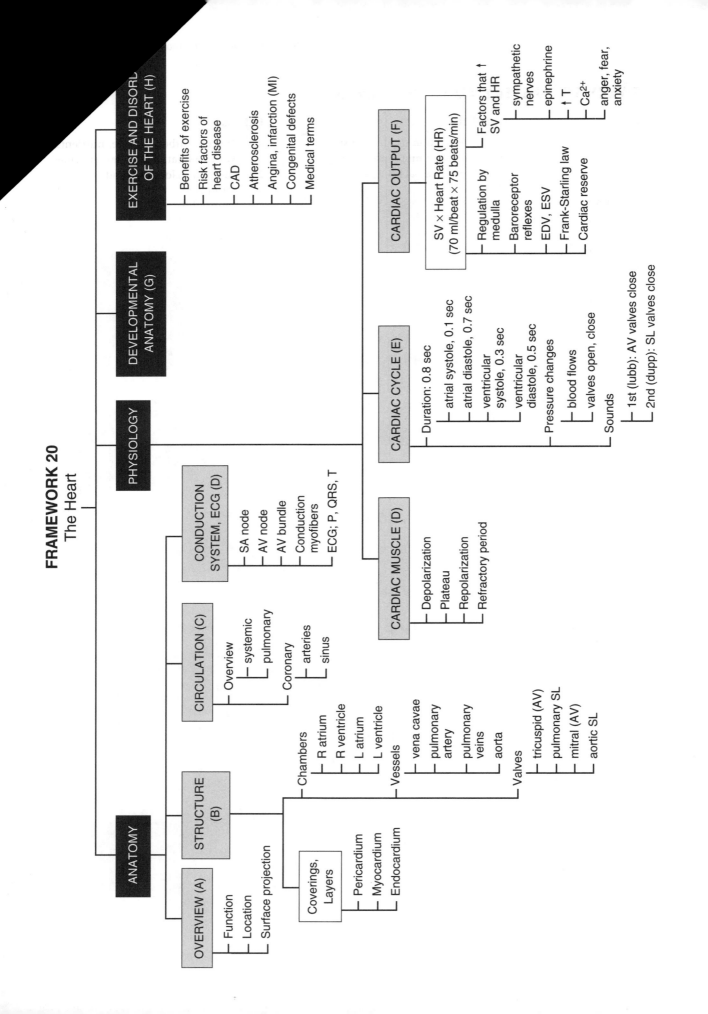

FRAMEWORK 20
The Heart

ANATOMY

OVERVIEW (A)
- Function
- Location
- Surface projection

STRUCTURE (B)

Coverings, Layers
- Pericardium
- Myocardium
- Endocardium

Chambers
- R atrium
- R ventricle
- L atrium
- L ventricle

Vessels
- vena cavae
- pulmonary artery
- pulmonary veins
- aorta

Valves
- tricuspid (AV)
- pulmonary SL
- mitral (AV)
- aortic SL

CIRCULATION (C)
- Overview
  - systemic
  - pulmonary
- Coronary
  - arteries
  - sinus

CONDUCTION SYSTEM, ECG (D)
- SA node
- AV node
- AV bundle
- Conduction myofibers
- ECG; P, QRS, T

PHYSIOLOGY

CARDIAC MUSCLE (D)
- Depolarization
- Plateau
- Repolarization
- Refractory period

CARDIAC CYCLE (E)
- Duration: 0.8 sec
  - atrial systole, 0.1 sec
  - atrial diastole, 0.7 sec
  - ventricular systole, 0.3 sec
  - ventricular diastole, 0.5 sec
- Pressure changes
  - blood flows
  - valves open, close
- Sounds
  - 1st (lubb): AV valves close
  - 2nd (dupp): SL valves close

CARDIAC OUTPUT (F)
- SV × Heart Rate (HR) (70 ml/beat × 75 beats/min)
  - Regulation by medulla
  - Baroreceptor reflexes
  - EDV, ESV
  - Frank-Starling law
  - Cardiac reserve
- Factors that ↑ SV and HR
  - sympathetic nerves
  - epinephrine
  - ↑ T
  - Ca^{2+}
  - anger, fear, anxiety

DEVELOPMENTAL ANATOMY (G)

EXERCISE AND DISORDERS OF THE HEART (H)
- Benefits of exercise
- Risk factors of heart disease
- CAD
- Atherosclerosis
- Angina, infarction (MI)
- Congenital defects
- Medical terms

C H A P T E R   **20**

The Cardiovascular System: The Heart

Make a fist, then open and squeeze tightly again. Repeat this about once a second as you read the rest of this Overview. Envision your heart as a muscular pump about the size of your fist. Unfailingly, your heart exerts pressure on your blood, moving it onward through the vessels to reach all body parts. The heart has one job; it is simply a pump. But this function is critical. Without the force of the heart, blood would come to a standstill, and tissues would be deprived of fluids, nutrients, and other vital chemicals. To serve as an effective pump, the heart requires a rich blood supply to maintain healthy muscular walls, a specialized nerve conduction system to synchronize actions of the heart, and intact valves to direct blood flow correctly. Heart sounds and EGG recordings as well as a variety of more complex diagnostic tools provide clues to the status of the heart.

Is your fist tired yet? The heart ordinarily pumps 24 hours a day without complaint and seldom reminds us of its presence. As you complete this chapter on the heart, keep in mind the value and indispensability of this organ. Start by studying the Chapter 20 Framework and refer to it frequently; note relationships among key terms in each section. As you begin your study of the heart, also carefully examine the Chapter 20 Topic Outline and Objectives; check off each one as you complete it.

TOPIC OUTLINE AND OBJECTIVES

A. Location and surface projection of the heart

1. Describe the location of the heart, and trace its outline on the surface of the chest.

B. Structure and function of the heart

2. Describe the structure of the pericardium and the heart wall.
3. Discuss the external and internal anatomy of the chambers of the heart.
4. Describe the structure and function of the valves of the heart.

C. Overview of circulation; coronary circulation

5. Describe the flow of blood through the chambers of the heart and through the systemic and pulmonary circulations.
6. Discuss the coronary circulation.

D. Cardiac muscle and the cardiac conduction system; electrocardiogram

7. Describe the structural and functional characteristics of cardiac muscle.
8. Explain the structural and functional features of the conduction system of the heart.
9. Describe how an action potential occurs in cardiac contractile fibers.
10. Describe the meaning of an electrocardiogram (ECG) and its diagnostic importance.

E. The cardiac cycle

11. Describe the phases, timing, and heart sounds associated with a cardiac cycle.

F. Cardiac output

12. Define cardiac output, and describe the factors that affect it.

G. Developmental anatomy

    13. Describe the development of the heart.

H. Exercise and the heart; disorders and medical terminology

    14. Explain the relationship between exercise and the heart.

WORDBYTES

Now become familiar with the language of this chapter by studying each wordbyte, its meaning, and an example of its use within a term. After you study the entire list, self-check your understanding by writing the meaning of each wordbyte on the line. As you continue through the *Learning Guide*, identify (and fill in) additional terms that contain the same wordbyte.

Wordbyte	Self-check	Meaning	Example(s)
angio-	_____	blood vessel	*angio*gram
auscult-	_____	listening	*auscult*ation
cardi-	_____	heart	*cardi*ologist
coron-	_____	crown	*coron*ary arteries
-cuspid	_____	point	tri*cuspid*
ectop-	_____	displaced	*ectop*ic pacemaker
endo-	_____	within	*endo*carditis
intercalat-	_____	inserted between	*intercalat*ed discs
-lunar	_____	moon	semi*lunar*
myo-	_____	muscle	*myo*cardial infarction
peri-	_____	around	*peri*cardium
pulmon-	_____	lung	*pulmon*ary artery
-sclerosis	_____	hard	arterio*sclerosis*
tri-	_____	three	*tri*cuspid
vascul-	_____	small vessel	cardio*vascul*ar

CHECKPOINTS

A. Location and surface projection of the heart (pages 636–638)

✓▲ **A1.** Do this exercise about heart functions.

    a. What is the primary function of the heart? _____

    b. Visualize a coffee cup; now imagine that sitting next to it is a gallon-size milk container. Keep in mind that a gallon holds about _____ quarts or approximately 4 liters. Now take a moment to imagine a number of containers lined up and holding the amount of blood pumped out by the heart.

        1. With each heartbeat, the heart pumps a stroke volume equal to about $^1/_3$ cup

        (70 mL) because 1 cup = about _____ mL.

        2. Each minute, the heart pumps enough blood to fill up 1.3 gallons (_____ liters).

        3. Within the 1440 minutes in a day, the heart would fill up about 1800 gallons

        (over _____ liters).

c. Reflecting on the work your heart is doing for you, what message would you give to your heart?

✓▲ **A2.** Closely examine Figure 20.1b in your text. Consider the location, size, and shape of your heart as you do this exercise. Trace its outline on your body.

a. Your heart lies in the _____ portion of your thorax, between your

two _____. About *(one-third? one-half? two-thirds?)* of the mass of your heart lies to the left of the midline of your body.

b. Your heart is about the size and shape of your _____.

c. The pointed part of your heart, called the _____, coincides with the

(superior? inferior?) left point, and is located in the _____ intercostal

space, about _____ cm (_____ inches) from the midline of your body.

d. The inferior surface of the heart lies on the _____. The right and left

borders of the heart face the _____.

e. The anterior surface of the heart faces the _____ and the

_____. Explain how the location of the heart accounts for the effectiveness of cardiopulmonary resuscitation (CPR).

B. Structure and function of the heart (pages 638–644)

✓▲ **B1.** Complete this Checkpoint on the structure of the heart wall and its coverings. Use answers in the box.

E.	**Endocardium**	**PC.**	**Pericardial cavity (site of pericardial fluid)**
FP.	**Fibrous pericardium**	**PP.**	**Parietal pericardium**
M.	**Myocardium**	**VP.**	**Visceral pericardium (epicardium)**

a. Arrange in order from most superficial to deepest.

_____ _____ _____ _____ _____ _____

b. Which layer of the heart is made of muscle that forms the bulk of the weight

of the heart? _____

c. In the condition known as pericarditis, where does fluid accumulate, causing

compression of the heart? _____

d. Which layer lines the inside of the heart and forms much of the tissue of valves?

_____

For extra review of the pericardium, look again at Checkpoint D2 in Chapter 4, page 00 in the *Learning Guide.*

A-1 _____   C _____   F _____

A-2 _____   D _____   G _____

A-3 _____   _____   H _____

B _____   E _____   I _____

Figure LG 20.1 Diagram of a frontal section of the heart. Letters follow the path of blood through the heart. Label, color, and draw arrows as directed in Checkpoints B2, B8, and D2.

✓▲**B2.** Refer to Figures LG 20.1 and 20.2. Identify all structures with letters A–I by writing labels on lines A–I on Figure LG 20.1.

✓▲**B3.** Check your understanding of heart structure by selecting the terms that fit descriptions below. Not all answers will be used.

A. Atria	**IVS.** Interventricular septum	**PV.** Pulmonary vein
CS. Coronary sulcus	**PA.** Pulmonary artery	**TC.** Trabeculae carnae
CT. Chordae tendineae	**PapM.** Papillary muscles	**V.** Ventricles
IAS. Interatrial septum	**PecM.** Pectinate muscles	

Label:
Anterior interventricular branch
Circumflex branch
Left coronary artery
Marginal branch
Right coronary artery
Posterior interventricular branch

Color
O Left coronary artery and branches
O Right coronary artery and branches

Figure LG 20.2 Anterior of the heart. Label and color as directed in Checkpoints B2, B8, and C4. Cardiac veins are solid black.

_____ a. Thin-walled chambers that receive blood from veins

_____ b. Blood vessel that carries blood that is rich in oxygen from lungs to left atrium

_____ c. Site of the fossa ovalis, the location of the foramen ovale in fetal life

_____ d. Strong wall separating the two ventricles

_____ e. Sites of most myocardium

_____ f. A groove separating left from right side of the heart; contains fat and coronary blood vessels

_____ g. Irregular ridges and folds of myocardium covered with smooth endocardium

_____ h. Strong tendons that anchor atrioventricular (AV) valves to ventricular muscle preventing eversion of the AV valves

_____ i. Muscles that give the anterior walls of atria a rough appearance

_____ j. Nipple-shaped muscles in ventricles that are attached to chordae tendineae

✓ **B4.** The human heart has *(2? 4? 6?)* valves. The primary function of heart valves is to

_____. Check your understanding of
heart valves by writing answers from the box next to related descriptions below.

A.	**Aortic semilunar**	P	**Pulmonary semilunar**
B.	**Bicuspid**	T.	**Tricuspid**

_____ a.   Also called the mitral valve

_____ b.   Prevents backflow of blood from right
         ventricle to right atrium

_____ c.   Prevents backflow from pulmonary trunk
         to right ventricle

_____ d.   Prevents backflow of blood into left
         atrium

_____ e.   Have half-moon-shaped leaflets or cusps
         (two answers)

_____ f.   Also called atrioventricular (AV) valves
         (two answers)

✓ **B5.** Which ventricle has a thicker wall? *(Right? Left?)* ventricle. The difference is related to
the fact that the left ventricle pumps blood to

_____ come

whereas the right ventricle pumps blood to _____. In which chamber
of the heart would you imagine greater pressure develops when that chamber
contracts? *(Right? Left?)* ventricle.

✓ **B6.** The fibrous skeleton of the heart consists of _____ tissue that

supports and anchors _____. Action potentials in atria *(can? cannot?)*
be transmitted across this fibrous skeleton. (More about the significance of this fact in
Checkpoints D1 and D2).

✓ **B7.** *A clinical correlation.* Name the microorganism that causes rheumatic fever (RF).

_____. What condition is likely to signal the presence of this microbe

and warn of possible effects (sequelae) upon the heart? _____.
Which parts of the heart are most likely to be affected?

Critical thinking. as you look at heart physiology, speculate about why these two
valves might be more susceptible. *Hint:* Refer to checkpoint B5 above.

✓▲ **B8.** Refer to Figures LG 20.1 and 20.2. and check your understanding of the pathway of
blood through the heart in this Checkpoint.
    a.   Draw arrows on Figure LG 20.1 to indicate direction of blood flow.
    b.   Color red the chambers of the heart and vessels that contain highly oxygenated blood;
        color blue the regions in which blood is low in oxygen and high in carbon dioxide.
    c.   On Figure LG 20.1, label the four valves that control blood flow through the heart.

d. On Figure LG 20.2 label the *ligamentum arteriosum*. In fetal life this is the site of the ductus *(arteriosus? venosus?)*, which connects the pulmonary *(artery? vein?)* to the

_____, thereby permitting blood to bypass the lungs.

O Vessels 1–3
O Vessel 4
O Vessels 5–7

Figure LG 20.3 Circulatory routes. Numbers 1–7 follow the pathway of blood through the systemic blood vessels. Color and label numbered vessels according to directions in Checkpoint C1. Letters A–I refer to Checkpoints C1–C2, and Chapter 21 Checkpoint F1, page LG 453.

C. Overview of circulation; coronary circulation (pages 644–648)

✓ **C1.** Refer to Figure LG 20.3 and complete this Checkpoint on blood vessels.

a. Label vessels 1–7. Note that numbers are arranged in the sequence of the pathway of blood flow. Use these labels:

Aorta and other large arteries	**Small artery**
Arteriole	**Small vein**
Capillary	**Venule**
Inferior vena cava	

b. Color vessels according to color code ovals.

c. The human circulatory system is called a(n) *(open? closed?)* system. State the significance of this fact.

▲ **C2.** On Figure LG 20.3, label the *pulmonary artery, pulmonary capillaries,* and a *pulmonary vein.* All blood vessels in the body other than pulmonary vessels are known as

_____ vessels because they supply a variety of body systems.

C3. Defend or dispute this statement: "The myocardium receives all of the oxygen and nutrients it needs from blood that is passing through its four chambers."

✓▲ **C4.** Refer to Figure LG 20.2 and complete this checkpoint.

a. Using the color code ovals on the figure, color the two coronary arteries and their main branches.

b. Label each of the vessels, using the list of labels on the figure.

c. Write another name for the coronary branch known as the "LAD."_____

d. The left coronary artery (and its branches) supplies blood to all of the chambers of

the heart *except* the _____.

✓▲ **C5.** The coronary sinus functions as *(an artery? a vein?)*. It collects blood that has passed

through coronary arteries and capillaries into _____ veins. The coronary sinus finally empties this blood into the *(right? left?) (atrium? ventricle?)*.

D. Cardiac muscle and the cardiac conduction system; electrocardiogram (pages 648–654)

✓ **D1.** Contrast cardiac muscle with skeletal muscle in this exercise.

a. Review these muscle types in the table in Chapter 10, Checkpoint E3, page LG 192.

b. In cardiac muscle, sarcoplasmic reticulum and transverse tubules are *(more? less?)* abundant than in skeletal muscle. As a result, cardiac muscle draws *(more? less?)* calcium from ECF during contraction.

   c. Unlike skeletal muscle, cardiac muscle contains desmosomes and gap junctions within intercalated discs. As a result, cardiac muscle fibers contract *(individually? as a unit?)*. When cardiac muscle contracts:

     A. Both atria and ventricles contract simultaneously as a unit.

     B. Atria and ventricles contract at different times as separate units.

   d. Identify the structural feature that causes atria and ventricles to contract as separate units.

✓ **D2.** In this exercise describe how the heart beats regularly and continuously.

   a. In embryonic life about _____% of cardiac muscle fibers become *autorhythmic*. State two functions of these specialized cells of the heart.

   b. Label the parts of the conduction system on Figure LG 20.1

   c. The normal pacemaker of the heart is the *(SA? AV?)* node. Each time the SA node "fires," impulses travel via the conduction system and also by the

     _____ junctions in _____ discs of cardiac muscle.

   d. What structural feature of the heart makes the AV node and AV bundle necessary for conduction from atria to ventricles?

   e. Through which part of the conduction system do impulses pass most slowly? *(SA node? AV node? AV bundle [of His]?)*. Based on the anatomy of this tissue, why does the rate of impulse conduction slow down?

     Of what advantage is this slowing?

✓ **D3.** Do this exercise about the pacemaker(s) of the heart.

   a. On its own, the SA node, located in the *(left? right?)* atrium, normally fires at about

     _____ to _____ times per minute. This rate is *(a little faster than? a little slower than? about the same as?)* a typical resting heart rate. At rest, *(acetylcholine? epinephrine?)* released from *(sympathetic? parasympathetic?)* nerves normally slows the pace of the

     SA node to modify the heart rate to about _____ beats per minute.

   b. How might this "normal pacemaker" be damaged? _____ If this occurs, then responsibility for setting the pace of the heart may be passed on to

     the _____ node, which fires at _____ to _____ times per minute.

   c. If both the SA and AV nodes fail, then autorhythmic fibers in _____

     may take over with a rate of only _____ to _____ beats per minute. Pacemakers at "other than the normal site" are known as _____ pacemakers and tend to be *(more? less?)* effective than the SA node in pacing the heart.

✓ **D4.** Describe the physiology of cardiac muscle contraction in this activity.
   a. Contractile fibers of the normal heart have a resting membrane potential of *(–70? –90?)* mV.
   b. Arrange in correct sequence the events in an action potential of cardiac muscle. The blank parentheses are for Activity D4c. One is done for you.

   __C (RD)__ → _____ (_____) → _____ or _____ (_____) → _____

   (_____ → _____ (_____)

   A. *Slow Ca²⁺ channels* open so Ca²⁺ enters muscle fibers. __P__
   B. Combined flow of Na⁺ and Ca²⁺ maintains depolarization for about 250 msec which is about 250X longer than depolarization in skeletal muscle.
   C. Na⁺ enters through *fast Na⁺ channels*; voltage rises rapidly to about +20 mV._____
   D. A second contraction cannot be triggered during this period, which is longer than

      for skeletal muscle. _____
   E. The presence of Ca²⁺ binding to troponin permits myocardial contraction via

      sliding of actin filaments next to myosin filaments. _____
   F. K⁺ channels open so K⁺ ions leave the fiber; meanwhile less Ca²⁺ enters as those

      channels are closing; voltage returns to resting level. _____
   c. Now label events A–F above by filling in parentheses. One is done for you. (For help refer to Figure 20.11 in your text.) Use these answers:
   d. State the significance of the prolonged depolarization (250 msec) of cardiac muscle

P.	Plateau	Rf.	Refractory
RD.	Rapid depolarization	Rp.	Repolarization

   compared to the brief depolarization (1 msec) of skeletal muscle.

   e. Explain the significance of the long refractory period of cardiac muscle.

   f. *A clinical correlation.* Calcium channel blockers [such as verapamil (Procardia)] tend to *(in? de?)*-crease contraction of the cardiac muscle (and of smooth muscle of coronary arteries). Write one or more condition(s) for which such a medication might be prescribed.

   Now name one chemical that enhances flow of Ca²⁺ through these channels to

   strengthen contraction of the heart. _____

D5. *A clinical correlation.* Describe how a patient's own skeletal muscle tissue can be used to assist the heart's pumping capacity.

✓ **D6.** Describe an ECG (or EKG) in this activity.
  a. What do the letters ECG (or EKG) stand for?

   An ECG is a recording of *(electrical changes associated with impulse conduction? muscle contractions?)* of the heart.
  b. On what parts of the body are leads (electrodes) placed for a 12-lead ECG?

  c. In Figure LG 20.4b, an ECG tracing using lead *(1? II? III?)* is shown. Label the following parts of that ECG on the figure: *P wave, P-R interval, QRS wave (complex), S-T segment, T wave.*
  d. *A clinical correlation.* Answer the following questions with names of parts of an ECG.

   1. Indicates atrial depolarization leading to atrial contraction _____
   2. Prolonged if the AV node is damaged, for example, by rheumatic fever, or in the

      condition described as "bundle (of His) block" _____

   3. Elevated in acute MI _____
   4. Elevated in hyperkalemia (high blood level of K^+) _____

   5. Flatter than normal in coronary artery disease _____

   6. Represents ventricular repolarization _____

   7. Represents ventricular depolarization _____

   8. Represents atrial repolarization _____

✓ **D7.** Do this exercise on arrhythmias.
  a. *(All? Not all?)* arrhythmias are serious.
  b. Conduction failure across the AV node results in an arrhythmia known as

   _____.

  c. Which is more serious? *(Atrial? Ventricular?)* fibrillation. Explain why.

  d. Excitation of a part of the heart other than the normal pacemaker (SA node) is

   known as a(n) _____ focus. Such excitations may be caused by ingestion of caffeine or by lack of sleep or by more serious problems such as ischemia.

Figure LG 20.4 Cardiac cycle. (a) Systole and diastole of atria and ventricles related to time. (b) ECG related to cardiac cycle. (c) Pressures in the aorta, atria, and ventricles. (d) Heart sounds related to cardiac cycle. (e) Volume of blood in ventricles. Label as directed in Checkpoints D6c and E2 to E4.

E. Cardiac cycle (Pages 654–657)

✓ **E1.** Complete the following overview of movement of blood through the heart.

a. Blood moves through the heart as a result of _____ and

_____ of cardiac muscle, as well as _____ and

_____ of valves. Valves open or close due to _____ within the heart.

b. Contraction of heart muscle is known as _____ whereas relaxation

of myocardium is called _____.

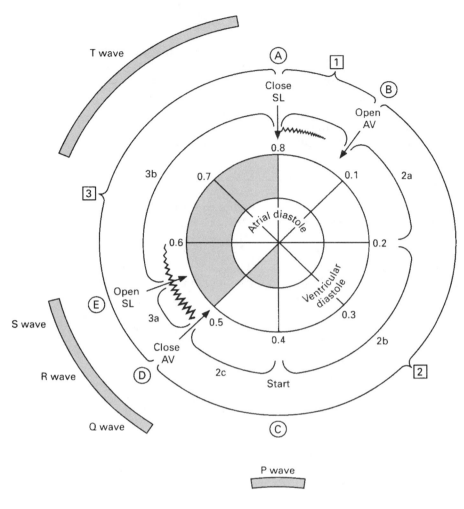

KEY

1   Isovolumetric relaxation
2a Rapid ventricular filling
2b Reduced ventricular filling (diastasis)
2c Atrial contraction
3a Isovolumetric contraction
3b Ventricular ejection

Figure LG 20.5 Cardiac cycle. Events of this cycle are shown on a special 0.8-sec "clock." Sectors are 0.1 sec each. Atrial activity is represented within inner circle and ventricular activity within outer circle. Shaded areas: systole; white areas of circle: diastole. Sounds are represented as ∧∨∧∨. Valve openings and closings and ECG correlations are also included. Brackets with key numbers 1 to 3b define important periods within the cycle. Refer to Checkpoints E2 and E3.

✓ **E2.** In this exercise, we will consider details of the cardiac cycle. Refer alternately to Figure LG 20.4 and Figure LG 20.5. Figure LG 20.4 will allow you to correlate pressure and volume changes within the heart to an ECG recording and to heart sounds. Figure LG 20.5 will help you to visualize events as they occur in a one-cycle "clock."

a. If the heart beats at 75 beats per minute, then *(54? 60? 75?)* complete cardiac cycles occur per 60-sec period. In other words, if your pulse is 75 (beats per minute), the

   duration of one cardiac cycle ("heartbeat") within you is _____ sec.

b. Now look at Figure LG 20.5 which represents one complete 0.8-sec cardiac cycle on a special 0.8-sec "clock." The clock is divided into eight sectors, or time periods, each of

   which is _____ sec in duration.

Note: In the following activities, it is critical to remember that the total time elapsed during one cardiac (clock) cycle is only 0.8 sec.

c. We could examine the cycle shown in Figure LG 20.5 at any point on the "clock." We will, however, begin our study with an overview of the cycle. Beginning at the top of the clock, the "12 o'clock" position, identify the three distinct phases—1, 2, and 3— that are bracketed on Figures LG 20.4 and LG 20.5.

   1. Phase 1 is a period in which the atria and ventricles are both *(contracting? relaxing?)*. In other words, they are both in *(systole? diastole?)*.

   2. Throughout phase 2 the ventricles *(eject? fill with?)* blood as they continue in

      (systole? diastole?) while atria go into _____ toward the end of Phase 2.

   3. Now the filled ventricles are ready to be stimulated (indicated by the _____ wave of the ECG) to contract and then eject the blood. That is exactly what hap-

      pens in phase _____.

   At the bottom of Figure 20.4, label these three bracketed phases.

d. Now let's look at details. Begin again at "12 o'clock" on Figure LG 20.5 and at the left side of LG Figure LG 20.4. As we come on to the scene, ventricles have just stopped contracting in the last cycle. Therefore ventricular pressure is *(rising? dropping?)*. When pressure within ventricles dips below that in the great arteries, blood fills

   semilunar valves and closes them (point _____). Notice that pressure within ventricles drops at a rapid rate now as ventricles continue to relax.

e. When intraventricular pressure becomes lower than that in atria, the force of blood within atria causes (atrioventricular) AV valves to *(open? close?)*. This happens at

   point _____.

f. Note (Figure LG 20.5) that AV valves now remain open all the way to point D, in other words, throughout the time of ventricular *(ejection ? filling?)*. Trace a pencil lightly along the filling curve shown on Figure LG 20.4e, starting at point B. Note that the curve is steepest just *(after B? before C?)*. In other words, the first part of ventricular diastole (just after the AV "floodgates" open) is the time when ventricles fill

   (rapidly? slowly?), whereas slower filling, known as _____, occurs later (just before point C).

g. Which wave of the ECG signals atrial systole? *(P? QRS? T?)* Duration of atrial systole is about *(0.1? 0.3? 0.5?)* sec. As a result of atrial contraction, pressure within the atria *(increases? decreases?)* (point C in Figure LG 20.4c), forcing the last 20–30% of all the blood that will enter ventricles into those chambers.

h. Point D marks the end of atrial systole, as well as the end of ventricular *(diastole? systole?)*. Ventricles are now filled to their maximum volume (see Figure LG 20.4e),

   which is about _____ mL of blood. This volume is known as the _____-

   _____ volume (EDV). Write EDV next to point E on Figure LG 20.4e.

i. The *(P? QRS? T?)* wave signals ventricular systole. Note that although the QRS wave itself happens over a very short period, the ventricular contraction that follows occurs over a period of about *(0.1? 0.3? 0.5?)* sec.

j. Trace a pencil lightly along the curve in Figure LG 20.4c showing ventricular pressure changes following the QRS wave. Notice that pressure there *(increases? decreases?)* *(slightly? dramatically?)*. In fact it quickly surpasses atrial pressure (at point

   _____), forcing the AV valves *(open? closed?)*.

k. A brief time later ventricular pressure becomes so great that it even surpasses pressure in the great arteries (pulmonary artery and aorta). This pressure forces blood against the undersurface of the _____ valves, *(opening? closing?)*

them. This occurs at point _____ on the figure.

l. Continued ventricular systole ejects blood from the heart into the great vessels. In the aorta this creates a systolic blood pressure of about *(15? 80? 120?)* mmHg.

m. Repolarization of ventricles (after the _____ wave) causes these chambers to go into *(systole? diastole?)*. Consequently, ventricular pressure begins to *(increase? decrease?)*. This completes one cardiac cycle.

n. The volume remaining in the ventricles at the end of systole is known as

_____-_____ volume (ESV). Write ESV next to point A to indicate that this much blood remains in the heart as a new cycle begins. ESV is

normally about _____ mL. The amount of blood ejected from the heart during

systole is known as _____ volume (SV) and equals about _____ mL. Show the calculation of stroke volume here:
End-diastolic volume (EDV) – End-systolic volume (ESV) = Stroke volume (SV)

_____ mL   – _____ mL   = _____ mL

o. Label points A, B, D, and E on Figure LG 20.4c to indicate which valves open or close at those points. Check your answers by referring to Figure LG 20.5.

p. *For extra review.* Refer to a figure of the heart, such as Figure LG 20.2, as you review this activity.

✓ **E3.** Summarize changes in ventricular volume during the cycle by relating events described in the key for Figure LG 20.5 with the curve in Figure LG 20.4e.

a. For example, the fact that the portion of the curve between points A and B is "flat" indicates that the volume of these chambers *(does? does not?)* change during this period. This is logical because the AV and SL valves are all *(open? closed?)* during this time.

This period (A to B) is known as the period of _____ relaxation.

b. Note that the "flat" portion of the curve in Figure LG 20.4e between points D and E is a time when ventricles are *(relaxing? contracting?)*. Which valves are closed during

this period? _____ In other words, ventricles have exerted enough

pressure to close *(AV? SL?)* valves but not enough pressure yet to open _____

valves. Name this period: isovolumetric _____.
(Note: Remember that the reason that periods A–B and D–E are called *isovolumetric* (iso = same) is that no blood enters or leaves ventricles because both sets of valves are *closed.)*

✓▲ **E4.** Do this exercise on heart sounds.

a. The first heart sound *(lubb? dupp?)* is produced by turbulence of blood at the

(opening? closing?) of the _____ valves. What causes the second sound?

b. Write *first* and *second* next to the parts of Figure LG 20.4d showing each of these sounds.

c. *A clinical correlation.* Using Figure 20.14 in your text, recognize why a stethoscope is placed at several points on the chest in order to best hear different heart sounds. If a stethoscope is placed at the level of about the fifth intercostal space (between ribs 5 and 6) on the left side of the sternum, the *(first? second?)* sound is better heard because turbulence from the *(mitral? aortic semilunar?)* valve is readily detected there. To best hear the second sound (related to SL valve closure), the stethoscope must then be moved in a more *(inferior? superior?)* direction on the chest.

E5. Explain what causes heart murmurs.

✓ **E6.** *Critical thinking.* Consider effects of a significant change in heart rate.
    a. Recall that a heart beating at about 75 beats/mm has a cardiac cycle of about 0.8 sec

        duration: ventricles contract for _____ sec and they relax and fill with blood for

        _____ sec.

    b. When the heart rate doubles, for example, to 150 beats/mm, the duration of the cardiac cycle *(doubles? is halved?)*, for example, to 0.4 sec. Which part of the cycle is especially affected? Ventricular *(systole? diastole?)* What are the consequences of this change?

F. Cardiac output (pages 657–662)

✓ **F1.** Determine the average cardiac output in a resting adult.
        Cardiac output = stroke volume × heart rate

        = _____ mL/stroke × _____ strokes/min

        = _____ mL/min (_____ liter/min)

✓ **F2.** At rest Dave has a cardiac output of 5 liters per minute. During a strenuous cross-country run, Dave's maximal cardiac output is 20 liters per minute. Calculate Dave's cardiac reserve.

$$\text{Cardiac reserve} = \frac{\text{maximal cardiac output}}{\text{cardiac output at rest}} =$$

Note that cardiac reserve is usually *(higher? lower?)* in trained athletes than in sedentary persons.

✓ **F3.** The two major factors (shown in Activity Fl) that control cardiac output are

        _____ and _____.

✓ **F4.** Do this activity about stroke volume (SV).
    a. Which statement is true about stroke volume (SV)? *(Hint:* Refer to Checkpoint E2n.)

        A. SV = ESV − EDV   B. SV = EDV − ESV   C. SV = $\dfrac{\text{ESV}}{\text{EDV}}$ D. SV = $\dfrac{\text{EDV}}{\text{ESV}}$

b. List the three factors that control stroke volume (SV):

_____ _____ _____

c. During exercise, skeletal muscles surrounding blood vessels squeeze *(more? less?)* blood back to the heart. This increase in venous return to the heart *(increases? decreases?)* end-diastolic volume (EDV). As a result, muscles of the heart are stretched *(more? less?)* so that preload (= stretching of the heart muscle) *(increases? decreases?)*.

d. Within limits, a stretched muscle contracts with *(greater? less?)* force than a muscle

that is only slightly stretched. This is a statement of the _____

_____ law of the heart. To get an idea of this, blow up a balloon slightly and then let it go. Then blow up a balloon quite full of air (much like the heart when stretched by a large venous return) and let that balloon go. In which stage do the walls of the balloon (heart) compress the air (blood) with greater force?

e. As a result of the Frank-Starling law, during exercise the ventricles of the normal heart contract *(more? less?)* forcefully, and stroke volume *(increases? decreases?)*.

f. State several reasons why venous return might be decreased, leading to reduction in stroke volume and cardiac output.

g. Elevated blood pressure or narrowing of blood vessels tends to _____-crease

afterload and therefore to _____-crease stroke volume and cardiac output. Identify two causes of increased afterload.

✓ **F5.** *The Big Picture: Looking Back.* Return to text references given below and answer questions related to cardiac function.

a. In Table 4.4, page 131 cardiac muscle cells appear *(parallel? branching?)*, which is a characteristic *(similar to? different from?)* skeletal muscle. What is the significance of this pattern of cardiac muscle tissue?

b. Refer to Table 17.4, page 560. Sympathetic nerves _____-crease heart rate and force of atrial and ventricular contraction and also *(dilate? constrict?)* coronary vessels. Therefore sympathetic nerves tend to _____-crease stroke volume and cardiac output.

c. Figure 17.2, page 551, shows that the vagus nerve is a *(sympathetic? parasympathetic?)* nerve. Its effects upon heart muscle, coronary vessels, and cardiac output are *(similar? opposite?)* to those of sympathetic nerves.

d. Table 18.8, page 592, states that epinephrine and norepinephrine have *(sympathomimetic? parasympathomimetic?)* effects. In other words, these hormones, produced by the adrenal *(cortex? medulla?)* have effects similar to those of the sympathetic division of the autonomic nervous system: they prepare the body for a

_____ response.

✓ **F6.** Summarize factors affecting heart rate (HR), stroke volume (SV), and cardiac output (CO) by completing arrows: ↑ (for increase) or ↓ (for decrease). The first one is done for you.
  a. ↑ exercise → ↑ preload
  b. Moderate ↑ in preload → I SV
  c. ↑ SV (if heart rate is constant) → I CO
  d. ↑ in heart contractility → I SV and CO
  e. Positively inotropic medication such as digoxin (Lanoxin) → I SV and CO
  f. ↑ afterload (for example, due to hypertension) → I SV and CO
  g. Hypothermia → I heart rate and CO
  h. ↑ heart rate (if SV is constant) → I CO
  i. ↑ stimulation of the cardioaccelerator center → I heart rate
  j. ↑ sympathetic nerve impulses → I heart strength (and SV) and I heart rate
  k. ↑ vagal nerve impulses → I heart rate
  l. Hyperthyroidism → I heart rate
  m. ↑ Ca^{2+} → I heart strength (and SV) and I heart rate
  n. ↓ SV (as in CHF) → ESV → excessive EDV and preload → I SV and CO

✓ **F7.** *A clinical correlation.* Do this exercise about Ms. S, who has congestive heart failure (CHF).
  a. Her weakened heart becomes overstretched much like a balloon (except much thicker!) that has been expanded 5000 times. Now her heart myofibers are stretched beyond the optimum length according to the Frank-Starling law. As a result, the

  force of her heart is _____-creased, and its stroke volume _____-creases (as shown in Activity F6n).
  b. Ms. S has right-sided heart failure, so her blood is likely to back up, distending vessels in *(lungs? systemic regions, such as in neck and ankles?)*; thus

  _____ edema results, with signs such as swollen hands and feet.
  c. Write a sign or symptom of left-sided heart failure.

  d. An intra-aortic balloon pump (IABP) is especially helpful for *(right? left?)*-sided CHF

  because the IABP _____-creases afterload and therefore can _____-crease stroke volume.

G. Developmental anatomy (page 663)

✓ **G1.** The heart is derived from _____-derm. The heart begins to develop during the *(third? fifth? seventh?)* week. Its initial formation consists of two endo-

  thelial tubes which unite to form the _____ tube.

✓ **G2.** Match the regions of the primitive heart below with the related parts of a mature heart listed below.

A.	**Atria**	**SV.**	**Sinus venosus**
BCTA.	**Bulbus cordis and truncus arteriosus**	**V.**	**Ventricle**

_____ a. Superior and inferior vena cava   _____ c. Parts of the fetal heart connected by foramen ovale

_____ b. Right and left ventricles

_____ d. Aorta and pulmonary trunk

H. Exercise and the heart; disorders and medical terminology (pages 662–666)

✓ **H1.** Circle answers that indicate benefits of physical conditioning, such as three to five aerobic workouts each week.

a. _____-crease in maximal cardiac output   d. _____-crease in resting heart rate

b. _____-crease in hemoglobin level of blood   e. _____-crease in blood pressure

c. _____-crease in capillary networks in   f. _____-crease in fibrinolytic activity
skeletal muscles

H2. List eight risk factors for heart disease. Circle five of those that can be modified by a healthy lifestyle.

a. _____   e. _____

b. _____   f. _____

c. _____   g. _____

d. _____   h. _____

✓ **H3.** Outline the sequence of changes in atherosclerosis in this learning activity.
a. Atherosclerosis involves damage to the walls of arteries. State several factors that may lead to damage of this lining.

b. The resulting fatty lesion in the arterial lining is known as an atherosclerotic

_____. The rough surface of plaque may snag platelets. These may then

cause _____ formation at the site, possibly leading to unwanted emboli.
c. Chemicals released in the plaque cause thickening of smooth muscle and collagen in the blood vessel wall. What are the ultimate effects of these changes?

✓ **H4.** Match diagnostic techniques and treatments in the box with descriptions below.

C.	**Catheterization**	**CABG.**	**Coronary artery bypass graft**
CA.	**Cardiac angiography**	**PTCA.**	**Percutaneous transluminal coronary angioplasty**

_____ a. Invasive procedure in which a long, slender tube is inserted into an artery or vein under x-ray observation

_____ b. Catheterization with use of contrast dye for visualization of coronary blood vessels or heart chambers

_____ c. Nonsurgical technique to increase blood supply to the heart as arterial plaque is compressed

_____ d. Surgical procedure utilizing a vessel from another part of the body (such as a vein from the thigh) to reroute blood around a blocked portion of a coronary artery

H5. Define *ischemia* and explain how it is related to *angina pectoris*.

✓ **H6.** *A clinical correlation*. Do this exercise about "heart attacks."

  a. "Heart attack" is a common name for a myocardial _____ (MI).

     What does the term *infarction* mean? _____
     In what other areas of the body might infarctions also occur? Cerebral infarction:

     _____ ; pulmonary infarction: _____ .
  b. List two or more immediate causes of an MI.

  c. List three treatments for an MI.

✓ **H7.** Match each congenital heart disease in the box with the related description below.

C.	Coarctation of aorta.	IVSD.	Interventricular septal defect
IASD.	Interatrial septal defect	PDA.	Patent ductus arteriosus

_____ a. Connection between aorta and pulmon-   _____ c. Septum between ventricles does not
         ary artery is retained after birth, allowing   develop properly.
         backflow of blood to right ventricle.
                                                _____ d. Aorta is abnormally narrowed.
_____ b. Foramen ovale fails to close.

✓ **H8.** Match disorders in the box with descriptions below.

CHF.	Congestive heart failure	PA.	Palpitation
CM.	Cardiomegaly	PT.	Paroxysmal tachycardia
CP.	Cor pulmonale		

_____ a. Enlarged heart   _____ d. Sudden period of rapid heart rate

_____ b. Right ventricular hypertrophy resulting   _____ e. Inability of the heart to function as an
         from lung disorders that create resistance   effective pump; blood backs up leading to
         to blood flow through pulmonary vessels   pulmonary or peripheral edema

_____ c. Fluttering of the heart

ANSWERS TO SELECTED CHECKPOINTS: CHAPTER 20

A1. (a) It is a pump. (b) 4 (one gallon = 4 quarts or 3.86 liters). (b1) 237. (b2) 5 (1.3 gallons = 5 liters). (b3) 7200 (1.3 gallons/mm × 1440 mm/day = 1800 gallons/day = over 7000 liters). (c) Perhaps: "Thanks!" "Great job!" "Amazing!" or "You deserve the best of care, and I intend to see that you get it."

A2. (a) Mediastinal, lungs; two-thirds. (b) Fist. (c) Apex, inferior, fifth, 9, 3.5. (d) Diaphragm; right and left lungs. (e) Sternum and ribs (or rib cartilages); compression of the heart between these bony structures anteriorly and the vertebrae posteriorly forces blood out of the heart.

B1. (a) FP PP PC VP M E. (b) M. (c) PC. (d) E.

B2. See Figure LG 20.1A

Ligamentum arteriosum

Pulmonary semilunar valve

AV bundle (of His)

Aortic semilunar valve

Bundle branches

Mitral valve

Left side of body

SA node

AV node

Right side of body

Tricuspid valve

Conduction myofibers (Purkinje fibers)

A 1 **Superior vena cava**	C **Right ventricle**	F **Pulmonary veins**
A 2 **Inferior vena cava**	D **Pulmonary trunk**	G **Left atrium**
A 3 **Coronary sinus**	**and arteries**	H **Left ventricle**
B **Right atrium**	E **Vessels in lungs**	I **Aorta**

Figure LG 20.1A Diagram of a frontal section of the heart.

B3. (a) A. (b) PV. (c) IAS. (d) IVS. (e) V. (f) CS. (g) TC. (h) CT. (i) PecM. (j) PapM.

B4. 4; ensure one-way flow of blood (or prevent back flow [or regurgitation] of blood into the chamber from which blood had just come). (a) B. (b) T. (c) P. (d) B. (e) A, P. (f) B, T.

B5. Left; all parts of the body except to lungs, only the lungs; left (usually develops pressure 4–5 times greater than that in the right ventricle).

B6. Dense fibrous connective, valves of the heart; cannot.

B7. Streptococcus; streptococcal sore throat ("strep throat"); valves, particularly the bicuspid (mitral) and the aortic semilunar. These two valves control blood flow through the left side of the heart, which is a higher-pressure system than the right side; therefore, valve damage is more likely, making these valves more vulnerable to effects of RF.

B8. (a) See Figure LG 20.1A. (b) Blue: A–D and first part of E; red: last part of E and F–I. (c) See Figure LG 20.A. (d) See Figure 20.4a in the text; arteriosus, artery, aorta.

C1. (a) 1, aorta; 2, small artery; 3, arteriole; 4, capillary; 5, venule; 6, small vein; 7, inferior vena cava. (b) See Figure 21.17 (page 691 of the text. (c) Closed; blood is carried from the heart to tissues and back to the heart within a closed system of tubes that contributes to hemodynamics, for example, by large arteries converting stored energy into kinetic energy of blood so that blood continues to move on through smaller vessels.

C4. (a) Left coronary artery: 4–6; right coronary artery: 1–3. (b) 1, right coronary artery; 2, posterior interventricular branch; 3, marginal branch; 4, left coronary artery; 5, circumflex branch; 6, anterior interventricular branch (c) Left anterior descending or anterior descending. (d) Right atrium.

C5. Vein; cardiac; right atrium.

D1. (a) See LG page 196. (b) Less; more. (c) As a unit; B. (d) The fibrous skeleton of the heart (page 642 of the text) separates atrial myocardium from ventricular myocardium.

D2. (a) One; pacemaker and impulse conduction. (b) See Figure LG 20.1A. (c) SA; gap, intercalated. (d) The fibrous skeleton (including AV valves) that completely separates atria from ventricles. (e) AV node; fibers have smaller diameter; affords time for ventricles to fill more completely before they contract.

D3. (a) Right, 90, 100; a little faster than; acetylcholine, parasympathetic, 75. (b) Ischemia as in myocardial infarction (MI); AV, 40, 50. (c) Conduction myofibers (Purkinje fibers), 20, 40; ectopic, less.

D4. (a) – 90. (b, c) C (RD) → A (P) → B or E (both P) → F (Rp) → D (Rf). (d) Myocardium can maintain contractions long enough to pump blood out of the heart. (e) Depolarization (causing contraction) cannot begin until after the refractory period (relaxation, providing time for the heart to fill with blood) is well underway. If the heart contracted all the time, it would never get to fill with blood, so no blood would be in the heart to be pumped out during contractions. (f) De; coronary artery disease (CAD), high blood pressure (hypertension), rapid heart rate (tachycardia); epinephrine.

D6. (a) An electrocardiogram (the recording) or electrocardiograph (the instrument); electrical changes associated with impulse conduction. (b) Arms, legs, and chest. (c) II; refer to Figures 20.12 and 20.13 of the text. (d1) P wave; (d2) P–R interval; (d3) S–T segment; (d4–d6) T wave; (d7) QRS complex; (d8) No wave because masked by QRS complex.

D7. (a) Not all. (b) Heart block. (c) Ventricular; atrial fibrillation reduces effectiveness of the atria by only about 20–30%, but ventricular fibrillation causes the heart to fail as a pump. (d) Ectopic.

E1. (a) Contraction, relaxation, opening, closing; pressure changes. (b) Systole, diastole.

E2. (a) 75; 0.8 (60 sec/75 cardiac cycles = 0.8 sec/cardiac cycle). (b) 0.1. (c1) Relaxing; diastole; (c2) fill with, diastole, systole; (c3) QRS; 3; phase 1: atrial and ventricular diastole (or isovolumetric relaxation); phase 2: ventricular diastole (and filling) with atrial contraction toward the end of this phase; phase 3: ventricular systole and atrial diastole. (d) Dropping; A. (e) Open; B. (f) Filling; after B; rapidly, diastasis. (g) P; 0.1; increases. (h) Diastole; 130; end-diastolic. (i) QRS; 0.3. (j) Increases dramatically; D, closed. (k) Semilunar, opening; E. (1) 120. (m) T, diastole; decrease. (n) End-systolic; 60; stroke, 70; 130−60 = 70. (o) A, closing of SL valves; B, opening of AV valves; D, closing of AV valves; E, opening of SL valves.

E3. (a) Does not; closed; isovolumetric. (b) Contracting; AV and SL; AV, SL; contraction.

E4. (a) Lubb, closing, AV; turbulence due to closing of SL valves. (b) Refer to Figure 20.13 in the text. (c) First, mitral; superior.

E6. (a) 0.3, 0.5. (b) Is halved, diastole; because the heart cannot adequately fill with blood during a very brief ventricular diastole, only a very small volume of blood is ejected from the heart during each cardiac cycle.

F1. CO = 70 ml/stroke × 75 strokes (beats)/min = 5250 ml/mm = 5.25 liters/min.

F2. Four; 20 liters per min/5 liters per mm = 4; higher.

F3. Stroke volume (SV), heart rate (HR).

F4. (a) B. (b) Preload, contractility, afterload. (c) More; increases; more, increases. (d) Greater; Frank-Starling; the more expanded balloon (heart) exerts greater force on air (blood). (e) More, increases. (f) Lack of exercise, loss of blood (hemorrhage), heart attack (myocardial infarction), or rapid heart rate (shorter diastole: see Checkpoint E6). (g) In, de; vasoconstriction of blood vessels, atherosclerosis.

F5. (a) Branching, different from; allows for more effective spreading of action potentials and contraction of the heart muscle as a unit, rather than as discrete muscle fibers. (b) In, dilate; in. (c) Parasympathetic; opposite. (d) Sympathomimetic; medulla, fight-or-flight.

F6. (a) ↑ exercise → ↑ preload. (b) Moderate ↑ in preload → ↑ SV. (c) ↑ SV (if heart rate is constant) → ↑ CO. (d) ↑ in heart contractility → ↑ SV and CO. (e) Positively inotropic medication such as digoxin (Lanoxin) → ↑ SV and CO. (f) ↑ afterload (for example, due to hypertension) → ↓ SV and CO. (g) Hypothermia → ↓ heart rate and CO. (h) ↑ heart rate (if SV is constant) → ↑ CO. (i) ↑ stimulation of the cardioaccelerator center → ↑ heart rate. (j) ↑ sympathetic nerve impulses → ↑ heart strength (and SV) and ↑ heart rate. (k) ↑ vagal nerve impulses → ↓ heart rate. (1) Hyperthyroidism → ↑ heart rate. (m) ↑ Ca^{2+} → ↑ heart strength (and SV) and ↑ heart rate. (n) ↓ SV (as is CHF) → ↑ ESV → excessive EDV and preload → ↓ SV and CO.

F7. (a) De, de. (b) Systemic regions, such as in neck and ankles; peripheral or systemic. (c) Difficulty breathing or shortness of breath (dyspnea), which is a sign of pulmonary edema. (d) Left, de, in.

G1. Meso; third; primitive heart.

G2. (a) SV. (b) V. (c) A. (d) BCTA.

H1. (a–c) In. (d–e) De. (f) In.

H3. (a) Cytomegalovirus, carbon monoxide from smoking, diabetes mellitus, prolonged hypertension, and fatty diet. (b) Plaque; clot (thrombus). (c) Narrowing of the arterial lumen, reducing blood flow and increasing risk of thrombus formation.

H4. (a) C. (b) CA. (c) PTCA. (d) CABG.

H6. Infarction, death of tissue because of lack of blood flow to that tissue; brain, lung. (b) Thrombus (clot), embolus (mobile clot), spasm of coronary artery. (c) Thrombolytic agents (such as t-PA), anticoagulants such as heparin, or surgery such as CABG or coronary angioplasty.

H7. (a) PDA. (b) IASD. (c) IVSD. (d) C.

H8. (a) CM. (b) CP. (c) PA. (d) PT. (e) CHF.

MORE CRITICAL THINKING: CHAPTER 20

1. Contrast structure and location of the AV valves with the SL valves.
2. Explain how and why interference in conduction through the AV node, the AV bundle, and bundle branches would alter the ECG and heart function.
3. Describe the structural differences that account for the functional differences between cardiac and skeletal muscle.
4. Relate opening and closing of heart valves to systole and diastole of atria and ventricles.
5. Contrast pulmonary edema with peripheral edema. Define each condition and relate them to different types of congestive heart failure (CHF). Also state two or more prominent signs or symptoms of each.
6. Describe components of a wellness lifestyle that are likely to enhance heart health.

MASTERY TEST: CHAPTER 20

Questions 1–3: Arrange the answers in correct sequence.

_____ _____ _____ _____ 1. Pathway of the conduction system of the heart:
    A. AV node
    B. AV bundle and bundle branches
    C. SA node
    D. Conduction myofibers (Purkinje fibers)

_____ _____ _____ _____ _____ 2. Route of a red blood cell supplying oxygen to myocardium of left atrium and returning to right atrium:
    A. Arteriole, capillary, and venule within myocardium
    B. Branch of great cardiac vein
    C. Coronary sinus leading to right atrium
    D. Left coronary artery
    E. Circumflex artery

_____ _____ _____ _____ _____ 3. Route of a red blood cell now in the right atrium:
    A. Left atrium
    B. Left ventricle
    C. Right ventricle
    D. Pulmonary artery
    E. Pulmonary vein

Questions 4–15: Circle the letter preceding the one best answer to each question.

4. All of the following are correctly matched *except:*
   A. Myocardium—heart muscle
   B. Visceral pericardium—epicardium
   C. Endocardium—forms heart valves which are especially affected by rheumatic fever
   D. Pericardial cavity—space between fibrous pericardium and parietal layer of serous pericardium

5. Which of the following factors will tend to decrease heart rate?
   A. Stimulation by cardiac nerves
   B. Release of the transmitter substance norepinephrine in the heart
   C. Activation of neurons in the cardioaccelerator center
   D. Increase of vagal nerve impulses
   E. Increase of sympathetic nerve impulses

6. The average cardiac output for a resting adult is about _____ per minute.
   A. 1 quart
   B. 5 pint
   C. 1.25 gallons (5 liters)
   D. 0.5 liter
   E. 2000 mL

7. When ventricular pressure exceeds atrial pressure, what event occurs first?
   A. AV valves open
   B. AV valves close
   C. Semilunar valves open
   D. Semilunar valves close

8. The second heart sound is due to turbulence of blood flow as a result of what event?
   A. AV valves opening
   B. AV valves closing
   C. Semilunar valves opening
   D. Semilunar valves closing

9. Choose the *false* statement about the circumflex artery.
   A. It is a branch of the left coronary artery.
   B. It provides the major blood supply to the right ventricle.
   C. It lies in a groove between the left atrium and left ventricle.
   D. Damage to this vessel would leave the left atrium with virtually no blood supply.

10. Choose the *false* statement about heart structure.
    A. The heart chamber with the thickest wall is the left ventricle.
    B. The apex of the heart is more superior in location than the base.
    C. The heart has four chambers.
    D. The left ventricle forms the apex and most of the left border of the heart.

11. All of the following are defects involved in Tetralogy of Fallot *except:*
    A. Ventricular septal defect
    B. Right ventricular hypertrophy
    C. Stenosed mitral valve
    D. Aorta emerging from both ventricles

12. Choose the *true* statement.
    A. The T wave is associated with atrial depolarization.
    B. The normal P-R interval is about 0.4 sec.
    C. Myocardial infarction means strengthening of heart muscle.
    D. Myocardial infarction is commonly known as a "heart attack" or a "coronary."

13. Choose the *false* statement.
    A. Pressure within the atria is known as intraarterial pressure.
    B. During most of ventricular diastole the semilunar valves are closed.
    C. During most of ventricular systole the semilunar valves are open.
    D. Diastole is another name for relaxation of heart muscle.

14. Which of the following structures are located in ventricles?
    A. Papillary muscles
    B. Fossa ovalis
    C. Ligamentum arteriosum
    D. Pectinate muscles

15. The heart is composed mostly of:
    A. Epithelium
    B. Muscle
    C. Dense connective tissue

Questions 16–25: Circle T (true) or F (false). If the statement is false, change the underlined word or phrase so that the statement is correct.

T F   16. The blood in the left chambers of the heart contains <u>higher</u> oxygen content than blood in the right chambers.

T F   17. At the point when intraarterial pressure surpasses ventricular pressure, semilunar valves <u>open</u>.

T F   18. The pulmonary <u>artery carries</u> blood from the lungs to the left atrium.

T F   19. The normal cardiac cycle <u>does not</u> require direct stimulation by the autonomic nervous system.

T F   20. During <u>about half</u> of the cardiac cycle, atria and ventricles are contracting simultaneously.

Questions 21–25: Fill-ins. Complete each sentence with the word or phrase that best fits.

_____ 21. Most ventricular filling occurs during atrial _____.

_____ 22. _____ is a cardiovascular disorder that is the leading cause of death in the U.S.

_____ 23. An ECG is a recording of _____ of the heart.

_____ 24. _____ is a term that means abnormality or irregularity of heart rhythm.

_____ 25. Are atrioventricular and semilunar valves ever open at the same

time during the cardiac cycle? _____

ANSWERS TO MASTERY TEST: CHAPTER 20

Arrange

1. C A B D
2. D E A B C
3. C D E A B

Multiple Choice

4. D	10. B
5. D	11. C
6. C	12. D
7. B	13. A
8. D	14. A
9. B	15. B

True–False

16 T
17. F. Close
18. F. Veins carry
19. T
20. F. No part

Fill-ins

21. Diastole
22. Coronary artery disease
23. Electrical changes or currents that precede myocardial contractions
24. Arrhythmia or dysrhythmia
25. No

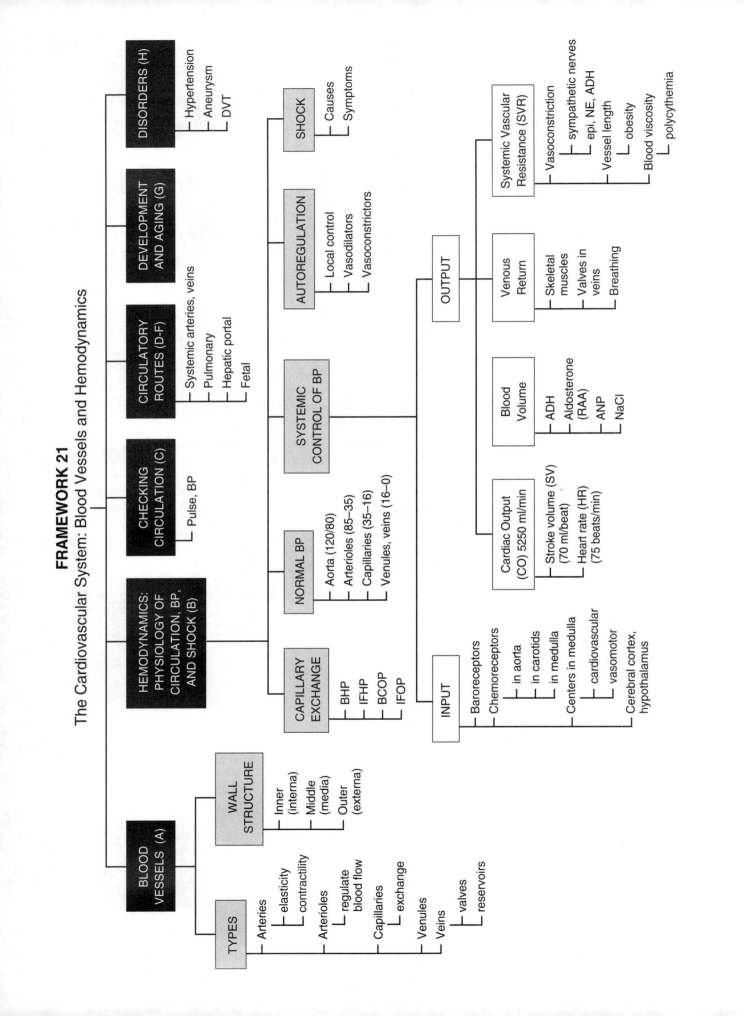

FRAMEWORK 21

The Cardiovascular System: Blood Vessels and Hemodynamics

BLOOD VESSELS (A)

TYPES
- Arteries
  - elasticity
  - contractility
- Arterioles
  - regulate blood flow
- Capillaries
  - exchange
- Venules
- Veins
  - valves
  - reservoirs

WALL STRUCTURE
- Inner (interna)
- Middle (media)
- Outer (externa)

HEMODYNAMICS: PHYSIOLOGY OF CIRCULATION, BP, AND SHOCK (B)

CAPILLARY EXCHANGE
- BHP
- IFHP
- BCOP
- IFOP

NORMAL BP
- Aorta (120/80)
- Arterioles (85–35)
- Capillaries (35–16)
- Venules, veins (16–0)

SYSTEMIC CONTROL OF BP

INPUT
- Baroreceptors
- Chemoreceptors
  - in aorta
  - in carotids
  - in medulla
- Centers in medulla
  - cardiovascular
  - vasomotor
- Cerebral cortex, hypothalamus

OUTPUT
- Cardiac Output (CO) 5250 ml/min
  - Stroke volume (SV) (70 ml/beat)
  - Heart rate (HR) (75 beats/min)
- Blood Volume
  - ADH
  - Aldosterone (RAA)
  - ANP
  - NaCl
- Venous Return
  - Skeletal muscles
  - Valves in veins
  - Breathing
- Systemic Vascular Resistance (SVR)
  - Vasoconstriction
    - sympathetic nerves
    - epi, NE, ADH
  - Vessel length
    - obesity
  - Blood viscosity
    - polycythemia

AUTOREGULATION
- Local control
- Vasodilators
- Vasoconstrictors

SHOCK
- Causes
- Symptoms

CHECKING CIRCULATION (C)
- Pulse, BP

CIRCULATORY ROUTES (D-F)
- Systemic arteries, veins
- Pulmonary
- Hepatic portal
- Fetal

DEVELOPMENT AND AGING (G)

DISORDERS (H)
- Hypertension
- Aneurysm
- DVT

C H A P T E R **21**

The Cardiovascular System: Blood Vessels and Hemodynamics

Blood is pumped by heart action to all regions of the body. The network of vessels is organized much like highways, major roads, side streets, and tiny alleys to provide interconnecting routes through all sections of the body. In city traffic, tension mounts when traffic is heavy or side streets are blocked. Similarly, pressure builds (arterial blood pressure) when volume of blood increases or when constriction of small vessels (arterioles) occurs. In a city, traffic may be backed up for hours, with vehicles unable to deliver passengers to destinations. A weakened heart may lead to similar and disastrous effects, backing up fluid into tissues (edema) and preventing transport of oxygen to tissues (ischemia, shock, and death of tissues).

Careful map study leads to trouble-free travel along the major arteries of a city. A similar examination of the major arteries and veins of the human body facilitates an understanding of the normal path of blood as well as clinical applications, such as the exact placement of a blood pressure cuff over the brachial artery or the expected route of a deep venous thrombus (pulmonary embolism) to the lungs.

As you continue your study of the cardiovascular system, carefully examine the Chapter 21 Topic Outline and Objectives; check off each one as you complete it. To organize your study of blood vessels and hemodynamics, glance over the Chapter 21 Framework now. Be sure to refer to the Framework frequently and note relationships among key terms in each section.

TOPIC OUTLINE AND OBJECTIVES

A. Anatomy of blood vessels and capillary exchange

1. Contrast the structure and function of arteries, arterioles, capillaries, venules, and veins.
2. Discuss the various pressures involved in the movement of fluids between capillaries and interstitial spaces.

B. Hemodynamics: factors affecting circulation and blood pressure; shock

3. Explain the factors that regulate the velocity and volume of blood flow.
4. Explain how blood pressure changes throughout the cardiovascular system, and describe the factors that determine mean arterial blood pressure.

5. Describe the factors that determine systemic vascular resistance.
6. Describe how blood pressure is regulated.
7. Define shock, and describe the four types of shock.

C. Evaluating circulation

8. Define pulse, and define systolic, diastolic, and pulse pressure.

D. Circulatory routes: systemic arteries

9. Describe the two main circulatory routes in adults.
10. Identify the four principal divisions of the aorta and locate the major arterial branches arising from each division.
11. Identify the two primary arterial branches of the ascending aorta.

12. Identify the three principal arteries that branch from the arch of the aorta.
13. Identify the visceral and parietal branches of the thoracic aorta.
14. Identify the visceral and parietal branches of the abdominal aorta.
15. Identify the two major branches of the common iliac arteries.

E. Circulatory routes: systemic veins

16. Identify the three systemic veins that return deoxygenated blood to the heart.
17. Identify the three major veins that drain blood from the head.
18. Identify the principal superficial veins that drain the upper limbs.
19. Identify the components of the azygos system of veins.
20. Identify the principal veins that drain the abdomen and the pelvis.
21. Identify the principal superficial and deep veins that drain the lower limbs.

F. Circulatory routes: hepatic portal, pulmonary, and fetal circulation

22. Identify the blood vessels and route of the hepatic portal circulation.
23. Identify the blood vessels and route of the pulmonary circulation.
24. Identify the blood vessels and route of the fetal circulation.

G. Developmental anatomy of blood vessels and blood; aging of the cardiovascular system

25. Describe the development of blood vessels and blood.
26. Explain the effects of aging on the cardiovascular system.

H. Disorders, medical terminology

WORDBYTES

Now become familiar with the language of this chapter by studying each wordbyte, its meaning, and an example of its use within a term. After you study the entire list, self-check your understanding by writing the meaning of each wordbyte on the line. As you continue through the *Learning Guide*, identify (and fill in) additional terms that contain the same wordbyte.

Wordbyte	Self-check	Meaning	Example(s)
hepat-	_____	liver	*hepat*ic artery
med-	_____	middle	tunica *med*ia
phleb-	_____	vein	*phleb*itis
port-	_____	carry	*port*al vein
retro-	_____	backward, behind	*retro*peritoneal
sphygmo-	_____	pulse	*sphygmo*manometer
tunica	_____	sheathe	*tunica* intima
vaso-	_____	vessel	*vaso*constriction
ven-	_____	vein	*ven*ipuncture

CHECKPOINTS

A. Anatomy of blood vessels and capillary exchange (pages 670–678)

A1. Review types of vessels in the human circulatory system by referring to Figure LG 20.3 and Chapter 20, Checkpoint C1, pages LG 413–414.

✓ **A2.** Check your understanding of the structure of blood vessels by matching terms in the box to related descriptions.

L.	**Lumen**	**TM.**	**Tunica media**
TE.	**Tunica externa**	**VV.**	**Vasa vasorum**
TI.	**Tunica intima**		

_____ a. The opening in blood vessels through which blood flows; narrows in atherosclerosis

_____ b. Formed largely of endothelium and basement membrane, it lines lumen; the only layer in capillaries

_____ c. Tiny blood vessels that carry nutrients to walls of blood vessels

_____ d. Composed of muscle and elastic tissue; contraction permits vasoconstriction

_____ e. Strong outer layer composed mainly of elastic and collagenous fibers

✓ **A3.** Match the names of vessels in the box with descriptions below. Answers may be used once, more than once, or not at all.

Aorta and other large arteries	**Medium-sized arteries**
Arterioles	**Small veins**
Capillaries	**Venules**
Inferior vena cava	

a. Sites of gas, nutrient, and waste exchange with tissues: _____

b. Play primary role in regulating moment-to-moment distribution of blood and in

   regulating blood pressure: _____

c. Sinusoids of liver are wider, more leaky versions of this type of vessel: _____

d. Reservoirs for about 60% of the volume of blood in the body; vasoconstriction (due to sympathetic impulses) permits redistribution of blood stored here (three answers):

   _____ _____ _____

e. Elastic arteries such as carotids or subclavians: _____

f. Flow of blood through here is called microcirculation: _____

g. Muscular arteries such as brachial or radial arteries: _____

A4. Contrast terms in the following pairs:

a. Structure of the tunica media of the following vessels:
   conducting arteries/distributing arteries

   artery/vein

b. Number of capillaries in *muscles and liver/tendons and ligaments*

c. Functions of *thoroughfare channel/true capillary*

d. Structure of *continuous capillary/fenestrated capillary*

✓ **A5.** Refer to Figure 21.4 in your text, and identify structural features of different types of capillaries and sinusoids that permit passage of materials across the vessel wall by four different routes. Select answers from the box.

Direct movement across endothelial membranes	**Intercellular clefts**
Fenestrations	**Pinocytic vesicles**

a. Structural features of very thin endothelial membrane with thin, incomplete, or even

absent basement membrane, for example, in sinusoids: _____

b. Gaps known as _____ between neighboring endothelial cells; may be large, for example, to permit passage of large proteins made in liver cells into the blood in liver sinusoids.

c. Pores known as _____ in plasma membranes of endothelial cells, for example, in cells of kidneys, intestinal villi, and choroid plexuses of the brain. The tight seal of the blood–brain barrier is attributed partly to lack of such pores.

d. _____ permitting endocytosis and exocytosis for transport of large molecules that cannot cross capillary walls in any other way.

✓ **A6.** Complete this exercise about the structure of veins.
a. Veins have *(thicker? thinner?)* walls than arteries. This structural feature relates to the fact that the pressure in veins is *(more? less?)* than in arteries. The pressure difference is demonstrated when a vein is cut; blood leaves a cut vein in *(rapid spurts? an even flow?)*.
b. Gravity exerts back pressure on blood in veins located inferior to the heart. To counteract this, veins contain _____ .
c. When valves weaken, veins become enlarged and twisted. This condition is called _____. This occurs more often in *(superficial? deep?)* veins. Why?

d. A vascular (venous) sinus has _____

_____

replacing the tunica media and tunica externa. So sinuses have the *(structure but not function? function but not structure?)* of veins. List two places where such sinuses are found.

A7. Define the following terms and explain how they are protective to the body:

a.　Anastomosis

b.　Collateral circulation

✓　**A8.** Complete this Checkpoint on distribution of blood. It may help to refer to Figure 21.6 in your text.

a.　At rest, *(arteries and arterioles? capillaries? venules and veins?)* contain most of the blood in the body. In which vessels is the smallest percentage of blood found in the resting person? *(Arteries and arterioles? Capillaries? Venules and veins?)*

b.　*(Sympathetic? Parasympathetic?)* nerve impulses to venules and veins cause vasoconstriction and release of blood from these vessels. Name two locations of venous "reservoirs" that can be activated to release blood when needed.

c.　State two examples of circumstances that might activate distribution of reservoir

　　blood. _____ _____

✓　**A9.** Identify the types of transport mechanisms involved in capillary exchange by doing this exercise. It may help to refer back to Checkpoint A5. Use these answers:

B.	**Bulk flow**	**D.**	**Diffusion**	**V.**	**Vesicular transport**

_____ a.　Passage of lipid-soluble chemicals, such as oxygen, carbon dioxide, and steroid hormones, through the phospholipid bilayer of capillary membrane endothelial cells

_____ b.　Transport of water-soluble substances, such as glucose or amino acids, through fenestrations or intercellular clefts

_____ c.　Movement of large plasma proteins, such as fibrinogen or albumin, made in liver cells through intercellular clefts in sinusoids to reach plasma

_____ d.　Movement of antibody proteins from maternal blood to fetal blood

_____ e.　Passive movements of large numbers of ions, molecules, and particles in the same direction, largely due to opposing forces of blood pressure and osmotic pressure

✓　**A10.** Recall that pressure is the principal factor causing blood to move through the vessels of the body. Pressure is also the major factor determining movement of substances across capillary membranes. Refer to Figure LG 21.1 and do this activity.

a.　Color arrows A–D on Figure LG 21.1 to identify the four pressures involved in capillary exchange.

b.　These pressures interact in a kind of tug-of-war that results in a net filtration

　　pressure (NFP) of _____ mm Hg. In other words, at the arterial end of the capillary, substances tend to move *(into? out of?)* the vessel.

○ BHP (blood hydrostatic pressure)
○ BCOP (blood colloid osmotic pressure)
○ IFHP (interstitial fluid hydrostatic pressure
○ IFOP (interstitial fluid osmotic pressure)

Figure LG 21.1 Practice diagram showing forces controlling capillary exchange and for calculation of NFP at arterial end of capillary. Numbers are pressure values in mm Hg. Color arrows as directed in Checkpoint A10.

c. Which value is substantially different at the venous end of capillaries? _____

Calculate the NFP there: _____ mm Hg

d. The near-equilibrium at the two ends of capillaries is based on _____'s law of the capillaries. Which system "mops up" the slight amount of fluid that is

not drawn back into veins? _____

e. Excessive amounts of fluid accumulating in interstitial spaces is the condition known

as _____. This may result from high blood pressure in which *(BCOP? BHP?)* is increased, loss of plasma protein in which *(BCOP? BHP?)* is decreased, or *(increase? decrease?)* in capillary permeability.

✓ **A11.** *Critical thinking.* On a random review of patient charts from a hospital, three clients were noted as having edema. Identify the physiological cause that was most likely the cause of each case of edema. Select from these answers:

↑	**BHP.**	**Increased blood hydrostatic pressure**
↑	**PC.**	**Increased permeability of capillaries**
↑	**ECV.**	**Increased extracellular volume**
↓	**PP.**	**Decreased concentration of plasma proteins**

_____ a. Paul P, age 72, has a diagnosis of congestive heart failure and experiences what he describes as "spells when I just can't get my breath."

_____ b. Tommy R, age 7, received second-degree burns on his left leg, which is swollen and seeping fluids.

_____ c. Elaine B, age 41, has kidney failure; she has gained 11 pounds since her last hemodialysis treatment 3 days ago. She reported that she "went overboard eating and drinking at a couple of parties over the weekend." (Two answers)

B. Hemodynamics: factors affecting circulation and blood pressure; shock (pages 678–689)

✓ **B1.** Check your understanding of circulation of blood in this Checkpoint.

   a. The typical human body contains about _____ liters of blood. At rest, this same amount of blood is pumped out by the heart each minute, as cardiac

   _____ (5 liters/minute).

   b. Now name the two factors that determine how rapidly and where this blood is distributed, for example, rapid delivery to brain and muscles, or slow delivery to intestine and skin. (See Checkpoint B4 for more on this topic.)

✓ **B2.** Do this exercise about blood flow.

   a. Figure 21.8 in your text shows that velocity of blood flow is greatest in *(arteries? capillaries? veins?)*. Flow is slowest in _____. As a result, ample time is available for exchange between capillary blood and tissues.

   b. As blood moves from capillaries into venules and veins, its velocity *(increases? decreases?)*, enhancing venous return.

   c. Although each individual capillary has a cross-sectional area which is microscopic, the body contains so many capillaries that their total cross-sectional area is enormous. (For comparison, look at the eraser end on one pencil; you are seeing the cross-sectional area of that one pencil. If you bind together 100 pencils with a strong rubber band and observe their eraser ends as a group, you see that the individual pencils contribute to a relatively large total cross-sectional area.) All of the capillaries

   in the body have a total cross-sectional area of about _____ cm^2.

   d. *(A direct? An indirect?)* relationship exists between cross-sectional area and velocity of blood flow. This is evident in the aorta also. This vessel has a cross section of only

   about _____ cm^2 and has a very *(high? low?)* velocity.

   e. *A clinical correlation.* Mr. Fey has an aortic aneurysm. The diameter of his aorta has increased from normal size to about 10 cm. The region of the aorta with the aneurysm therefore has a *(larger? smaller?)* cross-sectional area. One would expect the velocity of Mr. Fey's aortic blood flow to *(in? de?)*-crease. This change may put

   Mr. Fey at risk for _____.

   f. On the average, *(1? 5? 10?)* minute(s) is normally required for a blood cell to travel one complete circulatory circuit, for example, heart → lung → heart → foot → heart.

✓ **B3.** Complete this activity on blood pressure (BP) and mean arterial blood pressure (MABP).

   a. In the aorta and brachial artery, blood pressure (BP) is about 120 mm Hg immediately following ventricular contraction. This is called *(systolic? diastolic?)* BP. As ventricles

   relax (or go into _____), blood is no longer ejected into these arteries. However, the normally *(elastic? rigid?)* walls of these vessels recoil against

   blood, pressing it onward with a diastolic BP of _____ mm Hg.

   b. In a blood pressure (BP) of 120/80, the average of the two pressures (systolic and diastolic) is 100. However the MABP (average or mean blood pressure in arteries) for

   a BP of 120/80 is actually slightly less than 100; it is about _____.
   Reason: during the typical cardiac cycle (0.8 sec), the ventricles are in *(systole? diastole?)* for about two-thirds of the cycle and in systole for only about one-third of

the cycle (see Figure LG 20.5, page LG 419). As a result, the MABP is always slightly closer to the value of the *(systolic? diastolic?)* BP.

c. Try calculating MABP for a person with an arterial BP of 120/80. First write the

diastolic BP, _____ mm Hg. Then find the difference between systolic BP and

diastolic BP a value known as *pulse pressure*: 120 mm Hg – 80 mm Hg = _____ mm Hg. Finally, add one-third of pulse pressure to diastolic BP to arrive at MABP:

$BP_{diastolic}$ + ($^1/_3$ × pulse pressure) = MABP

_____ mm Hg + ($^1/_3$ × _____ mm Hg) = _____ mm Hg

✓ **B4.** Recall that in Checkpoint B1 you named two factors that determine distribution (or circulation) of cardiac output to the body. Closely examine these two factors in this Checkpoint.

a. One factor is _____ (such as blood pressure or MABP). Define blood

pressure: the pressure exerted by _____.
Picture the heart pumping harder and faster, much like a dam permitting more water (cardiac output) to pass over it so the river below swells; water (blood) presses forward and laps up against river banks (BP increases). So if cardiac output increases,

BP or MABP are likely to _____-crease. Finish the arrow to show the correct relationship:

↑ CO → | MABP. On the other hand, if CO decreases (as in hemorrhage or shock),

then MABP is likely to _____-crease.

b. Now visualize the effects of the second factor (resistance) on distribution of blood. Imagine yourself inside a red blood cell (much like a boat) surging forward at high velocity through the aorta or other large arteries. BP is high there (MABP of 93) as blood flows along freely. But as you enter a small artery or arteriole, your RBC encounters *(more? less?)* resistance to flow as the RBC repeatedly bounces up against and snags the walls of this narrow-diameter vessel. In other words, BP in any given blood vessel is *(directly? inversely?)* related to resistance to flow, while *(directly? inversely?)* related to cardiac output:

$$BP = \frac{CO}{R}$$

c. Because CO stays the same within the body at any given time, we can conclude that if resistance to flow is high through a vessel (such as an arteriole or capillary), that BP in that vessel will be *(high? low?)*. Referring to Figure 21.9 in your text, write the normal

ranges of BP values for the following types of vessels: arterioles, _____ mm Hg; capillar-

ies, _____ mm Hg; venules and veins, _____ mm Hg; venae cavae, _____ mm Hg.

d. On Figure 21.9 the steepest decline in the pressure curve occurs as blood passes through the *(aorta? arterioles? capillaries?)*. This indicates that the greatest resistance to flow is present in the *(aorta? arterioles? capillaries?)*.

✓ **B5.** Do this exercise on two very important and different relationships between resistance (R) and blood pressure.

a. In Checkpoint B4b, you saw that increase in resistance (R) *within a given blood vessel*

causes _____-crease in *BP in that vessel.* Therefore, BP (within a vessel) and R (in that same vessel) are inversely related. Show this by completing the arrow.

$\uparrow R_{\text{within a vessel}} \rightarrow | \ BP_{\text{in that same vessel}}$

For example, the RBC (boat) passing from an artery (wide river) into an arteriole (tributary) meets more resistance to flow in this narrow tributary, thereby _____-creasing BP *within that vessel*. So the increased resistance (R) blood meets as it passes through arterioles causes BP there to decrease from about 85 to 35 mm Hg.

b. However, several factors contribute to changes in *BP in systemic arteries* (usually measured in the brachial artery), for example, increase in BP from a normal

_____/_____ to a high BP of 162/98. One of these factors altering systemic BP is resistance. But here the resistance refers to inhibition of flow through many, many small systemic vessels, for example, by sympathetic nerve stimulation of arterioles

causing them to narrow. So the term SVR (_____ vascular

_____) is used to describe this resistance. Because less blood is engaged in runoff into arterioles, more blood stays in main arteries, thereby increasing systemic BP. Complete arrows to show this:

$\uparrow R_{\text{within many systemic arterioles, as in skin or abdomen}} \rightarrow | \ BP_{\text{in systemic arteries}}$

or

$\uparrow SVR \rightarrow | \ BP_{\text{systemic arterial}}$

Consider this analogy: if blood (water) cannot pass from an artery (wide river) into many arterioles (tributaries) because entranceways are blocked by vasoconstriction of arterioles (tributaries are clogged with silt and debris), then blood (water) will remain

in the main artery (wide river), _____-creasing BP *in main systemic arteries* (wide rivers).

c. In fact, systemic BP tends to increase if either cardiac output or systemic vascular resistance (SVR) or both of those factors increase. Show this by completing the equation:

$BP_{\text{systemic arterial}} = \underline{\hspace{2cm}} \times \underline{\hspace{2cm}}$

or

MABP = cardiac output × systemic vascular resistance

d. Describe three factors that may increase SVR leading to increased systemic BP:

1. _____-crease in blood viscosity, for example, due to _____

2. _____-crease in total length of blood vessels in the body, for example, due to

_____

3. _____-crease in diameter (or radius) of blood vessels, for example, by

_____ of arterioles

e. SVR is also called total _____ resistance because the vessels that are targeted to vasoconstrict (decrease radius) and create that resistance are those in

regions of the body (such as skin and _____) that could be considered more "peripheral," or less crucial, to survival.

✓ **B6.** *A clinical correlation.* Do this exercise about Mr. Tyler, who has been depressed since his wife died two years ago and has gained 50 pounds during that time.
   a. Which one of the following factors that increase systemic vascular resistance is most likely to be higher in Mr. Tyler?
      A. Blood viscosity
      B. Blood vessel length
      C. Blood vessel radius
      This factor *(does? does not?)* put Mr. Tyler at higher risk for hypertension.
   b. Because Mr. Tyler's stress level is often high, his *(sympathetic? parasympathetic?)* nerves are likely to be more active. Which of the three factors (A–C) above is most likely to be affected? *(A? B? C?)* If his arterioles decrease in diameter by 50%, the resistance to

      flow increases _____ times. This factor is likely to _____-crease his blood pressure.

   B7. Explain how these two factors increase venous return during exercise.
   a. Skeletal muscle pump (For help, refer to Chapter 20, Checkpoint F4c, page LG 423.)

   b. Respiratory pump

✓▲ **B8.** Imagine that your blood pressure (BP) is slightly lower than your body needs right now. Do this exercise to see how your body is likely to respond to increase your blood pressure.
   a. Refer to Figure LG 21.2 and fill in box 1 to describe types of receptors that trigger changes in blood pressure. Baroreceptors are sensitive to changes in

      _____ _____, and chemoceptors are activated by

      chemicals such as _____. These receptors are located in two locations:

      _____ and _____.
   b. Nerve messages from the carotid sinuses travel to the brain via cranial nerves

      _____, whereas messages from the aorta travel by cranial nerves _____. Write in these cranial nerves (C.N.) on the figure.
   c. This *input* reaches the cardiovascular center located within the _____ of the brain. Fill in box 3 to indicate this region of the brain. This center receives input from other parts of the brain also. Note three of these brain parts in box 2 in the figure.
   d. Consider the cardiovascular center as comparable to the control center of an automobile. The center has an accelerator portion (like a gas pedal) and an inhibitor portion

      (like the _____ of a car). When baroreceptors report that BP is too low (as if the car is running too slowly), the *(cardiostimulatory? cardioinhibitory?)* center must be activated; in the car analogy, you would *(step on the gas pedal? step on the brakes?).* At the same time, the cardioinhibitory center must be deactivated; continuing

      with the car analogy, what response do you make? _____
      What is the outcome for the car/heart?

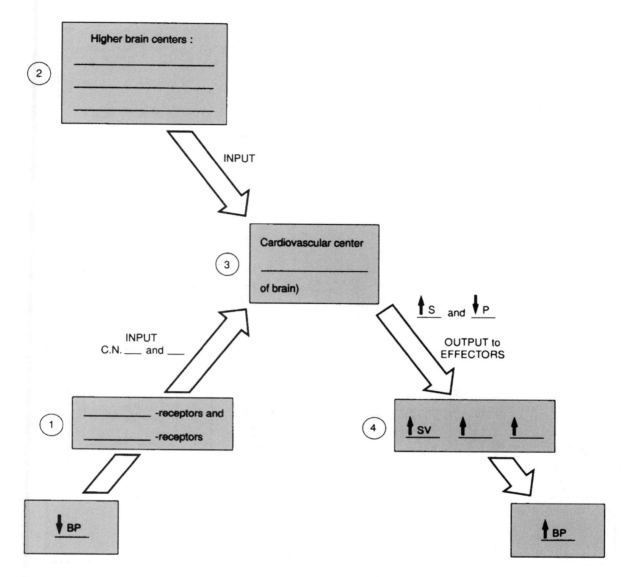

Figure LG 21.2 Overview of input and output regulating blood pressure. Complete as directed in Checkpoint B8.

e. In addition, a vasomotor portion of the cardiovascular center responds to lower blood pressure by *(stimulating? inhibiting?)* smooth muscle of blood vessels, resulting in

(*vasoconstriction? vasodilatation ?*). This helps to increase BP by _____-creasing systemic vascular resistance (SVR).

f. Summarize nervous output to effectors from the cardiovascular center by writing in box 4 in Figure LG 21.2 the three factors that are altered. The result of this output is

an attempt to restore homeostasis by _____-creasing your blood pressure.

g. Overall, Figure LG 21.2 demonstrates *(positive? negative?)* feedback mechanisms for regulating BP, because a decrease in BP initiates factors to increase BP.

✓ **B9.** On Figure LG 21.3 fill in blanks and complete arrows to show details of all of the nervous output, hormonal action, and other factors that increase blood pressure.

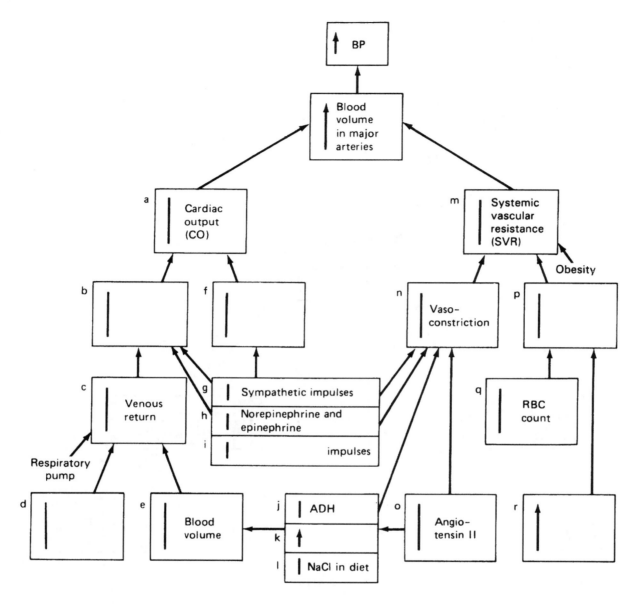

Figure LG 21.3 Summary of factors controlling blood pressure (BP).
Arrows indicate increase (↑) or decrease (↓). Complete as directed in Checkpoint B9.

✓ **B10.** *A clinical correlation.* Complete this exercise about blood pressure disorders.

  a. Hypertension means *(high? low?)* blood pressure. It may be caused by an excess of
    any of the factors shown in Figure LG 21.3. If directions of arrows in Figure LG 21.3
    are reversed, blood pressure can be *(increased? decreased?)*.

  b. Circle all the factors listed below that tend to decrease blood pressure.
    A. Increase in cardiac output, as by increased heart rate or stroke volume
    B. Increase in (vagal) impulses from cardioinhibitory center
    C. Decrease in blood volume, as following hemorrhage
    D. Increase in blood volume, by excess salt intake and water retention
    E. Increased systemic vascular resistance due to vasoconstriction of arterioles (blood
      prevented from leaving large arteries and entering small vessels)
    F. Decrease in sympathetic impulses to the smooth muscle of arterioles
    G. Decreased viscosity of blood via loss of blood protein or red blood cells, as by
      hemorrhage
    H. Use of medications called dilators because they dilate arterioles, especially in areas
      such as abdomen and skin

✓ **B11.** *For extra review.* Do the following Checkpoint on additional details on blood pressure regulation.

   a. Vasomotor _____ refers to a moderate level of vasoconstriction normally maintained by the medulla and sympathetic nerves. A sudden loss of sympathetic nerve output, for example, by head trauma or emotional trauma, can reduce this tone and cause BP to *(skyrocket? drop precipitously?)*. This may lead to

   _____ (fainting) or even to shock. This can also occur when pressure is exerted on the carotid sinus, for example by a tight collar.

   b. Smooth muscle of most small arteries and arterioles contains *(alpha? beta?)* receptors;

   in response to the neurotransmitter _____, these vessels vaso-*(constrict? dilate?)*. Sympathetic stimulation of most veins causes them to vaso-

   (constrict? dilate?), thereby _____-creasing venous return.

   c. Vasodilator medications therefore tend to _____-crease BP as blood pools in nonessential areas (such as GI tract or skin). Name two vasodilator chemicals that

   may be produced by the body: _____ from cells of the atria, and

   _____ from mast cells during inflammation.

   d. Baroreceptors involved with the aortic reflex primarily maintain *(cerebral? general systemic?)* BP, whereas receptors that trigger the carotid sinus reflex primarily control

   _____ BP.

   e. Carotid bodies and aortic bodies are sites of *(baro? chemo?)*-receptors that provide

   input for control of BP. Low oxygen (known as _____) or high

   carbon dioxide (called _____) will tend to cause *(vasoconstriction?*

   vasodilatation?) and consequently _____-crease BP. As a result, blood can move more rapidly to the lungs to pick up oxygen and give up carbon dioxide.

   f. Write ↑ or ↓ next to each hormone or other chemical listed below to indicate whether the chemical increases or decreases blood pressure.

   _____1. ADH

   _____2. Angiotensin II

   _____3. Aldosterone

   _____4. Epinephrine of NE

   _____5. ANP

✓ **B12.** Describe the process of autoregulation in this activity.

   a. Autoregulation is a mechanism for regulating blood flow through specific tissue areas. Unlike cardiac and vasomotor mechanisms, autoregulation occurs by *(local? autonomic nervous system?)* control.

   b. In other words, when a tissue (such as an active muscle) is hypoxic, that is, has a *(high? low?)* oxygen level, the cells of that tissue release *(vasoconstrictor? vasodilator?)*

   substances. These may include _____ acid, built up by active muscle. Other products of metabolism that serve as dilator substance include:

c. Compared to systemic vessels, pulmonary blood vessels respond *(similarly to? just the opposite from?)* hypoxic conditions. Arterioles in hypoxic parts of lungs will vaso-

_____. How can this mechanism be helpful to the body?

d. Recall from Chapter 19 that the initial step in hemostasis is *(vasodilation? vasospasm?)*. Name two chemicals that produce this effect.

✓ **B13.** Describe cause of shock and compensation for shock in this activity.
   a. Shock is said to occur when tissues receive inadequate supply to meet their demands for oxygen, nutrients, and waste removal. Shock resulting from decrease in circulat-

ing blood volume is known as _____ shock. Write two or more causes of this type of shock.

   b. Compensatory mechanisms activate *(sympathetic? parasympathetic)* nerves that

_____-crease both heart rate and systemic vascular resistance. These attempts to

increase BP normally work for blood losses up to _____% of total blood volume. One common indicator of this compensatory mechanism is *(warm, pink, and dry? cool, pale,*

and clammy?) skin related to vaso-_____ and _____-creased sweating.

   c. Chemical vasoconstrictors include _____, released from the adrenal

medulla; angiotensin II, formed when _____ is secreted by kidneys;

_____ made by the adrenal cortex; and the posterior pituitary

hormone, _____ .
   d. Which hormones cause water retention which helps to increase BP in shock?

   e. How does hypoxia help to compensate for shock?

✓ **B14.** *A clinical correlation.* Write arrows to indicate signs and symptoms of shock.
      (↑ = increase; ↓ = decrease)

   a. Sweat _____   d. Blood pH _____

   b. Heart rate (pulse) _____   e. Mental status _____

   c. BP _____   f. Urine volume _____

C. Evaluating circulation (pages 689–690)

C1. Define *pulse*.

▲ **C2.** In general, where may pulse best be felt? List six places where pulse can readily be palpated. Try to locate pulses at these points on yourself or on a friend.

✓ **C3.** Normal pulse rate is about _____ beats per minute. *Tachycardia* means *(rapid? slow?)*

pulse rate; _____ means slow pulse rate.

✓▲ **C4.** Describe the method commonly used for taking blood pressure by answering these questions about the procedure.
   a. Name the instrument used to check BP. _____

   b. The cuff is usually placed over the _____ artery. How can you tell that this artery is compressed after the bulb of the sphygmomanometer is squeezed?

   c. As the cuff is deflated, the first sound heard indicates _____ pressure because reduction of pressure from the cuff below systolic pressure permits blood to begin spurting through the artery during diastole.
   d. Diastolic pressure is indicated by the point at which the sounds _____. At this time blood can once again flow freely through the artery.
   e. Sounds heard while taking blood pressure are known as _____ sounds.

✓ **C5.** *A clinical correlation.* Mrs. Yesberger has a systolic pressure of 128 mm Hg and diastolic of 78 mm Hg. Write her blood pressure in the form in which BP is usually

expressed: _____ Determine her pulse pressure.

Would this be considered a normal pulse pressure? _____

D. Circulatory routes: systemic arteries (pages 690–709)

D1. On Figure LG 20.3 (page LG 413), label the *pulmonary artery, pulmonary capillaries,* and a *pulmonary vein.* Briefly contrast pulmonary and systemic vessels.

✓▲ **D2.** Use Figure LG 21.4 to do this learning activity about systemic arteries.

   a. The major artery from which all systemic arteries branch is the _____. It exits from the chamber of the heart known as the *(right? left?)* ventricle. The aorta

   can be divided into three portions. They are the _____ (N1 on the

   figure), the _____ (N2), and the _____ (N3).

   b. The first arteries to branch off the aorta are the _____ arteries. Label

   these (at O) on the figure. These vessels supply blood to the _____.

Right side
of body

Left side
of body

A

B

C

D

E

F

G

H

I

J

K

L

M

N

O

P

Q

R

S

T

U

V

W

X

2

1   3

Figure LG 21.4 Major systemic arteries. Identify letter labels as directed
in Checkpoints D2, D3, and D7–D9.

c. Three major arteries branch from the aortic arch. Write the names of these (at K, L, and M) on the figure, beginning with the vessel closest to the heart.

d. Locate the two branches of the brachiocephalic artery. The right subclavian artery

(letter _____) supplies blood to _____.

Vessel E is named the *(right? left?)* _____ artery. It supplies the right side of the head and neck.

e. As the subclavian artery continues into the axilla and arm, it bears different names (much as a road does when it passes into new towns). Label G and H. Note that H is

the vessel you studied as a common site for measurement of _____.

f. Label vessels I and J, and then continue the drawing of these vessels into the forearm and hand. Notice that they anastomose in the hand. Vessel *(I? J?)* is often used for checking pulse in the wrist. The deep palmar arch within the hand receives blood mainly from the *(radial? ulnar?)* artery, whereas the superficial palmar arch receives blood mainly from the *(radial? ulnar?)* artery. Both of the arches provide blood to

digital arteries that supply blood to _____.

g. Besides supplying blood to the upper limbs the subclavian arteries, each send a branch that ascends the neck through foramina in cervical vertebrae. This is the

_____ artery. The right and left vertebral arteries join at the base of the

brain to form the _____ artery. (Refer to D and C on Figure LG 21.5.)

✓▲ **D3.** The right and left common carotid arteries ascend the neck in positions considerably *(anterior? posterior?)* to the vertebral arteries. Find a pulse in one of your own

KEY

A. Anterior communicating cerebral artery (1)
B. Internal carotid arteries (cut) (2)
C. Basilar artery (1)
D. Vertebral arteries (2)

E. Anterior cerebral arteries (2)
F. Middle cerebral arteries (2)
G. Posterior communicating cerebral arteries (2)
H. Posterior cerebral arteries (2)

Figure LG 21.5 Arterial blood supply to the brain including vessels of the cerebral arterial circle (of Willis). View is from the undersurface of the brain. Numbers in parentheses indicate whether the arteries are single (1) or paired (2). Refer to Checkpoints D2, D4, and D5.

carotids. These vessels each bifurcate (divide into two vessels) at about the level of your mandible. Identify the two branches (A and B) on Figure LG 21.4. Which supplies the brain and eye? *(Internal? External?)* What structures are supplied by the external carotids? (See Exhibit 21.3 page 697 in the text, for help.)

✓▲**D4.** Refer to Figure LG 21.5 and answer these questions about the vital blood supply to the brain.
   a. An anastomosis, called the cerebral arterial circle of _____, is located just inferior to the brain. This circle closely surrounds the pituitary gland.
   b. The circle is supplied with blood by two pairs of vessels: anteriorly, the *(internal?*

   external?) carotid arteries and, posteriorly, the _____ arteries (joined to form the basilar artery).
   c. The circle supplies blood to the brain by three pairs of cerebral vessels, identified on

   Figure LG 21.5 by letter labels _____ _____, and _____. Vessels A and G complete

   the circle; they are known as anterior and posterior _____ arteries. Now cover the key and correctly label each of the lettered vessels.

✓ **D5.** *Critical thinking.* Relate symptoms to the interruption of normal blood supply to the brain as described in each case. Explain your reasons.
   a. A patient's basilar artery is sclerosed. Which part of the brain is most likely to be deprived of normal blood supply? *(Frontal lobes of cerebrum? Occipital lobes of cerebrum, as well as cerebellum and pons?)*
   b. Paralysis and loss of some sensation are noted on a patient's right side. Angiography points to sclerosis in a common carotid artery. Which one is more likely to be sclerosed? *(Right? Left?)* Explain.

   c. The middle cerebral artery which passes between temporal and frontal lobe (see broken lines on Figure LG 21.5) is blocked. Which sense is more likely to be affected? *(Hearing? Vision?)* Explain.

   d. Anterior cerebral arteries pass anteriorly from the circle of Willis and then arch backward along the medial aspect of each cerebral hemisphere. Disruption of blood flow in the right anterior cerebral artery will be most likely to cause effects in the *(right? left?) (hand? foot?)*. *Hint:* The affected brain area would include region Z in Figure LG 15.1 and/or region R in Figure LG 15.2, pages LG 291 and 293. Note that these regions are on the medial aspect of the right hemisphere.

✓▲**D6.** Identify which category (listed in the box) fits each artery named below.

PA.	**Parietal abdominal**		**VA.**	**Visceral abdominal**
PT.	**Parietal thoracic**		**VT.**	**Visceral thoracic**

_____ a. Esophageal arteries   _____ c. Renal arteries

_____ b. Lumbar and inferior phrenic arteries   _____ d. Superior phrenic arteries

✓▲ **D7.** Complete the table describing visceral branches of the abdominal aorta. Identify these on Figure LG 21.4 (vessels P to U). Refer also to Figure LG 24.0, page LG 000.

Artery	Letter	Structures Supplied

a. Celiac (three major branches)
  1.
  2.
  3.
		1. Liver
		2.
		3.

b.	Q	Small intestine and part of large intestine and pancreas
c.		Adrenal (suprarenal) glands
d. Renal	S	
e.		Testes (or ovaries)
f. Inferior mesenteric		

✓▲ **D8.** The aorta ends at about level _____ vertebra by dividing into right and left

_____ arteries. Label these at V on Figure LG 21.4.

✓▲ **D9.** Label the two branches (W and X on Figure LG 21.4) of each common iliac artery. Name structures supplied by each.
  a. Internal iliac artery

  b. External iliac artery

✓   **D10.** *For extra review.* Suppose that the right femoral artery were completely occluded. Describe an alternate (or collateral) route that would enable blood to pass from the right common iliac artery to the little toe of the right foot.

   Right common iliac artery →

▲ **D11.** *For extra review.* Identify locations of pulses on these three arteries on yourself: femoral, popliteal, dorsalis pedis.

E. Circulatory routes: systemic veins (pages 709–725)

✓▲ **E1.** Name the three main vessels that empty venous blood into the right atrium of the

heart. _____ _____ _____

✓▲ **E2.** Do this exercise about venous return from the head.
  a. Venous blood from the brain drains into vessels known as _____.
     These are *(structurally? functionally?)* like veins.
  b. Name the sinus that lies directly beneath the sagittal suture: _____
  c. Blood from all of the cranial vascular sinuses eventually drains into the

     _____ veins, which descend in the neck. These veins are positioned

     close to the _____ arteries.
  d. Where are the jugular arteries and carotid veins located?

✓ **E3.** Imagine yourself as a red blood cell currently located in the *left brachial vein*. Answer
     these questions.
  a. In order for you to flow along in the bloodstream to the *left brachial artery*, which of
     these must you pass through? *(A brachial capillary? The heart and a lung?)*
  b. In order for you to flow from the left brachial vein to the *right arm*, must you pass
     through the heart? *(Yes? No?)* Both sides of the heart, that is, right and left? *(Yes?
     No?)* One lung? *(Yes? No?)*

✓▲ **E4.** Once you are familiar with the pathways of systemic arteries, you know much about
     the veins that accompany these arteries. Demonstrate this by tracing routes of a drop
     of blood along the following deep vein pathways.
  a. From the thumb side of the left forearm to the right atrium

  b. From the medial aspect of the left knee to the right atrium

  c. From the right kidney to the right atrium

✓▲ **E5.** Contrast veins with arteries in this exercise.
  a. Veins have *(higher? lower?)* blood pressure than arteries and a *(faster? slower?)* rate of
     blood flow.
  b. Veins are therefore *(more? less?)* numerous, in order to compensate for slower blood
     flow in each vein. Many of the "extra" veins consist of vessels located just under the

     skin (in subcutaneous fascia); these are known as _____ veins.

✓▲ **E6.** On text Figures 21.26 to 21.28 (pages 717-725), locate the following veins. Identify the location at which each empties into a deep vein.

a. Basilic _____

b. Median cubital _____

c. Great saphenous _____

d. Small saphenous _____

E7. *A clinical correlation.* Write a short essay describing clinical significance of the great saphenous veins. Be sure to use these three terms: *varicose, intravenous,* and *bypass.*

▲ **E8.** Veins connecting the superior and inferior venae cavae are named

_____ veins. Identify the azygos system of veins on Figure 21.27 in the text. What parts of the thorax do they drain? Explain how these veins may help to drain blood from the lower part of the body if the inferior vena cava is obstructed.

✓▲ **E9.** Name the visceral veins that empty into the abdominal portion of the inferior vena cava. Then color them blue on Figure LG 26.1 (page LG 579). Contrast accompanying systemic arteries by coloring them red on that figure. Note on Figure LG 26.1 that the inferior vena cava normally lies on the *(right? left?)* side of the aorta. One way to remember this is that the inferior vena cava is returning blood to the *(right? left?)* side of the heart.

F. Circulatory routes: hepatic portal, pulmonary, and fetal circulation (pages 726–731)

✓▲ **F1.** Refer to Figure LG 20.3 (page LG 413) and do the following activity.
a. Label the following vessels: *gastric, splenic,* and *mesenteric vessels.* Note that all of these vessels drain directly into the vessel named *(7, inferior vena cava? H, portal vein to the liver?).* Label that vessel on Figure LG 20.3.
b. Label the vessel that carries oxygenated blood into the liver.
c. In the liver, blood from both the portal vein and hepatic artery mixes as it passes

through tortuous capillary-like vessels known as _____. On Figure LG 20.3 label a vessel that collects this blood and transports it to the inferior vena cava.

    d. What functions are served by the special hepatic portal circulation?

✓ **F2.** Contrast pulmonary circulation with systemic circulation in this activity Write P (pulmonary) or S (systemic).

    a. Which is considered a lower resistance system? List structural features that explain this lower resistance.

    b. Which arteries are likely to have a higher blood pressure (BP)? Write typical BP

       values for these vessels: aorta _____ mm Hg; pulmonary artery _____ mm Hg.

    c. Which capillaries normally have a hydrostatic pressure (BP) of about 10 mm Hg? Write one consequence of increased pulmonary capillary BP.

    d. Which system has its blood vessels constrict in response to hypoxia? Write one advantage and one disadvantage of this fact.

  F3. What aspects of fetal life require the fetus to possess special cardiovascular structures not needed after birth?

✓ **F4.** Complete the following table (on page LG 455) about fetal structures. (Note that the structures are arranged in order of blood flow from fetal aorta back to fetal aorta.)

✓ **F5.** For each of the following pairs of fetal vessels, circle the vessel with higher oxygen content.

    a. Umbilical artery/umbilical vein
    b. Femoral artery/femoral vein
    c. Lower abdominal aorta/thoracic portion of inferior vena cava

Fetal Structure	Structures Connected	Function	Fate of Structure After Birth
a.		Carries fetal blood low in oxygen and nutrients and high in wastes	
b. Placenta	(Omit)		
c.	Placenta to liver and ductus venosus		
d.		Branch of umbilical vein, bypasses liver	
e. Foramen ovale			
f.			Becomes ligamentum arteriosum

G. Developmental anatomy of blood vessels and blood; aging of the cardiovascular system (pages 731–732)

G1. Write a paragraph describing embryonic development of blood vessels. Include the following key terms: *15 to 16 days, mesoderm, mesenchyme, blood islands, spaces, endothelial, tunica.*

G2. List four sites of blood formation in the embryo and fetus.

✓ **G3.** Draw arrows next to each factor listed below to indicate whether it increases (↑) or decreases (↓) with normal aging.

_____ a. Size and strength of cardiac muscle cells   _____ c. Total cholesterol and LDLs in blood

_____ b. Cardiac output (CO)   _____ d. Systolic blood pressure

H. Disorders, medical terminology (pages 732–734)

✓ **H1.** Match types of hypertension listed in the box with descriptions below. One answer will be used twice; one answer will not be used.

Pri.	Primary	Sta–1.	Stage 1
Sec.	Secondary	Sta–2.	Stage 3
		Sta–4.	Stage 4

_____ a. Also known as essential or idiopathic (no known cause) hypertension

_____ b. Over 90% of cases of hypertension fit into this category

_____ c. Hypertension that has a known cause

_____ d. BP of 150/94 fits into this category

_____ e. BP of 200/110 fits into this category

✓ **H2.** Match each organ listed in the box with the related disorder that may cause secondary hypertension.

AC.	Adrenal cortex	AM.	Adrenal medulla	K.	Kidney

_____ a. Excessive production of renin, which catalyzes formation of angiotensin II, a powerful vasoconstrictor

_____ b. Pheochromocytoma, a tumor that releases large amounts of norepinephrine and epinephrine

_____ c. Release of excessive amounts of aldosterone, which promotes salt and water retention

H3. Describe possible effects of hypertension upon the following body structures:
a. Heart

b. Cerebral blood vessels

c. Kidneys

H4. Explain how each of the following therapeutic measures can help control hypertension.
a. Sodium restriction
b. Cessation of smoking

c. Exercise

d. Stress reduction

e. Vasodilators

f. A.C.E. inhibitors

g. Diuretics

H5. Define *aneurysm* and list four possible causes of aneurysms.

✓ **H6.** *A clinical correlation.* Explain how a deep vein thrombosis (DVT) of the left femoral vein can lead to a pulmonary embolism.

✓ **H7.** Match terms related to blood vessels with descriptions below. Use each answer once.

Angio.	**Angiogenesis**	**Occ.**	**Occlusion**
Art.	**Arteritis**	**Ortho.**	**Orthostatic hypotension**
C.	**Claudication**		

_____ a. Growth of new blood vessels, for example with weight gain, pregnancy, or to increase blood supply to tumor

_____ b. Inflammation of an artery

_____ c. Obstruction of a vessel, such as an artery, due to atherosclerotic plaque

_____ d. Pain upon exercise due to impaired circulation in limbs

_____ e. Significant drop in blood pressure upon standing up

ANSWERS TO SELECTED CHECKPOINTS: CHAPTER 21

A2. (a) L. (b) TI. (C) VV. (d) TM. (e) TE.

A3. (a) Capillaries. (b) Arterioles. (c) Capillaries. (d) Venules, small veins, inferior vena cava (actually all veins and venules). (e) Aorta and other large arteries. (f) Capillaries. (g) Medium-sized arteries.

A5. (a) Direct movement across endothelial membranes. (b) Intercellular clefts. (c) Fenestrations. (d) Pinocytic vesicles.

A6. (a) Thinner; less; an even flow. (b) Valves. (c) Vari-

cose veins; superficial; skeletal muscles around deep veins limit overstretching. (d) Surrounding dense connective tissue (such as the dura mater); function, but not structure; intracranial (such as superior sagittal venous sinus) and coronary sinus.

A8. (a) Venules and veins; capillaries. (b) Sympathetic; veins of skin and of abdominal organs such as liver and spleen. (c) Exercise, hemorrhage.

A9. (a–c) D. (d) V. (e) B.

A10. (a) A, IFOP; B, BHP; C, IFHP; D, BCOP. (b) +10; out of. (c) BHP is about 16 mm Hg; −9. (d) Starling; lymphatic. (e) Edema; BHP BCOP, increase.

A11. (a) ↑ BHP (pulmonary hypertension due to backlog of blood from a failing left ventricle. (b) ↓ PP (as plasma proteins leaked out of damaged vessels); also ↑ PC. (c) ↑ ECV and ↑ BHP.

B1. (a) 4–6; output. (b) Pressure (such as blood pressure and resistance to flow).

B2. (a) Arteries; capillaries. (b) Increases. (c) 4500–6000. (d) An indirect; 3 to 5, high. (e) Larger; de; clot formation (also shock because pooled blood there does not circulate to tissues). (f) 1.

B3. (a) Systolic; diastole; elastic, 70-80. (b) 93; diastole; diastolic. (c) 80; 120-80 = 40; 80 + ($1/3 \times 40 = 13$) = 93.

B4. (a) Pressure; blood on the wall of a blood vessel; in; ↑ CO → ↑ MABP; de. (b) More; inversely, directly. (c) Low; 85–35; 35–16; 16–0; close to 0. (d) Arterioles; arterioles.

B5. (a) De; ↑ $R_{\text{within a vessel}}$ → → ↓ $BP_{\text{in that same vessel}}$; de. (b) About 120/80; systemic (vascular) resistance; ↑ $R_{\text{(systemic arterioles skin or abdomen)}}$ → ↑ $BP_{\text{in systemic arteries}}$; or ↑ SVR → ↑ $BP_{\text{systemic arterial}}$; in. (c) $BP_{\text{systemic arteries}}$; = CO × SVR. (d1) In, dehydration or polycythemia. (d2) In, weight gain (or obesity), growth, or pregnancy. (d3) De, vasoconstriction. (e) Peripheral; digestive organs such as liver, spleen, stomach, intestine, as well as kidneys.

B6. (a) B; does. Recall that the body grows an average of about 300 km (200 miles) of new blood vessels for each extra pound of weight. (b) Sympathetic; C; 16; in.

B8. (a) Box 1: baro, chemo. Blood pressure; H^+, CO_2, and O_2; carotid arteries and aorta. (b) IX, X. (c) Box 3: medulla. Box 2: cerebral cortex, limbic system, and hypothalamus. (d) Brakes; cardiostimulatory, step on the gas pedal; take your foot off the gas; car goes faster/heart beats faster and with greater force of contraction. (e) Stimulating, vasoconstriction; in. (f) ↑ stroke volume (SV), ↑ heart rate (HR), and ↑ systemic vascular resistance (SVR); in. (g) Negative.

B9. (a) ↑ CO. (b) ↑ Stroke volume. (c) ↑ Venous return. (d) ↑ Exercise. (e) ↑ Blood volume. (f) ↑ Heart rate. (g) ↑ Sympathetic impulses. (h) ↑ Norepinephrine and epinephrine. (i) ↓ Vagal impulses. (j) ↑ ADH. (k) ↑ Aldosterone. (1) ↑ NaCl in diet. (m) ↑ Systemic vascular resistance (SVR). (n) ↑ Vasoconstriction. (o) ↑ Angiotensin II. (p) ↑ Viscosity of blood. (q) ↑ RBC count. (r) ↑ Plasma proteins.

B10. (a) High; decreased. (b) B C F G H.

B11. (a) Tone; drop precipitously; syncope. (b) Alpha, norepinephrine, constrict; constrict, in. (c) De; atrial natriuretic peptide (ANP), histamine. (d) General systemic, cerebral. (e) Chemo; hypoxia. hypercapnia, vasoconstriction, in. (fl–4) ↑ ; (f5) ↓.

B12. (a) Local. (b) Low, vasodilator; lactic; nitric oxide (EDRF), K^+, H^+, and adenosine. (c) Just the opposite; constrict; blood is shunted to parts of the blood that are well oxygenated. (d) Vasospasm; eicosanoids such as thromboxane A2 and prostaglandin F2a.

B13. (a) Blood; hypovolemic; loss of blood or other body fluids, or shifting of fluids into interstitial spaces (edema) as with bums. (b) Sympathetic, in: 10; cool, pale, and clammy, constriction, in. (c) Epinephrine, renin, aldosterone, ADH (vasopressin). (d) Aldosterone and ADH. (e) It causes local vasodilation.

B14. (a–b) ↑ . (c–f) ↓ .

C3. 60–100; rapid, bradycardia.

C4. (a) Sphygmomanometer. (b) Brachial; no pulse is felt or no sound is heard via a stethoscope. (c) Systolic. (d) Change or cease. (e) Korotkoff sounds.

C5. 128/78, 50 (128–78); yes, it is in the normal range.

D2. (a) Aorta; left; ascending aorta, arch of the aorta, descending aorta. (b) Coronary; heart. (c) K, brachiocephalic; L, left common carotid; M, left subclavian. (d) F; in general, the right extremity and right side of thorax, neck, and head; right common carotid. (e) G, axillary; H, brachial; blood pressure. (f) I, radial; J, ulnar; radial, I; ulnar; fingers. (g) Vertebral; basilar.

D3. Anterior; internal; external portions of head (face, tongue, ear, scalp) and neck (throat, thyroid).

D4. (a) Willis. (b) Internal, vertebral. (c) E, F, H; communicating.

D5. (a) Occipital lobes of cerebrum, as well as cerebellum and pons. (b) Left, because this artery supplies the left side of the brain. (Remember that neurons in tracts cross to the opposite side of the body, but arteries do not.) (c) Hearing, because it is perceived in the temporal lobe. (Visual areas are in occipital lobes). (d) Left foot.

D6. (a) VT. (b) PA. (C) VA. (d) PT.

D7.

Artery	Letter	Structures Supplied

a. Celiac (three major branches)
  1. Hepatic
 	
  2. Gastric
  3. Splenic
	P	1. Liver, **gallbladder and parts of stomach, duodenum, pancreas**
		2. **Stomach and esophagus**
		3. **Spleen, pancreas, and stomach**

b. **Superior mesenteric**	Q	Small intestine and part of large intestine and pancreas
c. **Adrenal (suprarenal)**	R	Adrenal (suprarenal) glands
d. Renal	S	**Kidneys**
e. **Gonadal (testicular or ovarian) Testes (or ovaries)**	T	Testes (or ovaries)
f. Inferior mesenteric	U	**Parts of colon and rectum**

D8. L–4, common iliac.

D10. Right external iliac artery → right lateral circumflex artery (right descending branch) → right anterior tibial artery → right dorsalis pedis artery → to arch and digital arteries in foot.

E1. Superior vena cava, inferior vena cava, coronary sinus.

E2. (a) Intracranial vascular (venous) sinuses; functionally. (b) Superior sagittal. (c) Internal jugular; common carotid. (d) There are no such vessels.

E3. (a) The heart and a lung. (b) Yes; yes; yes.

E4. (a) Left radial vein → left brachial vein → left axillary vein → left subclavian vein → left brachiocephalic vein → superior vena cava. (Note that there are left and right brachiocephalic veins, but only one artery of that name.) (b) Small veins in medial aspect of left knee → left popliteal vein → left femoral vein → left external iliac vein → left common iliac vein → inferior vena cava. (c) Right renal vein → inferior vena cava.

E5. (a) Lower, slower. (b) More; superficial.

E6. (a) Axillary. (b) Axillary. (c) Femoral. (d) Popliteal.

E9. See key, page 579 of the Guide. Right, right.

F1. (a) E, gastric; F, splenic; I, mesenteric; H, portal vein to the liver. (b) D, hepatic artery. (c) Sinusoids; G, hepatic vein (there are two but only one is shown on Figure LG 20.3. (d) While in liver sinusoids, blood is cleaned, modified, detoxified. Ingested nutrients and other chemicals are metabolized and stored.

F2. (a) P; pulmonary arteries have larger diameters and shorter length (a shorter circuit), thinner walls, less elastic tissue. (b) 5; 120/80; 25/8 (pulmonary BP is about one-fifth of systemic BP). (c) P; pulmonary edema with shortness of breath and impaired gas exchange. (d) P; an advantage is the diversion of blood to well-ventilated regions. One disadvantage is that chronic vasoconstriction

in persons with chronic hypoxia (as in emphysema or chronic bronchitis) may lead to cor pulmonale (see page 666 of your text).

F4.

Fetal Structure	Structures Connected	Function	Fate of Structure After Birth
a. Umbilical arteries (2)	Fetal internal iliac arteries to placenta	Carries fetal blood low in oxygen and nutrients and high in wastes	Medial umbilical ligaments
b. Placenta	(Omit)	Site where maternal and fetal blood exchange gases, nutrients, and wastes	Delivered as "afterbirth"
c. Umbilical vein (1)	Placenta to liver and ductus venosus	Carries blood high in oxygen and nutrients, low in wastes	Round ligament of the liver (ligamentum teres)
d. Ductus venosus	Umbilical vein to inferior vena cava	Branch of umbilical vein, bypasses liver	Ligamentum venosum
e. Foramen ovale	Right and left atria	Bypasses lungs	Fossa ovalis
f. Ductus arteriosus	Pulmonary artery to aorta	Bypasses lungs	Becomes ligamentum arteriosum

F5. (a) Umbilical vein. (b) Femoral artery. (c) Thoracic portion of inferior vena cava (because umbilical vein blood enters it via ductus venosus in upper abdomen).

G3. (a) ↓ (b) ↓ . (c) ↑ . (d) ↑ .

H1. (a-b) Pri. (c) Sec. (d) Sta-1. (e) Sta-3.

H2. (a) K. (b) AM. (c) AC.

H6. Embolus (a "clot-on-the-run") travels through femoral vein to external and common iliac veins to inferior vena cava to right side of heart. It can lodge in pulmonary arterial branches, blocking further blood flow into the pulmonary vessels.

H7. (a) Angio. (b) Art. (c) Occ. (d) C. (e) Ortho.

MORE CRITICAL THINKING: CHAPTER 21

1. Explain how the structure of the following types of blood vessels is admirably suited to their functions: elastic arteries, muscular arteries, arterioles, metarterioles, and capillaries.

2. Discuss advantages to health offered by collateral circulation. Include the terms anastomoses and end arteries in your discussion.

3. The typical human body contains about 4–6 liters of blood. Describe mechanisms that regulate where this blood goes at any given moment, for example, at rest versus during a period of vigorous exercise. Be sure to include the following terms in your discussion: metarteriole, precapillary sphincter, sympathetic, and vasoconstriction.

4. Describe all mechanisms that will help you to raise your blood pressure in response to a frightening experience. Be sure to include both nervous and hormonal responses.

5. Refer to Figure LG 21.3 and review related sections of Chapter 17 and 18 as you describe the specific mechanism by which each of the following medications or disorders lowers blood pressure: (a) angiotensin converting hormone inhibitor (ACE-I) medications; (b) beta-blocker medications; (c) alpha-blocker medications; (d) diuretics; (e) hemorrhage; and (f) anemia.

6. In most cases, arteries and veins are parallel in both name and locations, such as the brachial artery and brachial vein lying close together in the arm. Identify major arteries and veins that differ in names and locations.

MASTERY TEST: CHAPTER 21

Questions 1–2: Circle the letters preceding all correct answers to each question.

1. In the most direct route from the left leg to the left arm of an adult, blood must pass through all of these structures:
   A. Inferior vena cava
   B. Brachiocephalic artery
   C. Capillaries in lung
   D. Hepatic portal vein
   E. Left subclavian artery
   F. Right ventricle of heart
   G. Left external iliac vein

2. In the most direct route from the fetal right ventricle to the fetal left leg, blood must pass through all of these structures:
   A. Aorta   E. Ductus venosus
   B. Umbilical artery   F. Left ventricle
   C. Lung   G. Left common iliac artery
   D. Ductus arteriosus

Questions 3–9: Circle the letter preceding the one best answer to each question. Note whether the underlined word or phrase makes the statement true or false.

3. Choose the *false* statement.
   A. In order for blood to pass from a vein to an artery, it must pass through chambers of the heart.
   B. In its passage from an artery to a vein a red blood cell must ordinarily travel through a capillary.
   C. The wall of the femoral artery is thicker than the wall of the femoral vein.
   D. Most of the smooth muscle in arteries is in the intima.

4. Choose the *false* statement.
   A. Arteries contain valves, but veins do not.
   B. Decrease in the size of the lumen of a blood vessel by contraction of smooth muscle is called vasoconstriction.
   C. Most capillary exchange occurs by simple diffusion.
   D. Sinusoids are wider and more tortuous (winding) than capillaries.

5. Choose the *false* statement.
   A. Cool, clammy skin is a sign of shock that results from sympathetic stimulation of blood vessels and sweat glands.
   B. Nicotine is a vasodilator that helps control hypertension.
   C. Orthostatic hypotension refers to a sudden, dramatic drop in blood pressure upon standing or sitting up straight.
   D. Most hypertension is primary since it has no identifiable cause.

6. Choose the *false* statement.
   A. The vessels that act as the major regulators of blood pressure are arterioles.
   B. The only blood vessels that carry out exchange of nutrients, oxygen, and wastes are capillaries.
   C. Blood in the umbilical artery is normally more highly oxygenated than blood in the umbilical vein.
   D. At any given moment more than 50% of the blood in the body is in the veins.

7. Choose the *true* statement.
   A. The basilic vein is located lateral to the cephalic vein.
   B. The great saphenous vein runs along the lateral aspect of the leg and thigh.
   C. The vein in the body most likely to become varicosed is the femoral.
   D. The hemiazygos and accessory hemiazygos veins lie to the left of the azygos.

8. Choose the *true* statement.
   A. A normal blood pressure for the average adult is 160/100.
   B. During exercise, blood pressure will tend to increase.
   C. Sympathetic impulses to the heart and to arterioles tend to decrease blood pressure.
   D. Decreased cardiac output causes increased blood pressure.

9. Choose the *true* statement.
   A. Cranial venous sinuses are composed of the typical three layers found in walls of all veins.
   B. Cranial venous sinuses all eventually empty into the external jugular veins.
   C. Internal and external jugular veins do unite to form common jugular veins.
   D. Most parts of the body supplied by the internal carotid artery are ultimately drained by the internal jugular vein.

Questions 10–14: Circle the letter preceding the one best answer to each question.

10. Hepatic portal circulation carries blood from:
    A. Kidneys to liver
    B. Liver to heart
    C. Stomach, intestine, and spleen to liver
    D. Stomach, intestine, and spleen to kidneys
    E. Heart to lungs
    F. Pulmonary artery to aorta

11. Which of the following is a parietal (rather than a visceral) branch of the aorta?
    A. Superior mesenteric artery
    B. Bronchial artery
    C. Celiac artery
    D. Lumbar artery
    E. Esophageal artery

12. All of these vessels are in the leg or foot *except:*
    A. Saphenous vein   D. Dorsalis pedis artery
    B. Azygos vein   E. Popliteal artery
    C. Peroneal artery
13. All of these are superficial veins *except:*
    A. Median cubital   D. Great saphenous
    B. Basilic   E. Brachial
    C. Cephalic

14. Vasomotion refers to:
    A. Local control of blood flow to active tissues
    B. The Doppler method of checking blood flow
    C. Intermittent contraction of small vessels resulting in discontinuous blood flow through capillary networks
    D. A new aerobic dance sensation

Questions 15–17: Arrange the answers in correct sequence.

_____ _____ _____ _____   15. Route of a drop of blood from the right side of the heart to the left side of the heart:
    A. Pulmonary artery
    B. Arterioles in lungs
    C. Capillaries in lungs
    D. Venules and veins in lungs

_____ _____ _____ _____ _____   16. Route of a drop of blood from small intestine to heart:
    A. Superior mesenteric vein
    B. Hepatic portal vein
    C. Small vessels within the liver
    D. Hepatic vein
    E. Inferior vena cava

_____ _____ _____   17. Layers of blood vessels, from most superficial to deepest:
    A. Tunica interna
    B. Tunica externa
    C. Tunica media

Questions 18–20, use the following answers:

| A. **Aorta and other arteries** | C. **Capillaries** |
| B. **Arterioles** | D. **Venules and veins** |

_____ _____ _____ _____   18. Blood pressure in vessels, from highest to lowest:

_____ _____ _____ _____   19. Total cross-sectional areas of vessels, from highest to lowest:

_____ _____ _____ _____   20. Velocity of blood in vessels, from highest to lowest:

Questions 21–25: Fill-ins. Complete each sentence or answer the question with the word(s) or phrase that best fits.

_____   21. a weakened section of a blood vessel forming a balloonlike sac is

known as a(n) _____.

_____   22. Name the major factor that creates blood osmotic pressure (BCOP) that prevents excessive flow of fluid from blood to interstitial areas.

_____   23. Name the two blood vessels that are locations of baroreceptors and chemoreceptors for regulation of blood pressure.

_____   24. In taking a blood pressure, the cuff is first inflated over an artery. As the cuff is then slowly deflated, the first sounds heard indicate

the level of _____ blood pressure.

_____   25. Most arteries are paired (one on the right side of the body, one on the left). Name five or more vessels that are unpaired.

ANSWERS TO MASTERY TEST: CHAPTER 21

Multiple Answers

1. A C E F G
2. A D G

Multiple Choice

3. D	9. D
4. A	10. C
5. B	11. D
6. C	12. B
7. D	13. E
8. B	14. C

Arrange

15. A B C D
16. A B C D E
17. B C A
18. A B C D
19. C D B A
20. A B D C

Fill-ins

21. Aneurysm
22. Plasma protein such as albumin
23. Aorta and carotids
24. Systolic
25. Aorta, anterior communicating cerebral, basilar, brachiocephalic, celiac, gastric, hepatic, splenic, superior and inferior mesenteric, middle sacral

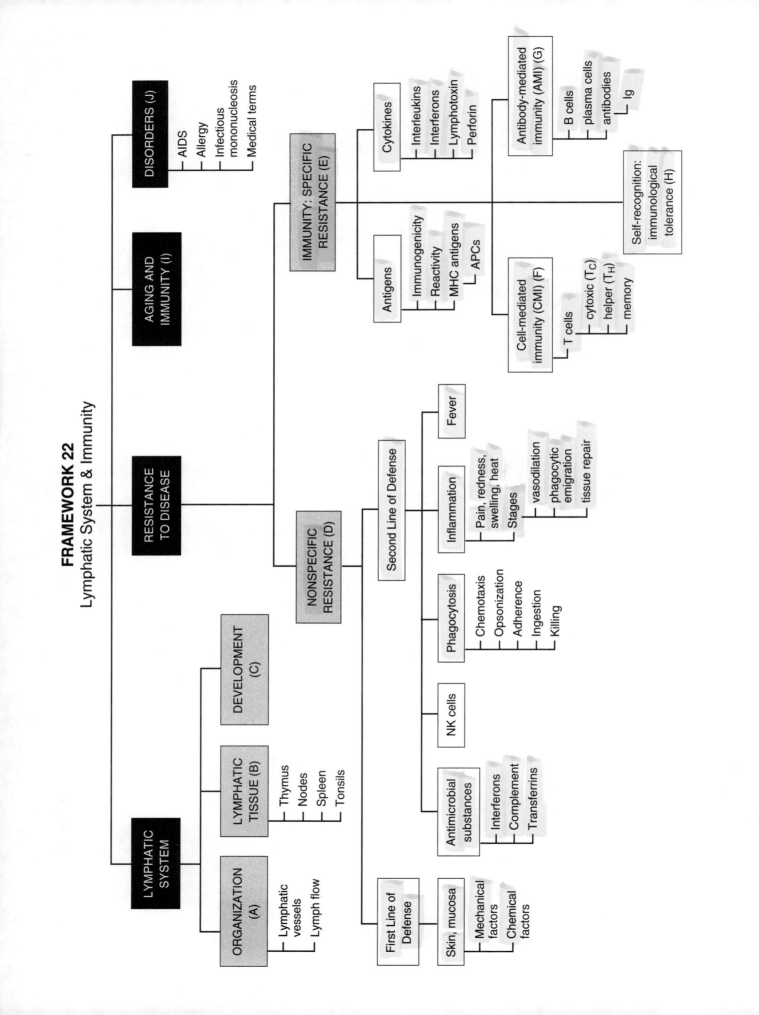

FRAMEWORK 22
Lymphatic System & Immunity

C H A P T E R **22**

The Lymphatic System, Nonspecific Resistance to Disease, and Immunity

The human body is continually exposed to foreign (nonself) entities: bacteria or viruses, pollens, chemicals in a mosquito bite, or foods or medications that provoke allergic responses. Resistance to specific invaders is normally provided by a healthy immune system made up of antibodies and battalions of T lymphocytes.

Nonspecific resistance is offered by a variety of structures and mechanisms, such as skin, mucus, white blood cells assisting in inflammation, and even a rise in temperature (fever) that limits the survival of invading microbes. The lymphatic system provides sentinels at points of entry (as in tonsils) and at specific sites (lymph nodes) near entry to body cavities (inguinal and axillary regions). The lymphatic system also wards off the detrimental effects of edema by returning proteins and fluids to blood vessels.

As you begin your study of this system, carefully examine the Chapter 22 Topic Outline and Objectives; check off each one as you complete it. To organize your study of the lymphatic system, nonspecific resistance to disease, and immunity, glance over the Chapter 22 Framework now. Be sure to refer to the Framework frequently and note relationships among key terms in each section.

TOPIC OUTLINE AND OBJECTIVES

A. The lymphatic system: vessels and flow

1. Describe the general components of the lymphatic system, and list its functions.
2. Describe the organization of lymphatic vessels.
3. Describe the formation and flow of lymph.

B. The lymphatic system: organs and tissues

4. List and describe the primary and secondary lymphatic organs and tissues.

C. Developmental anatomy of the lymphatic system

5. Describe the development of the lymphatic system.

D. Nonspecific resistance to disease

6. Describe the mechanisms of nonspecific resistance to disease.

E. Specific resistance to disease: immunity

7. Define immunity, and describe how T cells and B cells arise.
8. Explain the relationship between an antigen and an antibody.
9. Describe the role of antigen-presenting cells in processing exogenous and endogenous antigens.

F. Cell-mediated immunity

10. Describe the steps of a cell-mediated immune response.

G. Antibody-mediated immunity

11. Describe the steps of an antibody-mediated immune response.
12. Describe the chemical characteristics and actions of antibodies.

H. Self-recognition and immunological tolerance

13. Describe how self-recognition and immunological tolerance develop.

I. Aging and the immune system

14. Describe the effects of aging on the immune system.

J. Disorders, medical terminology

WORDBYTES

Now become familiar with the language of this chapter by studying each wordbyte, its meaning, and an example of its use within a term. After you study the entire list, self-check your understanding by writing the meaning of each wordbyte on the line. As you continue through the *Learning Guide*, identify (and fill in) additional terms that contain the same wordbyte.

Wordbyte	Self-check	Meaning	Example(s)
anti-	_____	against	*anti*body
auto-	_____	self	*auto*immune
axilla-	_____	armpit	*axilla*ry nodes
chyl-	_____	juice	cisterna *chyl*i
gen-	_____	to produce	anti*gen*, *gen*etic
hetero-	_____	other	*hetero*graft
iso-	_____	same	*iso*graft
lact-	_____	milk	*lact*eals
lymph	_____	clear water	*lymph*atic vessel
meta-	_____	beyond	*meta*stasis
-stasis	_____	to stand	meta*stasis*
xeno-	_____	foreign	*xeno*graft

CHECKPOINTS

A. The lymphatic system: vessels and flow (pages 738–742)

A1. Summarize the body's defense mechanisms against invading microorganisms and other potentially harmful substances. Be sure to use these terms in your paragraph: *pathogens, resistance, susceptibility.*

✓ **A2.** Write *S* next to specific immune responses and *NS* next to nonspecific responses.

_____ a. Barriers to infection, such as skin and mucous membranes

_____ b. Inflammation and phagocytosis

_____ c. Chemicals such as gastric or vaginal secretions, which are both acidic

_____ d. Antigen-antibody responses.

_____ e. Destruction of an intruder by a cytotoxic T lymphocyte

_____ f. Antimicrobial substances such as interferons (IFNs) or complement

✓ **A3.** Describe the functions of the lymphatic system in this Checkpoint.
  a. Lymphatic vessels drains two types of chemicals from spaces between cells; these

   are _____ and _____. They also transport

   _____ and *(water? fat?)*-soluble vitamins from the gastrointestinal (GI) tract to the bloodstream.
  b. The defensive functions of the lymphatic system are carried out largely by white blood

   cells named _____-cytes. *(B? T?)* lymphocytes destroy foreign substances

   directly, whereas B lymphocytes lead to the production of _____.

✓ **A4.** List the components of the lymphatic system.

A5. Describe relationships between terms in each pair.
  a. *Lymphatic capillary epithelium/one-way valve*

  b. *Anchoring filaments/edema*

  c. *Lacteal/chyle*

✓▲ **A6.** Answer the following questions about the largest lymphatics.
  a. In general, the thoracic duct drains *(three-fourths? one-half? one-fourth?)* of the body's lymph. This vessel starts in the lumbar region as a dilation known as the

   _____. This cisterna chyli receives lymph from which areas of the body?

b. In the neck the thoracic duct receives lymph from three trunks. Name these.

c. One-fourth of the lymph of the body drains into the _____ duct. The regions of the body from which this vessel collects lymph are:

d. The thoracic duct empties into the *(right? left?)* _____ *(artery? vein?)* close to its junction with the internal jugular vein. The right lymphatic duct empties into *(the same? a different?)* vein on the right side.

e. In summary, lymph fluid starts out originally in plasma or cells and circulates as

_____ fluid. Then as lymph it undergoes extensive cleaning as it passes through nodes. Finally, fluid and other substances in lymph are returned to

_____ in veins of the neck.

✓ **A7.** *Critical thinking.* Explain why a person with a broken left clavicle might exhibit these two symptoms:
a. Edema, especially of the left arm and both legs

b. Excess fat in feces (steatorrhea)

A8. Describe the general pathway of lymph circulation. *For extra review.* Refer to Chapter 1, Checkpoint D2 and Figure LG 1.1, page 8 in the *Learning Guide.*

✓ **A9.** The lymphatic system does not have a separate heart for pumping lymph. Describe two principal factors that are responsible for return of lymph from the entire body to major blood vessels in the neck. (Note that these same factors also facilitate venous return.)

B. The lymphatic system: organs and tissues (pages 742–747)

✓ **B1.** Write *P* next to descriptions of primary lymphatic organs and *S* next to descriptions of secondary lymphatic organs and tissues.

_____ a. Origin of B cells and pre-T cells

_____ b. Sites where most immune responses occur

_____ c. Includes lymph nodes, spleen, and lymphatic nodules

_____ d. Includes red bone marrow and thymus

✓▲ **B2.** Describe the thymus gland in this exercise.

   a. It is located in the *(neck? mediastinum?)*. Its position is just posterior to the

         _____ bone. Its size is relatively *(large? small?)* in childhood,

         _____-creasing in size after age 10 to 12 years.

   b. Like lymph nodes, the thymus contains a medulla surrounded by a

         _____. Within the cortex are densely packed _____-cytes.
The thymus is the site of maturation of *(B? T?)* cells *(only in childhood? throughout life?)*.

   c. The *(cortex? medulla?)* consists mostly of epithelial cells. What do these produce?

Figure LG 22.1 Structure of a lymph node. Arrows indicate the direction of lymph flow.
Label and color as directed on Checkpoint B3.

✓ **B3.** Complete this checkpoint on lymph node structure.

   a. Use terms in the box to label structures A-E on Figure LG 22.1. Then use color code
ovals to color each structure.

ALV.	**Afferent lymphatic vessel**	**MC.**	**Medullary cord**
LV.	**Efferent lymphatic vessel**	**TS.**	**Trabecular sinus**
GC.	**Germinal center**		

  b. Now write one answer from the box above on each line next to related descriptions.

    1. Vessels for entrance of lymph into node: _____

    2. Vessel for exit of lymph out of node: _____
    3. Region composed of B cells (that become plasma cells), antigen-presenting cells

       (APCs), and macrophages: _____
    4. Example of a channel for flow of lymph through the node: _____

    5. Tight strand of B cells and plasma cells: _____

    6. Regions of outer cortex: _____ _____

  B4. Describe how lymph is "processed" as it circulates through lymph nodes. How does
       this function of lymph nodes explain the fact that nodes ("glands") become enlarged
       and tender during infection?

✓▲ **B5.** Spread, or _____, of cancer may occur via the lymphatic system.
       Lymph nodes that have become cancerous tend to be enlarged, firm, and *(tender? non-
       tender?)*. Cancerous nodes also tend to be *(mobile? fixed via "tentacle-like extensions"?)*.

✓ **B6.** The term stroma refers to *(connective tissue scaffolding? functional cells?)* of an organ such

       as a lymph node, spleen, or liver. The term parenchyma refers to _____.

✓▲ **B7.** The spleen is located in the *(upper right? upper left? lower right? lower left?)* quadrant of

       the abdomen, immediately inferior to the _____. List three functions of:

  a. The white pulp _____

  b. The red pulp _____

✓▲ **B8.** Complete this exercise about lymphatic nodules.

  a. The acronym MALT refers to M_____   A_____

       L_____   T_____. Name four systems that contain MALT:

       _____ _____ _____ _____
  b. State advantages of the following locations of aggregations of lymphatic tissue:
     1. Peyer's patches

    2. Tonsils and adenoids

  c. Which tonsils are most commonly removed in a tonsillectomy? _____

    Which ones are called adenoids? _____

C. Developmental anatomy of the lymphatic system (page 747)

✓ **C1.** The lymphatic system derives from _____-derm, beginning about the *(fifth? seventh? tenth?)* week of gestation.

✓ **C2.** Lymphatic vessels arise from lymph sacs that form from early *(arteries? veins?)*. Name the embryonic lymph sac that develops vessels to each of the following regions of the body.
  a. Pelvic area and lower limbs: _____ lymph sac

  b. Upper limbs, thorax, head, and neck: _____ lymph sac. The left jugular lymph sac eventually develops into the upper portion of the main collecting

    lymphatic vessel, the _____ duct.

D. Nonspecific resistance to disease (pages 747-751)

D1. Briefly contrast the two types of resistance to disease: specific vs. nonspecific. (For help, refer back to Checkpoint A2.)

✓ **D2.** *A clinical correlation.* Match the following conditions or factors with the nonspecific resistance that is lost in each case.

Nonspecific Resistance Lost

_____ a. Cleansing of oral mucosa

_____ b. Cleansing of vaginal mucosa

_____ c. Flushing of microbes from urinary tract

_____ d. Lacrimal fluid with lysozyme

_____ e. Respiratory mucosa with cilia

_____ f. Closely packed layers of keratinized cells

_____ g. HCl production; pH 1 or 2

_____ h. Sebum production

Condition/Factor

A. Long-term smoking
B. Excessive dryness of skin
C. Use of anticholinergic medications that decrease salivation
D. Vaginal atrophy related to normal aging
E. Dry eyes related to decreased production of tears, as in aging
F. Decreased urinary output, as in prostatic hypertrophy
G. Skin wound
H. Partial gastrectomy (removal of part of the stomach)

✓ **D3.** *For extra review.* Describe nonspecific defenses more fully in this exercise.

  a. Interferons (IFNs) are proteins produced by cells infected with _____,

    such as cells named _____, _____, and _____.

    IFNs then stimulate *(infected? uninfected?)* cells to produce _____ proteins.

b. List the three types of interferons.

c. State two locations of complement proteins. _____ and _____
    Explain why these proteins are called complement. (More about complement in
    Checkpoint G6.)

d. NK cells, or _____ _____ cells, are a type of
    (lymphocyte? neutrophil?). About *(5-10? 60-70?)* percent of lymphocytes are NK cells.
    These cells attack a *(selected few? wide variety?)* of infectious microbes. One
    mechanisms utilized by NK cells involves their release of chemicals known as

    _____ that perforate plasma membranes of invading microbes.
e. The primary type of white blood cells that are involved in phagocytosis is the

    _____. Other leukocytes, known as _____, travel to

    infection sites and develop into highly phagocytic _____ cells. Name
    five organs that contain fixed macrophages.

✓ **D4.** Match each of the steps in phagocytosis with the related description. Note that
    descriptions are listed in chronological sequence.

A.	**Adherence**	**I.**	**Ingestion**
C.	**Chemotaxis**	**K.**	**Killing**

_____ a. Phagocytic cells "sniff' the delectable fra-
            grance of invading microbes and
            especially savor bacteria with the tasty
            complement coating. "Ah, dinner soon!"

_____ b. Phagocytes trap the invaders, possibly
            wedging them against a blood vessel or a
            clot, and then attach to the surface of the
            invader.

_____ c. Phagocytic tentacles called pseudopods
            surround the tasty morsels forming a
            culinary treat, the *phagosome*. "Slurp!"

_____ d. Lysosomal enzymes, oxidants, and diges-
            tive enzymes pounce on the microbial
            meal, a has-been within a short time.
            Residual bodies, like the pits of prunes, are
            discarded, and the phagocytic face smiles
            in contentment.

D5. Do this exercise on inflammation.
a. Defend or dispute this statement: "Inflammation is a process that can be both helpful
    and harmful."

b. List the four cardinal signs or symptoms of inflammation.

    _____ _____

    _____ _____

c. Briefly describe the three basic stages of inflammation.

1.

2.

3.

✓ **D6.** *For extra review.* Select the number that best fits each description of a phase of inflammation. Use these answers:

1. Vasodilation and increased capillary permeability
2. Phagocytic migration
3. Repair

_____ a. Brings defensive substances to the injured site and helps remove toxic wastes

_____ b. Histamine, prostaglandins, kinins, leukotrienes, and complement enhance this process

_____ c. Causes redness, warmth, and swelling of inflammation

_____ d. Localizes and traps invading organisms

_____ e. Occurs within an hour after initiation of inflammation; involves emigration, chemotaxis, and leukocytosis

_____ f. White blood cell and debris formation that may lead to abscess or ulcer

✓ **D7.** Identify the chemicals that fit the descriptions below.

C.	**Complement**	**LT.**	**Leukotrienes**
H.	**Histamine**	**PG.**	**Prostaglandins**
I.	**Interferon**	**T.**	**Transferrins**
K.	**Kinins**		

_____ a. Produced by damaged or inflamed tissues; aspirin and ibuprofen neutralize these pain-inducing chemicals

_____ b. Produced by mast cells and basophils (two answers)

_____ c. Made by some leukocytes and fibroblasts; stimulates cells near an invading virus to produce antiviral proteins that interfere with survival of the virus; effects are enhanced by fever

_____ d. Induces vasodilation, permeability, chemotaxis, and irritation of nerve endings (pain)

_____ e. Neutrophils are attracted by these chemicals present in the inflamed area (two answers)

_____ f. Proteins that inhibit bacterial growth by depriving microbes of iron

D8. Define these terms and relate each one to inflammation or to poor circulation:

a. Pus

b. Ulcer

c. Abscess

E. Specific resistance to disease: immunity (pages 751–756)

✓ **E1.** Contrast specific and nonspecific resistance in this exercise.

  a. *(Nonspecific? Specific?)* resistance is also known as *immunity* and is studied in the

    branch of science known as _____.

  b. Substances that are recognized as foreign and also provoke an immune response are
known as *(antigens? antibodies?)*. Immune responses involve memory which causes
a(n) *(slower and milder? even more rapid and vigorous?)* response to previously encoun-
tered antigens.

  c. *For extra review,* look back at Checkpoint A2.

✓ **E2.** Complete this exercise about cells that carry out immune responses.

  a. Name the two categories of cells that carry out immune responses.

    _____ and _____ Where do lymphocytes originate?

  b. Some immature lymphocytes migrate to the thymus and become *(B? T?)* cells. Here
T cells develop immunocompetence, meaning that these cells have the ability to

    _____.

    Some T cells become CD4+ cells and others become _____ cells, based on the type

    of _____ in their plasma membranes.

  c. Where do B cells mature into immune cells? _____

✓ **E3.** Contrast two types of immunity by writing AMI before descriptions of *antibody-mediated immunity* and CMI before those describing *cell-mediated immunity*.

_____ a.	Especially effective against microbes that enter cells, such as viruses and parasites	_____ c.	Involves plasma cells (derived from B cells) that produce antibodies
_____ b.	Especially effective against bacteria present in extracellular fluids	_____ d.	Utilizes killer T cells (derived from CD8 T cells) that directly attack the antigen
		_____ e.	Facilitated by helper T (CD4) cells

✓ **E4.** Do this exercise on antigens.

    a. An antigen is defined as "any chemical substance which, when introduced into the

       body," _____.
       In general, antigens are *(parts of the body? foreign substances?)*.

    b. Complete the definitions of the two properties of antigens.

        1. Immunogenicity: ability to _____ specific antibodies or specific T cells

        2. Reactivity: ability to _____ specific antibodies or specific T cells

    c. An antigen with both of these characteristics is called a(n) _____.

    d. A partial antigen is known as a _____. It displays *(immunogenicity?*

       reactivity?) but not _____. For example, for the hapten penicillin to

       evoke an immune response (immunogenicity), it must form a complete antigen by

       combining with a _____. Persons who have this particular protein

       are said to be _____ to penicillin.
    e. Describe the chemical nature of antigens.

    f. Explain why plastics used for valves or joints are not likely to initiate an allergic
       response and be rejected.

    g. Can an entire microbe serve as an antigen? *(Yes? No?)* List the parts of microbes that
       may be antigenic.

    h. If you are allergic to pollen in the spring or fall or to certain foods, the pollen or
       foods serve as *(antigens? antibodies?)* to you.
    i. Antibodies or specific T cells form against *(the entire? only a specific region of the?)*

       antigen. This region is known as the _____. Most antigens have
       (only one? only two? a number of?) antigenic determinant sites.

E5. Explain how human cells have the ability to recognize, bind to, and then evoke an
    immune response against over a billion different antigenic determinants. Describe
    two aspects.
    a. Genetic recombination

    b. Somatic mutations

✓ **E6.** Describe roles of major histocompatibility complex (MHC) antigens in this exercise.
    a. Describe the "good news" (how they help you) and the "bad news" (how they may be harmful) about MHC antigens.

    b. For what purpose is histocompatibility testing performed?

    c. Contrast two classes of MHC antigens according to types of cells incorporating the MHC into plasma membrane.
        MHC-I

        MHC-II

    d. Which of the following do T cells normally ignore?
        A. MHC with peptide fragment from foreign protein
        B. MHC with peptide fragment from self-protein
    e. If the body does fail to recognize its own tissues as "self," but instead considers them

        antigenic (foreign), the resulting condition is known as an _____ disorder.

✓ **E7.** Describe the processing and presenting of antigens in this activity.

    a. For an immune response to occur, either _____ or _____ cells must recognize the presence of a foreign antigen. *(B? T?)* cells can recognize antigens located in extracellular fluid (ECF) and not attached to any cells. However, for *(B? T?)* cells to recognize antigens, the antigens must have been processed by a cell, and then they must be presented in association with MHC-I or MHC-II self-antigens (as described in Checkpoint E6c).
    b. Define exogenous antigens.

        Cells that present these antigenic proteins are called APCs, or a_____ p_____ c_____. These cells are generally found *(in lymph nodes? at sites where antigens are likely to enter the body, such as skin or mucous membranes?)*. Name three types of APCs.

c. Place in correct sequence the six boxed codes that describe the steps APCs or other body cells take to initiate an immune response to intruder antigens. Use the lines provided.

Bind to MHC–II.	**Bind peptide fragment to MHC–II**
Exo/insert.	**Exocytosis and insertion of antigen fragment-MHC–II into APC plasma membrane.**
Fusion.	**Fusion of vesicles of peptide fragments with MHC–II**
Partial dig.	**Partial digestion of antigen into peptide fragments**
Phago/endo.	**Phagocytosis or endocytosis of antigen**
Present to T.	**Present antigen to T cell in lymphatic tissue**

1. _____ → 2. _____ → 3. _____

4. _____ → 5. _____ → 6. _____

d. Although only selected cells can present exogenous antigens, most cells of the body can present _____-genous antigens. State an example of endogenous antigens.

✓ **E8.** Match the name of each cytokine with the description that best fits.

Gamma-IFN.	**Gamma interferon**	**M.**	**Monokine**
IL–1.	**Interleukin-1**	**MMIF.**	**Macrophage migration inhibiting factor**
IL–2.	**Interleukin-2**		
IL–4.	**Interleukin-4**	**P.**	**Perforin**
LT.	**Lymphotoxin**	**TNF.**	**Tumor necrosis factor**

_____ a. General name for a cytokine made by a monocyte

_____ b. Made by activated helper T cells; stimulates B cells, leading to IgE production

_____ c. Known as T cell growth factor, it is made by helper T cells; needed for almost all immune responses; causes proliferation of cytotoxic T cells (as well as B cells) and activates NK cells

_____ d. Formerly called macrophage activating factor (MAF), stimulates phagocytosis in macrophages and neutrophils; enhances AMI and CMI responses

_____ e. Prevents macrophages from leaving the site of infection

_____ f. Destroys target cells by perforating their cell membranes

_____ g. Activates enzymes in the target cell that it kills by destruction of cell's DNA

_____ h. Produced largely by macrophages; induces fever (two answers)

E9. *A clinical correlation.* List one or more disorders for which each of these cytokines is used as a treatment:

a. Alpha interferon (Intron A)

b. Beta-interferon (Betaseron)

c. Interleukin-2 (IL-2)

F. Cell-mediated immunity (pages 756-760)

F1. Describe the stages of cell-mediated immunity (CMI) by completing the following outline that provides key terms. (*Suggestion*: Refer to Figure 22.14 in the text.)

a. Antigen recognition (car-starting analogy) and activation

    1. T cell receptors (TCRs)

    2. Costimulators

    3. (What is the meaning of *anergy*?)

b. Proliferation and differentiation of T cells in secondary lymphatic organs and tissues.

    1. Activated T cells

    2. Clones of three different T cells (See also Checkpoint F2)

c.　Elimination of the intruder by chemicals from T_C cells

　　1.　Perforin

　　2.　Lymphotoxin

　　3.　Gamma-IFN

✓　**F2.** Match T cells to descriptions of their roles. (*Suggestion*: Refer to Checkpoint E8 above.)

M. Memory T cells	T_C. Cytotoxic T cells	T_H. Helper T cells

_____ a.　Recognize antigen fragments associated with MHC–I molecules

_____ b.　Recognize antigen fragments associated with MHC–II molecules

_____ c.　Called T8 cells because developed from cells with CD8 protein; to become cytolytic, must be costimulated by IL–2 or other cytokines from T_H cells

_____ d.　Called T4 cells because developed from cells with CD4 protein; activated by APCs and costimulated by IL–l and IL–2; produce IL–2, IL–4, and IL–5

_____ e.　Programmed to recognize the original invader; can initiate dramatic responses to reappearance of the intruder

F3. Define *immunological surveillance*.

Name the three types of cells responsible for immunological surveillance.

This immune function is most helpful in eliminating *(all cancer cells? only cancer cells due to cancer-causing viruses?)*. Explain why transplant patients taking immunosuppressants are at risk for virus-associated cancers.

G. Antibody-mediated immunity (pages 760–765)

G1. Describe antibody-mediated immunity (AMI) in this activity.

a. Where do B cells perform their functions? *(At sites where invaders enter the body? In lymphatic tissue?)*

b. Now describe the steps in AMI, using these key terms or phrases as a guide.

  1. Activation of B cells

  2. Follicular dendritic cells

  3. Antigen-MHC–II

  4. Costimulation

  5. Plasma cells

  6. Four or five days

  7. Memory B cells

  8. Antibodies with identical structure to original antigen receptors

○ Heavy chain, constant   ○ Light chain, constant

○ Heavy chain, variable   ○ Light chain, variable

Figure LG 22.2 Diagram of an antibody molecule. Color and label as directed in Checkpoint G2.

✓ **G2.** Refer to Figure LG 22.2 and check your understanding of antibody structure in this Checkpoint.

a. Chemically, all antibodies are composed of glycoproteins named _____. They consist of *(2? 4? 8?)* polypeptide chains. The two heavy chains contain about

_____ amino acids; light chains contain about _____ amino acids. Crosshatch the *heavy chains* on the figure.

b. *(Constant? Variable?)* portions are diverse for different antibodies because these serve as binding sites for different antigens. *(Constant? Variable?)* portions *(differ? are identical?)* for all antibodies within the same class. Color constant and variable portions of both chains as indicated by color code ovals on the figure. Also label antibody *hinge region, binding sites, disulfide bonds,* and *carbohydrates.*

✓ **G3.** Because glycoproteins forming antibodies are involved in immunity, they are called

immunoglobulins, abbreviated _____. The different classes of Ig's are distinguishable by the *(constant? variable?)* portions of the antibody structure. Write the name of the related class of Ig next to each description.

_____ a. The first antibodies to be secreted after initial exposure to antigen, they are short lived; they destroy invading microbes by agglutination and lysis.

_____ b. The only type of Ig to cross the placenta, it provides specific resistance of newborns. Also significantly enhances phagocytosis and neutralizing of toxins in persons of all ages because it is the most abundant type of antibody.

_____ c. Found in secretions such as mucus, saliva, and tears and so protects against oral, vaginal, and respiratory infections.

_____ d. Located on mast and basophil cells and involved in allergic reactions, for example, to certain foods, pollen, or bee venom.

_____ e. Activate B cells to produce antibodies

G4. List five mechanisms that antibodies may use to attack and inactivate antigens.

G5. *A clinical correlation.* Define monoclonal antibodies (MAb's). These antibodies are

produced by _____ cells. Explain how a *hybridoma* cell is produced.

Describe use of monoclonal antibodies clinically.

✓ **G6.** Complete this Checkpoint on the role of complement in immunity. (Review Checkpoint D3c above.)
a. The complement system consists of more than 20 different *(proteins? lipids?)*. Name several of these.

b. The three overall effects of complement are:
1. *(Constriction? dilation?)* of blood vessels which _____-creases movement of WBCs into the infected area.
2. Coating of microbes, a process known as _____, which promotes phagocytosis.
3. Formation of the membrane attack complex (MAC) which leads to

_____ of the invading microbes.

✓ **G7.** Do this exercise on immunizations.
a. Once a specific antigen has initiated an immune response, either by infection or by a(n) *(initial? booster dose?)* immunization, the person produces some long-lived B and

T cells called _____.
b. The level of antibodies rises *(slowly? rapidly?)* after the initial immunization. This level

measured in serum is known as the antibody _____. Which antibody

titer rises first? *(IgA? IgG? IgM?)* Which rises next? _____
c. Upon subsequent exposure to the same antigen, such as during another infection or

by a _____ dose of vaccine, memory cells provide a *(more? less?)* intense response. Antibodies produced by this secondary response, also known as

_____ _____, have a *(higher? lower?)* affinity for the antigen compared to antibodies produced in the primary response.

✓ **G8.** *A clinical challenge.* Identify types of immunity described below by selecting from answers:

AAAI.	**Artificially acquired active immunity**
AAPI.	**Artificially acquired passive immunity**
NAAI.	**Naturally acquired active immunity**
NAPI.	**Naturally acquired passive immunity**

_____ a. As 31-year-old Bud tore apart an old shed, a rusty nail entered his left hand. Bud, who reported that he had not had a tetanus shot since he was "about 14," received a shot of tetanus immunoglobu-lins at the emergency room.

_____ b. Kelly has provided her 3-month-old baby Crystal with temporary immunity by the antibodies that crossed over the placenta during Kelly's pregnancy and also by antibodies in milk from breast feeding.

_____ c. Kim took her baby Jamie to the clinic for Jamie's regularly scheduled MMR (measles-mumps-rubella) immunization.

H. Self-recognition and immunological tolerance (pages 765–766)

✓ **H1.** The ability of your T cells to avoid reacting with your own body proteins is known as

immunological _____, B cells *(also? do not?)* display this tolerance.

Loss of such tolerance results in _____ disorders.

H2. Describe each of the processes listed below, which are steps in development of immunological tolerance involving T cells.

a. Positive selection

b. Deletion

c. Anergy

Which of the above steps *(a? b? c?)* appears to be the main mechanism that B cells

use to prevent responses to self-proteins? _____

H3. "The ability to recognize tumor antigens as nonself" is a definition of the characteristic of immunological *(surveillance? tolerance?)*. Describe how each of the following forms of therapy may help cancer patients whose own immunological surveillance is failing.

a. Adoptive cellular immunotherapy

b. Cytokine therapy

I. Aging and the immune system (pages 766–767)

✓ **I1.** Complete arrows to indicate immune changes that usually accompany aging.
I a. Number of helper T cells
I b. Risk of autoimmune conditions by production of antibodies against self
I c. Response to vaccines

I2. Write several health practices that can help you to maximize your resistance against disease.

J. Disorders, medical terminology (pages 767–770)

✓ **J1.** Answer these questions about AIDS.

a. AIDS refers to a_____ i_____-d_____

   s_____.

b. The causative agent for AIDS has been identified as the h_____

   i_____ v_____. What is the major effect of this virus?

c. List the groups of persons that have been most infected in the United States.

d. In the U.S. most AIDS patients are *(female? male?)*. Name the three groups of people

with greatest *increase* in incidence of HIV in the U.S. _____

_____ _____. Worldwide, _____ million people are

infected with HIV, and about *(10%? 25%? 50%?)* of them are women, and _____% children. Globally, HIV is largely transmitted by *(homosexual? heterosexual?)* contacts and by infected blood.

e. Like other viruses, the HIV virus *(does? does not?)* depend on host cells for replication.

The HIV virus is a _____-virus because its genetic code is carried in *(DNA? RNA?)*. What role does reverse transcriptase play in a retrovirus?

Name AIDS drugs that work by inhibiting this enzyme.

f. The covering over the HIV viral RNA contains glycoproteins known as *(CD4? GP120?)*; these serve as "docking proteins." These chemical protrusions permit the HIV virus to attach (or "dock") on *(CD4? GP120?)* receptors on human immune cells. Name two or more types of cells that may be infected by the HIV virus.

_____ _____

g. Once inside human cells, the HIV virus causes the human cells to become "factories" of over *(1,000? 1 million? 1 billion?)* HIV viruses per day. The T4 cells themselves *(are? are not?)* destroyed in the process.

h. The person is "HIV-positive" when *(T cells drop to under 200/mm³? anti-HIV antibodies are identified in the blood stream?)* The person is diagnosed as having AIDS when *(T cells drop to under 200/mm3? anti-HIV antibodies are identified in the blood stream?)*, or when opportunistic infections occur. Define this type of infection.

i. List five or more signs or symptoms of HIV infection.

j. Besides reverse transcriptase inhibitors, which drugs contribute to *triple therapy* (or

"the cocktail") available to HIV patients since 1995? _____ How do these help control the HIV virus?

k. Circle the four fluids through which the HIV virus has been found to be transmitted in sufficient quantities to be infective.

Blood   Breast milk   Mosquito venom   Saliva
Semen   Tears   Vaginal fluids

✓ **J2.** Match the four types of allergic responses in the box with the correct descriptions below

I. Type I	II. Type II	III. Type III	IV. Type IV

_____ a. Known as *cell-mediated* reactions, or delayed-type-hypersensitivity reactions, these lead to responses such as those that occur after exposure to poison ivy or after tuberculin testing (Mantoux test)

_____ b. Antigen-antibody *immune complexes* trapped in tissues activate complement and lead to inflammation, as in lupus (SLE) or rheumatoid arthritis (RA)

_____ c. Caused by IgG or IgM antibodies directed against RBCs or other cells, as in, for example, response to an incompatible transfusion; called a *cytotoxic* reaction

_____ d. Involves IgEs produced by mast cells and basophils; effects may be localized (such as swelling of the lips) or systemic, such as acute *anaphylaxis*

✓▲ **J3.** Match the condition in the box with the description below.

A. Autoimmune disease	LA. Lymphadenopathy
IM. Infectious mononucleosis	S. Splenomegaly

_____ a. Contagious infection of young women especially; B lymphocytes become infected by EBV virus and then enlarge to resemble monocytes

_____ b. Enlarged, tender lymph glands

_____ c. Enlarged spleen

_____ d. Multiple sclerosis (MS), rheumatoid arthritis (RA), insulin-dependent diabetes mellitus, and systemic lupus erythematosus (SLE) are examples.

✓ **J4.** Match types of transplants in the box with descriptions below.

ALLO. Allograph	AUTO. Autograft	Xeno. Xenograft

_____ a. Between animals of different species

_____ b. Between different parts of the same person, for example, healthy skin to regions of burned skin

_____ c. Between individuals of same species, but of different genetic backgrounds, such as a mother's kidney donated to her daughter or a blood transfusion

ANSWERS TO SELECTED CHECKPOINTS: CHAPTER 22

A2. (a-c) NS. (d-e) S. (f) NS.

A3. (a) Interstitial fluid and proteins; lipids (or fats), fat. (b) Lympho; T, antibodies.

A4. Lymph, lymph vessels, and lymph organs such as tonsils, spleen, and lymph nodes

A6. (a) Three-fourths; cisterna chyli; digestive and other abdominal organs and both lower limbs. (b) Left jugular, left subclavian, and left broncho-mediastinal. (c) Right lymphatic; right upper limb and right side of thorax, neck, and head. (d) Left subclavian vein; the same. (e) Interstitial; plasma.

A7. Fracture of this bone might block the thoracic duct, which enters veins close to the left clavicle. Lymph therefore backs up with these results: (a) Extra fluid remains in tissue spaces normally drained by vessels leading to the thoracic duct. (b) Lymph capillaries in the intestine normally absorb fat from foods. Slow lymph flow can decrease such absorption leaving fat in digestive wastes.

A9. Skeletal muscle contraction squeezing lymphatics which contain valves that direct flow of lymph. Respiratory movements.

B1. (a) P. (b) S. (c) S. (d) P.

B2. (a) Mediastinum; sternum; large, de. (b) Cortex; lympho; T, throughout life. (c) Medulla; thymic hormones.

B3. (a) A, ALV; B, TS; C, ELV; D, GC; E, MC. (b1) ALV; (b2) ELV. (b3) GC; (b4) TS; (b5) MC. (b6) TS, GC.

B5. Metastasis; nontender; fixed via "tentacle-like extensions" (hence the name *cancer*, meaning *crablike*).

B6. Connective tissue scaffolding; functional cells.

B7. Upper left, diaphragm. (a) Immune functions of B and T cells and phagocytosis of pathogens in blood. (b) Phagocytosis of defective or aging RBCs and platelets, storage of platelets, and hemopoiesis in fetal life.

B8. (a) Mucosa-associated lymphoid tissue; digestive, respiratory, urinary, and reproductive. (b1) Located in the wall of the small intestine, this tissue can protect against microbes found in food or present in excessive numbers in the GI tract; (b2) Because tonsils and adenoids surround the pharynx (throat), they protect against inhaled or ingested microbes. (c) Palatine; pharyngeals.

C1. Meso, fifth.

C2. Veins. (a) Posterior. (b) Jugular; thoracic.

D2. (a) C. (b) D. (c) F. (d) E. (e) A. (f) G. (g) H. (h)B.

D3. (a) Viruses; lymphocytes, macrophages, fibroblasts; uninfected, antiviral. (b) Alpha, beta, and gamma. (c) In blood plasma and on plasma membranes; when activated, they "complement" or enhance other defense mechanisms. (d) Natural killer, lymphocyte; 5–10; wide variety; perforins. (e) Neutrophil; monocytes, macrophages; skin, liver, lungs, brain, and spleen.

D4. (a) C. (b) A. (c) I. (d) K.

D6. (a–d) 1. (e) 2. (f) 3.

D7. (a) PG. (b) H, LT. (c) I. (d) K. (e) K, C. (f) T.

E1. (a) Specific, immunology. (b) Antigens; even more rapid and vigorous.

E2. (a) B cells (or B lymphocytes), T cells (or T lymphocytes); from stem cells in bone marrow. (b) T; perform immune functions against specific antigens if properly stimulated; CD8, antigen-receptor protein. (c) Bone marrow.

E3. (a) CMI. (b) AMI. (c) AMI. (d) CMI. (e) AMI and CMI

E4. (a) Is recognized as foreign and provokes an immune response; foreign substances. (b1) Stimulate production of; (b2) React with (and potentially be destroyed by). (c) Complete antigen, or immunogen. (d) Hapten; reactivity, immunogenicity; body protein; allergic. (e) Large, complex molecules, such as proteins, nucleoproteins, lipoproteins, glycoproteins, or complex polysaccharides. (f) They are made of simple, repeating subunits that are not likely to be antigenic. (g) Yes; flagella, capsules, cell walls, as well as toxins made by bacteria. (h) Antigens. (i) Only a specific region of

the; antigenic determinant (epitope); a number of.

E6. (a) Good news: they help T cells to recognize foreign invaders because antigenic proteins must be processed and presented in association with MHC antigens before T cells can recognize the antigenic proteins; bad news: MHC antigens present transplanted tissue as "foreign" to the body. causing it to be rejected. (b) Organ transplant such as kidney or liver or tissue transplant (such as bone marrow). (c) MHC–I molecules are in plasma membranes of all body cells except in RBCs; MHC–II antigens are present on APCs, thymic cells, and T cells activated by previous exposure to the antigen. (d) B (basis of the principle of self-tolerance). (e) Autoimmune.

E7. (a) B or T: B; T. (b) Antigens from intruders outside of body cells, such as bacteria, pollen, foods, cat hair; antigen presenting cells; at sites where antigens are likely to enter the body, such as skin or mucous membranes; macrophages, B cells, and dendritic cells (including Langerhans cells in skin). (c) 1. Phago/endo. 2. Partial dig. 3. Fusion. 4. Bind to MHC-II. 5. Exo/insert. 6. Present to T. (d) Endo; viral proteins made after a virus infects a cell and takes over cell metabolism.

E8. (a) M (b) IL–4. (c) IL–2. (d) Gamma-IFN. (e) MMIF. (f) P. (g) LT. (h) IL–1, TNF.

F2. (a) T_C. (b) T_H. (c) T_C. (d) T_H. (e) M.

Figure LG 22.2 Diagram of an antibody molecule.

G2. (a) Globulins; 4; 450, 220. See Figure LG 22.2A. (b) Variable; constant, are identical. See Figure LG 22.2A.

G3. Ig's; constant. (a) M. (b) G. (c) A. (d) E. (e) D.

G6. (a) Proteins; proteins C1–C9 and proteins B, D, and P (properdin). (b1) Dilation, in; (b2) Opsonization; (b3) Cytolysis (or bursting).

G7. (a) Initial, memory cells. (b) Slowly; titer; IgM; IgG. (c) Booster, more; immunological memory, higher.

G8. (a) AAPI. (b) NAPI. (c) AAAI.

H1. Tolerance: also; autoimmune.

I1. (a) ↓ .(b) ↑ .(c) ↓.

J1. (a) Acquired immunodeficiency syndrome.
(b) Human immunodeficiency virus: by destroying the immune system, the disease places the patient at high risk for opportunistic infections (normally harmless), such as PCP, and for other conditions such as KS. (c) Gay men who have not used safe sexual practices. IV drug users. hemophiliacs who received contaminated blood products before 1985. and heterosexual partners of HIV-infected persons. (d) Male; people of color, women, and teenagers; 40, 50; heterosexual. (e) Does; retro, RNA; directs DNA synthesis in host cells modeled on the HIV RNA code; AZT, ddl, and ddC. (f) GP120; CD4; CD4 (or T4 or helper T) cells, macrophages, and dendritic cells. (g) 1 billion; are. (h) Anti-HIV antibodies are identified in the blood stream; T cells drop to under 200/mm³; infections that do not occur in persons with normal immunity. (i) Initially, flu-like illness such as sore throat, swollen lymph glands, and night sweats; later, diarrhea, weight loss, serious respiratory infections, lymphoma, neuropathy, and dementia. (j) Protease inhibitors; they inhibit an enzyme (protease) needed to assemble the protein coat around newly-forming HIV viruses. (k) Blood, semen, vaginal fluids, and breast milk.

J2. (a) IV. (b) III. (c) II. (d) I.

J3. (a) IM. (b) LA. (c) S. (d) A.

J4. (a) XENO. (b) AUTO. (c) ALLO.

MORE CRITICAL THINKING: CHAPTER 22

1. Contrast lymphatic flow with blood circulation. Incorporate into your discussion the differences between (a) blood and lymph, (b) forces controlling flow of blood and of lymph, (c) lymph capillaries and blood capillaries, and (d) major lymph vessels and major blood vessels.

2. List the four cardinal signs or symptoms of inflammation. Then describe the processes that occur during inflammation that account for these signs or symptoms. Be sure to include roles of these chemicals: histamine, kinins, prostaglandins (PGs), and complement. Also discuss these processes: vasodilation, emigration, and leukocytosis.

3. Contrast nonspecific mechanisms for resistance known as the first line of defense and the second line of defense.

4. Contrast antigens and antibodies. Include several examples of each. Further distinguish endogenous from exogenous antigens, as well as complete and partial antigens.

5. Contrast antibody-mediated immunity (AMI) and cell-mediated immunity (CMI).

6. Contrast each of the following cytokines by describing types of cells producing the chemicals and also functions of each cytokine: interferons (alpha, beta, and gamma IFNs), interleukins (IL-I, IL–2, IL–4, IL–5), lymphotoxin, and perforin.

7. Differentiate the five classes of immunoglobulins.

8. Contrast helper T (T4) cells with cytotoxic T (T8) cells.

MASTERY TEST: CHAPTER 22

Questions 1–2: Arrange the answers in correct sequence.

_____ _____ _____ _____ _____ 1. Flow of lymph through a lymph node:
A. Afferent lymphatic vessel
B. Medullary sinus
C. Cortical sinus
D. Efferent lymphatic vessel

_____ _____ _____ _____ _____ 2. Activities in humoral immunity, in chronological order:
A. B cells develop in bone marrow or other part of the body.
B. B cells differentiate and divide (clone), forming plasma cells.
C. B cells migrate to lymphoid tissue.
D. B cells are activated by specific antigen that is presented.
E. Antibodies are released and are specific against the antigen that activated the B cell.

Questions 3–16: Circle the letter preceding the one best answer to each question. Take particular note of underlined terms in selected questions.

3. Both the thoracic duct and the right lymphatic duct empty directly into:
   A. Axillary lymph nodes
   B. Superior vena cava
   C. Cisterna chyli
   D. Subclavian arteries
   E. Subclavian veins

4. Which lymphatic trunk drains lymph primarily from the left upper limb?
   A. Left jugular trunk
   B. Left bronchomediastinal trunk
   C. Left subclavian trunk
   D. Left lumbar trunk

5. All of these are examples of nonspecific defenses *except:*
   A. Antigens and antibodies
   B. Saliva
   C. Complement
   D. Interferon
   E. Skin
   F. Phagocytes

6. All of the following correctly match lymphocytes with their functions *except:*
   A. Helper T cells: stimulate B cells to divide and differentiate
   B. Cytotoxic (killer) T cells: produce lymphokines that attract and activate macrophages and lymphotoxins that directly destroy antigens
   C. Natural killer (NK) cells: suppress action of cytotoxic T cells and B cells
   D. B cells: become plasma cells that secrete antibodies

7. All of the following match types of allergic reactions with correct descriptions *except:*
   A. Type I: anaphylactic shock, for example, from exposure to iodine or bee venom
   B. Type II: transfusion reaction in which recipient's IgG or IgM antibodies attack donor's red blood cells
   C. Type III: Involves Ag–Ab-complement complexes that cause inflammations, as in rheumatoid arthritis (RA)
   D. Type IV: cell-mediated reactions that involve immediate responses by B cells, as in the TB skin test

8. Choose the one *false* statement about T cells.
   A. T$_C$ cells are best activated by antigens associated with both MHC–I and MHC–II molecules.
   B. T$_H$ cells are best activated by antigens associated with MCH–II molecules.
   C. For T$_C$ cells to become cytolytic, they need costimulation by IL–2.
   D. Perforin is a chemical released by T$_H$ cells.

9. Choose the *false* statement about lymphatic vessels.
   A. Lymph capillaries are more permeable than blood capillaries.
   B. Lymphatics have thinner walls than veins.
   C. Like arteries, lymphatics contain no valves.
   D. Lymph vessels are blind-ended.

10. Choose the *false* statement about nonspecific defenses.
    A. Complement functions protectively by being converted into histamine.
    B. Histamine increases permeability of capillaries so that leukocytes can more readily reach the infection site.
    C. Complement and properdin are both proteins found in blood plasma.
    D. Opsonization enhances phagocytosis.

11. Choose the *false* statement about T cells.
    A. Some are called memory cells.
    B. They are called T cells because they are processed in the thymus.
    C. They are involved primarily in antibody-mediated immunity (AMI).
    D. Like B cells, they originate from stem cells in bone marrow.

12. Choose the *false* statement about lymph nodes.
    A. Lymphocytes are produced here.
    B. Lymph nodes are distributed evenly through out the body, with equal numbers in all tissues.
    C. Lymph may pass through several lymph nodes in a number of regions before returning to blood.
    D. Lymph nodes are shaped roughly like kidney (or lima) beans.

13. Choose the *false* statement about lymphatic organs.
    A. The palatine tonsils are the ones most often removed in a tonsillectomy.
    B. The spleen is the largest lymphatic organ in the body.
    C. The thymus reaches its maximum size at age 40.
    D. The spleen is located in the upper left quadrant of the abdomen.

14. Choose the *false* statement.
    A. Skeletal muscle contraction <u>aids</u> lymph flow.
    B. Skin is normally <u>more</u> effective than mucous membranes in preventing entrance of microbes into the body.
    C. Interferon is produced by <u>viruses</u>.
    D. An allergen is an <u>antigen</u>, not an antibody.
15. Choose the one *true* statement about AIDS.
    A. It appears to be caused by a virus that carries its genetic code in <u>DNA</u>.
    B. Persons at highest risk are <u>homosexual men and women</u>.
    C. Persons with AIDS experience a decline in <u>helper T cells</u>.
    D. AIDS is transmitted primarily through <u>semen and saliva</u>.

16. All of the following are cytokines known to be secreted by helper T cells except:
    A. Gamma interferon
    B. Interleukin–1
    C. Interleukin–2
    D. Interleukin–4

Questions 17–20: Circle T (true) or F (false). If the statement is false, change the underlined word or phrase so that the statement is correct.

T F   17. Antibodies are usually composed of <u>one light and one heavy</u> polypeptide chain.

T F   18. T_C cells are especially active against slowly growing bacterial diseases, some viruses, cancer cells associated with viral infections, and transplanted cells.

T F   19. Immunogenicity means the ability of an antigen to <u>react with</u> a specific antibody or T cell.

T F   20. A person with autoimmune disease produces <u>fewer than normal antibodies</u>.

Questions 21–25: Fill-ins. Complete each sentence with the word or phrase that best fits.

_____   21. _____ lymphocytes provide antibody-mediated immunity,

and _____ lymphocytes and macrophages offer cellular (cell-mediated) immunity.

_____   22. Helper T cells are also known as _____ cells.

_____   23. List the four fundamental signs or symptoms of inflammation.

_____   24. The class of immunoglobins most associated with allergy are

Ig _____.

_____   25. Three or more examples of mechanical factors that provide

nonspecific resistance to disease are _____.

ANSWERS TO MASTERY TEST: CHAPTER 22

Arrange

1. A C B D
2. A C D B E

Multiple Choice

3. E	10. A
4. C	11. C
5. A	12. B
6. C	13. C
7. D	14. C
8. D	15. C
9. C	16. B

True–False

17. F. Two light and two heavy; chains
18. T
19. F. Stimulate formation of
20. F. Antibodies against the individual's own tissues

Fill-Ins

21. B, T
22. T_H cells, T4 cells, and CD4 cells.
23. Redness, warmth, pain, and swelling
24. E
25. Skin, mucosa, cilia, epiglottis; also flushing by tears, saliva, and urine

FRAMEWORK 23
Respiratory System

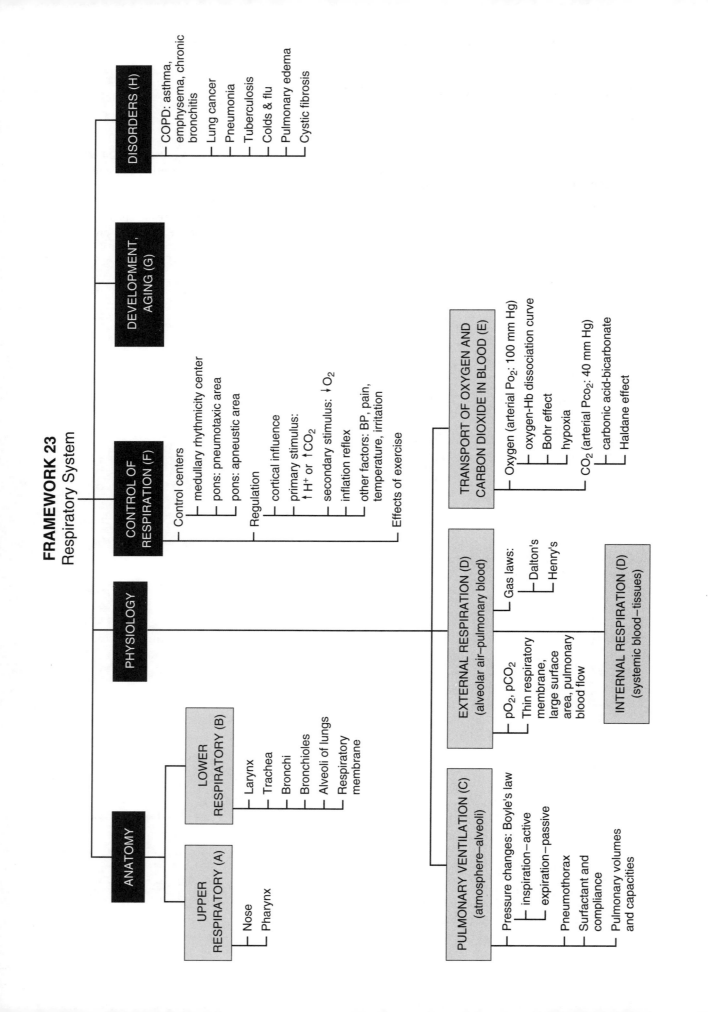

C H A P T E R **23**

The Respiratory System

Oxygen is available all around us. But to get an oxygen molecule to a muscle cell in the stomach or a neuron of the brain requires coordinated efforts of the respiratory system with its transport adjunct, the cardiovascular system. Oxygen traverses a system of ever-narrowing and diverging airways to reach millions of air sacs (alveoli). Each alveolus is surrounded by a meshwork of pulmonary capillaries, much like a balloon encased in a nylon stocking. Here oxygen changes places with the carbon dioxide wastes in blood, and oxygen travels through the blood to reach the distant stomach or brain cells. Interference via weakened respiratory muscles (as in normal aging) or extremely narrowed airways (as in asthma or bronchitis) can limit oxygen delivery to lungs. Decreased diffusion of gases from alveoli to blood (as in emphysema, pneumonia, or pulmonary edema) can profoundly affect total body function and potentially lead to death.

Take a couple of deep breaths and, as you begin your study of the respiratory system, carefully examine the Chapter 23 Topic Outline and Objectives. Check off each one as you complete it. To further organize your study of the respiratory system, glance over the Chapter 23 Framework. Be sure to refer to the Framework frequently and note relationships among key terms in each section.

TOPIC OUTLINE AND OBJECTIVES

A. Anatomy: upper respiratory system

1. Describe the anatomy and histology of the nose, pharynx, larynx, trachea, bronchi, and lungs.
2. Identify the functions of each respiratory system structure.

B. Anatomy: lower respiratory system

C. Physiology of respiration: pulmonary ventilation; lung volumes and capacities

3. Describe the events that cause inspiration and expiration
4. Define the various lung volumes and capacities

D. Physiology of respiration: exchange of oxygen and carbon dioxide

5. Explain Dalton's law and Henry's law
6. Describe the exchange of oxygen and carbon dioxide in external and internal respiration.

E. Physiology of respiration: transport of oxygen and carbon dioxide in blood

7. Explain how oxygen and carbon dioxide are transported in the blood.

F. Regulation of respiration; effects of exercise

8. Describe the various factors that regulate the rate and depth of respiration.
9. Describe the effects of exercise on the respiratory system.

G. Developmental anatomy and aging of the respiratory system

10. Describe the development of the respiratory system.
11. Describe the effects of aging on the respiratory system.

H. Disorders, medical terminology

WORDBYTES

Now become familiar with the language of this chapter by studying each wordbyte, its meaning, and an example of its use within a term. After you study the entire list, self-check your understanding by writing the meaning of each wordbyte on the line. As you continue through the *Learning Guide*, identify (and fill in) additional terms that contain the same wordbyte.

Wordbyte	Self-check	Meaning	Example(s)
atele-	_____	incomplete	*atele*ctasis
-baros	_____	pressure	hyper*baric*
dia-	_____	through	*dia*phragm
dys-	_____	bad, difficult	*dys*pnea
-ectasis	_____	dilation	atel*ectasis*
ex-	_____	out	expiration
hyper-	_____	over	*hyper*baric
in-	_____	in	*in*spiration
intra-	_____	within	*intra*pulmonic
-pnea	_____	breathe	a*pnea*
pneumo-	_____	air	*pneumo*thorax
pulmo-	_____	lung	*pulmo*nary
rhin-	_____	nose	*rhin*orrhea
spir-	_____	breathe	in*spir*ation

CHECKPOINTS

A. Anatomy: upper respiratory system (pages 775–780)

A1. Explain how the respiratory and cardiovascular systems work together to accomplish gaseous exchange among the atmosphere, blood, and cells.

✓ **A2.** Refer to Figure LG 23.1 and visualize the process of respiration in yourself as you do this exercise.

a. Label the first phase of respiration at (1) on the figure. This process involves transfer of air from your environment to *(alveoli? pulmonary capillaries?)* in your lungs.

Inflow of air rich in *(O_2? CO_2?)* is known as _____-spiration, whereas outflow of air

concentrated in *(O_2? CO_2?)* is known as _____-spiration.

b. Phase 2 of respiration involves exchange of gases between _____ in

Pulmonary
capillary

Alveolus

RA

LA

LV

RV

pO_2:_____ mm Hg

pCO_2:_____ mm Hg

pO_2:_____ mm Hg

pCO_2:_____ mm Hg

Inferior vena cava

Aorta

Cells

Systemic capillary

Figure LG 23.1 Phases of respiration and circulatory routes.
Numbers refer to Checkpoint A2. Fill in short lines according to Checkpoint D4.

● High in CO_2 (blue)

○ High in O_2 (red)

alveoli and _____ in pulmonary capillaries. This process is known as *(external? internal?)* respiration. Label phase 2 on the figure.

c. After oxygenated blood returns to your heart via your pulmonary *(artery? veins?)*, blood is pumped to your tissues where gas exchange occurs between

_____ and _____. This process is known as *(external? internal?)* respiration. Label phase 3 on the figure.

d. Blood high in CO, then returns to the heart where it passes directly into the pulmonary *(artery? vein?)* to reach the lungs. Here diffusion of CO_2 from pulmonary capillaries to alveoli occurs; this is another aspect of *(external? internal?)* respiration

(phase _____). Finally, you exhale this air, a process that is part of *(external*

respiration? pulmonary ventilation?) (phase _____).

   e. Which process takes place within tissue cells, utilizing O_2 and producing CO_2 as a waste product? *(Cellular? Internal?)* respiration.

   f. *For extra review,* color all blood vessels on Figure LG 23.1 according to color code ovals on the figure.

✓▲ **A3.** Identify parts of the respiratory system in this activity.

   a. Arrange respiratory structures listed in the box in correct sequence from first to last in the pathway of inspiration.

ADA.	Alveolar ducts and alveoli	MN.	Mouth and nose
BB.	Bronchi and bronchioles	P.	Pharynx
L.	Larynx	T.	Trachea

_____ → _____ → _____ → _____ → _____ → _____

   b. Write asterisks (*) next to structures forming the upper respiratory system.

   c. Most of the structures listed in the box above are parts of the *(conducting? respiratory?)* portion of the respiratory system.

✓▲ **A4.** Match each term related to the nose with the description that best fits.

External nares	Paranasal sinuses and meatuses
Internal nares (choanae)	Rhinoplasty
Nasal cartilages	Vestibule

   a. Mucous membrane-lined spaces and passageways: _____

   b. Nostrils: _____

   c. Anterior portion of nasal cavity: _____

   d. Openings from posterior of nose into nasopharynx: _____

   e. Repair of the nose: _____

   f. Known as septal, alar, and lateral nasal, these provide a structural framework of

       the nose: _____

✓▲ **A5.** Name the structures of the nose that are designed to carry out each of the following functions.

   a. Warm, moisten, and filter air

   b. Sense smell

   c. Assist in speech

✓▲ **A6.** Write *LP* (laryngopharynx), *NP* (nasopharynx), or *OP* (oropharynx) to indicate the location of each of the following structures.

_____ a. Adenoids   _____ d. Opening (fauces) from oral cavity

_____ b. Palatine tonsils   _____ e. Opening into larynx and esophagus

_____ c. Lingual tonsils

A7. On Figure LG 23.2, cover the key and identify all structures associated with the nose, palate, pharynx, and larynx. Color structures with color code ovals.

KEY

A. Conchae
B. External naris
O C. Hard palate
D. Oral cavity
E. Tongue
F. Lingual tonsil
O G. Epiglottis
H. Hyoid bone

O I. True vocal cords
O J. Thyroid cartilage
O K. Cricoid cartilage
L. Trachea
M. Paranasal sinuses
N. Internal naris
O. Opening of auditory tube

O P. Pharyngeal tonsil (adenoid)
Q. Nasopharynx
O R. Soft palate
O S. Palatine tonsil
T. Oropharynx
U. Laryngopharynx
V. Esophagus

Figure LG 23.2 Sagittal section of the right side of the head with the nasal septum removed. Color and identify labeled structures as directed in Checkpoint A7. Arrow refers to Checkpoint B2.

B. Anatomy: lower respiratory system (pages 780–790)

✓▲ **B1.** Match parts of the larynx with the descriptions that best fit.

A.	**Arytenoid cartilages**	RG.	**Rima glottidis**
C.	**Cricoid cartilage**	T.	**Thyroid cartilage**
E.	**Epiglottis**		

_____ a. Also known as the Adam's apple; longer in males than in females

_____ b. Includes a leaflike portion that covers the airway during swallowing

_____ c. A pair of triangular-shaped cartilages attached to muscles that control vocal cords

_____ d. Forms the inferior portion of the larynx

_____ e. Space between the vocal cords through which air passes during ventilation

B2. Explain how the larynx prevents food from entering the trachea. (*Hint:* Notice the arrow on Figure LG 23.2.)

B3. Tell how the larynx produces sound. Explain how pitch is controlled and what causes male pitch usually to be lower than female pitch.

✓▲ **B4.** Complete this Checkpoint about the trachea.
  a. The trachea is commonly known as the _____. It is located *(anterior? posterior?)* to the esophagus. About _____ cm (_____ inches) long, the trachea terminates at the _____, where the trachea leads to a Y-shaped intersection with the primary bronchi.

  b. The tracheal wall is lined with a _____ membrane and strengthened by 16–18 C-shaped rings composed of _____.

✓ **B5.** Describe the functional advantages provided by each of these parts of the trachea:
  a. Pseudostratified ciliated epithelium lining the mucosa

  b. C-shaped cartilage rings with the trachealis muscle completing the "ring" posteriorly

c. Sensitivity of the mucosa at the carina

✓▲ **B6.** *A clinical correlation.* Anna has aspirated a small piece of candy. Dr. Lennon expects to find it in the right bronchus rather than in the left. Why?

✓▲ **B7.** Bronchioles *(do? do not?)* have cartilage rings. How is this fact significant during an asthma attack?

▲ **B8.** Refer to Figure LG 23.3 and do this activity. Use Figure 23.8 (page 784 of your text) for help.
   a. Identify tracheobronchial tree structures J–N, and color each structure indicated with a color code oval.
   b. Color the two layers of the pleura and the diaphragm according to color code ovals. Also, cover the key and identify all parts of the diagram from A to I.

✓ **B9.** Write the correct term next to each description.

   a. Inflammation of the pleura: _____

   b. Excess fluid in the pleural space: _____

   c. Procedure to remove fluid from the pleural space: _____

✓▲ **B10.** Answer these questions about the lungs. (As you do the exercise, locate the parts of the lung on Figure LG 23.3.)
   a. The broad, inferior portion of the lung that sits on the diaphragm is called the

       _____ . The upper narrow apex of each lung extends just superior to

       the _____ . The costal surfaces lie against the _____ .

   b. Along the mediastinal surface is the _____ where the root of the

       lung is located. This root consists of _____ .
   c. Answer these questions with *right* or *left*.
       In which lung is the cardiac notch located? _____

       Which lung has just two lobes and so only two lobar bronchi? _____

       Which lung has a horizontal fissure? _____
       Write *S* (superior), *M* (middle), and *I* (inferior) on the three lobes of the right lung.

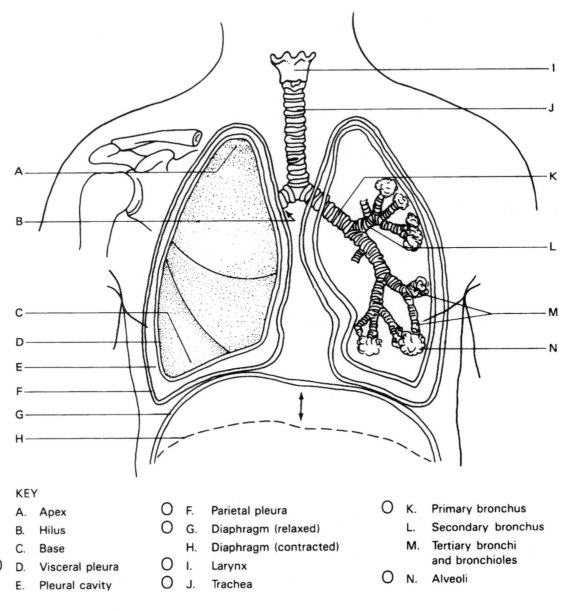

KEY

A. Apex
B. Hilus
C. Base
O D. Visceral pleura
E. Pleural cavity

O F. Parietal pleura
O G. Diaphragm (relaxed)
H. Diaphragm (contracted)
O I. Larynx
O J. Trachea

O K. Primary bronchus
L. Secondary bronchus
M. Tertiary bronchi
    and bronchioles
O N. Alveoli

Figure LG 23.3 Diagram of lungs with pleural coverings and bronchial tree. Color and identify labeled structures as directed in Checkpoints B8 and B10.

✓ **B11.** Describe the structures of a lobule in this exercise.

   a. Arrange in order the structures through which air passes as it is about to enter a

     lobule en route to alveoli. _____ _____ _____

A. Alveolar ducts	**R. Respiratory bronchiole**	**T. Tertiary bronchi**

   b. In order for air to pass from alveoli to blood in pulmonary capillaries, it must pass

     through the _____ _____ (respiratory) membrane. Identify structures in this pathway: A to E on Figure LG 23.4.

   c. *For extra review.* Now label layers 1–6 of the respiratory membrane in the insert in Figure LG 23.4.

Figure LG 23.4 Diagram of alveoli and pulmonary capillary. Insert enlarges respiratory membrane. Numbers and letters refer to Checkpoint B11. Complete partial pressures (in mm Hg) as directed in Checkpoint D4.

d. *A clinical correlation.* Normally the respiratory membrane is extremely *(thick? thin?).* Suppose pulmonary capillary blood pressure rises dramatically, pushing extra fluid out of blood. The presence of excess fluid in interstitial areas and ultimately in alveoli

is known as pulmonary _____ . Diffusion occurs *(more? less?)* readily through the fluid-filled membrane.

e. The alveolar epithelial layer of the respiratory membrane contains three types of cells. Forming most of the layer are flat *(squamous? septal? macrophage?)* (Type I alveolar) cells which are ideal for diffusion of gases. Septal (Type II alveolar) cells produce

an important phospholipid and lipoprotein substance called _____ , which reduces tendency for collapse of alveoli due to surface tension. Macrophage

cells help to remove _____ from lungs.

✓ **B12.** *The Big Picture: Looking Back.* Return to earlier chapters in your text and Guide as you contrast the dual blood supply to the lungs.

a. Which vessels supply oxygen-rich blood to the lungs? *(Bronchial? Pulmonary?)* arteries. These vessels carry blood directly from the *(aorta? right ventricle?)*. Review these on Figure 21.21 and Exhibit 21.4 in your text.

b. Which veins return blood carrying CO_2 from lung tissue toward the heart (Figure 21.27 and Exhibit 21.10 in your text, and Chapter LG 21, Checkpoint E8, page LG 453)?

_____ or _____ veins.

c. Refer to Chapter 20 in your *Learning Guide*, Checkpoints B8 and C2 (pages LG 412 and 414). Which vessel contains more highly oxygenated blood? Pulmonary *(artery? vein?)*.

d. Return to Chapter 21, Checkpoint F2, page 454, and answer these questions.
   1. Which system of vessels has higher blood pressure? *(Pulmonary? Systemic?)* One reason for this difference in blood pressure is that the extensive (lengthy) network of arterioles and capillaries in the *(pulmonary? systemic?)* circulation provides greater resistance to blood flow.
   2. Which vessels respond to hypoxia by vasoconstricting so that blood bypasses hypoxic regions and more efficiently circulates to oxygen-rich regions? *(Pulmonary? Systemic vessels, for example, those in skeletal muscle?)*

C. Physiology of respiration: pulmonary ventilation; lung volumes and capacities (pages 790–796)

✓ **C1.** Review the three phases of respiration in the box by matching them with the correct descriptions below.

E.	**External (pulmonary) respiration**	**PV.**	**Pulmonary ventilation**
I.	**Internal (tissue) respiration**		

_____ a. Movement of air between atmosphere and alveoli

_____ b. Exchange of gas between alveoli and blood

_____ c. Exchange of gas between blood and cells

✓ **C2.** Refer to Figure 23.15 in the text and describe the process of ventilation in this exercise.

a. In diagram (a), just before the start of inspiration, pressure within the lungs (called

_____) is _____ mm Hg. This is *(more than? less than? the same as?)* atmospheric pressure.

b. At the same time, pressure within the pleural cavity (called _____

pressure) is _____ mm Hg. This is *(more than? less than? the same as?)* alveolar and atmospheric pressure at sea level.

c. The first step in inspiration occurs as the muscles in the floor and walls of the thorax

contract. These are the _____ and _____ muscles. Note in diagram (b) (and also in Figure LG 23.3) that the size of the thorax *(increases? decreases?)*. Because the two layers of pleura tend to adhere to one another, the lungs will *(increase? decrease?)* in size also.

d. Increase in volume of a closed space such as the pleural cavity causes the pressure

there to *(increase? decrease?)* to _____ mm Hg. Because the lungs also increase in size

(due to pleural cohesion), alveolar pressure also *(increases? decreases?)* to _____ mm Hg. This inverse relationship between volume and pressure is a statement of

_____'s law.

e. A pressure gradient is now established. Air flows from high pressure area *(alveoli? atmosphere?)* to low pressure area *(alveoli? atmosphere?)*. So air flows *(into? out of?)* lungs. Thus *(inspiration? expiration?)* occurs. By the end of inspiration, sufficient air will have moved into the lungs to make pressure there equal to atmospheric

pressure, that is, _____ mmHg.

f. Inspiration is a(n) *(active? passive?)* process, whereas expiration is normally a(n)

(active? passive?) process resulting from _____ recoil. Identify two factors that contribute to this recoil.

During expiration, the diaphragm moves *(inf? sup?)*-eriorly and the sternum moves *(ant? post?)*-eriorly.

g. During labored breathing, accessory muscles help to force air out. Name two of these sets of muscles.

✓ **C3.** Answer these questions about compliance.

a. Imagine trying to blow up a new balloon. Initially, the balloon resists your efforts. Compliance is the ability of a substance to yield elastically to a force; in this case it is the ease with which a balloon can be inflated. So a new balloon has a *(high? low?)* level of compliance, whereas a balloon that has been inflated many times has *(high? low?)* compliance.

b. Similarly, alveoli that inflate easily have *(high? low?)* compliance. The presence of a

coating called _____ lining the inside of alveoli prevents alveolar walls from sticking together during ventilation and so *(increases? decreases?)* compliance.

c. Surfactant is made by type *(I? II? III?)* alveolar cells of the lungs, which produce this chemical especially during the final weeks before birth. A premature infant may lack

adequate production of surfactant; this condition is known as _____.

d. Collapse of all or part of a lung is known as _____ (See page 812 of the text.) Such collapse may result from lack of surfactant or from other factors. Name several of these factors.

✓ **C4.** Complete the arrows to indicate whether resistance to air flow increases (↑) or decreases (↓) in each case.

a. Inhalation, causing dilation of bronchi and bronchioles: |

b. Increase in sympathetic nerve impulses to relax smooth muscle of airways: |

c. COPD such as bronchial asthma or emphysema, with obstruction or collapse or airways: |

C5. Refer to Exhibit 23.1 in your text. Read carefully the descriptions of modified respiratory movements such as sobbing, yawning, sighing, or laughing, while demonstrating those actions yourself.

C6. Contrast these types of breathing patterns:
a. Eupnea/dyspnea (page 813 in the text)

b. Costal breathing/diaphragmatic breathing

C7. Each minute the average adult takes _____ breaths (respirations). Check your

own respiratory rate and write it here: _____ breaths per minute.

✓ **C8.** Of the total amount of air that enters the lungs with each breath, about *(99%? 70%? 30%? 5%?)* actually enters the alveoli. The remaining amount of air is much like the last portion of a crowd trying to rush into a store: it does not succeed in entering the alveoli during an inspiration but just reaches airways and then is quickly ushered out

during the next expiration. Such air is known as anatomic _____ and

constitutes about _____ mL of a typical breath.

✓ **C9.** Maureen, who weighs 150 pounds, is breathing at the rate of 15 breaths/minute. Her

tidal volume is 480 mL/breath. Her minute ventilation is _____ mL. Her alveolar

ventilation rate is likely to be about _____ mL/mm.

✓ **C10.** Match the lung volumes and capacities with the descriptions given. You may find it he helpful to refer to Figure 23.17 in the text.

ERV.	Expiratory reserve volume	RV.	Residual Volume
$FEV_{1.0}$	Forced expiratory volume in one second	TLC.	Total lung capacity
FRC.	Functional residual capacity	VC.	Vital capacity
IRV.	Inspiratory reserve volume	V_T.	Tidal volume

_____ a. The amount of air taken in with each inspiration during normal breathing is

called _____.

_____ b. At the end of a normal expiration the volume of air left in the lungs is called

_____. Emphysemics who have lost elastic recoil of their lungs cannot exhale adequately, so this volume will be large.

_____ c. Forced exhalation can remove some of the air in FRC. The maximum volume of air that can be expired beyond normal expiration is called _____. This volume will be small in emphysema patients.

_____ d. Even after the most strenuous expiratory effort, some air still remains in the lungs; this amount, which cannot be removed voluntarily, is called _____.

_____ e. The volume of air that represents a person's maximum breathing ability is called _____. This is the sum of ERV, V_T and IRV.

_____ f. Adding RV to VC gives _____.

_____ g. The excess air a person can take in after a normal inhalation is called _____.

_____ h. The amount of the vital capacity that can be forced out in 1 second is called _____.

✓ **C11.** Indicate normal values for each of the following.

    a. V_T =   _____ ml (_____ liter)

    b.   TLC   =   _____ ml (_____ liter)

    c.   VC   =   _____ ml (_____ liter)

    d.   FRC   =   _____ ml (_____ liter)

    e.   RV   =   _____ ml (_____ liter)

C12. State the medical and legal significance of minimal volume.

D. Physiology of respiration: exchange of oxygen and carbon dioxide (pages 796–799)

✓ **D1.** Check your understanding of gas laws by matching the correct law with the condition that it explains.

B. Boyle's law	D. Dalton's law	H. Henry's law

_____ a. If a patient breathes air highly concentrated in oxygen (as in a hyperbaric chamber), a higher percentage of oxygen will dissolve in the blood and tissues. This is the principle underlying the use of a hyperbaric chamber.

_____ b. The total atmospheric pressure (760 mm Hg) is due mostly to pressure caused by nitrogen, partly to pO_2 and slightly to pCO_2.

_____ c. Under high pN_2 more nitrogen dissolves in blood. The *bends* occurs when pressure decreases and nitrogen forms bubbles in tissue as it comes out of solution.

_____ d. As the size of the thorax increases, the pressure within it decreases.

(a)

| Nitrogen in atmosphere: 79% (600 mm Hg) | Oxygen in atmosphere: 21% (160 mm Hg) | Atmosphere (760 mm Hg) |

(b)

Figure LG 23.5 Atmospheric pressure. (a) Effect of gas molecules on column of mercury (Hg). (b) Partial pressures contributing to total atmospheric pressure. Color according to directions in Checkpoint D2.

✓ **D2.** Refer to Figure LG 23.5 and do this exercise about Dalton's law.

a. The atmosphere contains enough gaseous molecules to exert pressure upon a column

of mercury to make it rise about _____ mm. The atmosphere is said to have a pressure of 760 mm Hg.

b. Air is about _____% nitrogen and _____% oxygen. To show this, circle the nitrogen molecules in green and oxygen molecules in red in Figure LG 23.5b. Note

that carbon dioxide is not represented in the figure because only about _____% of air is CO_2.

c. Dalton's law explains that of the total 760 mm of atmospheric pressure, a certain amount is due to each type of gas. Determine the portion of the total pressure due to nitrogen molecules:

79% × 760 mm Hg = _____ mm Hg

This is the partial pressure of nitrogen, or _____.

d. Calculate the partial pressure of oxygen.

e. In Figure LG 23.5a, color the part of the Hg column due to N_2 pressure (up to the 600 mm Hg mark) *green*. Then color the portion of the Hg column due to O_2 pressure *red*.

D3. *A clinical correlation.* Hyperbaric oxygenation causes *(more? less?)* O_2 to dissolve in the patient's blood since the person is exposed to *(0.3-0.4? 3-4?)* atmosphere(s) of pressure.
Explain why this procedure kills anaerobic bacteria.

List several other clinical applications of this procedure.

✓ **D4.** Answer these questions about external and internal respiration.
  a. A primary factor in the diffusion of gas across a membrane is the difference in

    concentration of the gas (reflected by _____ pressures) on the two sides of the membrane. On the left side of Figure LG 23.4 write values for pO_2 (in mm Hg) in each of the following areas. (Refer to text Figure 23.18 for help.)
       Atmospheric air (Recall this value from Figure LG 23.5.)
       Alveolar air (Note that this value is lower than that for atmospheric pO_2 largely because some alveolar O_2 enters pulmonary blood.)
       Blood entering lungs
  b. Calculate the pO_2 difference (gradient) between alveolar air and blood entering lungs.

    _____ mm Hg – _____ mm Hg = _____ mm Hg
  c. Three other factors that increase exchange of gases between alveoli and blood are: *(large? small?)* surface area of lungs; *(thick? thin?)* respiratory membrane; and *(increased? decreased?)* blood flow through lungs (as in exercise).
  d. By the time blood leaves the lungs to return to heart and systemic arteries, its pO_2 is normally *(greater than? about the same as? less than?)* pO_2 of alveoli. Write the correct value on Figure LG 23.4.
  e. Now fill in all three pCO_2 values on Figure LG 23.4.
  f. *A clinical correlation.* Jenny, age 49, has had blood drawn from her radial artery to determine her arterial blood gases. Her pO_2 is 56 and her pCO_2 is 48. Are these

    typical values for a healthy adult? _____
  g. Fill in blood gas values (pO_2 and pCO_2) on the short lines on Figure LG 23.1, page LG 495.

✓ **D5.** With increasing altitude the air is "thinner," that is, gas molecules are farther apart, so atmospheric pressure is lower. Atop a 25,000-foot mountain, this pressure is only 282 mm Hg. Oxygen still accounts for 21% of the pressure. What is pO_2 at that level?

From this calculation you can see limitations of life (or modifications that must be made) at high altitudes. If atmospheric pO_2 is 59.2, neither alveolar nor blood pO_2 could surpass that level.

✓ **D6.** *A clinical correlation.* Match the clinical cases listed below with the factors (in box) affecting diffusion across the respiratory (alveolar-capillary) membrane. On each line that follows, write a rationale. One is done for you.

DD. **Diffusion distance**	**PP.** **Partial pressure of gases**
SA. **Surface area for gas exchange**	**SMW.** **Solubility and molecular weight of gases**

____**PP**____ a. Morphine has been prescribed for Ms. Iudica to alleviate her pain. Rationale: **Morphine slows the respiratory rate, so decreases alveolar pO_2.**

_____ b. Mr. Schmidt has interstitial viral pneumonia, with excessive accumulation of

fluid in interstitial spaces. Rationale: _____.

_____ c. Mrs. McLaughlin was diagnosed with emphysema nine years ago.

Rationale: _____.

_____ d. Rev. Wu has a left-sided congestive heart failure (CHF) with pulmonary edema. Arterial blood gas (ABG) analysis indicates her pO_2 is 54 mm Hg and

pCO_2 is 43 mm Hg. Rationale: _____.

E. Physiology of respiration: transport of oxygen and carbon dioxide in blood (pages 799–804)

✓ **E1.** Answer these questions about oxygen transport. Refer to Figure 23.19 in your text.

a. One hundred mL of blood contains about _____ mL of oxygen. Of this, about 19.7 mL

is carried as _____. Only a small amount of oxygen is carried in the dissolved state because oxygen has a *(high? low?)* solubility in blood or water.

b. Oxygen is attached to the _____ atoms in hemoglobin. The chemical

formula for oxyhemoglobin is _____. When hemoglobin carries all of the oxy-

gen it can hold, it is said to be fully _____. High pO_2 in alveoli will tend to *(increase? decrease?)* oxygen saturation of hemoglobin.

c. Refer to Figure 23.20 in your text. Note that arterial blood, with a pO_2 of about

100–105 mm Hg, has its hemoglobin _____% saturated with oxygen. (This may be

expressed as $SO_2 =$ _____%.)

d. List four factors that will enhance the dissociation of oxygen from hemoglobin so that oxygen can enter tissues. (*Hint:* Think of conditions within active muscle tissue.)

e. *For extra review.* Demonstrate the effect of temperature on the oxygen-hemoglobin dissociation curve. Draw a vertical line on Figure 23.22 of the text at $pO_2 = 40$ (the value for venous blood). Compare the percentage saturation of hemoglobin (SO_2) at these two body temperatures:

    38°C (100.4°F) _____% SO_2 43°C (109.4°F) _____% SO_2
    In other words, at a higher body temperature, for example in high fever, *(more? less?)* oxygen will be attached to hemoglobin, while *(more? less?)* oxygen will enter tissues to fuel metabolism.

f. By the time blood enters veins to return to the heart, its oxygen saturation of hemo-

    globin (SO_2) is about _____%. Note on Figure 23.20 of your text that although pO_2 drops from 105 in arterial blood to 40 in venous blood, oxygen saturation drops *(more? less?)* dramatically, that is, from 98% to 75%. Of what significance is this?

g. Fetal hemoglobin (Hb-F) carries 30% *(more? less?)* oxygen than adult hemoglobin (Hb-A). State the significance of this fact.

✓ **E2.** Carbon monoxide has about _____ times the affinity that oxygen has for hemoglobin. State the significance of this fact.

✓ **E3.** Complete this Checkpoint about transport of carbon dioxide.
a. Write the percentage of CO_2 normally carried in each of these forms: _____% is

    present in bicarbonate ion (HCO_3^-); _____% is bound to the globin portion of

    hemoglobin; _____% is dissolved in plasma.
b. Carbon dioxide (CO_2) produced by cells of your body diffuses into red blood cells

    (RBCs) and combines with water to form _____.
c. Carbonic acid tends to dissociate into two products. One is H^+ which binds to

_____. The other product is _____ (bicarbonate), which is carried in *(RBCs? plasma?)* in exchange for a *(K+? Cl−?)* ion that shifts into the RBC.

d. Now write the entire sequence of reactions described in (b) and (c). Be sure to include the enzyme that catalyzes the first reaction. Notice that the reactions show that increase in CO_2 in the body (as when respiratory rate is slow) will tend to cause a buildup of acid (H^+) in the body.

$$CO_2 + \underline{\quad\quad} \xrightarrow[\text{(enzyme)}]{} \underline{\quad\quad} \rightarrow H^+ + \underline{\quad\quad\quad\quad}$$

$$\qquad\qquad\qquad\qquad\qquad\qquad\quad \downarrow \qquad\qquad\qquad\qquad \downarrow$$

$$\qquad\qquad\qquad\qquad\qquad\qquad\; \text{Binds} \qquad\quad \text{Shifts to plasma}$$

$$\qquad\qquad\qquad\qquad\qquad\qquad\; \text{To Hb} \qquad\quad \text{in exchange for } Cl^-$$

E4. Note that as the red blood cells reach lung capillaries, the same reactions you just studied occur, but in reverse. Study Figure 23.24 b in the text carefully. Then list the major steps that occur in the lungs so that CO_2 can be exhaled.

F. Regulation of respiration; effects of exercise (pages 804–809)

✓ **F1.** Complete the table about respiratory control areas. Indicate whether the area is located in the medulla (M) or pons (P).

Name	M/P	Function
a.	M	Controls rhythm; consists of inspiratory and expiratory areas
b. Pneumotaxic		
c.		Prolongs inspiration and inhibits expiration

✓ **F2.** Answer these questions about respiratory control.

a. The main chemical change that stimulates respiration is increase in blood level of

_____, which is directly related to *(decrease in pO_2? increase in pCO_2?)* of blood.

b. Cells most sensitive to changes in blood CO_2 are located in the *(medulla? pons? aorta and carotid arteries?)*.

c. An increase in arterial blood pCO_2 is called _____. Write an arterial

pCO_2 value that is hypercapnic. _____ mm Hg *(Even slight? Only severe?)* hypercapnia will stimulate the respiratory system, leading to *(hyper? hypo?)*-ventilation.

d.   State two locations of chemoreceptors sensitive to changes in pO_2.

     (Even slight? Only large?) decreases in pO_2 level of blood will stimulate these chemore-
     ceptors and lead to hyperventilation. Give an example of a pO_2 low enough to evoke

     such a response. _____ mm Hg
e.   Increase in body temperature (as in fever), as well as stretching of the anal sphincter,

     will cause _____-crease in the respiratory rate.

f.   Take a deep breath. Imagine the _____ receptors in your airways
     being stimulated. These will cause *(excitation? inhibition?)* of the inspiratory and

     apneustic areas, resulting in expiration. This reflex, known as the _____
     reflex, prevents overinflation of the lungs.

g.   The term _____ means a brief cessation of breathing.

✓ **F3.** *A clinical correlation.* Identify types of hypoxia likely to be present in each situation.
         Use these answers: *anemic, histotoxic, hypoxic, stagnant.*

   a.   Cyanide poisoning: _____ hypoxia

   b.   Decreased circulation, as in circulatory shock: _____ hypoxia

   c.   Low hematocrit due to hemorrhage or iron-poor diet: _____ hypoxia

   d.   Chronic bronchitis or pneumonia: _____ hypoxia

✓ **F4.** Indicate whether the following factors will cause apnea (*A*), an increase in respiratory
         rate (*In*) or a decrease (*De*) in respiratory rate.

_____ a. Fever   _____ e. Sudden drop in blood pressure

_____ b. Prolonged somatic pain   _____ f. Activation of the limbic system during
                                                                 emotional anxiety
_____ c. Sudden, severe pain

_____ d. Sudden plunge into cold water   _____ g. Proprioceptive sensations that accompany
                                                                 exercise

✓ **F5.** Identify physiological adjustments made during exercise in this Checkpoint.

   a.   During exercise, blood flow to the lungs (known as pulmonary _____)

        increases as much as _____ times. The rate of diffusion of oxygen
        from alveolar air into pulmonary blood also rises significantly. This rate is known as

        oxygen _____ _____.
   b.   At the onset of moderate exercise, ventilation abruptly increases primarily in
        (rate? depth?), providing more oxygen to active muscles. This change in ventilation
        is due primarily to *(neural? chemical or physical?)* signals that alert the medulla.
   c.   The gradual increase in ventilation that follows is due to a slight increase in *(pO_2?*

        pCO_2?), a slight decrease in *(pO_2? pCO_2?)*, and a slight _____-crease in temperature
        that accompany exercise.

F6. Describe effects of smoking on respiratory efficiency. Include these terms in your description: *terminal bronchioles, hemoglobin, mucus, cilia,* and *elastic fibers.*

G. Developmental anatomy and aging of the respiratory system (pages 809–810)

✓ **G1.** Describe development of the respiratory system in this exercise.

    a. The laryngotracheal bud is derived from _____-derm. List structures formed from this bud.

    b. Identify portions of the respiratory system derived from mesoderm.

✓ **G2.** Describe possible effects of these age-related changes.
    a. Chest wall becomes more rigid as bones and cartilage lose flexibility.

    b. Decreased macrophage and ciliary action of lining of respiratory tract.

H. Disorders, medical terminology (pages 810–813)

✓ **H1.** Describe COPD in this Checkpoint.

    a. COPD stands for C_____ O_____ P_____

        D_____. In most cases, COPD *(is? is not?)* preventable. Name the two

        principal types of COPD: _____ and _____
    b. In emphysema, alveolar walls are broken down so that lungs contain *(more? fewer?)* alveoli that are *(larger? smaller?)* in size. As a result of alveolar destruction, the

        surface area of the respiratory membrane _____-creases, and the amount of gas

        diffusion _____-creases also.

c. This breakdown of alveolar walls is related to effects of smoking and decrease in an

enzyme known as alpha-1-_____. Since this enzyme normally inhibits proteases, then in the absence of the enzyme, proteases are *(more? less?)* active. One

protease, known as _____, is then free to destroy elastic fibers in alveolar walls so that alveoli remain inflated. Consequently, *(more? less?)* air is

trapped in lungs, leading to the classic "_____ chested" appearance of an emphysema patient.

d. Chronic bronchitis involves inflammation of bronchi with *(deficient? excessive?)* mucus production and edematous mucosa lining airways. As a result, airways are *(dilated? narrowed?)*.

e. Asthma (sometimes classified as a COPD) involved changes in mucus and mucosa much like chronic bronchitis. In addition, smooth muscle of airways undergoes broncho-*(dilation? spasm?)*. List several triggers for asthma attacks.

✓ **H2.** Lung cancer is the number *(1? 3? 10?)* cause of cancer death in the United States.

At the time of diagnosis, distant metastases are present in about _____% of patients.

Lung cancer is _____–_____ times more common in smokers than in nonsmokers. Secondhand smoke *(is? is not?)* known to lead to lung cancer deaths. The most

common form of lung cancer, known as _____ carcinoma, occurs in the *(alveoli? airways?)*, leading to blockage of airways, and inability to get air. (The American Lung Association's motto is, "If you can't breathe, nothing else

_____.")

✓ **H3.** Match the condition with the correct description.

A.	**Asphyxia**	**PE.**	**Pulmonary edema**
CF.	**Cystic fibrosis**	**PN.**	**Pneumonia**
COR.	**Coryza**	**SIDS.**	**Sudden infant death syndrome**
D.	**Dyspnea**	**TA.**	**Tachypnea**
E.	**Emphysema**	**TB.**	**Tuberculosis**

_____ a. Permanent inflation of lungs due to loss of elasticity; rupture and merging of alveoli, followed by their replacement by fibrous tissue

_____ b. Inherited disease in which ducts become obstructed with thick mucus; affects airways as well as digestive and reproductive organs

_____ c. Acute infection or inflammation of alveoli, which fill with fluid

_____ d. Oxygen starvation

_____ e. Caused by a species of *Mycobacterium*; lung tissue is destroyed and replaced with inelastic connective tissue

_____ f. Difficult or painful breathing; shortness of breath

_____ g. Rapid breathing rate

_____ h. Common cold; caused by rhinoviruses, and directly related to stress

_____ i. Abnormal accumulation of fluid in alveoli, for example, due to left ventricular failure

_____ j. A condition that may be prevented by having newborns sleep on their backs

H4. Explain how the abdominal thrust (Heimlich maneuver) helps to remove food that might otherwise cause death by choking.

ANSWERS TO SELECTED CHECKPOINTS: CHAPTER 23

A2. (a) Label (1) pulmonary ventilation; alveoli; O_2, in, CO_2, ex. (b) Air, blood; label (2) external respiration. (c) Veins, blood in systemic capillaries, tissue cells; label (3) internal respiration. (d) Artery, external, 2; pulmonary ventilation, 1. (e) Cellular. (f) Refer to Figure 21.17, p. 691 in your text.

A3. (a-b) MN* → P* → L → T → BB → ADA. (c) Conducting.

A4. (a) Paranasal sinuses and meatuses. (b) External nares. (c) Vestibule. (d) Internal nares (choanae). (e) Rhinoplasty. (f) Nasal cartilages.

A5. (a) Mucosa lining nose, septum, conchae, meati, and sinuses; lacrimal drainage; coarse hairs in vestibule. (b) Olfactory region lies superior to superior nasal conchae. (c) Sounds resonate in nose and paranasal sinuses.

A6. (a) NP. (b–d) OP. (e) LP.

B1. (a) T. (b) E. (c) A. (d) C. (d) RG.

B4. (a) Windpipe; anterior; 12 (4.5), carina. (b) Mucous, cartilage.

B5. (a) Protects against dust. (b) The open part of the cartilage ring faces posteriorly, permitting slight expansion of the esophagus during swallowing; the trachealis muscle provides a barrier preventing food from entering the trachea. (c) Promotes cough reflex, which helps deter food or foreign objects from moving from trachea into bronchi.

B6. The right bronchus is more vertical and slightly wider than the left.

B7. Do not; muscle spasms can collapse airways.

B9. (a) Pleurisy or pleuritis. (b) Pleural effusions. (c) Thoracentesis.

B10. (a) Base; clavicle; ribs. (b) Hilus; bronchi, pulmonary vessels, and nerves. (c) Left; left; right; refer to Figure 23.10, page 787 of the text.

B11. (a) T R A. (b) Alveolar-capillary. See KEY to Figure LU 23.4A. (c) 1, Surfactant; 2, alveolar epithelium, 3, epithelial basement membrane; 4, interstitial space; 5, capillary basement membrane; 6, capillary endothelium. (d) Thin; edema; less. (e) Squamous; surfactant; debris.

B12. (a) Bronchial; aorta. (b) Azygos, accessory hemiazygos. (c) Vein. (dl) Systemic; systemic; (d2) pulmonary.

C1. (a) PV. (b) E. (c) I.

C2. (a) Alveolar or intrapulmonic, 760; the same as. (b) Intrapleural, 756; less than. (c) Diaphragm, external intercostal; increases; increase. (d) Decrease, 754; decreases, 758; Boyle. (e) Atmosphere, alveoli; into; inspiration; 760. (f) Active, passive, elastic; recoil of elastic fibers stretched

during inspiration and inward pull of surface tension; sup, post. (g) Abdominal muscles and internal intercostals.

C3. (a) Low, high. (b) High; surfactant, increases. (C) II, respiratory distress syndrome (RDS). (d) Atelectasis; conditions creating external pressure on lungs, such as air in the pleural cavity (pneumothorax) or pleural effusions, or conditions that inhibit reinflation of lungs, such as bronchitis.

C4. (a) ↓ (b) ↓ (c) ↑ .

C8. 70% (350/500); dead space, 150 (or weight of the person).

C9. 7,200 mL/mm (= 15 breaths/min × 480 mL/breath); 4950 (= 15 breaths/min × 330 mL/breath).

C10. (a) V_T. (b) FRC. (c) ERV. (d) RV. (e) VC. (f) TLC.

C11. (a) 500 (0.5). (b) 6,000 (6). (c) 4,800 (4.8). (d) 2,400 (2.4). (e) 1,200 (1.2).

Dl. (a) H. (b) D. (c) H. (d) B.

D2. (a) 760. (b) 79 (or 78.6), 21 (or 20.9); 04. (c) 600 (or 597.4); pN_2. (d) 21 (or 20.9)% × 760 mm Hg = about 160 (158.8) mm Hg = pO_2.

D4. (a) Partial; see Figure LG 23.4A. (b) 105 − 40 = 65. (c) Large, thin, increased. (d) About the same as; 105. (e) See Figure LG 23.4A. (f) No. Typical radial arterial values are same as for alveoli or blood leaving lungs: $pO_2 = 100$, $pCO_2 = 40$. Jenny's values indicate inadequate gas exchange. (g) Labels for Figure LG 23.1: inferior vena cava, $pO_2 = 40$ mm Hg and $pCO_2 = 45$ mm Hg; aorta, $pO_2 = 100$–105 mm Hg and $pCO_2 = 40$ mm Hg.

Figure LG 23.4A
Diagram of alveoli and pulmonary capillary, Insert enlarges alveolar-capillary membrane.

D5. $282 \times 21\% = 59.2$ mm Hg.

D6. (b) DD; the excess fluid increases the thickness (diffusion distance) of the a-c membrane. (c) SA; emphysema destroys alveolar walls. (d) SMW. Because the a-c membrane is considerably more permeable to CO_2, she can exhale CO_2; so her pCO_2 is only slightly elevated above a normal of 40. Because O_2 exhibits lower solubility in the a–c membrane, her pO_2 (54) is considerably lower than normal (about 100).

E1. (a) 20; oxyhemoglobin; low. (b) Iron; HbO_2, saturated; increase. (c) 98; 98. (d) Increase in temperature, pCO_2, acidity, and BPG. (e) 62; 38; less, more. (f) 75; less. In the event that respiration is temporarily halted, even venous blood has much oxygen attached to hemoglobin and available to tissues. (g) More; this partially compensates for the inefficient fetal circulation and the low oxygen saturation of maternal blood.

E2. 200; oxygen-carrying capacity of hemoglobin is drastically reduced in carbon monoxide poisoning.

E3. (a) 70, 23, 7. (b) H_2CO_3 (carbonic acid). (c) Hemoglobin (as H•Hb); HCO_3^-, plasma, Cl^-. (d) CO_2 + H_2O $\xrightarrow{\text{carbonic anhydrase}}$ H_2CO_3 (carbonic acid) → H+ + HCO_3^- (bicarbonate).

F1.

Name	M/P	Function
a. **Medullary rhythmicity**	M	Controls rhythm; consists of inspiratory and expiratory areas
b. **Pneumotaxic**	P	**Limits inspiration and facilitates expiration**
c. **Apneustic**	P	Prolongs inspiration and inhibits expiration

F2. (a) H^+, increase in pCO_2, (b) Medulla. (c) Hypercapnia; any value higher than 40; even slight, hyper. (d) Aortic and carotid bodies, only large; usually below 60. (e) In. (f) Stretch; inhibition; inflation (Hering-Breuer). (g) Apnea.

F3. (a) Histotoxic. (b) Stagnant. (c) Anemic. (d) Hypoxic.

F4. (a–b) In. (c–d) A. (e–g) In.

F5. (a) Perfusion, five; diffusing capacity. (b) Depth; neural. (c) pCO_2, pO_2, in.

G1. (a) Endo; lining of larynx, trachea, bronchial tree, and alveoli. (b) Smooth muscle, cartilage, and other connective tissues of airways.

G2. (a) Decreased vital capacity, and so decreased pO_2, (b) Risk of pneumonia.

H1. (a) Chronic obstructive pulmonary disease; is (by avoiding smoking); emphysema and chronic bronchitis. (b) Fewer, larger; de, de. (c) Antitrypsin; more; elastase; more; barrel. (d) Excessive; narrowed. (e) Spasm; allergens such as pollen, food, emotional upset, or exercise (exercise-induced asthma).

H2. 1; 55; 10–30; is; bronchogenic, airways; matters.

H3. (a) E. (b) CF. (c) PN. (d) A. (e) TB. (f) D. (g) TA. (h) COR. (i) PE. (j) SIDS.

MORE CRITICAL THINKING: CHAPTER 23

1. Contrast location, structure, and function of each of these parts of the respiratory system: vestibule/conchae, nasopharynx/oropharynx, epiglottis/thyroid cartilage, primary bronchi/bronchioles, parietal pleura/visceral pleura.

2. Write job descriptions for the three types of cells that form alveoli.

3. Contrast Boyle's, Henry's, and Dalton's laws, and relate each law to your respiratory function.

4. All lung "capacities" consist of combinations of lung "volumes." Identify which volumes contribute to: vital capacity, total lung capacity, and functional residual capacity. Speculate as to why functional residual capacity is so named.

5. Lin has just had arterial blood gases (ABGs) drawn. Results indicate that her arterial pO_2 is 64 mm Hg and her pO_2 is 52 mm Hg. Determine if each of these values is high, normal, or low?

6. Contrast the Bohr effect with the Haldane effect on the gas-carrying capacity of hemoglobin.

7. Define and state one or more reasons for performing the following procedures: rhinoplasty, tracheostomy, intubation, bronchography, bronchoscopy, thoracentesis, hyperbaric oxygenation, and Heimlich (abdominal thrust) maneuver.

MASTERY TEST: CHAPTER 23

Questions 1–4: Arrange the answers in correct sequence.

_____ _____ _____ 1. From first to last, the steps involved in inspiration:
   A. Diaphragm and intercostal muscles contract.
   B. Thoracic cavity and lungs increase in size.
   C. Alveolar pressure decreases to 758 mm Hg.

_____ _____ _____ 2. From most superficial to deepest:
   A. Parietal pleura
   B. Visceral pleura
   C. Pleural cavity

_____ _____ _____ _____ _____ 3. From superior to inferior:
   A. Bronchioles
   B. Bronchi
   C. Larynx
   D. Pharynx
   E. Trachea

_____ _____ _____ _____ _____ _____ 4. Pathway of inspired air:
   A. Alveolar ducts
   B. Bronchioles
   C. Lobar bronchi
   D. Primary bronchi
   E. Segmental bronchi
   F. Alveoli

Questions 5–9: Circle the letter preceding the one best answer to each question.

5. All of the following terms are matched correctly with descriptions *except:*
   A. Internal nares: choanae
   B. External nares: nostrils
   C. Posterior of nose: vestibule
   D. Pharyngeal tonsils: adenoids

6. Which of these values (in mm Hg) would be normal for pO_2 of blood in the femoral artery?
   A. 40   D. 160
   B. 45   E. 760
   C. 100 F. 0

7. Pressure and volume in a closed space are inversely related, as described by _____ law.
   A. Boyle's   C. Dalton's
   B. Starling's D. Henry's

8. Choose the correct formula for carbonic acid:
   A. HCO_3^- D. HO_3C_2
   B. H_3CO_2 E. H_2C_3O
   C. H_2CO_3

9. A procedure in which an incision is made in the trachea and a tube inserted into the trachea is known as a(n):
   A. Tracheostomy
   B. Bronchogram
   C. Intubation
   D. Pneumothorax

Questions 10–20: Circle T (true) or F (false). If the statement is false, change the underlined word or phrase so that the statement is correct.

T F   10. In the chloride shift Cl⁻ moves into red blood cells in exchange for <u>H⁺</u>.

T F   11. When chemoreceptors sense <u>increase in pCO_2 or increase in acidity of blood H⁺</u> respiratory rate will normally be stimulated.

T F   12. Both increased temperature and increased acid content tend to cause oxygen to <u>bind more tightly to</u> hemoglobin.

T F   13. Under normal circumstances intrapleural pressure is <u>always less than atmospheric</u>.

T F   14. The pneumotaxic and apneustic areas controlling respiration are located in the <u>pons</u>.

T F   15. Fetal hemoglobin has a <u>lower</u> affinity for oxygen than maternal hemoglobin does.

T F   16. Most CO_2, is carried in the blood in the form of <u>bicarbonate</u>.

T F   17. The alveolar wall <u>does</u> contain macrophages that remove debris from the area.

T F   18. Intrapulmonic pressure means the same thing as <u>intrapleural</u> pressure.

T F   19. Inspiratory reserve volume is normally <u>larger than</u> expiratory reserve volume.

T F   20. The pO_2 and pCO_2 of blood leaving the lungs <u>are normally about the same</u> as pO_2 and pCO_2 of alveolar air.

Questions 21–25: Fill-ins. Complete each sentence with the word or phrase that best fits.

_____ 21. The process of exchange of gases between alveolar air and blood in pulmonary capillaries is known as _____.

_____ 22. Take a normal breath and then let it out. The amount of air left in your lungs is the capacity called _____ and it usually measures about _____ mL.

_____ 23. The Bohr effect states that when more H⁺ ions are bound to hemoglobin, less _____ can be carried by hemoglobin.

_____ 24. The epiglottis, thyroid, and cricoid cartilages are all parts of the _____.

_____ 25. _____ is a chemical that lowers surface tension and therefore increases inflatability (compliance) of lungs.

ANSWERS TO MASTERY TEST: CHAPTER 23

Arrange
1. A B C
2. A C B
3. D C E B A
4. D C E B A F

Multiple Choice
5. C
6. C
7. A
8. C
9. A

True–False
10. F, HCO₃⁻
11. T
12. F. Dissociate from
13. T
14. T
15. F. Higher
16. T
17. T
18. F. Alveolar
19. T
20. T

Fill-ins
21. External respiration (or pulmonary respiration, or diffusion)
22. Functional residual capacity (FRC), 2400
23. Oxygen
24. Larynx
25. Surfactant

FRAMEWORK 24
Digestive System

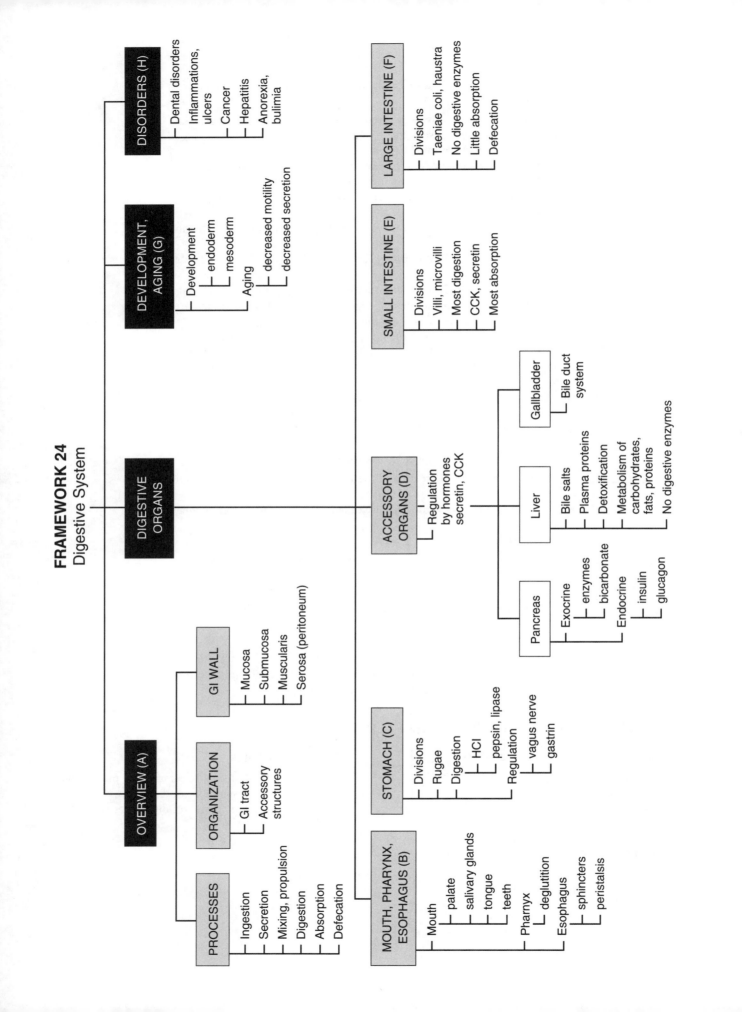

The Digestive System

Food comes in very large pieces (like whole oranges, stalks of broccoli, and slices of bread) that must fit into very small spaces in the human body (like liver cells or brain cells). The digestive system makes such change possible. Foods are minced and enzymatically degraded so that absorption into blood en route to cells becomes a reality. The gastrointestinal (GI) tract provides a passageway complete with mucous glands to help food slide along, muscles that propel food forward, muscles (sphincters and valves) that regulate flow, and secretions that break apart foods or modify pH to suit local enzymes. Accessory structures such as teeth, tongue, pancreas, and liver lie outside the GI tract; yet each contributes to the mechanical and chemical dismantling of food from bite-sized to right-sized pieces. The ultimate fate of the absorbed products of digestion is the story line of metabolism (in Chapter 25).

As you begin your study of the digestive system, carefully examine the Chapter 24 Topic Outline and Objectives. Check off each one as you complete it. You will also find food for thought in the Chapter 24 Framework. Be sure to refer to the Framework frequently and note relationships among key terms in each section.

TOPIC OUTLINE AND OBJECTIVES

A. Overview

1. Identify the organs of the digestive system.
2. Describe the basic processes of the digestive system.
3. Describe the layers that form the wall of the gastrointestinal tract.
4. Describe the peritoneum and its folds.

B. Mouth, pharynx, and esophagus

5. Identify the locations of the salivary glands, and describe the functions of their secretions.
6. Describe the structure and functions of the tongue.
7. Identify the parts of a typical tooth, and compare deciduous and permanent dentitions.
8. Describe the location and function of the pharynx.
9. Describe the location, anatomy, histology, and function of the esophagus.

C. Stomach

10. Describe the location, anatomy, histology, and function of the stomach.

D. Accessory organs: pancreas, liver, and gallbladder

11. Describe the location, anatomy, histology, and function of the pancreas.
12. Describe the location, anatomy, histology, and functions of the liver and gallbladder.
13. Describe the site of secretion and actions of gastrin, secretin, and cholecystokinin.

E. Small intestine

14. Describe the location, anatomy, histology, and function of the small intestine.

F. Large intestine

15. Describe the location, anatomy, histology, and functions of the large intestine.

G. Development and aging

16. Describe the development of the digestive system.
17. Describe the effects of aging on the digestive system.

H. Disorders and medical terminology

WORDBYTES

Now become familiar with the language of this chapter by studying each wordbyte, its meaning, and an example of its use within a term. After you study the entire list, self-check your understanding by writing the meaning of each wordbyte on the line. As you continue through the *Learning Guide*, identify (and fill in) additional terms that contain the same wordbyte.

Wordbyte	Self-check	Meaning	Example(s)
amyl-	_____	starch	*amyl*ase
-ase	_____	enzyme	malt*ase*
bucca-	_____	cheeks	*buccal* cavity
caec-, cec-	_____	blind	*cecum*
chole-, cholecyst-	_____	gallbladder	*cholecyst*itis
chym-	_____	juice	*chym*otrypsin
dent-	_____	tooth	*dent*ures
-ectomy	_____	removal of	append*ectomy*
entero-	_____	intestine	*entero*kinase
gastro-	_____	stomach	*gastr*in
gingiv-	_____	gums	*gingiv*itis
ileo-	_____	ileum	*ileo*cecal valve
jejun-	_____	empty	*jejun*oileostomy
lith-	_____	stone	chole*lith*iasis
or-	_____	mouth	*or*al
-ose	_____	sugar	lact*ose*
retro-	_____	behind	*retro*peritoneal
-rhea	_____	to flow	dia*rrhea*
stoma-	_____	mouth, opening	colo*stom*y
taen-	_____	ribbon	*taen*ia coli
vermi-	_____	worm	*vermi*form appendix

CHECKPOINTS

A. Overview (pages 818–823)

A1. Explain why food is vital to life. Give three specific examples of uses of foods in the body.

✓▲ **A2.** Cover the key and identify all digestive organs on Figure LG 24.1. Then color the structures with color code ovals; these five structures are all *(parts of the gastrointestinal tract? accessory structures?)*.

KEY

O A. Parotid gland
O B. Sublingual and
        submandibular glands
   C. Esophagus
O D. Liver
O E. Gallbladder
   F. Duodenum

   G. Colon, ascending
   H. Cecum
   I.   Vermiform appendix
   J. Mouth
   K. Pharynx
   L. Stomach
O M. Pancreas

   N. Colon, transverse
   O. Jejunum
   P. Ileum
   Q. Colon, descending
   R. Colon, sigmoid
   S. Rectum
   T. Anus

Figure LG 24.1 Organs of the digestive system. Color, draw arrows, and identify labeled structures as directed in Checkpoints A2, B2 and F3.

✓ **A3.** List the six basic activities of the digestive system.

_____   _____   _____

_____   _____   _____

✓ **A4.** _____ digestion occurs by action of enzymes (such as those in saliva)

and intestinal secretion, whereas _____ digestion involves action of
the teeth and muscles of the stomach and intestinal wall.

✓ **A5.** Color layers of the GI wall indicated by color code ovals on Figure LG 24.2.

Lumen

KEY
O Mucous membrane
O Muscularis
O Serous membrane
O Submucosa layer

Figure LG 24.2 Gastrointestinal (GI) tract and related gland seen
in cross section. Refer to Checkpoint A5 and color layers of the wall
according to color code ovals. Label muscularis layers.

✓▲ **A6.** Now match the names of layers of the GI wall with the correct description.

Muc.	**Mucosa**	**Ser.**	**Serosa**
Mus.	**Muscularis**	**Sub.**	**Submucosa**

_____ a. Also known as the peritoneum, it forms
mesentery and omentum

_____ b. Consists of epithelium, lamina propria,
and muscularis mucosae

_____ c. Connective tissue containing glands,
nerves, blood and lymph vessels

_____ d. Consists of an inner circular layer and an
outer longitudinal layer

A7. Describe the tissues listed below by completing this table. Identify the layer in which each tissue is located, type of tissue in each case, and functions of these tissues.

	Layer of GI Wall	Type of Tissue	Function(s)
a. Enteroendocrine tissue			Secretes hormones
b. MALT tissue		Lymphatic	
c. Submucosal plexus (of Meissner)	Submucosa and mucosa		
d. Myenteric plexus (of Auerbach)		ANS nerve tissue	

✓▲ **A8.** Match the names of these peritoneal extensions with the correct descriptions.

F.	**Falciform ligament**	M.	**Mesentery**
G.	**Greater omentum**	Meso.	**Mesocolon**
L.	**Lesser omentum**		

_____ a. Attaches liver to anterior abdominal wall

_____ b. Binds intestine to posterior abdominal wall; provides route for blood and lymph vessels and nerves to reach small intestine

_____ c. Binds part of large intestine to posterior abdominal wall

_____ d. "Fatty apron"; covers and helps prevent infection in small intestine

_____ e. Suspends stomach and duodenum from liver

A9. Define the following terms.

a. Enteric nervous system (ENS)

b. Peritonitis

Figure LG 24.3 Mouth (oral cavity). Complete as directed in Checkpoints B1 and B7.

○ Cuspids   ○ Molars

○ Incisors   ○ Premolars (bicuspids)

B. Mouth, pharynx, and esophagus (pages 823–832)

✓▲ **B1.** Refer to Figure LG 24.3 and label the following structures.

| Hard palate | Palatine tonsils | Uvula of soft palate |
| Lingual frenulum | Palatoglossal arch | Vestibule |

▲ **B2.** Identify the three salivary glands on Figure LG 24.1 and visualize their locations on yourself

✓▲ **B3.** Complete this exercise about salivary glands.
   a. Which glands are largest? *(Parotid? Sublingual? Submandibular?)*

   Which are inflamed by the viral infection known as mumps? _____
   b. Which secrete the thickest secretion due to presence of much mucus?

   _____
   c. About 1 to 1¹/₂ *(tablespoons? cups? liters?)* of saliva are secreted daily.
   d. State three or more functions of saliva.

   e. The pH of the mouth is appropriate for action of salivary amylase. This is about pH *(2? 6.5? 9?).*

f. List five or more chemicals that are in saliva.

g. List four or more factors that can stimulate increase in saliva production.

h. *(Sympathetic? Parasympathetic?)* nerves cause dryness of the mouth, whereas stimulation of saliva production occurs by *(sympathetic? parasympathetic?)* nerves.

▲ **B4.** Contrast terms in each pair:
a. *Intrinsic muscles of the tongue/extrinsic muscles of the tongue*

b. *Fungiform papillae/filiform papillae*

✓   **B5.** Arrange the following in correct sequence:

_____ _____ _____ a.   From most superficial to deepest:

A. Root	B. Neck	C. Crown

_____ _____ _____ b.   From most superficial to deepest within a tooth:

A. Enamel	B. Dentin	C. Pulp cavity

_____ _____ _____ c.   From hardest to softest:

A. Enamel	B. Dentin	C. Pulp cavity

✓▲ **B6.** List two functions of the periodontal ligament.

✓▲ **B7.** Refer to Figure LG 24.3 and do this exercise on dentitions.
   a. How many teeth (total) are in a complete set of deciduous (baby) teeth?

   _____ How many are in a complete set of permanent teeth? _____
   b. Color the teeth in the lower right of Figure LG 24.3 as indicated by color code ovals.
   c. Number the teeth (1–8) on the upper right of the diagram according to order of erup-
      tion. One (the first one to erupt) is done for you.
   d. Which teeth are most likely to become impacted? _____

✓ **B8.** Check your understanding of digestion by completing this Checkpoint.
   a. *(Deglutition? Mastication?)* is a term that means chewing. Food is formed into a soft
      mass known as a *(bolus? chyme?)*.
   b. Enzymes are used for *(chemical? mechanical?)* digestion of food. The main enzyme in

      saliva is salivary *(amylase? lipase?)*. This enzyme converts _____
      into smaller carbohydrates. Most starch *(is? is not?)* broken down by the time food
      leaves the mouth. What inactivates amylase in the stomach?

   c. Where is lingual lipase made? *(Salivary glands? Tongue? Tonsils?)* This enzyme starts

      the breakdown of _____.

✓ **B9.** The term *deglutition* means _____. Match each stage of deglutition
      with the correct description.

E. **Esophageal**	**P.** **Pharyngeal**	**V.** **Voluntary**

_____ a. Soft palate and epiglottis close off   _____ c. Peristaltic contractions push bolus
           respiratory passageways.   from pharynx to stomach.

_____ b. Tongue pushes food back into
           oropharynx.

✓▲ **B10.** Do this activity on the esophagus.

   a. The esophagus is about _____ cm (_____ in.) long. It ends inferiorly in the
      (stomach? duodenum?).
   b. The muscularis of the superior one-third of the esophagus consists of *(smooth?
      striated?)* muscle, whereas the inferior one-third consists of *(smooth? striated?)* muscle
      similar to that found in the stomach.
   c. Failure of the lower esophageal sphincter to close results in the stomach contents
      entering the esophagus and distending it and in irritation of the esophageal lining by
      the *(acidic? basic?)* contents of the stomach. The resulting condition is known as

      _____. Pain associated with either of these conditions may be

      confused with pain originating in the _____. Write four or more
      suggestions for controlling GERD.

✓▲ **B11.** *The Big Picture: Looking Back.* Refer to figures and exhibits in earlier text chapters that relate to digestive activity in the mouth.

    a. Exhibit 11.4 and Figure 11.7 (page 320) demonstrate the genioglossus, hyoglossus, and styloglossus. These are *(extrinsic? intrinsic?)* muscles of the tongue and are innervated by cranial nerve *(V? VII? XII?).*

    b. Table 17.4 (page 560) demonstrates that increased secretion of saliva is a function of the *(sympathetic? parasympathetic?)* division of the autonomic nervous system.

    c. Table 14.2 (page 475) describes cranial nerves. Which ones carry sensory nerves for taste as well as the autonomic nerves stimulating salivary glands? Cranial nerves

        _____ and _____. Pain in an upper tooth is transmitted by the *(mandibular?*

        maxillary? ophthalmic?) branch of the trigeminal nerve (cranial nerve _____).

    B12. Summarize digestion in the mouth, pharynx, and esophagus by completing parts *a* and *b* of Table LG 24.1 (next page).

C. Stomach (pages 832–840)

✓▲ **C1.** Refer to Figure LG 24.1 and complete this Checkpoint.

    a. Arrange these regions of the stomach according to the pathway of food from first to last.

Body	Fundus	Cardia	Pylorus

_____ → _____ →

_____ → _____

    b. Match the terms in the box with the correct description.

GC. Greater curvature	LC. Lesser curvature

_____ More lateral and inferior in location

_____ Attached to lesser omentum

_____ Attached to greater omentum

    c. In which condition does the pyloric sphincter fail to relax normally? *(Pyloric stenosis? Pylorospasm?)* Write one sign or symptom of this condition.

✓ **C2.** Complete this table about gastric secretions.

Name of Cell	Type of Secretion	Function of Secretion
a. Chief (zymogenic)		
b. Mucous		
c.	HCl	
d.	Gastrin	

Table LG 24.1 Summary of digestion.

Digestive Organs	Carbohydrate	Protein
a. Mouth, salivary glands	Salivary amylase: digests starch to maltose	
b. Pharynx, esophagus		
c. Stomach		
d. Pancreas		
e. Intestinal juices		
f. Liver	No enzymes for digestion of carbohydrates	
g. Large intestine	No enzymes for digestion of carbohydrates	No enzymes for digestion of proteins

Table LG 24.1 Summary of digestion. *(continued)*

Lipid	Mechanical	Other Functions
	Deglutition, peristalsis	
		1. Secretes intrinsic factor 2. Produces hormone stomach gastrin
Pancreatic lipase: digests about 80% of fats		
No enzymes for digestion of lipids		

✓ **C3.** How does the structure of the stomach wall differ from that in other parts of the GI tract?

   a. Mucosa

   b. Muscularis

✓ **C4.** Answer these questions about chemical digestion in the stomach.

   a. Pepsin is most active at very *(acid? alkaline?)* pH.

   b. State two factors that enable the stomach to digest protein without digesting its own cells (which are composed largely of protein).

   c. If mucus fails to protect the gastric lining, the condition known as

       _____ may result.

   d. Another enzyme produced by the stomach is _____, which digests

       _____. In adults it is quite *(effective? ineffective?)*. Why?

✓ **C5.** Answer these questions about control of gastric secretion.

   a. Name the three phases of gastric secretion: _____,

       _____, and _____. Which of these causes gastric

       secretion to begin when you smell or taste food? _____

   b. Gastric glands are stimulated mainly by the _____ nerves. Their fibers are *(sympathetic? parasympathetic?)*. Two stimuli that cause vagal impulses to

       stimulate gastric activity are _____ and _____.
       Three emotions that inhibit production of gastric secretions are

       _____, _____, and _____.

   c. Distention of the stomach triggers the _____ phase. Two effects of this stimulus are *(increase? decrease?)* in parasympathetic impulses and release of the hormone _____. What else triggers release of gastrin?

Gastrin travels through the blood to all parts of the body; its target cells are gastric glands and gastric smooth muscle, which are *(stimulated? inhibited?)*, and the pyloric and ileocecal sphincters, which are *(contracted? relaxed?)*.

d. *A clinical correlation.* Both acetylcholine, released by vagal nerves, and gastrin *(stimulate? inhibit?)* HCl production. Their effect is *(greater? less?)* in the presence of

histamine, which is made by _____ cells in the stomach wall. Describe the effect of H₂ blockers such as cimetidine (Tagamet) on the stomach (See page 832 of the text.).

e. When food (chyme) reaches the intestine, nerves initiate the _____ reflex, which *(stimulates? inhibits?)* further gastric secretion. Two hormones released by the intestine also inhibit gastric secretion as well as gastric motility, so the effect of these hormones is to *(stimulate? delay?)* gastric emptying. Name these two hormones:

_____ and _____.

✓ **C6.** Food stays in the stomach for about _____ hours. Which food type

leaves the stomach most quickly? _____ Which type stays in the

stomach longest? _____ Forcible expulsion of stomach contents

through the mouth is known as vomiting or _____. This process is

controlled by the vomiting center located in the _____. Prolonged vomiting *(can be? is never?)* serious. Explain why.

C7. Complete part *c* of Table LG 24.1, describing the role of the stomach in digestion.

D. Accessory organs: pancreas, liver, gallbladder (pages 840–847)

Dl. Refer to Figure LG 24.4 and color all of the structures with color code ovals according to colors indicated.

✓▲ **D2.** Complete these statements about the pancreas.

a. The pancreas lies posterior to the _____.
b. The pancreas is shaped roughly like a fish, with its head in the curve of the

_____ and its tail nudging up next to the _____.
c. The pancreas contains two kinds of glands. Ninety-nine percent of its cells produce *(endocrine? exocrine?)* secretions. One type of these secretions is an *(acid? alkaline?)* fluid to neutralize the chyme entering the small intestine from the stomach.

KEY

○ A. Gallbladder (green)
○ B. Liver (brown)
○ C. Right hepatic duct (green)
○ D. Common hepatic duct (green)
○ E. Cystic duct (green)
○ F. Common bile duct (green)
○ G. Accessory pancreatic duct (yellow)
○ H. Pancreatic duct (green and
         yellow striped)
    I. Duodenum

○ J. Hepatic vein (blue)
    K. Falciform ligament
○ L. Hepatic artery (red)
○ M. Portal vein (blue)
○ N. Spleen (purple)
○ O. Pancreas (yellow)
○ P. Aorta (red)
    Q. Common duct to duodenum
○ R. Superior mesenteric artery
         (red) and vein (blue)

Figure LG 24.4 Liver, gallbladder, pancreas, and duodenum, with associated blood vessels and ducts. (Stomach has been removed). Refer to Checkpoints D1, D2 and D6; color and draw arrows as directed.

d. One enzyme in the pancreatic secretions is trypsin; it digests *(fats? carbohydrates? proteins?)*. Trypsin is formed initially in the inactive form (trypsinogen) and is activated by *(HCl? NaHCO3? enterokinase?)*. Trypsin itself serves as an activator in the formation

of the protease named _____ and the peptidase named _____.

e. Most of the amylase and lipase produced in the body are secreted by the pancreas. Describe the functions of these two enzymes.

f. All exocrine secretions of the pancreas empty into ducts (_____ and _____ on

Figure LG 24.4). These empty into the _____.

g. *A clinical correlation.* In most persons the pancreatic duct also receives bile flowing

through the _____ (F on the figure). State one possible complication that may occur if a gallstone blocks the pancreatic duct (as at point H on the figure).

h. The endocrine portions of the pancreas are known as the _____

_____. List hormones they secrete. _____,

_____, _____, _____. Typical of all hormones, these pass into *(ducts? blood vessels?)*, specifically into vessels that empty

into the _____ vein.

D3. Fill in part *d* of Table LG 24.1, describing the role of the pancreas in digestion.

✓ **D4.** Describe regulation of the pancreas in this exercise.
a. The pancreas is stimulated by *(sympathetic? parasympathetic?)* nerves, specifically the

_____ nerves, and also by hormones produced by cells located in the

wall of the _____.
b. What is the chemical nature of chyme that stimulates release of CCK? Chyme that *(is acidic? contains partially digested fats and proteins?)* CCK then activates the pancreas to secrete fluid rich in *(HCO_3^-? digestive enzymes such as trypsin, amylase, and lipase?)*.
c. What is the chemical nature of chyme that stimulates release of secretion? Chyme that *(is acidic? contains partially digested fats and proteins?)* CCK then activates the pancreas to secrete fluid rich in *(HCO_3^-? digestive enzymes such as trypsin, amylase, and lipase?)*.

✓▲**D5.** Answer these questions about the liver.

a. This organ weighs about _____ kg (_____ pounds). It lies in the _____

quadrant of the abdomen. Of its two main lobes, the _____ is the largest.

b. The _____ ligament separates the right and left lobes. In the edge of this ligament is the ligamentum teres (round ligament of the liver), which is the

obliterated _____ vein.
c. Blood enters the liver via vessels named _____ and _____.

In liver lobules blood mixes in channels called _____ before leaving

the liver via vessels named _____.

✓▲**D6.** Complete Figure LG 24.4 as directed.
a. Identify the pathway of bile from liver to intestine by following the structures you colored green in that figure: C D F H. The general direction of bile is from *(superior to inferior? inferior to superior?)*.
b. Now draw arrows to show the direction of blood through the liver. In order to reach the inferior vena cava, blood must flow *(superiorly? inferiorly?)* through the liver.
c. Write letters of the three structures that form the *portal triad* on the undersurface of

the liver. _____ _____ _____

✓ **D7.** While blood is in liver sinusoids, hepatic cells have ample opportunity to act on this blood, modify it, and add new substances to it. Check your understanding of important functions of the liver in this Checkpoint.

  a. Bile secreted from the liver is composed largely of the pigment named _____

which is a breakdown product of _____ cells. Excessive amounts of

this pigment give skin a yellowish color, a condition known as _____.

Two functions of bile are _____ of fats and _____ of
fats (and fat-soluble vitamins).

  b. Name several plasma proteins synthesized by the liver.

  c. Identify two or more types of chemicals that are detoxified (metabolized) by the liver.

  d. List three types of cells phagocytosed by the liver.

  e. The liver can convert excess glucose to _____ and _____
and also reverse those processes. (More on this in Chapter 25.)

  f. The liver stores the four fat-soluble vitamins named _____, _____, _____,

and _____. It also stores a vitamin necessary for erythropoiesis, namely vitamin

_____. The liver also cooperates with kidneys and skin to activate vitamin _____.

  g. Note that the liver *(does? does not?)* secrete digestive enzymes.

D8. Complete part *f* of Table LG 24.1

✓ **D9.** Check your understanding of accessory structures of the digestive system by naming the organ that best fits each description below. Select answers in the box.

G.	**Gallbladder**	**L.**	**Liver**	**P.**	**Pancreas**

_____ a. Has both exocrine and endocrine functions

_____ b. Its primary functions are to store and concentrate bile

_____ c. Under influence of CCK, it contracts and ejects bile into the cystic duct

_____ d. Responds to CCK by secreting digestive enzymes

_____ e. Bilirubin, absorbed from worn-out red blood cells, is secreted into bile here

_____ f. Contains phagocytic cells known as stellate reticuloendothelial (Kupffer's) cells

_____ g. Site of carbohydrate, fat, and protein metabolism

_____ h. Removed in a cholecystectomy

✓ **D10.** Complete this activity about the "big three" hormones of the gut by selecting correct
answers describing functions of each hormone. Complete arrows as ↑ if the hormone
increases, stimulates, or promotes a certain activity; complete the arrow as ↓ if the
hormone decreases, inhibits, or slows an activity.

   a. Gastrin: I secretion of gastric juices, including HCl and pepsinogen; I gastric motility; I
contraction of the lower esophageal sphincter; and I contraction of the pyloric sphincter. In summary, gastrin I digestive activity of the stomach.

   b. CCK: I secretion of pancreatic juices rich in enzymes; I ejection of bile from gallbladder into bile duct system; I opening of hepatopancreatic ampulla (sphincter of Oddi)
so that bile can flow into duodenum; I gastric emptying (by I contraction of pyloric
sphincter and I satiety); I growth of pancreas; and I effects of secretin. The overall
effect of CCK is to I digestive activity in the small intestine.

   c. Secretin: I secretion of bicarbonate (in pancreatic juice and bile); I secretion of gastric
juice; I effects of CCK; I growth of pancreas. The overall effect of CCK is to I digestive
activity in the small intestine.

   d. In summary, gastrin tends to I gastric activity, and CCK and secretin tend to I
pancreas, liver, and gallbladder activity.

E. Small intestine (pages 847–857)

✓▲ **E1.** Describe the structure of the intestine in this Checkpoint.

   a. The average diameter of the small intestine is:

      A. 2.5 cm (1 inch)   B. 5.0 cm (2 inches)

   b. Its average length is about:

      A. 3 m (10 feet)   B. 6 m (20 feet)

   c. Name its three main parts (in sequence from first to last):

     _____ → _____ → _____

✓ **E2.** Match the correct term related to the intestine to the description that fits.

ALF.	**Aggregated lymphatic follicles (Peyer's patches)**	**MALT.**	**Mucosa-associated lymphoid tissue**
BB.	**Brush border**	**MV.**	**Microvilli**
CF.	**Circular folds (plicae circularis)**	**PC.**	**Paneth cells**
DG.	**Duodenal (Brunner's) glands**	**SC.**	**S cells and CCK cells**
L.	**Lacteal**	**V.**	**Villi**

_____ a. Fingerlike projections up to 1 mm high
that give the intestinal lining a velvety
appearance and increase absorptive
surface

_____ b. Microscopic fingerlike projections of
plasma membrane

_____ c. Circular folds that further increase
intestinal surface area

_____ d. Clusters of lymphatic follicles located
primarily in the wall of the ileum

_____ e. Submucosal glands that secrete protective alkaline fluids

_____ f. Lymphatic nodules within the mucous
membrane of the small intestine

_____ g. Fuzzy line marking the tips of microvilli;
contains digestive enzymes

_____ h. Lymphatic capillary found within each
villus

_____ l. Control level of microbes in intestine by
phagocytosis and secretion of lysozyme

_____ j. Enteroendocrine cells that secrete the
hormones secretin and CCK

E3. Name eight brush border enzymes.

These enzymes function primarily *(in the intestinal lumen? at the brush border?)*. Describe the action of one of these enzymes, α-*dextrinase.*

E4. Contrast segmentation and migrating motility complex (MMC), the two types of movement of the small intestine.

✓ **E5.** Write the main steps in the digestion of each of the three major food types.
   a. *Carbohydrates*
      Polysaccharides → _____ → monosaccharides
   b. *Proteins*
      Proteins → _____ → _____
   c. *Lipids*
      Neutral fats → emulsified fats → _____ _____

✓ **E6.** *A clinical correlation.* Persons with lactose intolerance lack the enzyme

   _____, which is necessary for digestion of lactose found in foods

   such as _____. Write a rationale for the usual symptoms of this condition.

E7. Complete the description of functions of intestinal juices by filling in part *e* of Table LG 24.1. Then review digestion of each of the three major food groups by reading your columns vertically.

✓ **E8.** *For extra review* of roles of GI organs in digestion, identify which chemicals in the list in the box are made by each of the following organs. Write one answer on each line provided.

Alb.	**Albumin**	**IF.**	**Intrinsic factor**
Amy.	**Amylase**	**Ins.**	**Insulin**
B.	**Bile**	**Lip.**	**Lipase**
CCK.	**Cholecystokinin**	**M.**	**Mucus**
Chol.	**Cholesterol**	**MLSD.**	**Maltase, lipase, sucrase, a-dextri-**
DA.	**Dipeptidases, aminopeptidases**		**nase**
E.	**Enterokinase**	**NP.**	**Nucleosidases, phosphatases**
FP.	**Fibrinogen and prothrombin**	**P.**	**Pepsinogen**
Gas.	**Gastrin**	**RD.**	**Ribonuclease, deoxyribonuclease**
Glu.	**Glucagon**	**S.**	**Secretin**
Gly.	**Glycogen**	**TCP.**	**Trypsinogen, chymotrypsinogen,**
HCl.	**Hydrochloric acid**		**procarboxypeptidase**

a. Salivary glands and tongue: _____ _____ _____

b. Pharynx and esophagus: _____

c. Stomach: _____ _____ _____ _____ _____ _____

d. Pancreas: _____ _____ _____ _____ _____

e. Small intestine: _____ _____ _____ _____ _____ _____ _____

f. Liver: _____ _____ _____ _____ _____

✓ **E9.** *For extra review* of secretions involved with digestion, match names of secretions listed for Checkpoint E8 with descriptions below. Blank lines follow some descriptions. On these indicate whether the secretion is classified as an enzyme (E), hormone (H), or neither of these (N). The first one is done for you.

__CCK__ a. Causes contraction of gallbladder and relaxation of the sphincter of the hepatopancreatic ampulla so that bile enters duodenum __**H**__

_____ b. Stimulates production of alkaline pancreatic fluids and bile _____

_____ c. Stimulates pancreas to produce secretions rich in enzymes such as lipase and amylase _____

_____ d. Promote growth and maintenance of pancreas; enhance effects of each other (two answers)

_____ e. Increases gastric activity (secretion, motility, and growth) _____

_____ f. Inhibit stomach motility and/or secretions (two answers)

_____ g. Precursors to protein-digesting enzymes (two answers)

_____ h. Activates pepsinogen _____

_____ i. The active form of this enzyme starts protein digestion in the GI tract

_____ j. Activates the inactive precursor to form trypsin

_____ k. Plasma proteins (two answers)

_____ l. A storage form of carbohydrate

_____ m. Most effective of this type of enzyme is produced by the pancreas; digests fats

_____ n. Emulsifies fats before they can be digested effectively _____

_____ o. Starch-digesting enzymes secreted by salivary glands and pancreas

_____ p. Intestinal enzymes that complete carbohydrate breakdown, resulting in simple sugars

_____ q. Digest DNA and RNA _____

_____ r. Brush border enzymes that carry out digestion on the surface of villi (three answers)

_____ s. Complete digestion of proteins into amino acids

✓ **E10.** Describe the absorption of end products of digestion in this exercise.

a. Almost all absorption takes place in the *(large? small?)* intestine. Absorption is a two-step process. Products of digestion must first be absorbed into *(blood or lymph capillaries? epithelial cells lining the intestine?)* and then be transported into *(blood or lymph capillaries? epithelial cells lining the intestine?)*.

b. Glucose and some amino acids move into epithelial cells lining the intestine by

   (primary'? secondary?) active transport coupled with active transport of _____. Fructose enters epithelial cells by *(diffusion? facilitated diffusion?)*. All simple sugars

   then enter capillaries by the process of _____.

c. Simple sugars, amino acids, and short-chain fatty acids are absorbed into *(blood?*

   lymph?) capillaries located in _____ in the intestinal wall. These

   vessels lead to the _____ vein and then to the _____ for storage or metabolism.

d. Long-chain fatty acids and monoglycerides first combine with _____ salts to form *(micelles? chylomicrons?)*. This enables fatty acids and monoglycerides to enter epithelial cells in the intestinal lining and soon enter lacteals leading to the *(portal vein? thoracic duct?)*. Most bile salts are ultimately *(eliminated in feces? recycled to*

   the liver?). This cycle is known as _____ circulation.

e. Aggregates of fats coated with _____ are known as chylomicrons. After traveling through lymph and blood, chylomicrons reach hepatocytes in the

   _____ or are stored in _____ where triglycerides are reformed.

f. About _____ liters (or _____ quarts) of fluids are ingested or secreted into the

   GI tract each day. Of this fluid, about _____ liters come from ingested food and

   about _____ liters derive from GI secretions. All but about 1 liter of the 9 liters is reabsorbed by *(facilitated diffusion? osmosis?)* into blood capillaries in the walls of the *(small? large?)* intestine. Several hundred milliliters of fluid are also reabsorbed each day into the *(small? large?)* intestine.

g. Normally about _____ mL of fluid exits each day in feces. When inadequate water

   reabsorption occurs, as in *(constipation? diarrhea?)*, then _____ such as Na$^+$ and Cl$^-$ are also lost also.

✓ **E11.** *The Big Picture: Looking Ahead.* Refer to Table 25.6 on page 906 in the text, and answer these questions.
  a. Which vitamins are absorbed with the help of bile? *(Water-soluble? Fat-soluble?)* Circle the fat-soluble vitamins: A B$_{12}$ C D E K
  b. Inadequate bile production or obstruction of bile pathways may lead to signs or symptoms related to deficiencies of fat-soluble vitamins. List several.

F. Large intestine (pages 857–863)

✓▲ **F1.** The large intestine is so named based on its *(diameter? length?)* compared to that of

the small intestine. The total length of the large intestine is about _____ m (_____ ft).

More than 90% of its length consists of the part known as the _____.

✓▲ **F2.** Arrange the parts of the large intestine listed in the box in correct sequence in the pathway of wastes

AnC.	Anal canal	R.	Rectum
AsC.	Ascending colon	SC.	Sigmoid colon
C.	Cecum	SF.	Splenic flexure
DC.	Descending colon	TC.	Transverse colon
HF.	Hepatic flexure		

_____ → _____ → _____ → _____ → _____ →

_____ → _____ → _____ → _____

▲ **F3.** Identify the regions of the large intestine in Figure LG 24.1. Draw arrows to indicate direction of movement of intestinal contents.

✓▲ **F4.** Contrast different portions of the GI tract by identifying structures or functions associated with each. Use these answers:

LI.	Large intestine	SI.	Small intestine	Sto.	Stomach

_____ a. Has thickened bands of longitudinal muscle known as taeniae coli

_____ b. Pouches known as haustra give this structure its puckered appearance

_____ c. Its fat-filled peritoneal attachments are known as epiploic appendages

_____ d. Bacteria here decompose bilirubin to urobilinogen, which gives feces its brown color

_____ e. Has rugae

_____ f. Has villi and microvilli

_____ g. (Vermiform) appendix is attached to the cecum here

_____ h. Diverticuli are outpouchings where the muscularis here has weakened

_____ i. Ileocecal valve is located here (two answers)

F5. Contrast terms in each pair:

a. *Gastroileal reflex/gastrocolic reflex*

b. *Haustral churning/mass peristalsis*

c. *Diarrhea/constipation*

✓ **F6.** Complete this activity describing chemicals associated with digestive wastes.

a. Flatus (gas) in the colon due to the gases _____,

_____, _____; these are products of bacterial
fermentation of *(carbohydrates? proteins? fats?)*.

b. Odors due to the chemicals _____ and _____; these
are products of bacterial breakdown of *(carbohydrates? proteins and amino acids? fats?)*.

c. Brown color due to the chemical _____, a product of bacterial

decomposition of _____.

✓ **F7.** Feces are formed by the time chyme has remained in the large intestine for about

_____ hours. List the chemical components of feces.

F8. Describe the process of defecation. Include these terms: *stretch receptors, rectal muscles,
sphincters, diaphragm,* and *abdominal muscles.*

✓ **F9.** Describe helpful effects of dietary fiber in this activity.

a. *(Soluble? Insoluble?)* fibers tend to speed up passage of digestive wastes. Potential
health benefits include lowering risk for a number of disorders, including

_____ (hard, dry stool) and colon cancer. Circle the two best sources of insoluble fiber in the following list.

apples   broccoli   oats and prunes   vegetable skins   wheat bran

b. Write one health benefit of soluble fiber.

c. Write three sources of soluble fiber from the list above.

F10. Complete part g of Table 24.1, describing the role of the large intestine in digestion.

G. Development and aging (pages 862–863)

✓ **G1.** Indicate which germ layer, endoderm (_E_) or mesoderm (_M_), gives rise to each of these structures.

_____ a. Epithelial lining and digestive glands of the GI tract

_____ b. Liver, gallbladder, and pancreas

_____ c. Muscularis layer and connective tissue of submucosa

G2. Identify which GI structures will form from each of the following embryonic structures.

a. Stomodeum

b. Proctodeum

c. Foregut

d. Hindgut

✓ **G3.** _A clinical correlation._ List physical changes with aging that may lead to a decreased desire to eat among the elderly population.
a. Related to the upper GI tract (to stomach)

b. Related to the lower GI tract (stomach and beyond)

H. Disorders, medical terminology (pages 863–866)

H1. Explain how tooth decay occurs. Include the roles of bacteria, dextran, plaque, and acid. List the most effective known measures for preventing dental caries.

H2. Briefly describe these disorders, stating possible causes of each.

a. Periodontal disease

b. Peptic ulcer

✓ **H3.** Match the terms with the descriptions.

A.	Anorexia nervosa	Dys.	Dysphagia
B.	Bulimia	E.	Eructation
Cho.	Cholecystitis	F.	Flatus
Colit.	Colitis	Hem.	Hemorrhoids
Colos.	Colostomy	Hep.	Hepatitis
Con.	Constipation	Htb.	Heartburn
Dia.	Diarrhea	P.	Peptic ulcer

_____ a. Incision of the colon, creating artificial anus

_____ b. Inflammation of the liver

_____ c. Inflammation of the colon

_____ d. Burning sensation in region of esophagus and stomach; probably due to gastric contents in lower esophagus

_____ e. Frequent defecation of liquid feces

_____ f. Inflammation of the gallbladder

_____ g. Infrequent or difficult defecation

_____ h. Craterlike lesion in the GI tract due to acidic gastric juices, Helicobacter pylori bacteria, and/or drugs such as NSAIDs

_____ i. Excess air (gas) in stomach or intestine, usually expelled through anus

_____ j. Binge-purge syndrome

_____ k. Loss of appetite and self-imposed starvation

_____ l. Difficulty in swallowing

_____ m. Belching

ANSWERS TO SELECTED CHECKPOINTS: CHAPTER 24

A2. Accessory structures.

A3. Ingestion, secretion, motility, digestion, absorption, defecation

A4. Chemical, mechanical.

A5.

Figure LG 24.2A Gastrointestinal (GI) tract and related gland seen in cross section.

A6. (a) Ser. (b) Muc. (c) Sub. (d) Mus.

A8. (a) F. (b) M. (c) Meso. (d) G. (e) L.

B1.

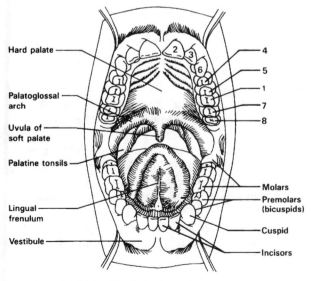

Figure LG 24.3A Mouth (oral cavity).

B3. (a) Parotid; parotid. (b) Sublingual. (c) Liters.
(d) Dissolving medium for foods, lubrication, waste removal medium, and source of antibodies,

lysozyme. (e) 6.5. (f) Water, Cl^-, HCO_3^-, HPO_4^{2-}, urea, uric acid, lysozyme, salivary amylase.
(g) Smell, sight, touch, or memory of food; nausea; swallowing of irritating food; nerve impulses. (h) Sympathetic, parasympathetic.

B5. (a) C B A. (b) A B C. (c) A B C.

B6. Anchors tooth to bone and acts as a shock absorber.

B7. (a) 20; 32. (b–c) See Figure LG 24.3A. (d) Third molars or "wisdom teeth."

B8. (a) Mastication; bolus. (b) Chemical; amylase; starch; is not; acidic pH of the stomach. (c) Tongue; triglycerides (or fats).

B9. Swallowing. (a) P. (b) V. (c) E.

B10. (a) 25 (10); stomach. (b) Striated, smooth.
(c) Acidic, gastroesophageal reflux disease (GERD); heart (heartburn); avoid smoking, alcohol, certain foods, large meals, and lying down just after meals; take H2-blocker medications.

B11. (a) Extrinsic, XII. (b) Parasympathetic. (c) VII, IX; maxillary, V.

C1. (a) Cardia → fundus → body → pylorus. (b) GC; LC; GC. (c) Pylorospasm; vomiting.

C2.

Name of Cell	Type of Secretion	Function of Secretion
a. Chief (zymogenic)	Pepsinogen Gastric lipase	Precursor of pepsin Splits triglycerides
b. Mucous	Mucus	Protects gastric lining from acid and pepsin
c. Parietal	HCl Intrinsic factor	Activates pepsinogen to pepsin; inhibits gastrin secretion; stimulates secretion of secretin and CCK; kills microbes in food; facilitates absorption of vitamin B_{12}
d. Enteroendocrine (G) cells	Gastrin	Stimulates secretion of HCl and pepsinogen; contracts lower esophageal sphincter; increases gastric motility and relaxes pyloric and ileocecal sphincters

C3. (a) Arranged in rugae. (b) It has three, rather than two, layers of smooth muscle; the extra one is an oblique layer located inside the circular layer.

C4. (a) Acid. (b) Pepsin is released in the inactive state (pepsinogen); mucus protects the stomach lining from pepsin. (c) Ulcer. (d) Gastric lipase, emulsified fats, as in butter; ineffective; its optimum pH is 5 or 6, and most fats have yet to be emulsified (by bile from liver).

C5. (a) Cephalic, gastric, intestinal; cephalic. (b) Vagus;

parasympathetic; sight, smell, or thought of food and presence of food in the stomach; anger, fear, anxiety. (c) Gastric; increase, gastrin; presence of partially digested proteins, alcohol, and caffeine in the stomach; stimulated, relaxed. (d) Stimulate; greater, mast; H_2 blockers decrease HCl production which could otherwise lead to "acid stomach" or peptic ulcer. (e) Enterogastric, inhibits; delay; secretin, CCK.

C6. Two to four; carbohydrates; fats (triglycerides); emesis; medulla; can be; loss of HCl can lead to alkalosis (increase in blood pH).

D2. (a) Stomach. (b) Duodenum, spleen. (c) Exocrine; alkaline. (d) Proteins; enterokinase; chymotrypsin (and elastase), carboxypeptidase. (e) Amylase digests carbohydrates, including starch; lipase digests lipids. (f) G, H; duodenum. (g) Common bile duct; pancreatic proteases such as trypsin, chymotrypsin, and carboxypeptidase may digest tissue proteins of the pancreas itself. (h) Pancreatic islets (of Langerhans); insulin, glucagon, somatostatin, and pancreatic polypeptide; blood vessels, portal.

D4. (a) Parasympathetic, vagus, small intestine. (b) Contains partially digested fats and proteins; digestive enzymes such as trypsin, amylase, and lipase. (c) Is acidic; HCO_3.

D5. (a) 1.4, 3.0; upper right; right. (b) Falciform; umbilical. (c) Hepatic artery, portal vein; sinusoids. hepatic veins.

D6. (a) Superior to inferior. (b) Draw arrows from L and M toward B and then J; superiorly. (c) F, L, and M

D7. (a) Bilirubin, red blood; jaundice; emulsification, absorption. (b) Clotting proteins (prothrombin and fibrinogen), albumin, and globulins. (c) Products of protein digestion, such as ammonia; medications such as sulfa drugs or penicillin; and steroid hormones. (d) Worn-out RBCs, WBCs, and some bacteria. (e) Glycogen, fat. (f) A, D, E, K: B_{12}, D. (g) Does not.

D9. (a) P. (b–c) G. (d) P. (e–g) L. (h) G.

D10. (a) All ↑ except ↓ contraction of pyloric sphincter. (b) All ↑ except ↓ gastric emptying. (c) All ↑ except, ↓ gastric juice. (d.) ↑; ↑.

E1. (a) A. (b) B. (c) Duodenum → jejunum → ileum.

E2. (a) V. (b) MV. (c) CF. (d) ALF. (e) DG. (f) MALT. (g) BB. (h) L. (i) PC. (j) SC.

E5. (a) Shorter-chain polysaccharides such as a-dextrins (5–10 glucoses), trisaccharides (such as maltotriose), or disaccharides (such as maltose).

(b) Shorter-chain polypeptides → dipeptides and amino acids. (c) Fatty acids, monoglycerides, and glycerol.

E6. Lactase, milk and other dairy products; gas (flatus) and bloating because bacteria ferment the undigested lactose present in the GI tract.

E8. (a) Amy, Lip, M. (b) M. (c) Gas, HCl, IF, Lip, M, P. (d) Amy, Glu, Ins, Lip, RD, TCP. (e) CCK, DA, E, M, MLSD, NP, S. (f) Alb, B, Chol, FP, Gly.

E9. (a) CCK, H. (b) S, H. (b) CCK, H. (d) CCK and S. (e) G, H. (f) CCK and S. (g) P, TCP. (h) HCl, N. (i) P. (j) E. (k) Alb and FP. (l) Gly. (m) Lip. (n) B, N. (o) Amy. (p) MLSD. (q) NP. (r) MLSD, DA, and NP. (s) DA.

E10. (a) Small; epithelial cells lining the intestine, blood or lymph capillaries. (b) Secondary, Na^+; facilitated diffusion; facilitated diffusion. (c) Blood, villi; portal, liver. (d) Bile, micelles; thoracic duct; recycled to the liver; enterohepatic. (e) Protein; liver, adipose. (f) 9.3 (9.8); 2, 7; osmosis, small; large. (g) 100–200; diarrhea, electrolytes.

E11. (a) Fat-soluble: A D E K. (b) Examples: A, night blindness and dry skin; D, rickets or osteomalacia due to decreased calcium absorption; K, excessive bleeding.

F1. Diameter: 1.5 (5.0); colon.

F2. C → AsC → HF → TC → SF → DC → SC → R → AnC.

F4. (a–d) LI. (e) Sto. (f) SI. (g–h) LI. (i) SI, LI.

F6. (a) Methane, carbon dioxide, hydrogen; carbohydrates. (b) Indole, skatole; proteins and amino acids. (c) Stercobilin, bilirubin.

F7. 3–10: water, salts, epithelial cells, bacteria and products of their decomposition, as well as undigested foods.

F9. (a) Insoluble; constipation; vegetable skins and wheat bran. (b) Lowers blood cholesterol level; apples, broccoli, oats and prunes.

G1. (a) E. (b) E. (c) M.

G2. (a) Oral cavity. (b) Anus. (c) Pharynx, esophagus, stomach, and part of the duodenum. (d) Remainder of duodenum through most of the transverse colon.

G3. (a) Decreased taste sensations, gum inflammation (pyorrhea) leading to loss of teeth and loose-fitting dentures, and difficulty swallowing (dysphagia). (b) Decreased muscle tone and neuromuscular feedback may lead to constipation.

H3. (a) Colos. (b) Hep. (c) Colit. (d) Htb. (e) Dia. (f) Cho. (g) Con. (h) P. (i) F. (j) B. (k) A. (l) Dys. (m) E.

MORE CRITICAL THINKING: CHAPTER 24

1. Contrast the superior and inferior portions of the esophagus, stomach, and small and large intestines with regard to (a) types of epithelium in the mucosa, and (b) types of muscle in the muscularis.
2. Place in correct sequence each of these structures in the pathway of food through the GI tract. Then briefly describe each structure: anal canal, ascending colon, cecum, duodenum, esophagus, hepatic flexure, ileocecal sphincter, oropharynx, pyloric sphincter, sigmoid colon.
3. Contrast functions of chemicals within each grouping: amylase/trypsin/lipase; gastrin/CCK/secretin.
4. Describe nine functions of the liver. Determine which ones are essential to survival.
5. Identify with which digestive organs the following structures are associated: dentin, lacteal, frenulum linguae, uvula, pyloric region, ileum, brush border, villi, cecum, haustra, and sinusoids.
6. Identify with which digestive organs the following conditions are associated: periodontal membrane disease, pyorrhea, cholecystitis, cirrhosis, hepatitis A-E, diverticulosis, appendicitis, and achalasia.

MASTERY TEST: CHAPTER 24

Questions 1–10: Circle the letter preceding the one best answer to each question.

1. Which of these organs is not part of the GI tract but is an accessory organ?
   A. Mouth   D. Small intestine
   B. Pancreas   E. Esophagus
   C. Stomach
2. The main function of salivary and pancreatic amylase is to:
   A. Lubricate foods
   B. Help absorb fats
   C. Digest polysaccharides to smaller carbohydrates
   D. Digest disaccharides to monosaccharides
   E. Digest polypeptides to amino acids
3. Which enzyme is most effective at pH 1 or 2?
   A. Gastric lipase   D. Salivary amylase
   B. Maltase   E. Pancreatic amylase
   C. Pepsin
4. Choose the *true* statement about fats.
   A. They are digested mostly in the stomach.
   B. They are the type of food that stays in the stomach the shortest length of time.
   C. They stimulate release of gastrin.
   D. They are emulsified and absorbed with the help of bile.
5. Which of the following is a hormone, not an enzyme?
   A. Gastric Lipase   C. Pepsin
   B. Gastrin   D. Trypsin

6. Which of the following is under only nervous (not hormonal) control?
   A. Salivation
   B. Gastric secretion
   C. Intestinal secretion
   D. Pancreatic secretion
7. All of the following are enzymes involved in protein digestion *except:*
   A. Amylase   D. Pepsin
   B. Trypsin   E. Chymotrypsin
   C. Carboxypeptidase
8. All of the following chemicals are produced by the walls of the small intestine *except:*
   A. Lactase   D. Trypsin
   B. Secretin   E. Brush border enzymes
   C. CCK
9. Migrating motility complex (MCC) is a form of _____ in the small intestine.
   A. Peristalsis   C. Haustral churning
   B. Segmentation   D. Mastication
10. Choose the *false* statement about layers of the wall of the GI tract.
    A. Most large blood and lymph vessels are located in the submucosa.
    B. The mesenteric plexus is part of the muscularis layer.
    C. Most glandular tissue is located in the layer known as the mucosa.
    D. The mucosa layer forms the peritoneum.

Questions 11–15: Arrange the answers in correct sequence.

_____ _____ _____   11. Arrange from greatest to least in number in a
                             complete set of adult dentition:
                             A. Molars
                             B. Premolars
                             C. Cuspids (canines)

_____ _____ _____ _____   12. GI tract wall, from deepest to most superficial:
                                    A. Mucosa
                                    B. Muscularis
                                    C. Serosa
                                    D. Submucosa

_____ _____ _____ _____ _____   13. Pathway of chyme:
                                           A. Ileum
                                           B. Jejunum
                                           C. Cecum
                                           D. Duodenum
                                           E. Pylorus

_____ _____ _____ _____ _____   14. Pathway of bile:
                                           A. Bile canaliculi
                                           B. Common bile duct
                                           C. Common hepatic duct
                                           D. Right and left hepatic ducts
                                           E. Hepatopancreatic ampulla and duodenum

_____ _____ _____ _____ _____   15. Pathway of wastes:
                                           A. Ascending colon
                                           B. Transverse colon
                                           C. Sigmoid colon
                                           D. Descending colon
                                           E. Rectum

Questions 16–20: Circle T (true) or F (false). If the statement is false, change the underlined word or phrase so that the statement is correct.

T F 16.   In general, the sympathetic nervous system <u>stimulates</u> salivation and
            secretions of the gastric and intestinal glands.
T F 17.   The principal chemical activity of the stomach is to begin digestion of <u>protein</u>.
T F 18.   The esophagus produces <u>no digestive enzymes or mucus</u>.
T F 19.   Cirrhosis and hepatitis are diseases of the <u>liver</u>.
T F 20.   The greater omentum <u>connects the stomach to the liver</u>.

Questions 21–25: Fill-ins. Complete each sentence with the word or phrase that best fits.

_____   21. Stomach motility is _____-creased by CCK and _____-creased
                               by stomach gastrin and parasympathetic nerves.

_____   22. Mumps involves inflammation of the _____ salivary glands.

_____   23. Most absorption takes place in the _____, although some

                               substances, such as _____, are absorbed in the stomach.

_____   24. The clinical specialty that deals with diseases of the stomach and
                               intestines is _____.

_____   25. Epithelial lining of the GI tract, as well as the liver and pancreas,

                               are derived from _____-derm.

ANSWERS TO MASTERY TEST: CHAPTER 24

Multiple Choice

1. B
2. C
3. C
4. D
5. B
6. A
7. A
8. D
9. A
10. D

Arrange

11. A B C
12. A D B C
13. E D B A C
14. A D C B E
15. A B D C E

True–False

16. F. Inhibits
17. T
18. F. No digestive enzymes, but it does produce mucus
19. T
20. F. Connects the stomach and duodenum to the transverse colon and drapes over the transverse colon and coils of the small intestine.

Fill-ins

21. De, in
22. Parotid
23. Small intestine; alcohol
24. Gastroenterology
25. Endo

FRAMEWORK 25
Metabolism

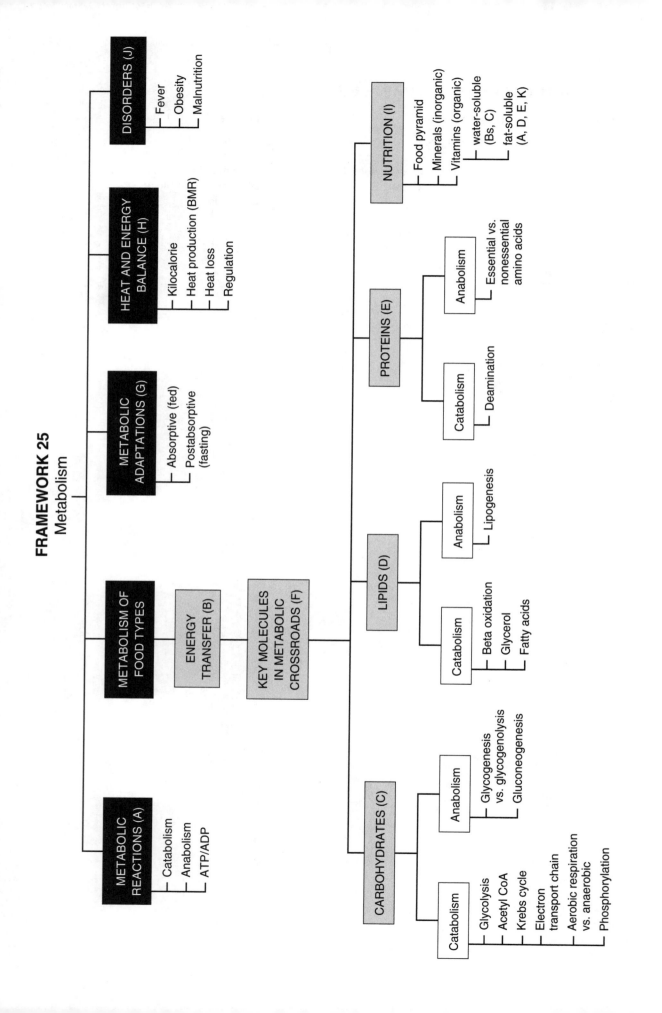

C H A P T E R **25**

Metabolism

Ingested foods, once digested, absorbed, and delivered, are used by body cells. This array of biochemical reactions within cells is known as metabolism. Products of digestion may be reassembled, as in formation of human protein from the amino acids of meat or beans, or in synthesis of fats from ingested oils. These reactions are examples of anabolism. The flip side of metabolic currency is catabolism, as in the breakdown of complex carbohydrates stored in liver or muscle to provide simple sugars for quick energy. Catabolism releases energy that maintains body heat and provides ATP to fuel activity. Vitamins and minerals play key roles in metabolism—for example, in enzyme synthesis and function.

The Chapter 25 Framework provides an organizational preview of the metabolism of the major food groups. Study the key terms and concepts it presents. As you begin this chapter, carefully examine the Chapter 25 Topic Outline and check off each objective as you complete it.

TOPIC OUTLINE AND OBJECTIVES

A. Metabolic reactions

1. Explain the role of ATP in anabolism and catabolism

B. Energy transfer

2. Describe oxidation-reduction reactions.
3. Explain the role of ATP in metabolism.

C. Carbohydrate metabolism

4. Describe the fate, metabolism, and functions of carbohydrates.

D. Lipid metabolism

5. Describe the lipoproteins that transport lipids in the blood.
6. Describe the fate, metabolism, and functions of lipids.

E. Protein metabolism

7. Describe the fate, metabolism, and functions of proteins.

F. Summary; key molecules in metabolism

8. Identify the key molecules in metabolism, and describe the reactions and the products they may form.

G. Metabolic adaptations

9. Compare metabolism during the absorptive and postabsorptive states.

H. Heat and energy balance

10. Define basal metabolic rate (BMR), and explain several factors that affect it.
11. Describe the factors that influence body heat production.
12. Explain how normal body temperature is maintained by negative feedback loops involving the hypothalamic thermostat.

I. Nutrition

13. Describe how to select foods to maintain a healthy diet.
14. Compare the sources, functions, and importance of minerals and vitamins in metabolism.

WORDBYTES

Now become familiar with the language of this chapter by studying each wordbyte, its meaning, and an example of its use within a term. After you study the entire list, self-check your understanding by writing the meaning of each wordbyte on the line. As you continue through the *Learning Guide*, identify (and fill in) additional terms that contain the same wordbyte.

Wordbyte	Self-check	Meaning	Example(s)
ana-	_____	upward	*ana*bolism
calor-	_____	heat	*calor*ie
cata-	_____	downward	*cata*bolism
de-	_____	remove, from	*de*amination
-gen	_____	to form	gluconeo*gen*esis
-lysis	_____	destruction	hydro*lysis*
neo-	_____	new	gluco*neo*genesis

CHECKPOINTS

A. Metabolic reactions (pages 871–872)

✓ **A1.** Write brief descriptions of the three possible "fates" of foods absorbed by the GI tract.

a. _____

b. _____

c. _____

✓ **A2.** Complete this table comparing catabolism with anabolism.

Process	Definition	Releases or Uses Energy	Examples
a. Catabolism			
b. Anabolism			

A3. ATP stands for a_____ t_____ p_____.
Explain how it functions as the "energy currency" (or "money machine cash") of the cell. (For help, refer to Chapter 1, Checkpoint E16, page 35 of the *Learning Guide*.)

✓ **A4.** *For extra review*, refer to LG Chapter 10, Checkpoint D1c, page LG 188, and notice the difference between the structure of ATP and ADP. ADP has its *(1? 2? 3?)* phosphate "trailers" attached to the adenosine "truck" by one low-energy and one high-energy bond. ATP has *(1? 2? 3?)* phosphate "trailers" attached to the adenosine "truck" by one low-energy and *(1? 2? 3?)* high-energy bond(s). Therefore, *(ATP? ADP?)* contains more energy that can be utilized for cell activities.

✓ **A5.** About *(1–2%? 10–20%? 40%? 65–70%?)* of the energy released in catabolism is avail-

able for cellular activities. The rest of the energy is converted to _____.

B. Energy transfer (pages 872–874)

✓ **B1.** Complete this overview of catabolism (energy production).
  a. Organic nutrients such as glucose are rich in hydrogen; energy is contained within the C —— H bonds. In catabolism, much of the energy in glucose is ultimately released

  and stored in the high-energy molecule named _____. Write the chemical formulas to show this overall catabolic conversion:

  _____ → _____ + _____ + energy stored in ATP.
  Glucose   *carbon dioxide*   *water*

  Note that the inorganic compound carbon dioxide is energy-*(rich?; poor?)* because it lacks C —— H bonds.
  b. The reaction shown is an oversimplification of the entire catabolic process, which actually entails many steps. Each pair of hydrogen atoms (2H) removed from glucose consists of two components: one hydrogen ion or proton lacking orbiting electrons, expressed as *(H⁺? H⁻?)*, and a hydrogen nucleus with two orbiting electrons, known as a *(hydride? hydroxyl?)* ion, and expressed as *(H⁺? H⁻?)*.
  c. Hydrogens removed from glucose must first be trapped by H-carrying coenzymes,

  such as _____, _____, or _____; all of these coenzymes are derivatives of a

  vitamin named _____. Ultimately the Hs combine with oxygen to

  form _____. In other words, glucose is dehydrogenated (loses H) while oxygen is reduced (gains H).
  d. Catabolism involves a complex series of stepwise reactions in which hydrogens are transferred. Because oxygen is the final molecule that serves as an oxidizer, catabolism is often described as biological *(reduction? oxidation?)*.
  e. Along the way, energy from the original C–H bonds of glucose (and present in the

  electrons of H) is trapped in _____. This process is known as

  _____ because it involves the addition of a phosphate to ADP. A more complete expression of the total catabolism of glucose (shown briefly in (a) above) is included below. Write the correct labels under each chemical: *oxidized, oxidizer, reduced, reducer:*

  Glucose   + O_2 + ADP → CO_2 + H_2O + ATP
  (_____)   (_____)   (_____)   (_____)

✓ **B2.** Check your understanding of oxidation-reduction reactions and generation of ATP involved in metabolism by circling the correct answer in each case.
  a. Which molecule is more reduced and therefore contains more chemical potential energy (can release more energy when catabolized)?
    A. Lactic acid ($C_3H_6O_3$)
    B. Pyruvic acid ($C_3H_4O_3$)

b. Which is more oxidized and therefore now contains less chemical potential energy since it has given up Hs to some coenzyme?
   A. Lactic acid ($C_3H_6O_3$)
   B. Pyruvic acid ($C_3H_4O_3$)
c. Which is more likely to happen when a molecule within the human body is oxidized?
   A. The molecule gives up Hs
   B. The molecule gains Hs
d. Which coenzyme is more reduced, indicating that it has received Hs when some other molecule was oxidized?
   A. NAD^+
   B. $NADH + H^+$
e. Which molecule contains more chemical potential energy (can release more energy when catabolized)?
   A. ATP
   B. ADP
f. Which is an example of oxidative phosphorylation?
   A. Creatine-P + ADP → creatine + ATP
   B. Generation of ATP by energy released by the electron transfer chain
g. Which is the substrate in the following example of substrate-level phosphorylation? Creatine-P + ADP → creatine + ATP
   A. ATP
   B. Creatine

C. Carbohydrate metabolism (pages 874–886)

✓ **C1.** Answer these questions about carbohydrate metabolism.
a. The story of carbohydrate metabolism is really the story of _____ metabolism because this is the most common carbohydrate (and in fact the most common energy source) in the human diet.
b. What other carbohydrates besides glucose are commonly ingested? How are these converted to glucose?

c. Just after a meal, the level of glucose in the blood *(increases? decreases?)*. Cells use some of this glucose; by _____ glucose, they release energy.
d. List three or more mechanisms by which excess glucose is used or discarded.

e. Increased glucose level of blood is known as _____. This can occur after a meal containing concentrated carbohydrate or in the absence of the hormone _____ because this hormone *(facilitates? inhibits?)* entrance of glucose into cells. Lack of functional insulin occurs in the condition known as _____.

f. Name two types of cells that do not require insulin for glucose uptake so that glucose entry is always possible (provided blood glucose level is adequate).

g. As soon as glucose enters cells, glucose combines with _____ to form glucose-6-phosphate. This process is known as _____ and is

catalyzed by enzymes called _____. What advantage does this
process provide?

✓ **C2.** Describe the process of glycolysis in this exercise.

   a. Glucose is a _____-carbon molecule (which also has hydrogens and oxygens in it).

   During glycolysis each glucose molecule is converted to two _____-carbon mole-

   cules named _____. *(A lot? A little?)* energy is released from glucose
   during glycolysis. This process requires *(one? many?)* step(s) and *(does? does not?)*
   require oxygen.
   b. The initial steps of glycolysis also involve _____-ation of glucose,
   along with conversion of this carbohydrate to another six-carbon sugar,

   _____. These preliminary steps in glycolysis are catalyzed by the

   enzyme _____, called the key regulator of glycolysis. This enzyme
   is most active when the levels of *(ADP? ATP?)* are high and *(ADP? ATP?)* are low
   (indicating the need for generation of ATP by glycolysis). When ATP levels are
   already high, then cellular glucose is instead used for *(ana? cata?)*-bolism of *(glycogen?*
   $CO_2 + H_2O?$*)*.
   c. The phosphorylated fructose is broken down into two *(2? 3? 6?)*-carbon molecules

   named glucose 3-phosphate (_____) and dihydroxyacetone phosphate. The G 3–P
   molecules are then *(oxidized? reduced?)*, giving up their hydrogens to *(FAD? NAD?)*.

   d. The end result of glycolysis is formation of two _____ acids and a
   net gain of *(2? 3? 4?)* ATP molecules.
   e. The fate of pyruvic acid depends on whether sufficient _____ is
   available. If it is, pyruvic acid undergoes chemical change in phases 2 and 3 of
   glucose catabolism (see below). Those processes *(do? do not?)* require oxygen; that is,
   they are *(aerobic? anaerobic?)*.
   f. If the respiratory rate cannot keep pace with glycolysis, insufficient oxygen is
   available to break down pyruvic acid. In the absence of sufficient oxygen, pyruvic

   acid is temporarily converted to _____. This is likely to occur during

   active _____.
   g. Several mechanisms prevent the accumulation of an amount of lactic acid that might

   be harmful. The liver changes some back to _____. Excess pCO_2
   caused by exercise *(stimulates? inhibits?)* respiratory rate, so more oxygen is available
   for breakdown of the pyruvic acid.

✓ **C3.** Review glucose catabolism up to this point and summarize the remaining phases
   by doing this exercise. Refer to Figure LG 25.1. A six-carbon compound named

   (a) _____ is converted to two molecules of (b) _____

Figure LG 25.1 Summary of glucose catabolism, described in three phases.
Phase 1 (glycolysis) occurs in the cytosol. Phase 2 (Krebs cycle) takes place in the matrix
of mitochondria. Phase 3 (electron transport) takes place on inner mitochondrial membrane.
Letters refer to Checkpoint C3.

via a number of steps, together called (c) _____, occurring in the

(d) _____ of the cell.

In order for pyruvic acid to be further broken down, it must undergo a

(e) _____ step. This involves removal of the one-carbon molecule,

(f) _____. The remaining two-carbon (g) _____

group is attached to a carrier called (h) _____, forming

(i) _____. (This compound hooks on to a four-carbon compound called (j) _____. The result is a six-carbon molecule of

(k) _____. The name of this compound is given to the cycle of

reactions called the (1) _____ cycle, occurring in the

(m) _____.

Two main types of reactions occur in the Krebs cycle. One is the decarboxylation

reaction, in which (n) _____ molecules are removed, and hydrogens removed at these steps combine with coenzymes FAD or NAD. Notice these locations on Figure 25.5 of your text. (What happens to the CO_2 molecules that are removed?) As a result, the six-carbon citric acid is eventually shortened to regenerate oxaloacetic acid, giving the cyclic nature to this process.

The other major type of reaction (oxidation–reduction) involves removal of

(o) _____ atoms during oxidation of compounds such as isocitric and a-ketoglutaric acids. (See Figure 25.5 in your text.) These hydrogens are carried off

by two enzymes named (p) _____ and "trapped" for use in phase 3. In phase 3, hydrogen atoms on coenzymes NAD and FAD (and later coenzyme Q) are

ionized to (q) _____ and (r) _____. Electrons are shuttled along a chain of

(s) _____. (Refer to Checkpoint C4.) During electron transport, energy in these electrons (derived from hydrogen atoms in ingested foods) is

"tapped" and ultimately stored in (t) _____.

Finally, the electrons, depleted of some of their energy, are reunited with hydrogen

ions and with (u) _____ to form (v) _____.
Notice why your body requires oxygen—to keep drawing hydrogen atoms off of nutrients (in biological oxidation) so that energy from them can he stored in ATP for use in cellular activities.

✓ **C4.** Now describe details of the electron transport chain in this activity.
  a. The first step in this chain is transfer of high-energy electrons from NADH + H⁺ to the coenzyme *(FMN? Q?)*. As a result this carrier is *(oxidized? reduced?)* to

  _____. From which vitamin is this carrier derived? _____
  b. As electrons are passed on down the chain, $FMNH_2$ is *(oxidized? reduced?)* back to

  _____.

  c. Which mineral is the portion of cytochromes on which electrons are ferried?

  _____ Cytochromes are positioned toward the *(beginning? end?)* of the transport chain.
  d. Circle two other minerals that are known to be involved in the electron transport chain and are hence necessary in the diet.
    Calcium   Copper   Iodine   Sulfur   Zinc
  e. The last carrier in the chain is *(coenzyme Q? cytochrome a₃?)*. It passes e⁻ to

  _____, the final oxidizing molecule. This is a key step in the process of *(aerobic? anaerobic?)* cellular respiration.

f. Where are the electron carriers located? Circle the correct answer.

> **MM.**   **Mitochondrial matrix**
>
> **IMM.**   **Inner mitochondrial membrane (cristae)**
>
> **OMM.**   **Outer mitochondrial membrane**

g. The carriers are grouped into *(two? three? four?)* complexes. Each complex acts to pump *(protons = H+? electrons = e⁻?)* from the *(matrix? space between inner and outer mitochondrial membranes?)* into the _____ .

h. As a result, a gradient of H+ is set up across the *(inner? outer?)* mitochondrial membrane, resulting in potential energy known as the _____ force. How do these protons (H+) travel back into the mitochondrial matrix?

i. The channels through which H+ diffuses contain the enzyme known as ATP _____ . Because the formation of the resulting ATP is driven by an energy source that consists of a gradient and subsequent diffusion of a *chemical* (specifically _____), this mechanism is known as _____ generation of ATP.

✓ **C5.** Summarize aerobic respiration in this activity.

a. Complete the chemical equation showing the overall reaction for aerobic respiration.

$$\underset{\text{Glucose}}{\underline{\hspace{1cm}}} + 6\,\underset{\text{Oxygen}}{\underline{\hspace{1cm}}} + 36\text{–}38\ \text{ADPs} + 36\text{–}38\ \text{P} \rightarrow 6\,\underset{\substack{\text{Carbon}\\\text{dioxide}}}{\underline{\hspace{1cm}}} + 6\,\underset{\text{Water}}{\underline{\hspace{1cm}}} + \underline{\hspace{1cm}}\ \text{ATPs}$$

b. Fill in the correct number of high-energy molecules (ATPs or GTPs) yielded from processes in aerobic respiration.

1. Substrate-level phosphorylation via oxidation of glucose to pyruvic acid in glycolysis: _____ ATPs.

2. Oxidative phosphorylation in electron transport chain (ETS) of six reduced NADs (NADH + 2H+) resulting from glycolysis that takes place under aerobic conditions: _____ ATPs.

3. Oxidative phosphorylation in electron transport chain (ETS) of two reduced NADs (NADH + 2H+) yielded from transition step (formation of acetyl CoA): _____ ATPs.

4. Oxidative phosphorylation in electron transport chain (ETS) of six reduced NADs (NADH + 2H+) yielded from the Krebs cycle: _____ ATPs.

5. Oxidative phosphorylation in electron transport chain (ETS) of two reduced FADS yielded from the Krebs cycle: _____ ATPs. (Notice that each FADH₂ yields fewer ATPs than NADH + 2H+ because FADH₂ enters the ETS at a lower level.)

6. Substrate-level phosphorylation via oxidation via succinyl CoA to succinic acid in the Krebs cycle: _____ GTPs.

7. Grand total of ATPs or GTPs resulting from complete catabolism of glucose by glycolysis, Krebs cycle, electron transport, and resulting phosphorylation of ADP

   to ATP (or GDP to GTP): _____ ATPs and/or GTPs

✓ **C6.** Most of the 36 to 38 ATPs generated from the total oxidation of hydrogens derive from *(glycolysis? the Krebs cycle?)*. Of those generated by the Krebs cycle, most are formed from the oxidation of hydrogens carried by *(NAD? FAD?)*. About what percentage of the energy originally in glucose is stored in ATP after glucose is completely catabolized? *(10%? 40%? 60%? 99%?)*

✓ **C7.** Complete the exercise about the reaction shown below.

   (1)
   Glucose ⇌ Glycogen
   (2)

   a. You learned earlier that excess glucose may be stored as glycogen. In other words,

      glycogen consists of large branching chains of _____. Name the process of glycogen formation by labeling (1) in the chemical reaction.

   b. Between meals, when glucose is needed, glycogen can be broken down again to release glucose. Label (2) above with the name of this process. [Note that a number of steps are actually involved in both processes (1) and (2).]

   c. Which of these two reactions is anabolic? _____

   d. Where is most (75%) of glycogen in the body stored'? _____

   e. Identify a hormone that stimulates reaction (1). Write its name on the upper arrow in the reaction above. As more glucose is stored in the form of glycogen, the blood

      level of glucose _____-creases. Now write below the lower arrow the names of two hormones that stimulate glycogenolysis.

   f. In order for glycogenolysis to occur, a _____ group must be added to glucose as it breaks away from glycogen. The enzyme catalyzing this reaction is

      known as _____. Does the reverse reaction [shown in (2)] require phosphorylation also? *(Yes? No?)*

C8. Define *gluconeogenesis* and briefly discuss how it is related to other metabolic reactions.

Name three hormones that stimulate gluconeogenesis.

✓ **C9.** Circle the processes at left and the hormones at right that lead to increased blood glucose level (hyperglycemia).

Processes   **Hormones**
Glycogenesis   Insulin
Glycogenolysis   Glucagon
Glycolysis   Epinephrine
Gluconeogenesis   Cortisol
Thyroid hormone
Growth hormone

For extra review. Refer to Figures LG 18.3 and LG 18.6, pages 360 and 366 of the *Learning Guide*.

D. Lipid metabolism (pages 886–889)

✓ **D1.** Most lipids are *(polar? nonpolar?)*, and therefore are hydro-*(philic? phobic?)*. Check your understanding of mechanisms that make lipids water-soluble in blood by completing this Checkpoint. Select answers from the box that best fit descriptions below.

A.	**Apoproteins**	**LDLs.**	**Low-density lipoproteins**
C.	**Chylomicrons**	**VLDLs.**	**Very low-density lipoproteins**
HDLs.	**High-density lipoproteins**		

_____ a. Designated by letters A, B, C, D, and E, these proteins surround an inner core of lipids within lipoproteins. These chemicals can serve as "docking" molecules that bind to receptors on body cells.

_____ b. Named so because these molecules are composed mainly (close to 50%) of protein which has high density; prevent accumulation of cholesterol in blood, so called the "good" cholesterol.

_____ c. Named so because these molecules are composed of small amounts (25%) of protein but large amounts (50%) of cholesterol (which has low density). These

carry 75% of total blood cholesterol, and are called the "bad" cholesterol since excessive numbers of these form atherosclerotic plaques in arteries.

_____ d. Made largely (50%) of triglycerides, their numbers increase with a high-fat diet, leading ultimately to high blood levels of LDLs.

_____ e. Formed almost exclusively (85%) of triglycerides derived from dietary fats, they travel through lymphatic pathways in the GI tract, and then into blood to reach adipose or muscle tissue.

✓ **D2.** *Critical thinking*. Complete this exercise on cholesterol in diet and health.
   a. Many foods bear the label "no cholesterol." Does a diet of such foods assure a low blood cholesterol?

   b. Total cholesterol level should be below _____ mg/deciliter, and the level of LDLs

   (which make up most of total cholesterol) should be less than _____ mg/dL. "Total cholesterol to HDL" ratio should be *(less than? greater than?)* 3:1. Two ways to improve

   (lower) this ratio are to _____-crease total cholesterol level and to _____-crease HDL level. Regular aerobic exercise can achieve both of those goals.

c. Complete the table below as you evaluate and comment on each client's cholesterol profile. (Note that total cholesterol includes VLDLs also, so that LDLs + HDLs do not add up to total cholesterol.)

Client	Total cholesterol*	LDL*	HDL*	Cholesterol: HDL ratio
1. Dave Comment:	296	226	40	296/40 = 7.4:1
2. Juan Comment:	198	138	45	198/45 = _____
3. Denise Comment:	202	160	38	_____
4. Jerry Comment:	222	140	74	_____

* mg/dL blood

✓ **D3.** About 98% of all of the energy reserves stored in the body are in the form of *(fat? glycogen?)*. About *(10%? 25%? 50%?)* of the stored triglycerides (fats) in the body are located in subcutaneous tissues. Is the fat stored in these areas likely to be "the same fat" that was located there two years ago? *(Yes? No?)* Explain.

✓ **D4.** Do this exercise about fat metabolism.
   a. The initial step in catabolism of triglycerides is their breakdown into

   _____ and _____, a process called _____-lysis

   and catalyzed by enzymes called _____-ases.

   b. Glycerol can then be converted into _____ (G 3-P); this molecule can then enter glycolytic pathways to increase ATP production. If the cell does not

   need to generate ATP, then G 3-P can be converted into _____.

   This anabolic step is an example of gluco-_____.

c. Recall that fatty acids are long chains of carbons with attached hydrogens and a few oxygens. Two-carbon pieces are "snipped off" of fatty acids by a process called

_____, occurring in the _____.

d. These two-carbon pieces (_____) attach to coenzyme _____.

The resulting molecules, named _____ may enter the Krebs cycle. Complete catabolism of one 16-carbon fatty acid can yield a net of *(36–38? 129?)* ATPs.

e. When large numbers of acetyl CoAs form. they tend to pair chemically:

Acetic acid   +   acetic acid → _____
(2C)   (2C)   (4C)

The presence of excessive amounts of these acids in blood *(raises? lowers?)* blood pH. Because these acids are called "keto" acids, this condition is known as

_____.

f. Slight alterations of acetoacetic acid leads to formation of β-hydroxybutyric acid

and _____. Collectively, these three chemicals are known as

_____ bodies. Formation of ketone bodies is known as

_____; it takes place in the _____.

g. Circle the cells that can use acetoacetic acid for generation of ATP:
   Brain cells   Hepatic cells
   Cardiac muscle cells   Cells of the renal cortex

h. Explain why the blood level of acetoacetic acid is normally maintained at a low level.

i. Ketogenesis occurs when cells are forced to turn to fat catabolism. State two or more reasons why cells might carry out excessive fat catabolism leading to ketogenesis and ketosis.

j. *A clinical correlation.* Explain why a person whose diabetes is out of control might have sweet breath.

E. Protein metabolism (pages 889–891)

✓ **E1.** Which of the three major nutrients (carbohydrates, lipids, and proteins) fulfills each function?

a. Most direct source of energy; stored in body least: _____

b. Constitutes almost all of energy reserves, but also used for body-building:

_____

c. Usually used least for fuel; used most for body structure and for regulation:

_____

✓ **E2.** Throughout your study of systems of the body, you have learned about a variety of roles of proteins. List at least six functions of proteins in the body, using one or two words for each function. Include some structural and some regulatory roles.

_____   _____   _____

_____   _____   _____

✓ **E3.** List names of hormones that enhance transport of amino acids into cells or otherwise promote protein synthesis (anabolism).

Figure LG 25.2 Metabolism of protein.
Lowercase letters refer to Checkpoint E4.

✓ **E4.** Refer to Figure LG 25.2 and describe the uses of protein.

Two sources of proteins are shown: (a) _____ and (b) _____.

Amino acids from these sources may undergo a process known as (c) _____ to form new body proteins. If other energy sources are used up, amino acids may

undergo catabolism. The first step is (d) _____, in which an amino

group (NH_2) is removed and converted (in the liver) to (e) _____, a

component of (f) _____ which exits in urine. The remaining portion

of the amino acid may enter (g) _____ pathways at a number of points (Figure 25.13 in your text). In this way, proteins can lead to formation of

(h) _____.

E5. Contrast essential amino acids with nonessential amino acids.

How many amino acids are considered essential to humans? _____

✓ **E6.** Phenylketonuria (PKU) is a metabolic disorder in which the amino acid

_____ cannot be converted into tyrosine. A screening test for this

disorder *(is? is not?)* available. The sweetener _____ (Nutrasweet) contains phenylalamine, so children with PKU should not consume this.

F. Key molecules at metabolic cross roads (pages 891–893)

✓ **F1.** Key molecules in metabolism provide "crossroads" among the three different

categories of nutrients: carbohydrates, _____, and _____.
These key molecules provide pathways that permit you to put on weight (fat) from eating excessive carbohydrates or even proteins, and allow you to be energized

(generate _____) by eating any of the three food types. Refer to Figure LG 25.14 in the text and identify roles of each of these key molecules (in the box) in metabolism.

Acetyl CoA. **Acetyl coenzyme A**	**PA.** **Pyruvic acid**
G 6-P. **Glucose 6-phosphate**	

_____ a. Two carbon molecule formed from pyruvic acid when oxygen is available; leads into Krebs cycle and electron transfer system

_____ b. Entrance molecule for fatty acids and ketone bodies into Krebs cycle; also the key molecule in conversion of excessive carbohydrates to fat

_____ c. Formed from beta-oxidation of fatty acids during fat catabolism; excessive amounts of these (as in diabetics lacking adequate insulin or persons who are starving or dieting excessively) can lead to ketone body formation and potentially lethal ketoacidosis (see page 889 of the text)

_____ d. Three carbon molecule that feeds into aerobic metabolism if oxygen is available, otherwise can form lactic acid

_____ e. Similar to amino acid alanine except for one amino group; molecule by which amino acids (from protein in foods) can provide energy (via catabolic pathways)

_____ f. Molecule into which glucose is converted (reversible reaction) when glucose enters any body cell

_____ g. Can form oxaloacetic acid to feed into the Krebs cycle or lead into gluconeogenesis, resulting in synthesis of carbohydrates (glucose) from proteins (alanine)

_____ h. Molecule by which glycogen can enter into pathways to be used as fuel by liver or muscle cells, as in "carbo loading" before a marathon race (see page 883 of the text)

_____ i. Precursor for ribose portion of RNA and deoxyribose portion of DNA

✓ **F2.** *For extra review.* Focus on interconversions of carbohydrates and fats in this Checkpoint.
   a. Fill in the blanks in the pathway of conversion of excessive carbohydrates to body fat. Use the same answers as for Checkpoint F1.

   Glycogen → _____ → _____ → _____ →
   Fatty acid portion of fats

   b. Mammals, including humans, *(can? cannot?)* convert acetyl CoA into pyruvic acid. As a result, fatty acids *(can? cannot?)* be used to form glucose. Gluconeogenesis involving fats can occur, however, by the following pathway:

   The _____ portion of triglycerides → glyceraldehyde 3-phosphate

   (G 3-P) → _____.

G. Metabolic adaptations (pages 893–898)

✓ **G1.** Contrast these two metabolic states by writing *A* if the description refers to the absorptive state and *P* if it refers to the postabsorptive state.

_____ a. Period when the body is fasting and is challenged to maintain an adequate blood glucose level

_____ b. State during which glucose (stored in glycogen in liver and muscle) is released

_____ c. State during which most systems (excluding nervous system) switch over to use of fatty acids as energy sources

_____ d. Time when the body is absorbing nutrients from the GI tract

_____ e. State during which the principal concern is formation of stores of glucose (as glycogen) and fat

_____ f. State when insulin is released under stimulation from glucose-dependent insulinotropic peptide (GIP); insulin then facilitates transport of glucose and amino acids into cells

_____ g. Period dominated by anti-insulin hormones, such as glucagon

✓ **G2.** Do this exercise about maintenance of glucose level between meals.
   a. The normal blood glucose level in the postabsorptive state is about *(30–50? 70–110? 150–170? 600–700?)* mg glucose/100 mL blood.

b. Now list four sources of glucose that may he called upon during the postabsorptive state.

_____

_____

_____

_____

c. Can fatty acids serve as a source of glucose? *(Yes? No?)* Explain.

d. Fats can be used to generate ATP by being broken down into _____, which can then enter the Krebs cycle. Therefore, fatty acids are said to provide a glucose-*(sparing? utilizing?)* mechanism between meals.

✓ **G3.** Identify the hormones (in the box) that stimulate each metabolic process listed below. Write one answer on each line. (Refer to Tables 25.3 and 25.4 in your text for help.)

Cor.	**Cortisol**	**IGF.**	**Insulinlike growth factors**
Epi.	**Epinephrine**	**Ins.**	**Insulin**
Glu.	**Glucagon**	**Thy.**	**Thyroid hormone (T3/T4)**

a. Glycogenolysis: _____ _____

b. Glycogenesis: _____

c. Gluconeogenesis: _____ _____

d. Protein synthesis: _____ _____ _____

e. Lipogenesis: _____

f. Lipolysis: _____ _____ _____ _____

g. Protein breakdown: _____

✓ **G4.** *For extra review,* identify which processes (a-g) in Checkpoint G3 are *anabolic* and which are *catabolic.*
Anabolic (*Hint:* terms may include "-genesis" or *creation*) _____ _____ _____
_____
Catabolic (*Hint:* terms may include "-lysis" or *destruction*) _____ _____ _____

✓ **G5.** *For extra review,* identify the hormones (in the box for Checkpoint G3) that are

hyperglycemic, that is, tend to increase blood glucose. _____ _____ _____ _____
These four hormones *(mimic insulin? are anti-insulin?)* in action, and they are more active during the *(absorptive? postabsorptive?)* state since the body requires more glucose then. During this period, *(sympathetic? parasympathetic?)* nerves stimulate

release of the hormone _____ from the adrenal medulla.

✓ **G6.** *A clinical correlation.* Describe metabolic changes during fasting or starvation.
a. Contrast these terms: fasting/starving.

b. State the major factor that determines the life span possible without food.

c. Explain what accounts for the "wasting" appearance of people who are starving.

d. Explain why extremely high blood levels of ketone bodies (which provide up to 2/3 of the energy to the brain after many weeks of starvation) may be dangerous and even lethal for example, leading to coma and death.

H. Heat and energy balance (pages 898–902)

H1. Define basal metabolic rate (BMR).

✓ **H2.** Contrast *calorie* and *kilocalorie*.

If you ingest, digest, absorb, and metabolize a slice of bread, the energy released from the bread equals about 80 to 100 *(cal? kcal?)*.

✓ **H3.** Catabolism of foods *(uses? releases?)* energy. Most of the heat produced by the body comes from catabolism by *(oxidation? reduction?)* of nutrients.

✓ **H4.** Discuss BMR in this exercise.

a. BMR may be determined by measuring the amount of _____ consumed within a period of time, because oxygen is necessary for the metabolism of foods. Normally, for each liter of oxygen consumed, the body metabolizes enough

food to release about _____ kcal.

b. Suppose you consume 15 liters of oxygen in an hour. Your body would release

_____ kcal of heat in that hour. At this metabolic rate, you would need to

consume _____ kcal per day to meet your metabolic needs.

c. A BMR of 20% below the standard value is likely to be due to *(hyper? hypo?)*-thyroid.

✓ **H5.** Do this activity on regulation of body temperature.

    a. Which temperature is normally higher? *(Core? Shell?)* Explain how an increase in core temperature, for example, above 112–114°F can be fatal.

    b. Circle the factor in each pair that is likely to lead to increased metabolic rate with increased body temperature.

        1. Age: 12 years old   52 years old

        2. Body temperature: 98.6°F   103.6°F

        3. State of activity: Running 2 miles   Typing at computer

        4. Adrenal medulla or thyroid hormones:

          Increased epinephrine or thyroid hormone

          Decreased epinephrine or thyroid hormone

        5. Specific dynamic action related to ingestion of: Protein   Carbohydrate

H6. Write the percentage of heat loss by each of the following routes (at room temperature). Then write an example of each of these types of heat loss. One is done for you.

    a. Radiation _____   _____

    b. Evaporation _____   _____

    c. Convection __**15**__   **Cooling by draft while taking a shower**_____

    d. Conduction _____   _____

✓ **H7.** During an hour of active exercise Bill produces 1 liter of sweat. The amount of heat loss (cooling) that accompanies this much evaporation is _____ Cal. On a humid day, *(more? less?)* of Bill's sweat occurs, so Bill would be cooled *(more? less?)* on such a day.

✓ **H8.** The body's "thermostat" is located in the preoptic area of the *(eyes? hypothalamus? medulla? skin?)*. If the body temperature needs to be raised, nerve impulses from the preoptic area are sent to the heat-*(losing? promoting?)* center, which is primarily *(parasympathetic? sympathetic?)*. An example of such a response is vaso-*(constriction? dilation?)* of blood vessels in skin, resulting in heat *(loss or release? conservation?)*.

H9. Explain why elderly persons are at greater risk for hypothermia.

Write seven or more signs or symptoms of hypothermia.

✓ **H10.** Complete this Checkpoint about regulation of food intake.

     a. The site within the brain that is the location of feeding and satiety centers is the (*medulla? hypothalamus?*). When the satiety center is active, an individual will feel (*hungry? satiated or full?*).

     b. Circle all of the answers that will keep the feeding center active and enhance the desire to eat.

         ↓ Glucose in blood

         ↓ Amino acids in blood

         ↓ Fats entering intestine (with related ↓ in CCK release)

         ↓ Environmental temperature

         ↓ Stretching of stomach

I. Nutrition (pages 902–907)

I1. List three primary functions of nutrients.

✓ **I2.** List the six principal classes of nutrients.

Circle the nutrient needed in greatest quantities

✓ **I3.** Fill in the number of calories needed each day by typical persons in each category.

_____ a. Active women, teen girls, and most men   _____ c. Elderly adults

_____ b. Active men and teen boys   _____ d. Children

✓ **I4.** Fill in the percentage of calories needed each day in each category of nutrients.

     a. Carbohydrates: _____%

     b. Fats: no more than _____% with no more than _____% of total calories in the form of saturated fats

     c. Proteins: _____%

✓ **I5.** *Critical thinking.* Keep in mind that carbohydrates and proteins each provide about 4 Cal/g and fats provide about 9 Cal/g. Considering the information in Checkpoint I4, determine the number of grams of each of the three major nutrient groups appropriate for a healthy diet of 2000 Cal/day. One is done for you.

     a. Carbohydrates

b. Fats

c. Proteins **2000 Cal × (0.12–0.15) = 240–300 Cal/day. At 4 Cal/g = *60–75g/day*.**

16. Write seven guidelines for healthy eating.

✓ **17.** Using the Food Pyramid (page 903 in the text), identify the number of servings suggested per day for each food group listed.

_____ a. Bread, cereal, rice, pasta

_____ b. Vegetables

_____ c. Fruits

_____ d. Meat, fish, poultry, dry beans, eggs, nuts

_____ e. Milk, yogurt, and cheese

_____ f. Fats, oils, sweets

✓ **18.** Define *minerals*.

Minerals make up about _____% of body weight and are concentrated in the

_____.

✓ **19.** Study Table 25.5 in your text. Then check your understanding of minerals by doing this matching exercise

Ca.	Calcium	Fe.	Iron	Na.	Sodium
Cl.	Chlorine	I.	Iodine	P.	Phosphorus
Co.	Cobalt	K.	Potassium	S.	Sulfur
F.	Fluorine	Mg.	Magnesium		

_____ a. Main anion in extracellular fluid, part of HCl in stomach; component of table salt

_____ b. Involved in generation of nerve impulse, helps to regulate osmosis, acts in buffer systems

_____ c. Most abundant mineral in the body. found mostly in bones and teeth; necessary for normal muscle contraction and for blood clotting

_____ d. Important component of hemoglobin and cytochromes

_____ e. Main cation inside of cells; used in nerve transmission

_____ f. Essential component of thyroxin

_____ g. Constituent of vitamin B_{12}, so necessary for red blood cell formation

_____ h. Important component of amino acids, vitamins, and hormones

_____ i. Improves tooth structure

_____ j. Found mostly in bones and teeth; important in buffer system and in ATP
processes; component of DNA and RNA

✓ I10. Circle the correct answer. The primary purpose of vitamins is:
A. Synthesis of body structures
B. Regulation of body activities

✓ I11. Vitamins are _(organic? inorganic?)_. Most vitamins _(can? cannot?)_ be synthesized in the
body. In general, what are the functions of vitamins?

✓ I12. Contrast the two principal groups of vitamins, and list the main vitamins in each
group.
a. Fat-soluble _____

b. Water-soluble _____

I13. Defend or dispute this statement: "Most persons who eat a balanced diet do need to
take vitamin or mineral supplements."

✓ I14. Select the vitamin that fits each description.

A	B$_1$	B$_2$	B$_{12}$	C	D	E	K

_____ a. This serves as a coenzyme that is essential
for blood clotting, so it is called the anti-
hemorrhagic vitamin; synthesized by
intestinal bacteria.

_____ b. Its formation depends upon sunlight on
skin and also on kidney and liver activation; necessary for calcium absorption.

_____ c. Riboflavin is another name for it; a component of FAD; necessary for normal
integrity of skin, mucosa, and eye.

_____ d. This vitamin acts as an important
coenzyme in carbohydrate metabolism;
deficiency leads to beriberi.

_____ e. Formed from carotene, it is necessary for
normal bones and teeth; it prevents night
blindness.

_____ f. This substance is also called ascorbic acid;
deficiency causes anemia, poor wound
healing, and scurvy.

_____ g. This coenzyme, the only B vitamin not
found in vegetables, is necessary for normal erythropoiesis; absorption from GI
tract depends on intrinsic factor.

_____ h. Also known as tocopherol, it is necessary
for normal red blood cell membranes:
deficiency is associated with sterility in
some animals.

J. Disorders (pages 908–909)

✓ **J1.** Arrange in order the events believed to occur in the production of fever of 39.4°C

(103°F) and recovery from this state. _____ _____ _____ _____ _____ _____

A. 39.4°C (103°F) temperature is reached and maintained.

B. Prostaglandins (PGs) cause the hypothalamic "thermostat" to be reset from normal 37.5°C (98.6°F) to a higher temperature such as 39.4°C (103°F).

C. A "chill" occurs as skin feels cool (due to vasoconstriction) and shivering occurs— attempts to raise body temperature to the 39.4°C (103°F) designated by the hypothalamus.

D. Source of pyrogens is removed (for example, bacteria are killed by antibiotics), lowering hypothalamic "thermostat" to 37.5°C (98.6°F).

E. "Crisis" occurs. Obeying hypothalamic orders, the body shifts to heat loss mechanisms such as sweating and vasodilation, returning body temperature to normal.

F. Infectious organisms cause phagocytes to release interleukin-l, which then causes hypothalamic release of PGs.

J2. In what ways is a fever beneficial?

✓ **J3.** Define obesity, and state several causes of obesity.

J4. *A clinical correlation.* Which condition involves deficiency in both protein and calories? *(Kwashiorkor? Marasmus?)* Explain why an individual with kwashiorkor may have a large abdomen even though the person's diet is inadequate.

ANSWERS TO SELECTED CHECKPOINTS: CHAPTER 25

A1. (a) Immediate energy supply. (b) Building blocks. (c) Storage for future use for (a) or (b).

A2.

Process	Definition	Releases or Uses Energy	Examples
a. Catabolism	Breakdown of complex organic compounds into smaller ones	Energy released as heat or stored in ATP	Glycolysis, Krebs cycle, glycogenolysis
b. Anabolism	Synthesis of complex organic molecules from smaller ones	Uses energy (ATP)	Synthesis of protein, fats, or glycogen

A4. 2; 3, 2; ATP.

A5. 40%; heat.

B1. (a) ATP; $C_6H_{12}O_6 \rightarrow CO_2 + H_2O$ + energy stored in ATP: poor. (b) H^+, hydride, H^-. (c) NAD^+, $NADP^+$, FAD; niacin; H_2O. (d) Oxidation. (e) ATP; phosphorylation;

Glucose + O_2 + ADP →
(reducer) (oxidizer)

CO_2 + H_2O + ATP
(oxidized) (reduced)

B2. (a) A (b) B. (c) A. (d) B. (e) A. (f) B. (g) B.

C1. (a) Glucose. (b) Starch, sucrose, lactose; by enzymes located mostly in liver cells. (c) Increases, oxidizing. (d) Stored as glycogen, converted to fat or protein and stored, excreted in urine. (e) Hyperglycemia; insulin, facilitates; diabetes mellitus. (f) Neurons and liver cells. (g) Phosphate; phosphorylation, kinases: cells can utilize the trapped glucose since phosphorylated glucose cannot pass out of most cells.

C2. (a) 6; 3, pyruvic acid; a little; many, does not. (b) Phosphoryl, fructose; phosphofructokinase; ADP, ATP; ana, glycogen. (c) 3, G 3-P; oxidized, NAD. (d) Pyruvic, 2. (e) Oxygen; do, aerobic. (f) Lactic acid; exercise. (g) Pyruvic acid; stimulates.

C3. (a) Glucose. (b) Pyruvic acid. (c) Glycolysis. (d) Cytosol. (e) Transition. (f) CO_2. (g) Acetyl. (h) Coenzyme A. (i) Acetyl coenzyme A. (j) Oxaloacetic acid. (k) Citric acid. (1) Citric acid (or Krebs). (m) Matrix of mitochondria. (n) CO_2, which are exhaled. (o) Hydrogen. (p) NAD and FAD. (q) H^+. (r) Electrons (e^-) (or H^-). (s) Cytochromes or electron transfer system. (t) ATP. (u) Oxygen. (v) Water.

C4. (a) FMN: reduced. $FMNH_2$; B_2. (riboflavin). (b) Oxidized, FMN. (c) Iron (Fe); end. (d) Copper and sulfur. (e) Cytochrome a_3; oxygen; aerobic. (f) IMM. (g) Three; protons = H^+, matrix, space between inner and outer mitochondrial membranes. (h) Inner, proton motive; across special channels in the inner mitochondrial membrane. (i) Synthetase; H, chemiosmotic.

C5. (a) $C_6H_{12}O_6$ + $6O_2$: + 36 or 38 ADPs
Glucose   Oxygen

+ 36 or 38 (P) → $6CO_2$ + $6H_2O$ +
Carbon   Water

36–38 ATPs. (b1) 2; (b2) 4–6. (b3) 6. (b4) 18. (b5) 4. (b6) 2. (b7) 36–38.

C6. The Krebs cycle: NAD; 40.

C7. (a) Glucose; I: glycogenesis. (b) 2: Glycogenolysis. (c) Glycogenesis. (d) Muscles. (e) Insulin; de; glucagon and epinephrine. (f) Phosphate; phosphorylase; yes.

C9. Processes: glycogenolysis and gluconeogenesis; hormones: all except insulin.

D1. Nonpolar, phobic. (a) A. (b) HDLs. (c) LDLs. (d) VLDLs. (e) C.

D2. (a) No. Cholesterol comes not only from diet, but also from synthesis of cholesterol, for example, from saturated fatty acids (as in palm oil, coconut oil, or partially hydrogenated margarine), within the liver. (b) 200, 130; less than; de, in. (c1) Dave's seriously high levels of cholesterol, LDLs, and his ratio of 7.4 put him at almost four times the risk of a heart attack as Juan with a cholesterol of 198. (c2) Juan's total cholesterol is good, but his ratio of 198/45 = 4.5 puts him at increased risk. (c3) Denise's LDL level is high; raising her HDLs would improve her 202/38 ratio of 5.3. (c4) Jerry, at age 74, still runs six miles most days, so keeps his HDLs high (normal range is 39–96). His is the best ratio: 222/72 = 3.0, even though his total cholesterol is borderline-high. (He's working on cutting back on saturated fats.)

D3. Fat: 50%: no. it is continually catabolized and resynthesized.

D4. (a) Fatty acids. glycerol, lipo, lip. (b) Glyceraldehyde 3-phosphate; glucose; neogenesis. (c) Beta oxidation, matrix of mitochondria. (d) Acetic acid, A; acetyl CoA (or acetyl coenzyme A); 129. (e) Acetoacetic acid; lowers; ketoacidosis. (f) Acetone; ketone; ketogenesis, liver. (g) Cardiac muscle cells, cells of the renal cortex, and (during starvation) brain cells. (h) Acetoacetic acid is normally catabolized for ATP production as fast as the liver generates it. (i) Examples of reasons: starvation, fasting diet, lack of glucose in cells due to lack of insulin (diabetes mellitus), excess growth hormone (GH) which stimulates fat catabolism. (j) As acetone (formed in ketogenesis) passes through blood in pulmonary vessels. some is exhaled and detected by its sweet aroma.

E1. (a) Carbohydrates. Exception: cardiac muscle and cortex of the kidneys utilize acetoacetic acid from fatty acids preferentially over glucose. (b) Lipids. (c) Proteins.

E2. See Exhibit 2.6 in your text.

E3. Insulin (as well as insulinlike growth factors, IGFs), thyroid hormones (T3 and T4), estrogen, and testosterone.

E4. (a) Worn-out cells as in bone or muscle. (b) Ingested foods. (c) Anabolism. (d) Deamination. (e) Ammonia. (f) Urea. (g) Glycolytic or Krebs cycle. (h) $CO_2 + H_2O$ + ATP.

E6. Phenylalanine; is; aspartame.

Fl. Proteins and lipids; ATP. (a–c) Acetyl-CoA. (d-e). PA. (f-i) G 6-P.

F2. (a) G 6-P. PA, acetyl CoA. (b) Cannot; cannot; glycerol, glucose.

G1. (a–c) P (d–f) A. (g) P.

G2. (a) 70–110. (b) Liver glycogen (4-hour supply); muscle glycogen and lactic acid during exercise; glycerol from fats; as a last resource, amino acids from tissue proteins. (c) No. See Checkpoint F2b on page 563. (d) Acetic acids (and then acetyl CoA); sparing.

G3. (a) Glu Epi. (b) Ins. (c) Cor Glu. (d) IGF Ins Thy. (e) Ins. (f) Cor Epi IGF Thy. (g) Cor.

G4. Anabolic: b-e. Catabolic: a, f, and g.

G5. Cor Glu Epi Thy (or all except Ins and IGF); are anti-insulin; postabsorptive; sympathetic, epinephrine.

G6. (a) Fasting is short term (a few hours or days) food deprivation; starvation refers to long periods of up to months without food. (b) The amount of adipose tissue (and also availability of water to prevent dehydration). (c) When dietary glucose is not available, amino acids from muscles and fatty acids from adipose stores are used for ATP production. (d) Glucose (normally the primary fuel for the nervous system: See page 895 of the text) is not available, and fatty acids cannot pass the blood-brain barrier. Ketone bodies can be used as energy sources, but high levels lead to ketoacidosis (a form of metabolic acidosis) which depresses the central nervous system.

H2. (a) A kilocalorie is the amount of heat required to raise 1000 grams (= 1000 mL = I liter) of water 1°C; a kilocalorie (kcal) = 1000 calories (cal); kcal.

H3. Releases; oxidation.

H4. (a) Oxygen; 4.8. (b) 72 (= 4.8 kcal/liter oxygen × 15 liters/hr); 1728 (= 72 kcal/hr × 24 hrs/day. (c) Hypo.

H5. (a) Core; hyperthermia denatures (changes chemical configurations of) proteins. (b1) 12 years old; (b2) 103.6°F; (b3) Running 2 miles; (b4) Increased epinephrine or thyroid hormone; (B5) Protein.

H7. 580 (= 0.58 cal/ml water × 1000 ml/liter); less, less.

H8. Hypothalamus; promoting, sympathetic; constriction, conservation.

H10. (a) Hypothalamus; satiated or full. (b) Satiety; all answers.

I2. Carbohydrates, proteins, lipids, minerals, vitamins, and water; water.

I3. (a) 2200. (b) 2800. (c) 1600. (d) 2200.

I4. (a) 50–60. (b) 30, 10. (c) 12–15.

I5. (a) 2000 Cal × (0.50–0.60) = 1000–1200 Cal/day. At 4 Cal/g = 250–300 g/day. (b) 2000 Cal × (0.30) = less than 600 cal/day. At 9 Cal/g = 600/9 = 67 g/day of total fat: a maximum of 10% should be saturated fat = 22 g/day.

I7. (a) 6–11. (b) 3–5. (c) 2–4. (d) 2–3. (e) 2–3. (f) Sparingly.

I8. Inorganic substances; 4, skeleton.

I9. (a) Cl. (b) Na. (c) Ca. (d) Fe. (e) K. (f) I. (g) Co. (h) S. (i) F. (j) P.

I10. B.

I11. Organic; cannot; most serve as coenzymes, maintaining growth and metabolism.

I12. (a) A D E K. (b) B complex and C.

I14. (a) K. (b) D. (c) B_2. (d) B_1. (e) A. (f) C. (g) B_{12}. (h) E.

J1. F B C A D E.

J4. Marasmus; protein deficiency (due to lack of essential amino acids) decreases plasma proteins. Blood has less osmotic pressure, so fluids exit from blood. Movement of these into the abdomen (ascites) increases the size of the abdomen. Fatty infiltration of the liver also adds to the abdominal girth.

MORE CRITICAL THINKING: CHAPTER 25

1. Contrast anabolism and catabolism. Give one example of each process in metabolism of carbohydrates, lipids, and proteins.

2. Explain the significance of B-complex vitamins in normal metabolism. Include roles of niacin (B_1), which forms coenzymes NAD^+ and $NADP^+$, riboflavin (B_2), which forms coenzymes FAD and FMN, and pantothenic acid, which forms coenzyme A.

3. Describe interconversions of carbohydrates, lipids, and proteins, including roles of the three key molecules, acetyl CoA, pyruvic acid, and G 6-P.

4. Explain why red blood cells derive all of their ATP from glycolysis. (Hint: which organelles do they lack?)

5. When you wake up at 7 AM and have not eaten since dinner last night, which hormones are helping you to maintain an adequate blood level of glucose? How do they accomplish this?

6. Lil is outside on a snowy day without a warm coat. Her skin is pale and chilled and she begins to shiver. Explain how these responses are attempts of her body to maintain homeostasis of temperature. What other responses might raise her body temperature?

7. Describe guidelines for healthy eating and discuss roles foods may play in disorders such as cancer and diabetes

MASTERY TEST: CHAPTER 25

Questions 1–12: Circle the letter preceding the one best answer to each question.

1. Which of these processes is anabolic?
   A. Pyruvic acid → CO_2 + H_2O + ATP
   B. Glucose → pyruvic acid + ATP
   C. Protein synthesis
   D. Digestion of starch to maltose
   E. Glycogenolysis

2. Which of the following can be represented by the equation Glucose → pyruvic acids + small amount ATP?
   A. Glycolysis
   B. Transition step between glycolysis and Krebs cycle
   C. Krebs cycle
   D. Electron transport system

3. Which hormone is said to be hypoglycemic because it tends to lower blood sugar?
   A. Glucagon   C. Growth hormone
   B. Glucocorticoids D. Insulin
   E. Epinephrine

4. The complete oxidation of glucose yields all of these products *except:*
   A. ATP   C. Carbon dioxide
   B. Oxygen   D. Water

5. All of these processes occur exclusively or primarily in liver cells except one, which occurs in virtually all body cells. This one is:
   A. Gluconeogenesis
   B. Beta oxidation of fats
   C. Ketogenesis
   D. Deamination and conversion of ammonia to urea
   E. Krebs cycle

6. The process of forming glucose from amino acids or glycerol is called:
   A. Gluconeogenesis   D. Deamination
   B. Glycogenesis   E. Glycogenolysis
   C. Ketogenesis

7. All of the following are parts of fat absorption and fat metabolism *except:*
   A. Lipogenesis   D. Ketogenesis
   B. Beta oxidation   E. Glycogenesis
   C. Chylomicron formation

8. Which of the following vitamins is water soluble? A, C, D, E, K?

9. At room temperature most body heat is lost by:
   A. Evaporation   C. Conduction
   B. Convection   D. Radiation

10. Choose the *false* statement about vitamins.
    A. They are organic compounds.
    B. They regulate physiological processes.
    C. Most are synthesized by the body.
    D. Many act as parts of enzymes or coenzymes.

11. Choose the *false* statement about temperature regulation.
    A. Some aspects of fever are beneficial.
    B. Fever is believed to be due to a "resetting of the body's thermostat."
    C. Vasoconstriction of blood vessels in skin will tend to conserve heat.
    D. Heat-producing mechanisms which occur when you are in a cold environment are primarily parasympathetic.

12. Which of the following processes involves a cytochrome chain?
    A. Glycolysis
    B. Transition step between glycolysis and Krebs cycle
    C. Krebs cycle
    D. Electron transport system

Questions 13–14. Arrange the answers in correct sequence.

_____ _____ _____ _____ _____ _____   13.   Steps in complete oxidation of glucose:
- A. Glycolysis takes place.
- B. Pyruvic acid is converted to acetyl CoA.
- C. Hydrogens are picked up by NAD and FAD; hydrogens ionize.
- D. Krebs cycle releases CO_2 and hydrogens.
- E. Oxygen combines with hydrogen ions and electrons to form water.
- F. Electrons are transported along cytochromes, and energy from electrons is stored in ATP.

_____ _____ _____   14.   Using the Food Pyramid as a guide, arrange these food groups according to number of servings suggested per day, from greatest to least:
- A. Meat, fish, poultry, dry beans, eggs, and nuts
- B. Vegetables
- C. Bread, cereal, rice, pasta

Questions 15–20: Circle T (true) or F (false). If the statement is false, change the underlined word or phrase so that the statement is correct.

T F   15.   Complexes of carrier molecules utilized in the electron transport system are located in the mitochondrial matrix.

T F   16.   Catabolism of carbohydrates involves oxidation, which is a process of addition of hydrogens.

T F   17.   Nonessential amino acids are those that are not used in the synthesis of human protein.

T F   18.   The complete oxidation of glucose to CO_2, and H_2O yields 4 ATPs.

T F   19.   Anabolic reactions are synthetic reactions that release energy.

T F   20.   Carbohydrates, proteins, and vitamins provide energy and serve as building materials.

Questions 21–25: Fill-ins. Complete each sentence with the word or phrase that best fits.

_____   21.   Catabolism of each gram of carbohydrate or protein results in

release of about _____ kcal, whereas each gram of fat leads to

about _____ kcal.

_____   22.   Pantothenic acid is a vitamin used to form the coenzyme named

_____.

_____   23.   _____ is the mineral that is most common in extracellular fluid (ECF); it is also important in osmosis, buffer systems, and nerve impulse conduction.

_____   24.   Aspirin and acetaminophen (Tylenol) reduce fever by inhibiting

synthesis of _____ so that the body's "thermostat" located in

the _____ is reset to a lower temperature.

_____   25.   In uncontrolled diabetes mellitus, the person may go into a state

of acidosis due to the production of _____ acids resulting from

excessive breakdown of _____.

ANSWERS TO MASTERY TEST: CHAPTER 25

Multiple Choice

1. C
2. A
3. D
4. B
5. E
6. A
7. E
8. C
9. D
10. C
11. D
12. D

Arrange

13. A B D C F E
14. C B A

True–False

15. F. Mitochondrial inner (cristae) membrane
16. F. Oxidation, removal of hydrogens
17. F. Can be synthesized by the body
18. F. 36 to 38
19. F. Synthetic reactions that require energy
20. F. Carbohydrates and proteins but not Vitamins

Fill-Ins

21. 4, 9
22. Coenzyme A
23. Sodium (Na^+)
24. Prostaglandins (PGs), hypothalamus
25. Acetoacetic (keto-), fats

FRAMEWORK 26
Urinary System

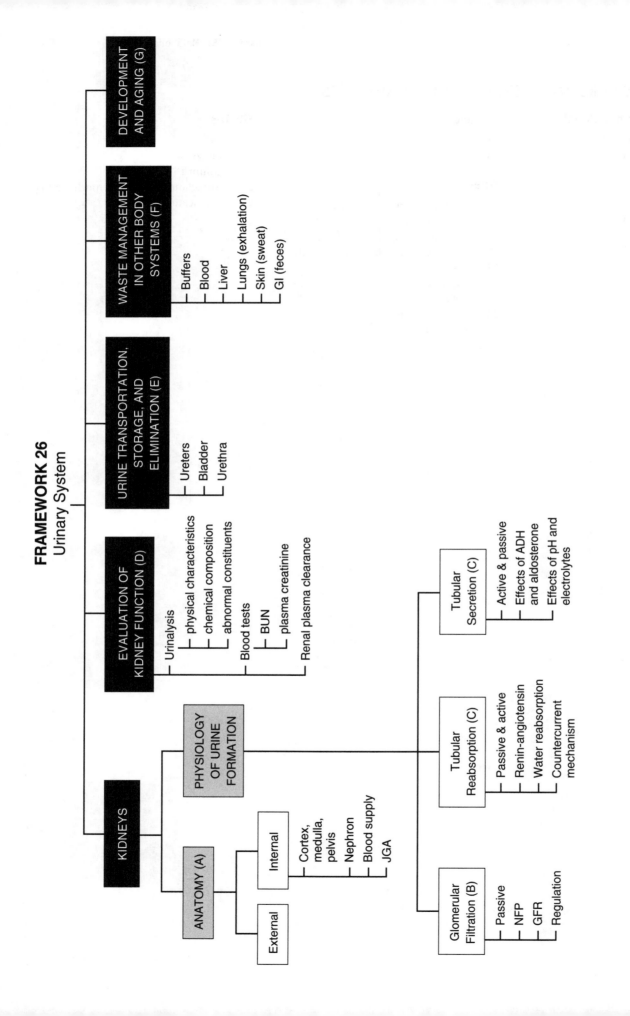

C H A P T E R **26**

The Urinary System

Wastes like urea or hydrogen ions (H$^+$ are products of metabolism that must be removed or they will lead to deleterious effects. The urinary system works with lungs, skin, and the GI tract to eliminate wastes. Kidneys are designed to filter blood and selectively determine which chemicals stay in blood or exit in the urine. The process of urine formation in the kidney requires an intricate balance of factors such as blood pressure and hormones (ADH and aldosterone). Examination of urine (urinalysis) can provide information about the status of the blood and possible kidney malfunction. A healthy urinary system requires an exit route for urine through ureters, bladder, and urethra.

Start your study of the urinary system with a look at the Chapter 26 Framework and familiarize yourself with the key concepts and terms. Examine the Chapter 26 Topic Outline and Objectives. Check off each one as you complete it.

TOPIC OUTLINE AND OBJECTIVES

A. Overview; kidney anatomy

1. List the functions of the kidneys.
2. Describe the external and internal gross anatomical features of the kidneys.
3. Trace the path of blood flow through the kidneys.
4. Describe the structure of renal corpuscles and renal tubules.

B. Kidney physiology: glomerular filtration

5. Identify the three basic tasks accomplished by nephrons and collecting ducts, and indicate where each task occurs.
6. Describe the filtration membrane.
7. Discuss the pressures that promote and oppose glomerular filtration.

C. Kidney physiology: tubular reabsorption, and tubular secretion

8. Describe the routes and mechanisms of tubular reabsorption and secretion.
9. Describe how specific segments of the renal tubule and collecting duct reabsorb water and solutes.

10. Describe how specific segments of the renal tubule and collecting duct secrete solutes into urine.
11. Describe how the renal tubule produces dilute and concentrated urine.

D. Evaluation of kidney function

12. Define urinalysis and describe its importance.
13. Define renal plasma clearance and describe its importance.

E. Urine transportation, storage, and elimination

14. Describe the anatomy, histology, and physiology of the ureters, urinary bladder, and urethra.

F. Waste management in other body systems

15. Describe the ways that body wastes are handled.

G. Development and aging of the urinary system

16. Describe the development of the urinary system.
17. Describe the effects of aging on the urinary system.

H. Disorders

WORDBYTES

Now study the following parts of words that may help you better understand terminology in this chapter.

Wordbyte	Self-check	Meaning	Example(s)
af-	_____	toward	*af*ferent
anti-	_____	against	*anti*diuretic
azot-	_____	nitrogen-containing	*azot*emia
calyx	_____	cup	major *calyx*
cyst-	_____	bladder	*cyst*ostomy
densa	_____	dense	macula *densa*
ef-	_____	out	*ef*ferrent
-ferent	_____	carry	af*ferent*
-genic	_____	producing	myo*genic*
glomus	_____	ball	*glom*erular
insipid	_____	without taste	diabetes *insipid*us
juxta-	_____	next to	*juxta*glomerular
macula	_____	spot	*macula* densa
myo-	_____	muscle	*myo*genic
recta	_____	straight	vasa *recta*
ren-	_____	kidney	*ren*al vein
-ulus	_____	small	glomer*ulus*
urin-	_____	urinary	*urin*alysis

CHECKPOINTS

A. Overview; kidney anatomy (pages 914–924)

✓ **A1.** Select terms in the box that best label descriptions of different kidney functions designed to maintain homeostatic balance of blood and other body fluids.

pH	pH	P.	Pressure (BP)
E.	Electrolyte concentration	V.	Volume
G.	Glucose	W.	Wastes

_____ a. Regulation of Ca^+, Cl^-, K^+ and Na^+

_____ b. Mechanisms utilizing renin and changes in resistance of renal blood vessels

_____ c. Mechanisms utilizing ADH and aldosterone

_____ d. Alterations in elimination or conservation of H^+ or HCO_3^-

_____ e. Deamination of the amino acid glutamine which facilitates gluconeogenesis

_____ f. Removal of ammonia, urea, uric acid, and creatine

KEY

	A.	Pyramid	L.	Celiac artery
	B.	Papilla	M.	Left adrenal (suprarenal vein)
	C.	Calyx	N.	Superior mesenteric artery
	D.	Renal column	O.	Left renal artery and vein
O	E.	Pelvis	P.	Left spermatic or ovarian artery and vein
	F.	Medulla	Q.	Aorta
O	G.	Cortex	R.	Inferior mesenteric artery
O	H.	Ureter	S.	Inferior vena cava
	I.	Ureteral opening	T.	Left common iliac artery and vein
	J.	Bladder	U.	Left internal iliac artery and vein
O	K.	Urethra	V.	Left external iliac artery and vein

Figure LG 26.1 (Left side of diagram) Organs of the urinary system.
Structures A to G are parts of the kidney. Identify and color as directed in Checkpoint A2.
(Right side of diagram) Blood vessels of the abdomen. Color as directed in Chapter 21,
Checkpoint E9, page 453 of the Learning Guide.

✓▲ **A2.** Identify the organs that make up the urinary system on Figure LG 26.1 and answer
the following questions about them.
   a. The kidneys are located at about *(waist? hip?)* level, between _____ and _____

   vertebrae. Each kidney is about _____ cm (_____ in.) long. Visualize kidney
   location and size on yourself.
   b. The kidneys are in an extreme *(anterior? posterior?)* position in the abdomen. They are

   described as _____ because they are posterior to the peritoneum.
   c. Identify the parts of the internal structure of the kidney on Figure LG 26.1. Then
   color all structures as indicated by color code ovals.

✓▲ **A3.** Select terms that match each description related to kidneys. Choose from these terms. Not all terms will be used.

Adipose capsule	**Renal calyx**	**Renal fascia**
Nephrology	**Renal capsule**	**Renal papilla**
Parenchyma	**Renal columns**	**Renal pyramids**

_____ a. Cuplike structure that drains urine from papillary ducts and empties in renal pelvis

_____ b. Outermost layer of connective tissue that anchors kidneys to abdominal wall

_____ c. Study of kidneys

_____ d. Functional part of organ, such as kidney

_____ e. Makes up most of renal medulla

_____ f. Portion of renal cortex located between renal pyramids

✓ **A4.** About what percentage of cardiac output passes through the kidneys each minute?

_____% This amounts to about _____ mL/min.

✓▲ **A5.** Follow the pathway that blood takes through kidneys by naming vessels 1–8 on Figure LG 26.2. Some arrows are shown; add more to reinforce your understanding of the pathway of blood.

1. _____   2. _____   3. _____

4. _____   5. _____   6a. _____

6b. _____   7. _____

Which of these vessels extends most deeply into the renal medulla? _____

✓▲ **A6.** Explain what is unique about the blood supply of kidneys in this activity.
   a. In virtually all tissues of the body, blood in capillaries flows into vessels named

       _____. However, in kidneys, blood in glomerular capillaries flows

       directly into _____ _____.
   b. Which vessel has a larger diameter? *(Afferent? Efferent?)* arteriole. State one effect of this difference.

✓▲ **A7.** Do this exercise about the functional unit of the kidney, a nephron.
   a. On Figure LG 26.2, label parts A–E of the nephron, with letters arranged according to the flow of urine. Several structure E's empty into a single structure F, known as a

       _____ _____ .

   b. Which one of those structures forms part of the renal corpuscle? _____
       The other part of the renal corpuscle is a cluster of capillaries known as a

       _____ and numbered *(4? 6b?)* in the figure.

Figure LG 26.2 Diagram of a nephron. Letters refer to Checkpoints A7 and A9. Numbers refer to Checkpoints A5 and A7. The boxed area refers to Checkpoint A10.

✓ **A8.** Contrast two types of nephrons by writing *C* for cortical nephrons and *JM* for juxtamedullary nephrons.

_____ a. Also known as short-looped nephrons because these lie only in the renal cortex

_____ b. Constitute about 15–20% of nephrons

_____ c. Have a thin (squamous epithelial) layer in ascending limb of loop of Henle

_____ d. Permit significant variations in concentration of urine

✓ **A9.** Do this exercise on microscopic structure of the nephron.
   a. As water and other substances are "cleaned" (filtered) out of glomerular blood, they

   collect in the _____, which lies between _____ and parietal layers of the capsule. The *(parietal? visceral?)* layer of the glomerular (Bowman's) capsule forms the outermost boundary of the renal capsule.
   b. Identify the part of the renal tubule on the figure that consists of cuboidal epithelium with a brush border of many microvilli that increase the absorptive surface area.

   _____ Which part consists of simple squamous epithelium?

   _____
   c. In which regions are *principal cells* located? *(B and C? E and F?)* These cells are

sensitive to two hormones. Name them. _____ and

_____. The role of intercalated cells—also located in E and F—is to

regulate _____.

✓▲ **A10.** Describe the juxtaglomerular apparatus (JGA) in this exercise.
    a. As shown in the boxed structure in Figure LG 26.2, the JGA consists of *(two? four?)* types of cells. One type is located in the final portion of the *(ascending? descending?)*

        limb of the loop of Henle. These cells, known as _____ _____

        cells, monitor concentration of _____ in the tubular lumen.

    b. The second type of cell is a _____ _____

        _____ cell located in the wall of the *(afferent arteriole? glomerulus?)*.

        These cells are known as _____ cells; they secrete the chemical

        known as _____, which regulates _____.

    A11. Christian has undergone a right *nephrectomy*. How is his remaining kidney likely to compensate?

B. Kidney physiology: glomerular filtration (pages 924–929)

✓ **B1.** Do this exercise on urine formation.
    a. List the three steps of urine formation.

        _____ _____ _____

    b. Which of these steps involve passage of substances out of blood and into the

        "forming urine"? _____ _____ Which step involves selective return of valuable chemicals from "forming urine" back into blood?

        _____

    c. What percentage of plasma entering nephrons is filtered out of glomerular blood and passed into the capsular space? About *(1%? 15–20%? 48–60%? 99%?)*. This

        percentage is known as the filtration _____. The actual amount of

        glomerular filtrate is about *(180? 60? 1–2?)* liters/day in males, and about _____

        liters/day in females. In fact, about _____ times the total volume of blood plasma is filtered out of blood in the course of a day. So blood is "cleaned" by kidneys many times each day.
    d. Of this filtrate, about _____ liters/day of fluid are reabsorbed in the

        "second step" of urine formation. This is about _____% of the glomerular filtrate.

        So the total amount of urine eliminated from kidneys daily is usually about _____ liters (or quarts).

✓▲ **B2.** Complete this Checkpoint on the filtration membrane.
  a. Walls of glomerular capillaries are *(more? less?)* "leaky" than other capillaries of the body. State a functional advantage of this feature.

  b. The three portions of the sandwichlike membrane consists of:

    1. Glomerular capillary _____ cells with large pores called *(filtration slits? fenestrations?)* between these cells. The size of these pores prevents

      _____ and _____ from passing out of blood.

    2. An acellular layer known as the _____ _____
    3. Modified simple squamous epithelial cells known as *(pedicels? pococytes?)* that are part of the *(visceral? parietal?)* layer of the glomerular (Bowman's) capsule. Tiny "feet" called *(pedicels? pococytes?)* wrap around glomerular capillaries leaving spaces known as called *(filtration slits? fenestrations?)*. These spaces are small enough to *(permit? prohibit?)* filtration of most plasma proteins such as albumin out of blood.
  c. The structure of the glomerular-endothelial membrane allows only certain substances to *pass freely* out of blood and into filtrate ("forming urine"). Circle those substances that do pass uninhibited into filtrate:

Albumin (and other	**Glucose**	**Small proteins**
large proteins)	**Na+ and other ions**	**Urea**
Amino acids	**Platelets**	**Water**
Ammonia	**Red blood cells**	**White blood cells**

✓ **B3.** Refer to Figure LG 26.3. Now describe the first step in urine production in this checkpoint. Glomerular filtration is a process of *(pushing? pulling?)* fluids and solutes out of

  _____ and into the fluid known as _____. The direction of glomerular filtration is shown by the thick arrow from W to X on the figure.
  b. *(Most substances? Only a few types of substances?)* are forced out of blood in the process of glomerular filtration. Refer to Figure LG 26.3, area W (plasma in glomeruli) and area X (filtrate just beyond the capsule). Compare the compositions of these two fluids (described in columns two and three of Table 26.3 in your textbook). Note that

  during filtration all solutes are freely filtered from blood except _____. Why is so little protein filtered?

  c. Blood pressure (or glomerular blood hydrostatic pressure or GBHP) in glomerular

  capillaries is about _____ mm Hg. This value is *(higher? lower?)* than that in other capillaries of the body. This extra pressure is accounted for by the fact that the diameter of the efferent arteriole is *(larger? smaller?)* than that of the afferent arteriole. Picture three garden hoses connected to each other, representing the afferent arteriole, glomerular capillaries, and efferent arteriole. The third one (efferent arteriole) is

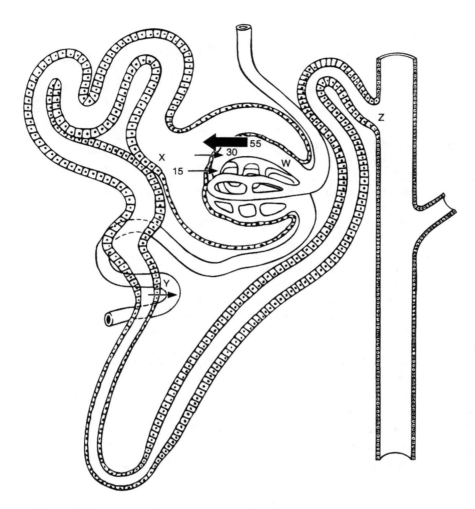

Figure LG 26.3 Diagram of a nephron with renal tubule abbreviated.
Numbers refer to Checkpoint B4; letters refer to Checkpoints B3, C1, and C2.

extremely narrow; it creates such resistance that pressure builds up in the first two,
fluids will be forced out of pores in the middle (glomerular) hose.
d. Describe three structural features of nephrons that enhance their filtration capacity.

✓ **B4.** Glomerular blood hydrostatic pressure (GBHP) is not the only force determining the
amount of filtration occurring in glomeruli.
a. Name two other forces (shown on Figure 26.9 in your textbook).

_____ _____

b. Normal values for these three forces are given on Figure LG 26.3. Write the formula
for calculating net filtration pressure (NFP); draw an arrow beneath each term to
show the direction of the force in Fig. LG 26.3. Then calculate a normal NFP using
the values given.

NFP   = glomerular blood hydrostatic pressure – _____ – _____

Normal NFP   = _____ mm Hg – _____ mm Hg – mmHg

                 = _____ mmHg

Note that blood (hydrostatic) pressure pushes *(out of? into?)* glomerular blood and is
largely counteracted by the other two forces.

c. *For extra review* of calculation of NFP, look back at Chapter 21, Checkpoint A10 on LG page 437.

✓ **B5.** *A clinical correlation.* Show the effects of alterations of these pressures in a pathological situation. Determine NFP of the patient whose pressure values are shown below.
    Glomerular blood hydrostatic pressure = 44
    Blood colloid osmotic pressure = 30
    Capsular filtrate hydrostatic pressure = 15

    NFP = _____
    Note which value(s) is(are) abnormal and suggest causes.

✓ **B6.** Define GFR in this exercise.
    a. To what do the letters GFR refer?

        g _____ f _____ r _____
    b. Fill in the blanks with normal values for GFR:

        a. Females: _____ mL/min × 1440 min/day* = 151,200 mL/day = 151.2 liters/day**

        b. Males: _____ mL/min × 1440 min/day = _____ mL/day = _____ liters/day
        *1440 min/day = 60 minutes/hour × 24 hours/day
        **rounded to 150 liters/day, on page 925 in your text
    c. *A clinical correlation.* Describe two causes of decreased GFR. GFR is likely to decrease

        if blood pressure in kidneys _____-creases, for example, during stress when

        sympathetic nerves decrease the diameter of the _____-ferent arteriole. Imagine
        that the first of the three garden hoses in Checkpoint B3 is narrowed, so *(more? less?)*
        blood flows into the glomerulus (middle hose). Either enlarged prostate gland or

        kidney _____ lodged in the ureter back up urine into renal tubules
        of kidneys, opposing glomerular blood pressure, and this will decrease GFR also.
    d. What consequences may be expected to accompany an abnormally low GFR?

✓ **B7.** List three categories of mechanisms that control GFR and ultimately help to regulate blood pressure in the body:

    _____ _____ _____
    One autoregulation mechanism involves vasoconstriction of afferent arterioles

    (leading to _____-creased GFR) in response to stretching of these arterioles due to
    renal blood flow accompanying high blood pressure. This means of protecting
    kidneys from overwhelming blood flow is called the *(myogenic? tubuloglomerular feedback?)* mechanism.

✓ **B8.** *A clinical correlation.* Mrs. K's blood pressure has dropped slightly. Explain how another renal autoregulation mechanism functions to ensure adequate renal blood flow.
    a. When Mrs. K's systemic blood pressure decreases, her NFP and GFR are likely to *(increase? decrease?)*, causing filtrate to move more *(rapidly? slowly?)* through the renal tubules. As a result, *(more? less?)* fluid and salts are resorbed in the proximal convoluted tubules and loop of Henle.

    b. Consequently, the amount of fluid slowly passing the macula densa is *(high? low?)* in volume and concentration of salts. This condition causes JGA cells to produce *(more? fewer?)* of certain vasoconstrictor substances. As afferent arterioles then dilate, *(more? less?)* blood flows into glomerular capillaries, causing NFP and GFR to *(increase? decrease?)*.

    c. This restoration of homeostasis by renal autoregulation occurs by a *(positive? negative?)* feedback mechanism. The name autoregulation indicates that neither the autonomic nervous system nor any chemical made outside the kidneys is involved. The control *(is? is not?)* intrinsic to kidneys.

✓ **B9.** During stress the *(sympathetic? parasympathetic?)* nerves to kidneys tend to cause vaso-constriction of the *(afferent? efferent?)* arteriole. (Look back at Checkpoint B6c.)

The resulting _____-crease in GFR leads to _____-creased urine output. This

helps to _____-crease blood volume and blood pressure, providing blood to brain, heart, and skeletal muscle for "fight-or-flight" responses.

✓ **B10.** Contrast hormonal effects on GFR in this Checkpoint.

    a. Angiotensin II _____-creases GFR by *vaso-(constricting? dilating?)* both afferent and efferent arterioles.

    b. ANP, secreted by the _____ when blood volume there _____-creases,

relaxes glomerular _____ cells, thereby _____-creasing the surface

area of glomerular capillaries. ANP therefore _____-creases GFR and urinary

output. This negative feedback mechanism ultimately _____-creases blood volume returning to the heart.

C. Kidney physiology: tubular reabsorption, and tubular secretion (pages 929–942)

✓ **C1.** Refer to Figure LG 26.3 and do this exercise about formation of urine.

    a. We have already seen that glomerular filtration results in movement of substances from *(blood to filtrate/forming urine? filtrate/forming urine to blood?)*, as shown by the arrow between W and X. If this were the only step in urine formation, all the substances in the filtrate would leave the body in urine. Note from Table 26.3 in your

text that the body would produce _____ liters of urine each day, and it would contain many valuable substances. Obviously some of these "good" substances must be drawn back into blood and saved.

    b. Recall that blood in most capillaries of the body flows into vessels named

_____, but blood in glomerular capillaries flows into

_____ and then into _____. This unique arrangement permits blood to recapture some of the substances indiscriminately pushed out during filtration. This occurs during the second step of urine formation, called

_____. This process moves substances in the *(same? opposite?)* direction as glomerular filtration, as shown by the direction of the arrow at area Y of Figure LG 26.3. The *(proximal? distal?)* convoluted tubule is the site of most reabsorption.

✓ **C2.** Refer to Table 26.3 on page 930 in your text, and notice the two columns on the far right. Write "Y" above REABSORBED: RETURNED TO BLOOD and "Z" above URINE EXCRETED. Then refer to related areas Y and Z on Figure LG 26.3. Discuss the results of tubular reabsorption in this learning activity. (*Hint:* cross reference with Figure 26.21, page 942 in the text also.)

a. Which solutes are 100% reabsorbed into area Y so that virtually none remains in urine (area Z)?

b. About what percentage of water that is filtered out of blood is reabsorbed back into

blood? _____% Identify two ions that are about 99% reabsorbed. _____ _____

c. Which ion is 100% reabsorbed but still shows up in urine because some of this ion is

secreted (step 3 in urine formation) from blood to urine? _____

d. *(Urea? Uric acid?)* is about 90% reabsorbed, whereas _____ is about 50% reabsorbed.

e. Identify the solute that is filtered from blood and not reabsorbed.

_____ State the clinical significance of this fact.

f. Tubular reabsorption is a *(nonselective? discriminating?)* process that enables the body to save valuable nutrients, ions, and water.

✓ **C3.** Refer to Figure 26.11 (page 930) and do this exercise on reabsorption routes.

a. Reabsorption occurs by two possible routes: *between* renal tubule cells, the method

known as _____-cellular reabsorption, or across tubule cells, the process called

_____-cellular reabsorption.

b. Identify the apical and basolateral portions of renal tubule cells on Figure 26.11. The apical portion of these cells faces the *(lumen containing "forming urine"? peritubular capillaries?)*, whereas the bases of cells lie closer to the *(lumen containing "forming urine"? peritubular capillaries?)*. Now arrange in correct sequence the steps of

reabsorption of Na⁺ in the proximal convoluted tubules: _____ _____ _____

A. Sodium pumps actively expel Na⁺ from basolateral portions of tubule cells, a process utilizing about large amounts of ATP (See Checkpoint C4).

B. Na⁺ passively diffuses from negatively charged fluid within the lumen into apical portions of tubular cells.

C. Na⁺ diffuses into pericapillary blood.

Reabsorption of Na⁺ is extremely important since more Na⁺ is filtered from glomeruli

than any other substance except _____, and Na⁺ reabsorption is closely linked to *(facultative? obligatory?)* water reabsorption—which normally

accounts for about _____% of water reabsorption.

C4. *For extra review.* Describe reabsorption mechanisms in this exercise.

a. About _____% of total resting metabolic energy is used for reabsorption of Na+ in renal tubules by *(primary? secondary?)* active transport mechanisms. Explain how active transport of Na+ also promotes osmosis of water and diffusion of substances such as K+, Cl−, HCO_3^-, and urea.

b. Reabsorption of glucose, amino acids, and lactic acid occurs almost completely in the *(proximal? distal?)* convoluted tubule, and the process utilizes Na+ *(symporters? antiporters?)*. This mechanism is an example of *(primary? secondary?)* active transport because it is the sodium pump, not the symporter, that uses ATP. If blood concentra-

tion of a substance such as glucose is abnormally high, the renal _____ is surpassed. In other words, the blood level of glucose is greater than the T_m,

(or _____ _____).

✓ **C5.** *A clinical correlation.* Eileen and Marlene are both "spilling sugar" in urine; that is,

they have _____-uria. Discuss possible causes of glycosuria in each case, considering their blood sugar levels.

a. Eileen: glucose of 340 mg/l00 mL plasma

b. Marlene: glucose of 80 mg/100 mL plasma

✓ **C6.** Identify the part of the renal tubule that best fits the description. Use these answers.

A.	Ascending limb of the loop of Henle	**DCT.**	Distal convoluted tubule
CD.	Collecting duct	**PCT.**	Proximal convoluted tubule
D.	Descending limb of the loop of Henle		

_____ a. This portion of the renal tubule with a prominent brush border permits tubular reabsorption of most filtered water and almost 100% of filtered nutrients (such as glucose).

_____ b. This is the first site where osmosis of water is not necessarily coupled to reabsorption of other filtrates.

_____ c. Little or no water is reabsorbed here because these cells are impermeable to water.

_____ d. Variable amounts of Na^+ and water may be reabsorbed here under the influence of aldosterone and ADH. (two answers)

_____ e. Absorption of almost all HCO_3^- and at least half of Na^+, K^+, Cl^-, and urea occurs here.

✓ **C7.** The third step in urine production is _____. It involves movement of substances from *(blood to urine? urine to blood?)*. In other words, tubular secretion is movement of substances in *(the same? opposite?)* direction as movement occurring in filtration. List four or more substances that are secreted by the process of tubular secretion.

✓ **C8.** Do this exercise on hormonal regulation of urine volume and composition.

   a. Aldosterone and ADH are hormones that act on _____ cells located in *(PCT and loop of Henle? last part of DCT as well as collecting ducts?)*. Aldosterone is made by the *(adrenal medulla? adrenal cortex? hypothalamus?)*. This hormone stimulates production of a protein that helps pump *(Na^+? K^+?)* and also water from urine to blood. So aldosterone tends to *(in? de?)*-crease blood pressure.

   b. ADH causes principal cells to become *(more? less?)* permeable to water; as a result,

   reabsorption of water _____-creases, leading to _____-crease of blood pressure.

   c. ADH is made by the _____ and released from the *(anterior? posterior?)* pituitary. It is regulated by a *(positive? negative?)* feedback mechanism because

   decreased body fluid leads to _____-crease of ADH release and retention of body fluid.

   d. Diabetes insipidus involves _____-crease of ADH and _____-crease of urinary output.

✓ **C9.** Do this exercise on regulation of blood level of potassium ion (K^+).
   a. What percentage of filtered K^+ is normally reabsorbed in PCT, loop of Henle, and

   DCT? *(Hint: Refer to Table 26.3 in the text)* _____%
   b. Explain why kidneys might need to secrete variable amounts of K^+ from blood to urine.

   c. When kidney reabsorption of the cation Na^+ increases, secretion of K^+ (also a cation)

   _____-creases. Aldosterone *(enhances? inhibits?)* such processes, causing lowered blood K^+ levels. On the other hand, aldosterone deficiency may lead to *(low? high?)* blood levels of K^+, and even cardiac arrest.

✓ **C10.** Explain how tubular secretion helps to control pH. Refer to figures in the text and do this exercise.

a. Jenny s respiratory rate is slow, and her arterial pH is 7.33. Her blood therefore

contains high levels of _____. Her blood is therefore somewhat *(acidic? alkaline?)*. Explain why.

b. Jenny's kidneys can assist in restoring normal pH (Figure 26.13a, b). The presence of acid (H^+) stimulates the PCT cells to eliminate H^+. In order for this to happen, another cation present in the tubular lumen must exchange places with H^+. What

ion enters kidney tubule cells? _____ It joins HCO_3^- to form _____. This process requires *(antiporters? symporters?)*.

c. Figure 26.17a shows another possible mechanism for reducing acidity of blood. Inter-calated cells located in *(PCT? collecting ducts?)* pump H^+ out into the lumen, causing urine to be up to *(10? 1000?)* times more acidic than blood. This pump *(does? does not?)* require ATP; the pump is an example of *(primary? secondary?)* active transport.

The leftover HCO_3^- (from H_2CO_3) may now enter peritubular blood, _____-creasing blood pH even more.

d. In addition, when blood is acidic, more ammonia (_____) is produced as a byprod-uct of *(lipid? protein?)* metabolism. H^+ in the collecting duct may then be buffered by

combining with ammonia to form ammonium (_____). which is excreted into urine (Text Figure 26.17b) p 600.

✓ **C11.** Now describe chemicals involved in other mechanisms that regulate GFR by completing this checkpoint. Select the chemical that matches each description. Use the following list of answers. (It may help to refer to Figure LG 18.5, page 366 in the *Learning Guide*.)

ACE.	Angiotensin converting enzyme	Ang II.	Angiotensin II
ADH.	Antidiuretic hormone	ANP.	Atrial natriuretic peptide
Ald.	Aldosterone	Epi.	Epinephrine
Ang I.	Angiotensin I	R.	Renin

_____ a. A vasoconstrictor that also stimulates thirst and release of aldosterone and ADH

_____ b. Produced by adrenal cortex, it causes collecting ducts to reabsorb more Na^+ and water

_____ c. Released from the posterior pituitary, this hormone also increases blood volume

_____ d. This hormone increases GFR; it opposes actions of renin, aldosterone, and ADH and suppresses secretion of those chemicals

_____ e. Released under sympathetic stimulation; leads to arteriolar vasoconstriction and decreased GFR

_____ f. Of all chemicals in the above list, the one that tends to decrease blood pressure

✓ **C12.** State an example of a situation in which your body needs to produce dilute urine in order to maintain homeostasis.

Now describe how your body can accomplish this by doing the following learning activity.

a. To form dilute urine, kidney tubules must reabsorb *(more? fewer?)* solutes than usual. Two factors facilitate this. One is permeability of the ascending limb to Na$^+$, K$^+$, and Cl$^-$ *(and also to? but not to?)* water. Thus solutes enter interstitium and then capillaries, but water stays in urine.

b. The second requirement for dilute urine is a *(high? low?)* level of ADH. The function

   of ADH is to _____-crease permeability of distal and collecting tubules to water. Less ADH forces more water to stay in urine, leading to a *(hyper? hypo?)*-osmotic urine.

✓ **C13.** State two reasons why it might be advantageous for the body to produce a concentrated urine.

Now refer to Figure LU 26.4 and describe mechanisms for concentrating urine in this exercise.

a. To help you clearly differentiate microscopic parts of the kidney, first color yellow the renal tubules (containing urine that is forming), color red the capillaries (containing blood), and color green all remaining areas (the interstitium). (Recall that kidneys are composed of over one million nephrons. The interstitium is all of the matrix between nephrons.) Also draw arrows to show route of "forming urine" within the renal tubule.

b. Now find [1] on the figure. This marks the endpoint of the figure where concentrated urine exits from the nephron. Such urine is known as *(hypo? hyper?)*-tonic (or

   _____-osmotic) urine. To produce a small amount of urine that is

   highly concentrated in wastes, the rate of reabsorption must _____-crease. Therefore more fluid from the urine forming within the lumen of the collecting duct will pass back into blood. (See [3] on the figure.)

c. This leads to the key question of how to increase tubular reabsorption. The answer lies in establishment of a solute gradient with the highest concentration in the *(outer? inner?)* medullary interstitium (IMI), around the inner portions of the long-loop juxtamedullary tubules. Solute concentration in this region may reach a maximum

   of *(300? 600? 1200?)* mOsm, as compared to only _____ mOsm in the cortical and outer medullary interstitium. (Note these values on the figure.)

d. Because the final segments of the collecting ducts reside in the inner medullary interstitium (IMI) region, urine in these ducts is exposed to this significantly hypertonic environment, shown at [2] on the figure. Much like a desert (or a dry sponge or blotter) surrounding a small stream of fluid, this "dry" (hypertonic) interstitium acts to draw water out of the urine. Ultimately, the fluid in the interstitium will pass into

   _____ blood ([3] on the figure).

e. So how can this hypertonic interstitium be developed? Two mechanisms are

   involved. The first depends on the relative _____ of different portions of the renal tubule to water and to several solutes. The second is the

   _____ mechanism (discussed in Checkpoint C14).

f. Consider the first mechanism. On Figure LG 26.4, find the thick portion of the tubule (next to the [4]). This is part of the *(ascending? descending?)* tubule. The thick

Figure LG 26.4 The concentration of urine in the nephron. Numbers refer to checkpoint C13. Color as directed in Checkpoints C13 and C14.

(squamous? columnar?) epithelium here causes this region to be *(permeable? imperme-*

able?) to water in the absence of ADH. However, ions such as Na⁺, _____ and

_____ can be moved (by _____-porters) from the tubule into interstitium and
then to blood here. Show this by writing names of these chemicals in circles on the
figure. (Note that water is forced to stay inside the ascending limb.)

g. As a result, the fluid in the outer medullary and cortical interstitium becomes *(more?*
less?) concentrated (or hypertonic). Many of these ions then move into long capillar-

ies known as vasa _____ ([5] on the figure), carrying ions deep into
the inner medullary interstitium (IMI) ([3]) and increasing hypertonicity.

h. As the "desert" (IMI) exerts an osmotic (pulling) force, water is drawn out of urine
into IMI and then into capillary blood ([2] and [3] on the figure). Of course, this

mechanism depends on the hormone _____, which permits normally impermeable
collecting tubule cells to develop water pores and therefore to become more water
permeable. By itself this mechanism is effective in increasing urine concentration.

i. But urine concentration can be enhanced even further by the role played by urea.
Concentration of urea within these tubules is increased as water moves out of the
tubule lumen ([2] on the figure). Because this final portion of the collecting tubule is
(permeable? impermeable?) to urea, this solute can diffuse out into the IMI, further
contributing to the "desert-like" hypertonicity. Show this by writing "urea" near the
[6]. The urea factor increases renal capability for water reabsorption and concentration of urine even more.

j. In summary, we have discussed a urine-concentrating mechanism based on
differences of specific regions of renal tubule in permeability to water and urea. The
resulting concentration gradient consists of a relatively hypotonic renal cortex or
outer medullary region (300 mOsm) with an inner medullary interstitium (IMI)

highly concentrated in the waste _____, as well as ions _____

_____, and _____. This IMI "desert" (or blotter) is ready to pull (osmose) water

out of urine. But this can happen only when the hormone _____ is released as the

signal that the body needs to "save fluid." ADH then causes _____
cells in collecting ducts to become more permeable, allowing water to leave them and
move into IMI and then into capillary blood. The final outcome (at [1] on the figure)
is a *(large? small?)* volume of *(dilute? concentrated?)* urine.

✓ **C14.** Refer to Figure LG 26.4 and check your understanding of the countercurrent mechanism in this checkpoint.

a. The term *countercurrent* derives from the fact that the "forming urine" in the descending limbs of renal tubules runs parallel and in the *(same? opposite?)* direction(s) to that
in the ascending limb. This is also true of blood flow in the comparable regions of the
vasa recta. Draw arrows in the vasa recta to show direction of blood flow there.

b. This mechanism, like that discussed in Checkpoint C13 leads to production of *(dilute?*
concentrated?) urine. The countercurrent mechanism *(also? does not?)* depend(s) on the
concentration gradient in the interstitium. The *(ascending? descending?)* limb is highly

permeable to water but not to solutes, whereas the _____ limb
exhibits just the reverse situation: it is permeable to *(water? solutes?)* *(and? but not to?)*
water. Show this on Figure LG 26.4 by coloring large, hollow arrows to indicate correct locations of water or solute reabsorption.

c. A consequence of movement of water out of the descending limb ([8] in the figure) is increased concentration of solutes (_____ mOsm) in the renal tubule as that fluid rounds the bend at the "hairpin" ([9] in the figure).

d. Explain why urine becomes more and more dilute, finally falling to *(100? 300? 600?)* mOsm by the top of the ascending tubule.

e. But as urine flows into the DCT and collecting tubules, water will be pulled out (into the IMI "desert") at times when ADH is present. So in the presence of this hormone,

it is possible to produce highly concentrated urine (_____ mOsm) that is *(two? four?*

ten?) times the osmolality of blood plasma and glomerular filtrate (_____ mOsm).

f. The next time you finish a vigorous exercise session, pause for a moment to reflect upon the amazing intricacy of these mechanisms that allow you to limit your fluid loss in urine because you have just eliminated so much fluid from your

_____ glands!

✓ **C15.** Do this exercise on actions of diuretics.

a. A chemical that mimics the action of ADH is called a(n) *(antidiuretic? diuretic?)*. Such

a chemical _____-creases urinary output.

b. A medication that increases urinary output is called a(n) _____.

Removal of extra body fluid in this way is likely to _____-crease blood pressure.

c. Most diuretics cause loss of *(K+? Na+?)* from blood plasma; therefore patients must be sure to ingest adequate food or a supplement containing this ion.

✓ **C16.** *For extra review,* match the names of diuretics with descriptions below.

Thi.	**Thiazides (Diurel)**
Alc.	**Alcohol**
Caf.	**Caffeine (as in coffee, tea, or cola)**
Fur.	**Furosemide (Lasix)**
Spir.	**Spironolactone (aldactone)**

_____ a. Unlike most diuretics, this diuretic is potassium-sparing because it inhibits action of aldosterone; therefore clients do not normally need to take K+ supplements

_____ b. Inhibits secretion of ADH

_____ c. Promote NaCl (and water) loss from distal convoluted tubule (DCT)

_____ d. Naturally occurring diuretic that inhibits Na+ (and water) reabsorption

_____ e. The most potent diuretic drugs; known as loop diuretics

C17. *A clinical correlation.* Mrs. P is taking furosemide (Lasix) to lower her blood pressure. Answer the following questions.

a. Why is this diuretic called a *loop diuretic?*

b. How does the action of this diuretic lower blood pressure?

c. Mrs. P's nurse reminds her to eat a banana or orange each day. Why?

d. Some time later Mrs. P is given a prescription for spironolactone (Aldactone). Speculate as to why Mrs. P's doctor changed her prescription.

D. Evaluation of kidney function (pages 943–945)

✓ **D1.** Review factors that influence urine volume in this checkpoint.

a. *Blood pressure.* When blood pressure drops, _____ cells of the kidney

release _____, which leads to formation of _____.
This substance *(increases? decreases?)* blood pressure in two ways: directly, because it
serves as a *(vasoconstrictor? vasodilator?)*, and indirectly, because it stimulates release of

_____, which causes retention of both _____ and _____. Increase
in fluid volume will *(increase? decrease?)* blood pressure.

b. *Blood concentration.* The hypothalamic hormone _____ helps *(conserve? eliminate?)*
fluid. When body fluids are hypertonic, ADH is released, causing *(increased? decreased?)* tubular reabsorption of water back into blood.

c. *Temperature.* On hot days skin gives up *(large? small?)* volumes of fluid via sweat
glands. As blood becomes more concentrated, the posterior pituitary responds by
releasing *(more? less?)* ADH, so urine volume *(increases? decreases?)*.

d. *Diuretic medications* tend to _____-crease urinary output and _____-crease blood
pressure. For extra review, return to Checkpoint C15.

D2. Describe the following characteristics of urine. Suggest causes of variations from the
normal in each case.

a. Volume: _____ mL/day _____

b. Color: _____ _____

c. Turbidity: _____ _____

d. Odor: _____ _____

e. pH: _____ _____

f. Specific gravity: _____ _____

D3. The following substances are not normally found in urine. Explain what the presence
of each might indicate.
a. Albumin

b. Bilirubin

c. Casts

D4. Contrast the following pairs of conditions. Describe what is present in urine in each case and possible causes.
a. *Glucosuria/acetonuria*

b. *Hematuria/pyuria*

c. *Vaginitis due to Candida albicans/vaginitis due to Trichomonas vaginalis*

✓ **D5.** *A clinical correlation.* Of the following factors circle the ones that are likely to increase above normal in a patient with kidney failure. State the rationale for your answer.

BUN   Plasma creatinine   GFR   Renal plasma clearance of glucose

✓ **D6.** For a chemical to work effectively in renal clearance tests for evaluation of renal function, that chemical *(must? must not?)* be readily filtered and *(must? must not?)* be readily absorbed. A chemical is often used for testing renal clearance _____.

D7. Compare and contrast *hemodialysis* and *continuous ambulatory peritoneal dialysis* (CAPD) as methods of cleansing the blood.

E. Urine transportation, storage, and elimination (pages 945–947)

✓▲ **E1.** Arrange in correct sequence the structures through which urine flows once it is formed in collecting ducts:

MajC.	Major calyx	Ureter.	Ureter
MinC.	Minor calyx	Ureth.	Urethra
PapD.	Papillary duct	UrinB.	Urinary Bladder
RenP.	Renal pelvis		

Collecting duct → _____ → _____ →

_____ → _____ → _____ →

_____ → _____ → outside the body

✓▲ **E2.** Which organs in the list above (Checkpoint E1) are singular in the human body; in other words, only one of that organ is normally present in each person?

_____

✓▲ **E3.** Refer to Figure LG 26.1, page LG 579, and complete the following exercise about the ureters.

a. Ureters connect _____ to _____. Ureters are about

_____ cm (_____ in.) long. What causes urine to flow through these tubes?

b. What prevents urine from moving (retrograde) back into the ureters during contraction of the urinary bladder?

c. Kidneys and ureters are located in the *(anterior? posterior?)* of the abdominopelvic

cavity. They are said to be _____-peritoneal.

d. Ureters enter the urinary bladder at two corners of the *(detrusor? trigone?)* muscle.

The third corner marks the opening of the _____.

e. Both the ureters and the urinary bladder are lined with _____ epithelium. What advantage does this type of epithelium offer in these organs?

✓ **E4.** In the micturition reflex, *(sympathetic? parasympathetic?)* nerves stimulate the

_____ muscle of the urinary bladder and cause relaxation of the internal sphincter. What stimulus initiates the micturition reflex?

✓▲ **E5.** Refer to Figures LG 28.2 and 28.3 (pages 632 and 635 in your *Guide*), and complete the following exercise about the bladder and the urethra.
  a. The urinary bladder is located in the *(abdomen? true pelvis?)*. Two sphincters lie just inferior to it. The *(internal? external?)* sphincter is under voluntary control.

  b. Urine leaves the bladder through the _____. In females the length of

this tube is _____ cm (_____ in.); in males it is about _____ cm (_____ in.). Write in correct sequence (according to the pathway of urine) the three portions of the male urethra:

_____ → _____ → _____
In addition to urine, what other fluid passes through the male urethra?

_____

✓ **E6.** Ureters, urinary bladder, and urethra are lined with _____ membrane. What is the clinical significance of that fact?

E7. Contrast *urinary incontinence* and *retention*. List possible causes of each.

F. Waste management in other body systems (pages 947–949)

  F1. Complete this list of body structures and tissues, besides urinary organs, that perform excretory functions. Then list the substances they eliminate or detoxify.

Structure or tissue	Substances eliminated or made less toxic
1. _____	Excess H^+ is bound _____
2. _____	_____
3. Liver _____	_____
4. _____	CO_2, heat, and some water eliminated _____
5. Sweat Glands _____	_____
6. GI tract (via anus) _____	_____

✓ **F2.** Which of these "waste management" systems acts most like garbage trucks or sewer

lines to transport waste to kidneys for removal there? _____

G. Development and aging of the urinary system (page 949)

✓ **G1.** Do this exercise about urinary system development.

a. Kidneys form from _____-derm, beginning at about the _____ week of gestation.

b. Which develops first? *(Pro? Meso? Meta?)*-nephros Which extends most superiorly in

location? _____-nephros. Which one ultimately forms the kidney? _____-nephros.

c. The urinary bladder develops from the original _____, which is derived

from _____-derm.

✓ **G2.** GFR is likely to _____-crease by about 50% between age 40 and 70. This change is a function of:

a. _____-creased renal blood flow (related to decreased cardiac output)

b. _____-creased kidney mass.

✓▲ **G3.** Two other common problems among older adults are inability to control voiding

(_____) and increased frequency of _____ (UTIs). Explain why UTIs may occur more often in elderly males.

Nocturia may also occur more among the aged population. Define *nocturia*.

H. Disorders, clinical terminology (pages 949–951)

✓ **H1.** Kidney stones are known as renal _____. Lithotripsy is a *(surgical? nonsurgical?)* technique for destroying kidney stones by high-frequency

_____ waves. Suggest two or more health practices that may help you to avoid formation of kidney stones.

✓ **H2.** List three or more signs or symptoms of urinary tract infections (UTIs).

H3. Discuss and contrast two types of renal failure in this learning activity.

a. *(Acute? chronic?)* kidney failure is an abrupt cessation (or almost) of kidney function. List several causes of acute renal failure (ARF).

b. *(Acute? chronic?)* renal failure is progressive and irreversible. Describe changes that occur during the three stages of CRF.

✓ **H4.** Match each term in the box with descriptions below.

A.	Azotemia		E.	Enuresis
C.	Cystitis		GN.	Glomerulonephritis
			PCD.	Polycystic disease

_____ a. Inherited disorder in which kidneys (and other organs such as liver or pancreas) develop multiple fluid-filled cavities

_____ b. Inflammation of the urinary bladder

_____ c. Bed wetting

_____ d. Inflammation of the kidney involving glomeruli; may follow strep infection

_____ e. Presence of nitrogen-containing chemicals such as creatinine or urea in blood

ANSWERS TO SELECTED CHECKPOINTS: CHAPTER 26

A1. (a) E. (b) P. (c) V. (d) pH. (e) G. (f) W.

A2. (a) Waist, T12, L2; 10-12 (4-5). (b) Posterior; retroperitoneal. (c) See KEY to Figure LG 26.1.

A3. (a) Calyx. (b) Renal fascia. (c) Nephrology. (d) Parenchyma. (e) Renal pyramids. (f) Renal columns.

A4. 20-25; 1200.

A5. 1, Arcuate artery; 2, interlobular artery; 3, afferent arteriole; 4, glomerulus; 5, efferent arteriole; 6a, peritubular capillaries; 6b, vasa recta; 7, interlobular vein; 8, arcuate vein. Vasa recta.

A6. (a) Venules; efferent arterioles. (b) Afferent; increases glomerular blood pressure. (See Checkpoint B3c.)

A7. (a) A, glomerular (Bowman's) capsule; B, proximal convoluted tubule; C, descending limb of the loop of Henle; D, ascending limb of the loop of Henle; E, distal convoluted tubule; collecting duct (F). (b) A, glomerular (Bowman's) capsule; glomerulus, 4.

A8. (a) C. (b–d) JM.

A9. (a) Capsular (Bowman's) space, visceral; parietal.

(b) B (proximal convoluted tubule), C (descending limb of the loop of Henle), and the first part of D (ascending limb of the loop of Henle). (c) E and F; ADH and aldosterone; blood and urine pH.

A10. (a) Two; ascending; macula densa, NaCl. (b) Modified smooth muscle, afferent arteriole; juxtaglomerular (JG); renin, blood pressure.

B1. (a) Glomerular filtration, tubular reabsorption, tubular secretion. (b) Glomerular filtration and tubular secretion; tubular reabsorption. (c) 15–20%; fraction; 180; 150; 65. (d) 178–179 in males and 148–149 in females; 99; 1–2 (= 180 – 178 or 179).

B2. (a) More; blood passing through glomerular vessels in kidneys can be cleaned out. (b1) Endothelial, fenestrations, blood cells and platelets; (b2) Basal lamina; (b3) Podocytes, visceral; pedicels, filtration slits; prohibit. (c) Amino acids, ammonia, glucose, Na^+ and other ions, small proteins, urea, and water.

B3. (a) Pushing, blood, filtrate. (b) Most substances; protein molecules. Those molecules are too large to pass through healthy endothelial-capsular membranes, and the basement membrane restricts passage of proteins. (c) 60; higher; smaller. (Refer to Checkpoint A6.) (d) Glomerular capillaries are long, thin, and porous and have high blood pressure.

B4. (a) Blood colloid osmotic pressure (BCOP) and capsular hydrostatic pressure (CHP). (b) NFP = GBHP – CHP – BCOP. NFP = 55 – 15 – 30; 10; out of.

B5. NFP = –1 mm Hg = 44 – 30 – 15. The NFP of –1 indicates that no net filtration would occur; no urine (anuria) would result. In fact some fluids would pass from filtrate back into blood due to the low glomerular blood pressure possibly related to hemorrhage.

B6. (a) Glomerular filtration rate. (b) 105, 151.2 (or 150); 125, 180. (c) De, af; less; stones. (d) Additional water and waste products (such as H^+ and urea) normally eliminated in urine are retained in blood, possibly resulting in hypertension and acidosis.

B7. Autoregulation, neural and hormonal mechanisms; de; myogenic.

B8. (a) Decrease, slowly; more. (b) Low; fewer; more, increase. (c) Negative; is.

B9. Sympathetic, afferent; de, de, in.

B10. (a) De, constricting. (b) Atria of the heart, in, mesangial, in; in; de.

C1. (a) Blood to filtrate forming urine: 180. (b) Venules; efferent arterioles, capillaries: tubular reabsorption; opposite.

C2. (a) Glucose and bicarbonate (HCO_3^-). (b) 99; Na^+ and Cl^-. (c) K^+. (d) Uric acid, urea. (e) Creatinine; creatinine clearance is often used as a measure of glomerular filtration rate (GFR). One hundred percent of the creatinine (a breakdown product of muscle protein) filtered or "cleared" out of blood will show up in urine because 0% is reabsorbed. (f) Discriminating.

C3. (a) Para, trans. (b) Lumen containing "forming urine"; peritubular capillaries; B A C; water; obligatory, 90.

C5. Glycos. (a) Eileen's hyperglycemia could be caused by ingestion of excessive amounts of carbohydrates or by hormonal imbalance, such as deficiency in effective insulin (diabetes mellitus) or excess growth hormone. Eileen's glucose level of 340 mg/l00 ml plasma exceeds the renal threshold of about 200 mg/100 ml plasma because kidneys have a limited transport maximum (T_m) for glucose. (b) Although Marlene's blood glucose is in the normal (70–110 mg/per 100 ml) range, she may have a rare kidney disorder with a reduced transport maximum (T_m) for glucose.

C6. (a) PCT. (b) D. (c) A. (d) DCT and CD. (e) PCT.

C7. Tubular secretion; blood to urine; the same. K^+, H^+, ammonium (NH_4^+), creatinine, penicillin, and para-aminohippuric acid.

C8. (a) Principal, last part of DCT as well as collecting ducts; adrenal cortex; Na^+ in. (b) More, in, in. (c) Hypothalamus, posterior; negative, in. (d) De, in.

C9. (a) 100. (b) Dietary K^+ deficit or loss of K^+ accompanying tissue trauma or use of certain medications (such as most diuretics) leads to need for increased plasma K^+ levels via reabsorption. Dietary K^+ excesses (such as some salt substitutes) require extra elimination (secretion) of K^+. (c) In; enhances; high.

C10. (a) CO_2; acidic; increasing levels of CO_2 tend to cause increase of H^+: $CO_2 + H_2O \rightarrow H_2CO_3 \rightarrow H^+ + HCO_3^-$. (Review content on transport of CO_2 in Chapter 23.) (b) Na^+; $NaHCO_3^-$ (sodium bicarbonate); antiporters. (c) Collecting ducts, 1000; does, primary; in. (d) NH_3, protein; NH_4.

C11. (a) Ang II. (b) Ald. (c) ADH. (d) ANP. (e) Epi. (f) ANP.

C12. During periods of high water intake or decreased sweating. (a) More; but not to. (b) Low; in; hypo.

C13. Water retention by kidneys raises blood pressure as plasma volume is increased and also helps prevent or reverses dehydration, for example, when drinking water is unavailable or when extra water loss occurs via sweating or fever. (a) Refer to Figure LG 26.4A. (b) Hyper, hyper; in. (c) Inner; 1200, 300. (d) Capillary. (e) Permeability; countercurrent. (f) Ascending; columnar, impermeable; K^+, Cl^-; sym; see Figure LG 26.4A. (g) More; recta. (h) ADH. (i) Permeable; see Figure LG 26.4A. (j) Urea, Na^+, K^+, Cl^-; ADH; principal; small, concentrated.

Figure LG 26.4A The concentration of urine in the nephron.

C14. (a) Opposite; see Figure LG 26.4A. (b) Concentrated; also; descending, ascending, solutes, but not to; see Figure LG 26.4A. (c) 1200. (d) 100; see Checkpoint C13f. (e) 1200, four, 300. (f) Sweat.

C15. (a) Antidiuretic; de. (b) Diuretic (or "water pill"); de. (c) K$^+$.

C16. (a) Spir. (b) Alc. (c) Thi. (d) Caf. (e) Fur.

D1. (a) Juxtaglomerular, renin, angiotensin II; increases, vasoconstrictor, aldosterone, Na$^+$, H$_2$O; increase. (b) ADH, conserve; increased. (c) Large; more, decreases. (d) In, de.

D5. BUN and plasma creatinine both increase in blood because less urea and creatinine are filtered from blood. Because GFR decreases, kidneys "clear out" less glucose from blood (renal plasma clearance of glucose decreases), resulting in hyperglycemia.

D6. Must, must not; creatinine.

E1. PapD → MinC → MajC → RenP → Ureter → UrinB → Ureth.

E2. Urin B, Ureth.

E3. (a) Kidneys, urinary bladder; 25–30 (10–12); gravity, hydrostatic pressure, and peristaltic contractions. (b) No anatomical valve is present, but filling of the urinary bladder compresses urethral openings, thus forcing urine through the open urethral sphincter. (c) Posterior; retro. (d) Trigone; urethra. (e) Transitional; the epithelium stretches and becomes thinner as these organs fill with urine.

E4. Parasympathetic, detrusor; stretching the walls of the bladder by more than 200–400 mL (about 0.8–1.6 cups) of urine.

E5. (a) True pelvis; external. (b) Urethra; 3.8 (1.5), 20 (8); prostatic → membranous → spongy; semen.

E6. Mucous; microbes can spread infection from the exterior of the body (at urethral orifice) along mucosa to kidneys.

F2. #2 blood in the list on pages 947–9.

G1. (a) Meso, third. (b) Pro; pro; meta. (c) Cloaca (urogenital sinus portion), endo.

G2. De. (a) De. (b) De.

G3. Incontinence of bladder; urinary tract infections; increased incidence of prostatic hypertrophy leads to urinary retention; excessive urination at night.

H1. Calculi; nonsurgical, shock; drink plenty of water (8–10 glasses/day) to keep urine dilute, avoid excessive calcium, and discuss with a dietitian and/or urologist a diet that minimizes extremes of urine pH.

H2. Burning with urination or painful urination (dysuria); frequency of urination due to irritation of the urinary tract; if the infection spreads up the ureters to kidneys (located at waist level), low back pain may result.

H4. (a) PCD. (b) C. (c) E. (d) GN. (e) A.

MORE CRITICAL THINKING: CHAPTER 26

1. Discuss the significance of the structure and arrangement of blood vessels within a nephron.

2. Describe how blood pressure is increased by mechanisms initiated by renin and aided by angiotensin I and II, aldosterone, and ADH.

3. Tamika, an accident victim, was admitted to the hospital 3 hours ago. Her chart indicates that she had been hemorrhaging at the scene of the accident. Joyce's nurse monitors her urinary output and notes that it is currently 12 mL/hr. (Normal output is 30–60 mL/hr.) Explain why a person in such severe stress is likely to have decreased urinary output (oliguria).

4. Describe several mechanisms that allow you to: (a) assure adequate blood flow into kidneys; (b) prevent loss of glucose and amino acids in your urine; and (c) produce more concentrated urine (conserving water) when you produce large amounts of sweat.

5. Who is more likely to be at higher risk for urinary tract infections (UTIs)? Explain why in each case. (a) 30-year-old women or men. (b) 30-year-old men or 70-year-old men.

6. Discuss signs and symptoms that are likely to be present in a person who has chronic renal failure (CRF).

MASTERY TEST: CHAPTER 26

Questions 1–6: Arrange the answers in correct sequence.

_____ _____ _____   1. From superior to inferior:
   A. Ureter
   B. Bladder
   C. Urethra

_____ _____ _____ _____   2. From most superficial to deepest:
   A. Renal capsule
   B. Renal medulla
   C. Renal cortex
   D. Renal pelvis

_____ _____ _____ _____ _____   3. Pathway of glomerular filtrate:
   A. Ascending limb of loop
   B. Descending limb of loop
   C. Collecting tubule
   D. Distal convoluted tubule
   E. Proximal convoluted tubule

_____ _____ _____ _____ _____   4. Pathway of blood:
   A. Arcuate arteries
   B. Interlobular arteries
   C. Renal arteries
   D. Afferent arteriole
   E. Interlobar arteries

_____ _____ _____ _____ _____   5. Pathway of blood:
   A. Afferent arteriole
   B. Peritubular arteries and vasa recta
   C. Glomerular capillaries
   D. Venules and veins
   E. Efferent arteriole

_____ _____ _____ _____ _____ _____   6. Order of events to restore low blood pressure to normal:
   A. This substance acts as a vasoconstrictor and stimulates aldosterone.
   B. Renin is released and converts angiotensinogen to angiotensin I.
   C. Angiotensin I is converted into angiotensin II.
   D. Decrease in blood pressure or sympa thetic nerves stimulate juxtaglomeru- lar cells.
   E. Under influence of this hormone Na$^+$ and H$_2$O reabsorption occurs.
   F. Increased blood volume raises blood pressure.

Questions 7–12: Circle the letter preceding the one best answer to each question.

7. Which of these is a normal constituent of urine?
   A. Albumin
   B. Urea
   C. Glucose
   D. Casts
   E. Acetone

8. Which is a normal function of the bladder?
   A. Oliguria
   B. Nephrosis
   C. Calculi
   D. Micturition

9. Which parts of the nephron are composed of simple squamous epithelium?
   A. Ascending and descending limbs of loop of the nephron
   B. Glomerular capsule (parietal layer) and descending limb of loop of the nephron
   C. Proximal and distal convoluted tubules
   D. Distal convoluted and collecting tubules

10. Choose the one *false* statement.
    A. An increase in glomerular blood hydrostatic pressure (GBHP) causes a decrease in net filtration pressure (NFP).
    B. Glomerular capillaries have a higher blood pressure than other capillaries of the body.
    C. Blood in glomerular capillaries flows into arterioles, not into venules.
    D. Vasa recta pass blood from the efferent arteriole toward venules and veins.

11. Choose the one *true* statement about renal corpuscle structure and function.
    A. Afferent arterioles offer more resistance to flow than do efferent arterioles.
    B. An efferent arteriole normally has a larger diameter than an afferent arteriole.
    C. The juxtaglomerular apparatus (JGA) consists of cells of the proximal convoluted tubule (PCT) and the afferent arteriole.
    D. Most water reabsorption normally takes place across the proximal convoluted tubule (PCT).

12. Which one of the hormones listed here causes increased urinary output and naturesis?
    A. ANP
    B. ADH
    C. Aldosterone
    D. Angiotensin II

Questions 13–20: Circle T (true) or F (false). If the statement is false, change the underlined word or phrase so that the statement is correct.

T F   13. Antidiuretic hormone (ADH) <u>decreases</u> permeability of distal and collecting tubules to water, thus <u>increasing</u> urine volume.

T F   14. A ureter is a tube that carries urine from <u>the bladder to the outside of the body</u>.

T F   15. Diabetes insipidus is a condition caused by a defect in production of <u>insulin</u> or lack of sensitivity to this hormone by principal cells in collecting ducts.

T F   16. A person taking diuretics is likely to urinate <u>less</u> than a person taking no medications.

T F   17. Blood colloid osmotic pressure (BCOP) is a force that tends to <u>push substances from blood into filtrate</u>.

T F   18. The thick portion of the ascending limb of the loop of the nephron <u>is</u> permeable to Na⁺ and Cl⁻ but not to water.

T F   19. The countercurrent mechanism permits production of <u>small volumes of concentrated urine</u>.

T F   20. In infants 2 years old and under, <u>retention</u> is normal.

Questions 21–25: Fill-ins. Complete each sentence with the word or phrase that best fits.

_____ 21. The glomerular capsule and its enclosed glomerular capillaries

constitute a _____.

_____ 22. A _____ consists of a renal corpuscle and a renal tubule.

_____ 23. A toxic level of urea in blood is know as _____.

_____ 24. Sympathetic impulses cause greater constriction of _____

arterioles with resultant _____-crease in GFR and _____-crease
in urinary output.

_____ 25. A person with chronic renal failure is likely to be in acidosis

because kidneys fail to excrete _____, may be anemic because

kidneys do not produce _____, and may have symptoms of

hypocalcemia because kidneys do not activate _____.

ANSWERS TO MASTERY TEST: CHAPTER 26

Arrange

1. A B C
2. A C B D
3. E B A D C
4. C E A B D
5. A C E B D
6. D B C A E F

Multiple Choice

7. B   10. A
8. D   11. D
9. B   12. A

True–False

13. F. Increases; decreasing
14. F. A kidney to the bladder
15. F. ADH
16. F. More
17. F. Pull substances from filtrate
    into blood
18. T
19. T
20. F. Incontinence

Fill-ins

21. Renal corpuscle
22. Nephron
23. Uremia
24. Afferent, de, de
25. H, erythropoietic factor, vitamin D

FRAMEWORK 27
Fluid, Electrolyte, Acid–Base Balance

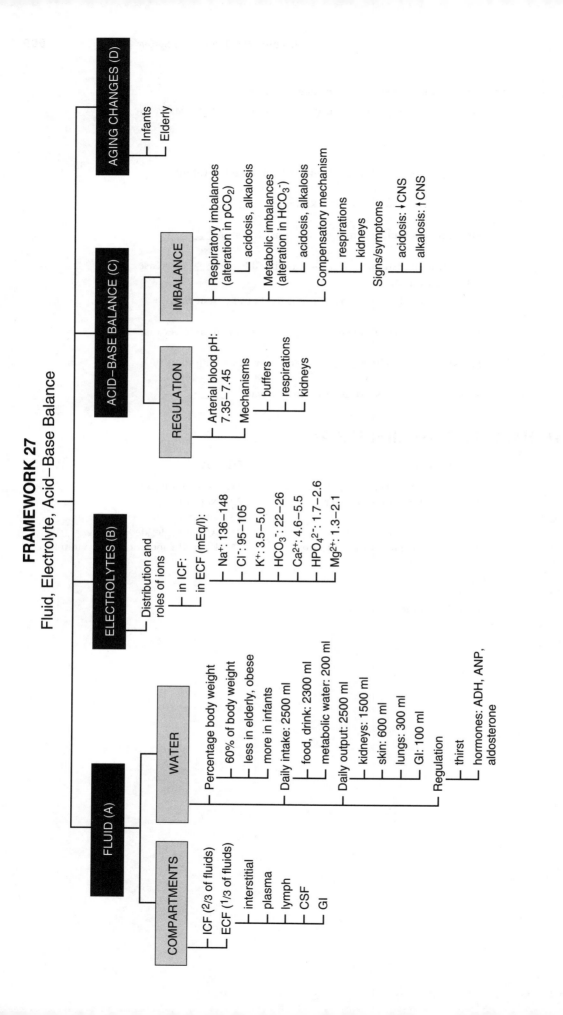

FLUID (A)

COMPARTMENTS
- ICF (2/3 of fluids)
- ECF (1/3 of fluids)
  - interstitial
  - plasma
  - lymph
  - CSF
  - GI

WATER
- Percentage body weight
  - 60% of body weight
  - less in elderly, obese
  - more in infants
- Daily intake: 2500 ml
  - food, drink: 2300 ml
  - metabolic water: 200 ml
- Daily output: 2500 ml
  - kidneys: 1500 ml
  - skin: 600 ml
  - lungs: 300 ml
  - GI: 100 ml
- Regulation
  - thirst
  - hormones: ADH, ANP, aldosterone

ELECTROLYTES (B)

- Distribution and roles of ions
  - in ICF:
  - in ECF (mEq/l):
    - Na$^+$: 136–148
    - Cl$^-$: 95–105
    - K$^+$: 3.5–5.0
    - HCO$_3^-$: 22–26
    - Ca^{2+}: 4.6–5.5
    - HPO$_4^{2-}$: 1.7–2.6
    - Mg^{2+}: 1.3–2.1

ACID–BASE BALANCE (C)

REGULATION
- Arterial blood pH: 7.35–7.45
- Mechanisms
  - buffers
  - respirations
  - kidneys

IMBALANCE
- Respiratory imbalances (alteration in pCO$_2$)
  - acidosis, alkalosis
- Metabolic imbalances (alteration in HCO$_3^-$)
  - acidosis, alkalosis
- Compensatory mechanism
  - respirations
  - kidneys
- Signs/symptoms
  - acidosis: ↓CNS
  - alkalosis: ↑CNS

AGING CHANGES (D)
- Infants
- Elderly

C H A P T E R **27**

Fluid, Electrolyte, and Acid–Base Homeostasis

For the maintenance of homeostasis, levels of fluid, ions or electrolytes, acids and bases must be kept within acceptable limits. Without careful regulation of pH, for example, nerves will become overactive (in alkalosis) or underfunction, leading to coma (in acidosis). Alterations in blood levels of electrolytes can be lethal: excess K^+ can cause cardiac arrest, and decrease in blood Ca^{2+} can result in tetany of the diaphragm and respiratory failure. Systems you studied in Chapter 23 (respiratory) and Chapter 26 (urinary), along with buffers in the blood, normally achieve fluid, electrolyte, and acid–base balance.

As you begin your study of fluids, electrolytes, and acid–base balance, carefully examine the Chapter 27 Topic Outline and Objectives; check off each one as you complete it. Be sure to refer to the Framework frequently and note relationships among key terms in each section.

TOPIC OUTLINE AND OBJECTIVES

A. Fluid compartments and fluid balance

1. Compare the locations of intracellular fluid (ICF) and extracellular fluid (ECF), and describe the various fluid compartments of the body.
2. Describe the sources of water and solute gain and loss, and explain how each is regulated.

B. Electrolytes in body fluids

3. Compare the electrolyte composition of the three fluid compartments: plasma, interstitial fluid, and intracellular fluid.
4. Discuss the functions of sodium, chloride, potassium, bicarbonate, calcium, phosphate, and magnesium ions, and explain the regulation of their concentrations.

C. Acid–base balance

5. Compare the roles of buffers, exhalation of carbon dioxide, and kidney excretion of H^+ in maintaining pH of body fluids.
6. Define acid–base imbalances, describe their effects on the body, and explain how they are treated.

D. Aging and fluid, electrolyte, and acid–base balance

7. Describe the changes in fluid, electrolyte, and acid–base balance that may occur with aging.

WORDBYTES

Now become familiar with the language of this chapter by studying each wordbyte, its meaning, and an example of its use within a term. After you study the entire list, self-check your understanding by writing the meaning of each wordbyte on the line. As you continue through the *Learning Guide*, identify (and fill in) additional terms that contain the same wordbyte.

Wordbyte	Self-check	Meaning	Example(s)
extra-	_____	outside	*extra*cellular
hyper-	_____	above	*hyper*ventilation
hypo-	_____	below	*hypo*tonic
intra-	_____	within	*intra*cellular
-osis	_____	condition of	alkal*osis*

CHECKPOINTS

A. Fluid compartments and fluid balance (pages 956–961)

✓ **A1.** Do this exercise about body fluid by answering questions about 40-year-old Alex, who is 193 cm (6′ 4″) tall and weighs 100 kg (220 pounds).

   a. Fill in the expected weight in each category:

   1. _____ kg of solids (such as fats, protein, and calcium)

   2. _____ kg of fluid within cells

   3. _____ kg of fluid in interstitial fluid

   4. _____ kg of fluid in blood plasma

   b. The total amount of fluid in Alex's body amounts to about _____ kg, whereas the

   amount of his extracellular fluid is about _____ kg.

✓ **A2.** About 67% of body fluids are *(inside of? outside of?)* cells. Check your understanding of fluid compartments by circling all fluids that are considered *extracellular fluids (ECF):*

Blood plasma
Aqueous humor and vitreous humor of the eyes
Endolymph and perilymph of ears
Fluid within liver cells
Fluid immediately surrounding liver cells

Lymph
Pericardial fluid
Synovial fluid
Cerebrospinal fluid (CSF)

For extra review, refer to Figure LG 1.1, page 8 of the *Learning Guide.*

✓ **A3.** *A clinical correlation.* Is a 45-year-old woman who has a water content of 55% likely to be in fluid balance? Explain.

✓ **A4.** Circle the person in each pair who is more likely to have a higher proportion of
    water content.
    a. 37-year-old woman/67-year-old woman
    b. Female/male
    c. Lean person/obese person

✓ **A5.** Daily intake and output (I and 0) of fluid usually both equal _____ mL (_____

    liters). Of the intake, about _____ mL are ingested in food and drink, and _____
    mL result from catabolism of foods (discussed in Chapter 25).

✓ **A6.** Now list four systems of the body that eliminate water. Indicate the amounts lost by
    each system on an average day. One is done for you.
    a. Integument (skin): 600 mL/day

    b. _____ : _____ mL/day

    c. _____ : _____ mL/day

    d. _____ : _____ mL/day
        From this exercise, it is clear that the largest route for elimination of fluid is normally
        via: (circle answer)

        exhalation from lungs   feces   sweat   urine

        On a day when you exercise vigorously, how would you expect routes of fluid output
        to change?

✓ **A7.** Do this exercise on relationships between water and solutes.
    a. Define osmolarity of body fluids.

    b. Write two categories of changes in the body that can increase blood osmolarity.

    c. Identify the hormones that alter blood levels of NaCl and water:
        1. Aldosterone promotes *(excretion? retention?)* of Na^+ and Cl^- by kidneys.
        2. Angiotension II ultimately has the *(same? opposite?)* effect on blood levels of NaCl
            compared to aldosterone. Explain why.

        Angiotensin is formed when kidneys produce the chemical named _____
        in response to several factors such as (see Figure LG 18.5, page LG 366):

3. The name ANP indicates that this hormone promotes *(excretion? retention?)* of Na^+

and Cl^- by kidneys since the "N" in ANP stands for _____ (or loss of

Na^+). ANP is released from the _____ when they are stretched by a large volume of blood. Natriuresis will then lead to loss of both NaCl and

_____, returning blood volume to normal.
4. The name ADH indicates that this hormone promotes *(promotes? opposes?)* diuresis which is defined as a *(large? small?)* volume of urine. A high level of ADH there-fore causes a *(high? low?)* urine output. ADH causes insertion of water-channel

proteins called _____ into cells in collecting tubules of kidneys, causing

a(n) _____-crease in permeability of those cells to water. Excessive amounts of ADH can lead to so much water retention that blood osmolarity (and Na^+ concen-

tration) _____-creases significantly. (See Checkpoint A10).

✓ **A8.** Describe mechanisms for replacement of fluid that result from reduction of the body's fluid volume.
a. Loss of fluid volume is a condition known as *(de? re? over?)*-hydration. This condition

stimulates thirst in three ways. One is _____-creased blood volume (and blood pres-

sure), which stimulates release of _____ from kidneys and leads to

production of _____ II. This chemical plays several roles in increas-ing blood pressure, one of which is to *(stimulate? inhibit?)* thirst receptors located in

the _____.
b. Identify the other two factors that stimulate thirst receptors during dehydration:

_____-crease in blood osmotic pressure, as well as _____-creased saliva production.
c. List three causes of dehydration.

In these cases, is it advisable to begin fluid replacement before or after onset of thirst? *(Before? After?)*

✓ **A9.** Circle the factors that increase fluid output:
A. ADH
B. Aldosterone
C. Diarrhea
D. Increase in blood pressure (BP)
E. Fever
F. Hyperventilation
G. Atrial natriuretic peptide (ANP)

✓ **A10.** Do this exercise on movement of water between compartments.
a. List several examples of causes of low blood levels of Na^+ (hyponatremia).

Interstitial fluid   Cell   Blood plasma   Interstitial fluid   Cell   Blood plasma

(a)   (b)

○ **Interstitial fluid to ICF**   ○ **Plasma to interstitial fluid**

Figure LG 27.1 Diagrams of fluid compartments (a) Direction of fluid shift resulting from Na+ deficit. Color according to Checkpoint A10 (b) Write in electrolytes according to Checkpoint B5.

b. Show on Figure LG 27.1a the shift of fluid resulting from loss of Na⁺. Color arrows 1 and 2 according to color code ovals. Note that this figure also shows effects likely to occur in the large intestine if a *(hypertonic? hypotonic or tap water?)* enema is administered.

✓ **A11.** In this section on regulation of water balance, you have seen that electrolyte levels of the body *(do? do not?)* affect fluid levels, and fluid levels *(do? do not?)* affect electrolyte levels. Explain.

B. Electrolytes in body fluids (pages 961–965)

B1. List four general functions of electrolytes.

_____

_____

_____

_____

B2. Define each term: *mole, mmole/liter, mEq/liter.*

✓ **B3.** Circle all of the answers that are classified as electrolytes:
    A. Proteins in solutions   D. K^+ and Na^+
    B. Anions   E. H^+
    C. Glucose

✓ **B4.** Circle the electrolytes that are anions.
    Ca^{2+} HCO_3^- (bicarbonate)   Lactate Cl^-
    HPO_4^{2-} (phosphate) Na^+ H^+ K^+

✓ **B5.** On Figure LG 27.1b write the chemical symbols for electrolytes that are found in greatest concentrations in each of the three compartments. Write one symbol on each line provided. Use the following list of symbols. (*Hint:* Refer to Figure 27.6 in the text.)

Cl^-.	**Chloride**	Mg^{2+}.	**Magnesium**
HCO_3^-.	**Bicarbonate**	Na^+.	**Sodium**
HPO_4^{2-}.	**Phosphate**	$Protein^-$.	**Protein**
K^+.	**Potassium**		

✓ **B6.** Analyze the information in Figure LG 27.1b by answering these questions.
    a. Circle the two compartments that are most similar in electrolyte concentration.
        cell   blood plasma   interstitial fluid (IF)
    b. List two major differences in electrolyte content of these two similar compartments.

    c. Most body protein is located in which compartment? _____
    d. Protein anions in intracellular fluid are more likely to be bound to the cation *(Na^+? K^+?)*.

B7. Check your understanding of the functions and disorders related to major electrolytes by completing Table LG 27.1 (next page).

✓ **B8.** *A clinical correlation.* Norine is having her serum electrolytes analyzed. Complete this exercise about her results.
    a. Her Na^+ level is 128. Norine's value indicates that she is likely to be in a state of *(hyper? hypo?)*-natremia. Write two or more signs or symptoms of this electrolyte imbalance.

    b. A diagnosis of hypochloremia would indicate that Norine's blood level of _____

    ion is lower than normal. A normal range for serum chloride (Cl^-) level is _____

    mEq/liter. One cause of low chloride is excessive _____.
    c. A normal range for K^+ in serum is 3.5–5.0 mEq/liter. This range is considerably *(higher? lower?)* than that for Na^+. This is reasonable since K^+ is the main cation in

(ECF? ICF?), yet lab analysis of serum electrolytes is examining K$^+$ in *(ECF? ICF?)*.

d. Norine's potassium (K$^+$) level is 5.6 mEq/liter, indicating that she is in a state of

hyper_____. If Norine's serum K$^+$ continues to increase, for example to 8.0 mEq/l, her most serious risk, possibly leading to death, is likely to be

_____.

e. A normal range for Mg^{2+} level of blood is 1.3–2.1 mEq/liter. Norine's electrolyte report indicates a value of 1.1 mEq/liter for Mg^{2+}. This electrolyte imbalance is

known as _____. List three or more factors related to nutritional state that could lead to this condition.

f. Norine's Ca^{2+} level is 3.9 mEq/]iter. A normal range is 4.5–5.5 mEq/liter. This

electrolyte imbalance is known as _____. This condition is often associated with *(decreased? elevated?)* phosphate levels, because blood levels of calcium and phosphate are *(directly? inversely?)* related. Low blood levels of both Ca^{2+} and Mg^{2+} cause overstimulation of the central nervous system and muscles. List two symptoms of these electrolyte imbalances.

Table LG 27.1 Summary of major blood electrolytes

Electrolyte (Name and Symbol)	Normal Range (SerumLevel), in mEq/liter	Principal Functions	Signs and Symptoms of Imbalances:	
			Hyper-	Hypo-
a.	136–142			
b.				Hypochloremia: muscle spasms, alkalosis, ↓ respirations, coma
c.		Most abundant cation in ICF; helps maintain fluid volume of cells; functions in nerve transmission		
d.			Hypercalcemia: weakness, lethargy, confusion, stupor, and coma	
e.	1.7–2.6	Helps form bones, teeth; important components of DNA, RNA, ATP, and buffer system		
f. Magnesium (Mg^{2+})				

✓ **B9.** *The Big Picture: Looking Back* and *A clinical correlation*. Refer to figures in earlier chapters of the *Learning Guide* or the text and complete this Checkpoint about regulation and imbalances of electrolytes.

a. On Figure LG 3.2, page LG 46, you identified the major anions and cations inside of cells and in fluids surrounding cells. When significant tissue trauma occurs, injured cells are likely to release their contents, leading to increased blood levels of the major ICF cation *(K+? Na+?)*. This condition is know as hyper-*(calc? kal? natr?)*-emia.

b. In Figure LG 10.1, page LG 181, you identified sarcoplasmic reticulum (SR) in muscle tissue. Which electrolyte is sequestered inside of SR and released into sarcoplasm surrounding thick and thin myofilaments only when this ion is needed to trigger muscle contraction? *(Ca²⁺? Cl⁻? K⁺ Na⁺?)*

c. On Figure LG 12.2, page LG 230, you demonstrated the major anions and cations in and around nerve cells. Which major ECF cation is permitted to pass through voltage-gated channels in neuron plasma membranes to initiate an action potential? *(K+? Na+?)*

d. On Figure LG 12.3, page LG 232, you identified that the voltage of a resting

membrane potential is typically _____ mV. The major ICF cation in neurons is *(K+? Na+?)*. In hypokalemia, nerve cells (as well as body fluids) would be deficient in K⁺ and so would be more likely to have a resting membrane potential of *(–65? –80?)*. This means a greater stimulus would be necessary to raise the membrane potential to

threshold of about _____ mV. In hypokalemia, nerves and muscles are therefore likely to be *(hyper? hypo ?)*-active. Write two or more signs or symptoms of hypokalemia related to nerves or muscles.

e. Figure LG 18.4, page LG 365, shows that *(calcitonin? PTH?)* enhances movement of calcium into bones; this occurs by stimulation of osteo-*(blasts? clasts?)* in bone. PTH

_____-creases calcium in blood by drawing calcium from several sources. Name

three. _____ _____ _____ In fact, excessive PTH can lead to *(hyper? hypo?)*-calcemia.

f. Figure LG 18.5, page LG 366, emphasizes the effect of aldosterone on blood levels of several electrolytes. Aldosterone tends to increase blood levels of *(H+? K+? Na+ ?)* and decrease blood levels of *(H+? K+? Na+ ?)*. An overproduction of aldosterone (as in primary aldosteronism) is likely to lead to *(hypo? hyper?)*-natremia and *(hypo? hyper?)*-kalemia.

g. The same figure points out that ADH _____-creases reabsorption of water from urine into blood. The more water reabsorbed, the *(more? less?)* concentrated Na⁺ will be in blood. What effect would overproduction of ADH (as in the syndrome of inappropriate ADH production [SIADH]) have upon blood level of Na⁺? Leads to *(hypo? hyper?)*-natremia. What effect would deficient production of ADH (as in diabetes insipidus) have upon blood level of Na⁺? Urine output would greatly

_____-crease, leading to dehydration and *(hypo? hyper?)*-natremia.

h. Refer to text Figure 23.24 (page 803) and complete the chemical formulas showing formation of bicarbonate.

$$CO_2 + H_2O \rightarrow \underset{\text{(carbonic acid)}}{\underline{\hspace{3cm}}} \rightarrow \underset{\text{(bicarbonate)}}{\underline{\hspace{3cm}}} + H^+$$

In fact, this equation indicates how most (_____%) of CO_2 in blood is transported as bicarbonate. Bicarbonate also serves as a buffer for *(acids? bases?)*. An arterial blood level of bicarbonate of 32 mEq/ml is *(high? normal? low?)*. Excessive blood levels of

bicarbonate can be reduced as bicarbonate is excreted by _____.

✓ **B10.** *A clinical correlation.* Edema is an abnormal increase of *(intracellular fluid? Interstitial fluid? plasma?)*. Describe three or more factors that may lead to edema.

C. Acid–base balance (pages 965–969)

✓ **C1.** Complete this Checkpoint about acid–base regulation.
  a. The pH of blood and other ECF should be maintained between _____ and _____.
  b. Name the three major mechanisms that work together to maintain acid–base balance.

  _____   _____   _____

  c. List four important buffer systems in the body.

  _____   _____

  _____   _____

✓ **C2.** Refer to Figure LG 27.2 and complete this exercise about buffers.
  a. In Figure LG 27.2, you can see why buffers are sometimes called "chemical sponges." They consist of an anion (in this case bicarbonate, or HCO_3^-) that is combined with a cation, which in this buffer system may be either H^+ or Na^+. When H^+ is drawn to

  the HCO_3^- "sponge," the weak acid _____ is formed. When Na^+

  attaches to HCO_3^-, the weak base _____ is present. As you color H^+ and Na^+, imagine these cations continually trading places as they are "drawn in or squeezed out of the sponge."
  b. Buffers arc chemicals that help the body to cope with *(strong? weak?)* acids or bases that are easily ionized and cause harm to the body. When a strong acid, such as hydrochloric acid (HCl), is added to body fluids, buffering occurs, as shown in Figure LG 27.2b. The easily ionized hydrogen ion (H^+) from HCl is "absorbed by the sponge" as Na^+ is released from the buffer/sponge to combine with Cl^-. The two products,

  _____ and _____, are weak (less easily ionized).
  c. Now complete Figure LG 27.2c by writing the names of products resulting from buffering of a strong base, H^+ (and its arrow) and Na^+ (and its arrow) to show this exchange reaction.
  d. When the body continues to take in or produce excess strong acid, the concentration

  of the _____ member of this buffer pair will decrease as it is used up in the attempt to maintain homeostasis of pH.
  e. Though buffer systems provide rapid response to acid–base imbalance, they are limited since one member of the buffer pair can be used up, and they can convert only strong acids or bases to weak ones; they cannot eliminate them. Two systems of the

  body that can actually eliminate acid or basic substances are _____

  and _____.

a. H$^+$ and Na$^+$: only one of you
can join HCO$_3^-$ at any one time.

b. Na$^+$ is squeezed out
as H$^+$ is drawn in.

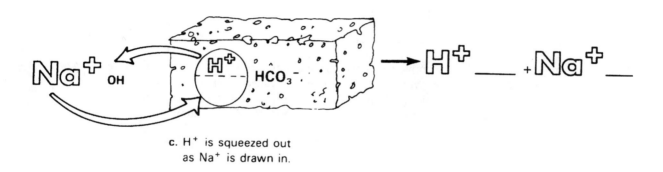

c. H$^+$ is squeezed out
as Na$^+$ is drawn in.

O H$^+$ and arrows showing movement of H$^+$

O Na$^+$ and arrows showing movement of Na$^+$

Figure LG 27.2 Buffering action of carbonic acid—sodium bicarbonate system.
Complete as directed in Checkpoint C2.

✓ **C3.** *Far extra review.* Check your understanding of the carbonic acid-bicarbonate buffer in
this Checkpoint. When a person actively exercises, *(much? little?)* CO$_2$ forms, leading
to H$_2$CO$_3$ production. Consequently, *(much? little?)* H$^+$ will be added to body tissues.
In order to buffer this excess H$^+$ (to avoid drop in pH and damage to tissues), the
(H$_2$CO$_3$? NaHCO$_3$?) component of the buffer system is called upon. The excess H$^+$

then replaces the _____ion of NaHCO$_3$, in a sense "tying up" the potentially
harmful H$^+$.

✓ **C4.** Phosphates are more concentrated in *(ECF? ICF?)*. (For help, see Figure 27.6 in the
text.) Name one type of cell in which the phosphate buffer system is most important.

_____

C5. Write the reaction showing how phosphate buffers strong acid.

✓ **C6.** Which is the most abundant buffer system in the body? _____ Circle the part of the amino acid shown below that buffers acid. Draw a square around the part of the amino acid that buffers base.

$$NH_2 \text{———} C \text{———} COOH$$

with R above C and H below C.

C7. Explain how the hemoglobin buffer system buffers carbonic acid, so that an acid even weaker than carbonic acid (that is, hemoglobin) is formed.
For extra review. See Chapter 23, Checkpoint E3, pages 509–510 in the *Learning Guide* and Figure 23.24 in the text.

✓ **C8.** Complete this Checkpoint about respiration and kidneys related to pH.
 a. Hyperventilation will tend to *(raise? lower?)* blood pH because, as the person exhales CO_2, less CO_2 is available for formation of _____ acid and free hydrogen ion.
 b. A slight decrease in blood pH will tend to *(stimulate? inhibit?)* the respiratory center and then will *(increase? decrease?)* respirations.
 c. The kidneys regulate acid–base balance by altering their tubular secretion of _____ or elimination of _____ ion.

✓ **C9.** An acid–base imbalance caused by abnormal alteration of the respiratory system is classified as *(metabolic? respiratory?)* acidosis or alkalosis. Any other cause of acid–base imbalance, such as urinary or digestive tract disorder, is identified as *(metabolic? respiratory?)* acidosis or alkalosis.

✓ **C10.** *Critical thinking.* Indicate which of the four categories of acid–base imbalances listed in the box is most likely to occur as a result of each condition described below.

MAcid.	Metabolic acidosis	RAcid.	Respiratory acidosis
MAlk.	Metabolic alkalosis	RAlk.	Respiratory alkalosis

_____ a. Decreased blood level of CO_2, as a result of hyperventilation

_____ b. Decreased respiratory rate in a patient taking overdose of morphine

_____ c. Excessive intake of antacids

_____ d. Prolonged vomiting of stomach contents

_____ e. Ketosis in uncontrolled diabetes mellitus

_____ f. Decreased respiratory minute volume in a patient with emphysema or fractured rib

_____ g. Excessive loss of bicarbonate from the body, as in prolonged diarrhea or renal dysfunction

✓ **C11.** Refer to the table below and do this exercise on diagnosis of acid–base imbalances.

Patient	Arterial pH	pCO_2 (mm Hg)	HCO_3^- (m Eq/liter)
Normal	acidosis 7.35–7.45 alkalosis	alkalosis 35–45 acidosis	acidosis 22–26 alkalosis
Patient W	← 7.29	52* →	30** →
Patient X	7.50 →	← 28*	23**
Patient Y	← 7.31	← 32**	← 19*
Patient Z	7.44	53** →	35* →

Arrows indicate which values have increased above normal (→) or decreased below normal (←). Absence of an arrow means the value is normal. * indicates *primary value* associated with the initial acid–base problem. ** indicates *secondary value* which may be due to compensatory mechanisms for acid–base imbalances.

a. First refer to normal values for arterial blood pH, pCO_2, and HCO_3^-. Acidosis is indicated by arterial blood pH less than *(7.35? 7.45?)*. Keep in mind that acidity and pH

are *(directly? inversely?)* related; that is, as acidity of blood increases, its pH _____-creases. Which patients have blood pH that indicate acidosis? Patients *(W? X? Y? Z?)*

b. Look at the blood chemistry values for patient W, a 61-year-old man with diagnoses of emphysema and pneumonia. The table indicates that a pCO_2 value greater than *(35? 45?)* may be an indicator of acidosis. Patients with chronic respiratory problems retain CO_2, as indicated by this patient's pCO_2 of 52, which is *(higher? lower?)* than normal. The arrow drawn to the right of the 52 indicates that such high CO_2 levels have led this patient into acidosis.

c. Patient X is a 27-year-old woman who has just arrived in the emergency room. She is hyperventilating and in a state of high anxiety after an accident that has critically injured her child. Hyperventilation tends to *(increase? decrease?)* blood pCO_2, as

shown by her value of _____ mm Hg. The arrow next to the 28 points left to show the effect (alkalosis) of this lowered blood pCO_2.

d. Both patients W and X have acid–base imbalances caused by changes in respiration. Therefore patient W's imbalance is classified as *(respiratory? metabolic?)* acidosis, and

patient X's imbalance is known as _____ _____.

e. In respiratory acid–base imbalances, the value that is altered is always *(pCO_2?* HCO_3^-?)*. For clarity, call this the *primary* value associated with respiratory acid–base imbalances; this value is marked by * in the table. The alteration is very logical; for

example, increased pCO_2 is associated with _____-osis because CO_2 leads to production of carbonic acid, and decreased pCO_2 indicates

_____-osis because less carbonic acid is made.

f. In order to maintain homeostasis, the body will try to compensate for acid–base imbalances by altering the *secondary* (or "other") value *in the same direction* as the primary value, so that the two values are more matched or in balance. The secondary value is marked with ** in the table. Where compensation is occurring, you will see the arrow for the secondary value (**) going in the same direction as the primary value (*). For respiratory acidosis, in which pCO_2 (the primary value) _____-creases,

we can expect compensatory mechanisms to _____-crease the secondary value (HCO_3^-), possibly to a value greater than normal, in an attempt to make pH more alkaline. It is even possible that a person in respiratory acidosis could move into metabolic alkalosis due to overcompensation, for example, by reduced kidney excretion of sodium bicarbonate ($NaHCO_3$).

g. Does it appear that patient W has done some compensating for his respiratory acidosis? *(Yes? No?)* This is indicated by his HCO_3^- value, which is *(within? outside?)* the normal range. This patient has *(fully? partially?)* compensated respiratory acidosis because his pH is still abnormal.

h. If patient W had his HCO_3^- value within the normal range, this would indicate *(full? partial? no?)* compensation. If patient W's values were pH of 7.39, pCO_2 of 52 mm Hg, and HCO_3^- of 30, this return to the normal pH range would indicate *(fully? partially? un-?)* compensated respiratory acidosis.

i. Patient X's HCO_3^- value (23 mEq) is *(high? normal? low?)*, indicating that she has *(fully? partially? un-?)* compensated respiratory alkalosis. Although she may eventu-

ally excrete HCO_3^- from her _____ to match the decreased pCO_2 level, this has not yet occurred because her anxiety and hyperventilation have been relatively short-term.

j. Patient Y is in diabetic ketoacidosis, is lethargic, and is hyperventilating. The lowered HCO_3 level indicates that the patient is using bicarbonate to buffer ketoacids resulting from uncontrolled diabetes. The primary value (*) affected is *(pCO_2? HCO_3^-?)*, indicating *(respiratory? metabolic?)* acid–base imbalance. What factor indicates partial compensation?

k. Patients with ketoacidosis or other metabolic acidosis (such as from excessive aspirin

intake) may even move into _____ alkalosis by compensatory hyperventilation. In that case, the pCO_2 would continue to drop, and pH would return to normal and go right on to a value greater than *(7.35? 7.45?)*.

l. Patient Z has a history of peptic ulcer and had been vomiting for days preadmission. The patient states that she had taken a "bunch of antacids" for her ulcer. Does her pH indicate a pH imbalance? *(Yes? No?)* Are her other two values normal? *(Yes? No?)* What type of acid–base imbalances have normal pH but abnormal pCO_2 and HCO_3^-? *(Fully? Partially? Un-?)* compensated imbalances.

m. Patient Z's ulcer, vomiting (loss of gastric HCl), and ingestion of antacids, as well as the level of her pH (7.44) so close to the alkaline end of the normal range, all suggest

metabolic _____-osis. Her compensatory mechanisms apparently include *(hypo? hyper?)*-ventilation, with retention of CO_2 which has helped to *(raise? lower?)* her pH back into the normal range.

n. Complete this Checkpoint by writing complete diagnoses (including degree of compensation) for each patient in the table.

✓ **C12.** Acidosis has the effect of *(stimulating? depressing?)* the central nervous system (CNS). Write one or more signs of acidosis.

Alkalosis is more likely to cause *(spasms? lethargy?)* because this acid–base imbalance

_____ the CNS.

D. Aging and fluid, electrolyte, and acid–base balance (pages 969–970)

✓ **D1.** Complete this tablet contrasting fluids and electrolytes in infants and adults. Select correct answers from those given in *italics*.

Factor (answers)	Infant	Adult
a. Percentage of total body mass that is water: *(55–60% 70–90%)*		
b. Metabolic rate (with resulting metabolic wastes) *(Higher? Lower?)*	Higher	
c. Ability of kidneys to concentrate urine (retaining water) *(Greater? Less?)*		
d. Body surface area (which permits loss of fluid by evaporation) *(Greater? Less?)*		
e. Respiration rate (which permits loss of fluid in expiration) *(Greater? Less?)*		
f. Concentration of K^+ (if higher, leads towards acidosis) *(Higher? Lower?)*		

D2. *Critical thinking.* Older adults may experience increased risk for changes in fluid, electrolytes, and acid–base as a result of normal aging changes. Write possible causes and signs or symptoms associated with each imbalance. Check your answers against Table 27.1 in the text.

Imbalance	Possible causes	Signs or symptoms
a. Hypokalemia		
b. Hyponatremia		
c. Hypernatremia		
d. Acidosis (respiratory or metabolic)		

ANSWERS TO SELECTED CHECKPOINTS: CHAPTER 27

A1. (a1) 40; (a2) 40; (a3) 16: (a4) 4. (b) 60, 20.

A2. Inside of; all answers except fluid within liver cells.

A3. The percentage is within the normal range; however, more information is needed because the fluid must be distributed correctly among the three major compartments: cells (ICF) and ECF in plasma and interstitial fluid. For example, excessive distribution of fluid to interstitial space occurs in the fluid imbalance known as edema.

A4. (a) 37-year-old woman. (b) Male. (c) Lean person.

A5. 2500 (2.5); 2300, 200.

A6. (Any order) (a) Kidney, 1500. (b) Lungs, 300. (c) GI, 100. Urine. With exercise, output of fluid increases greatly via sweat glands and somewhat via respirations. To compensate, urine output decreases, and intake of fluid increases.

A7. (a) The degree of concentration or osmotic pressure of body fluids. (b) Loss of fluid (dehydration) or gain of electrolytes (such as by a high intake of salty foods). (c1) Retention. (c2) Same, since angiotensin II promotes release of aldosterone from the adrenal cortex; renin; dehydration or low renal artery blood pressure, low blood Na^+ level or high blood K^+ level. (c3) Excretion, natriuretic; atria; water. (c4) Opposes, large; low; aquaporin-2, in; de.

A8. (a) De: de, renin, angiotensin; stimulate, hypothalamus. (b) In. de. (c) Sweating. vomiting. diarrhea; before.

A9. C–G.

A10. (a) Excessive amounts of ADH, intake of large amounts of water or decreased intake of Na^+; (see causes of hyponatremia in Table 27.1 of the text.) (b) 1, Shift of fluid leading to overhydration or water intoxication of brain cells, possibly leading to convulsions and coma; 2, shift of fluid (from plasma to replace IF that entered cells), possibly leading to hypovolemic shock. Hypotonic or tap water.

A11. Do, especially Na^+ (the major cation in ECF), which "holds" (retains) water. Note (in Checkpoint A7c4) that excessive water retention dilutes blood levels of Na^+, possibly leading to hyponatremia.

B3. All except C (glucose).

B4. Cl^-, HCO_3^- (bicarbonate), HPO_4^{2-} (phosphate), lactate.

B5.

Figure LG 27.1A
Diagram of fluid compartments.

B6. (a) Blood plasma and interstitial fluid. (b) Plasma contains more plasma protein, and slightly less sodium and less chloride than IF. (c) Intracellular fluid (ICF). (d) K^+.

B8. (a) Normal serum Na^+ ranges from 136 to 148; hypo; examples: headache, low blood pressure, tachycardia, and weakness, possibly leading to confusion, stupor, and coma if her serum Na^+ continues to drop. (b) Cl^-; 95–105; emesis, dehydration or use of certain diuretics. (c) Lower; ICF, ECF. (d) Kalemia; ventricular fibrillation resulting in cardiac arrest. (e) Hypomagnesemia; malnutrition, malabsorption, diarrhea, alcoholism, or excessive lactation. (f) Hypocalcemia; elevated, inversely; tetany, convulsions.

B9. (a) K^+; kal. (b) Ca^{2+}. (c) Na^+. (d) –70; K^+; –80; –55; hypo; flaccid paralysis, fatigue, changes in cardiac muscle and G1 muscle function. (e) Calcitonin, blasts; in; bones, intestine, kidneys; hyper. (f) Increases Na^+ and decreases H^+ and K^+, hyper, hypo. (g) In; less; hypo; in, hyper. (h) H_2CO_3, HCO_3^-; 70; acids; high; kidneys.

B10. Interstitial fluid; hypertension (which increases filtration at arteriolar ends of capillaries), excessive NaCl intake (which causes fluid retention), congestive heart failure (which backs up blood into venules and capillary beds), or blockage of lymph flow (for example, by cancer metastasized to lymph nodes).

C1. (a) 7.35–7.45. (b) Buffers, respiration, and kidney excretion. (c) Carbonic acid-bicarbonate, phosphate, hemoglobin-oxyhemoglobin, protein.

C2. (a) Carbonic acid, or H_2CO_3; sodium bicarbonate, or $NaHCO_3$. (b) Strong; the salt sodium chloride (NaCl), the weak acid H_2CO_3 (H•HCO_3). (c) Water (H_2O or H•OH) and sodium bicarbonate ($NaHCO_3$). (d) $NaHCO_3$ or weak base. (e) Respiratory (lungs), urinary (kidneys).

C3. Much; much; $NaHCO_3$; Na^+.

C4. ICF; kidney cells.

C6. Protein;

$$\text{Buffers acid} - \!\!\left(\!NH_2\!\right)\!\! - C \,\boxed{\!-\!COOH\!} - \text{Buffers base}$$

with R above C and H below C.

C8. (a) Raise, carbonic. (b) Stimulate, increase. (c) H^+, HCO_3^-.

C9. Respiratory; metabolic.

C10. (a) RAlk. (b) RAcid. (c) MAlk. (d) MAlk. (e) MAcid. (f) RAcid. (g) MAcid.

C11. (a) 7.35; inversely; de: W and Y. (b) 45 (except when pCO_2 is high due to hypoventilation as compensation for metabolic alkalosis: see Patient Z); higher. (c) Decrease. 28. (d) Respiratory, respiratory alkalosis. (e) pCO_2: acid. alkal. (f) In, in. (g) Yes: outside; partially. (h) No: fully. (i) Normal, un: kidneys. (j) HCO_3^-, metabolic; pCO_2 is decreased also related to hyperventilation. (k) Respiratory; 7.45. (1) No; no; fully. (m) Alkal; hypo. lower. (n) Patient W: respiratory acidosis. partially compensated; X: respiratory alkalosis uncompensated: Y: metabolic

acidosis, partially compensated; Z: metabolic alkalosis, fully compensated.

C12. Depressing; lethargy, stupor, coma; spasms, stimulates.

D1. (a) 70-90%; 55-60%. (b) Higher; lower. (c-e) Greater; less. (f) Higher; lower.

MORE CRITICAL THINKING: CHAPTER 27

1. Contrast the chemical composition of intracellular fluid with that of plasma and interstitial fluid.
2. Explain the importance of maintaining fluid balance among the three major fluid compartments of the body.
3. Explain why changes in fluids result in changes in concentration of electrolytes in interstitial and intracellular fluids. Be sure to state which two electrolytes normally exert the greatest effect on this balance.
4. Describe effects of ADH, ANP, and aldosterone on fluid and electrolyte balance.
5. Describe how you might be able to distinguish two persons with acid–base imbalances: one in acidosis and one in alkalosis.
6. Contrast causes of respiratory acidosis and metabolic acidosis. State several examples of causes.

MASTERY TEST: CHAPTER 27

Questions 1–13: Circle T (true) or F (false). If the statement is false, change the underlined word or phrase so that the statement is correct.

T F   1. Hyperventilation tends to raise pH.
T F   2. Chloride is the major intracellular anion.
T F   3. Under normal circumstances fluid intake each day is greater than fluid output.
T F   4. During edema there is increased movement of fluid out of plasma and into interstitial fluid.
T F   5. Parathyroid hormone causes an increase in the blood level of calcium.
T F   6. When aldosterone is in high concentrations, sodium is conserved (in blood) and potassium is excreted (in urine).
T F   7. Elderly persons are at lower risk compared to young adults for acidosis and sodium imbalances.
T F   8. About two-thirds of body fluids are found in intracellular fluid (ICF) and one-third of body fluids are found in extracellular fluid (ECF).
T F   9. The body of an infant contains a higher percentage of water than the body of an adult.
T F   10. Starch, glucose, HCO_3^- Na$^+$, and K$^+$ are all electrolytes.
T F   11. Protein is the most abundant buffer system in the body.
T F   12. Sodium plays a much greater role than magnesium in osmotic balance of ECF because more mEq/liter of sodium than magnesium are in ECF.
T F   13. The pH of arterial blood is higher than that of venous blood.

Questions 14–20: Circle the letter preceding the one best answer to each question.

14. Three days ago Ms. Weldon fractured several ribs in an auto accident, and her breathing has been shallow since then. Her arterial blood pH is 7.29, pCO_2 is 50 mm Hg, and HCO_3^- is 30 mEq/liter.

Her acid-base imbalance is most likely to be:
A. Respiratory acidosis
B. Respiratory alkalosis
C. Metabolic acidosis
D. Metabolic alkalosis

15. Ms. Weldon's acid–base imbalance (question 14) appears to be:
    A. Fully compensated
    B. Partially compensated
    C. Uncompensated

16. Mr. Wong is in cardiac care post-myocardial infarction. His arterial blood pH is 7.31, pCO_2 is 39 mm Hg, and HCO_3^- is 20 mEq/liter. His acid–base imbalance is most likely to be:
    A. Metabolic acidosis, fully compensated
    B. Metabolic acidosis, partially compensated
    C. Metabolic acidosis, uncompensated

17. On an average day the greatest volume of fluid output is via:
    A. Feces   C. Expiration
    B. Sweat   D. Urine

18. Which term refers to a lower than normal blood level of sodium?
    A. Hyperkalemia   D. Hyponatremia
    B. Hypokalemia   E. Hypercalcemia
    C. Hypernatremia

19. Which two electrolytes are in greatest concentration in ICF?
    A. Na^+ and Cl^- D. K^+ and HPO_4^{2-}
    B. Na^+ and K^+ E. Na^+ and HPO_4^{2-}
    C. K^+ and Cl^- F. Ca^{2+} and Mg^{2+}

20. All of the following factors will tend to increase fluid output except:
    A. Fever
    B. Vomiting and diarrhea
    C. Hyperventilation
    D. Increased glomerular filtration rate
    E. Decreased blood pressure

Questions 21–25: Fill-ins. Complete each sentence with the word or phrase that best fits.

_____ 21. List four factors that increase production of aldosterone.

_____ 22. _____ are chemicals that have at least one ionic bond and dissociate into positive and negative ions.

_____ 23. To compensate for acidosis, kidneys will increase secretion of

_____, and in alkalosis kidneys will eliminate more _____.

_____ 24. Normally blood serum contains about _____ mEq/liter of Na+

and _____ mEq/liter of K+.

_____ 25. The three main mechanisms used by the body to compensate

for acid–base imbalances are _____,

_____, and _____.

ANSWERS TO MASTERY TEST: CHAPTER 27

True–False	Multiple Choice	Fill-ins
1. T	14. A	21. Decreased blood volume or cardiac output, decrease of Na^+ or increase of K in ECF, or physical stress.
2. F. Extracellular anion	15. B	
3. F. Equals	16. C	
4. T	17. D	
5. T	18. D	22. Electrolytes
6. T	19. D	23. H^+, HCO_3^-.
7. F. Higher	20. E	24. 136 to 148; 3.5 to 5.0
8. T		25. Buffers, respirations, and renal output of H^+ or HCO_3^-.
9. T		
10. F. Not all. (Starch and glucose are nonelectrolytes.)		
11. T		
12. T		
13. T		

FRAMEWORK 28
The Reproductive Systems

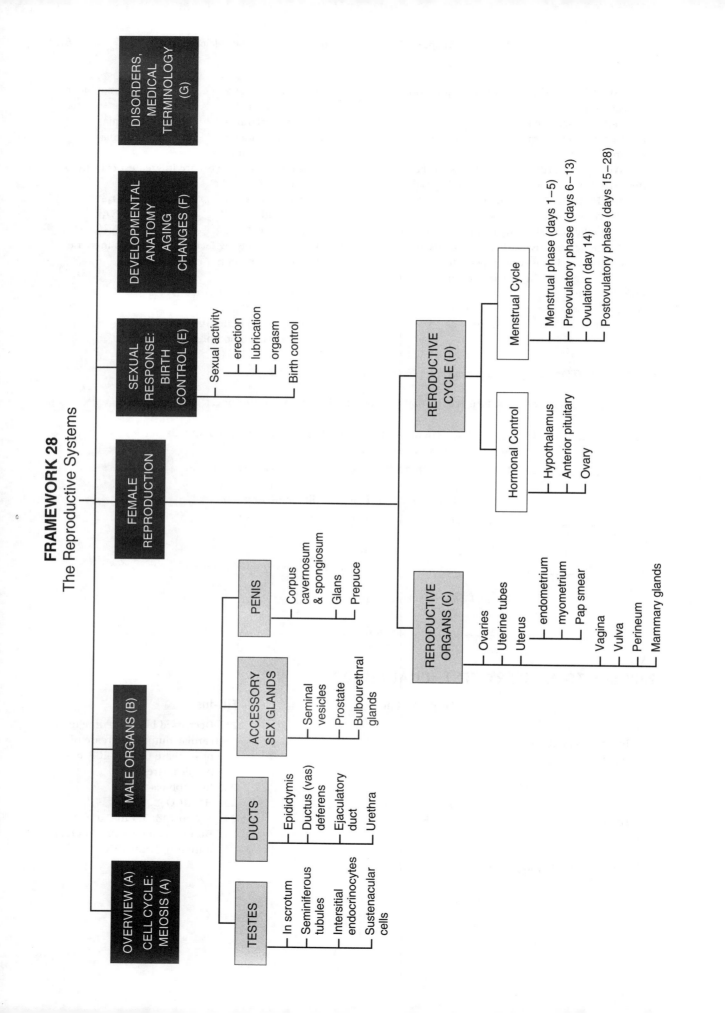

C H A P T E R **28**

The Reproductive Systems

Unit V introduces the systems in females and males that provide for continuity of the human species, not simply homeostasis of individuals. The reproductive systems produce sperm and ova that may unite and develop to form offspring. In order for sperm and ova to reach a site for potential union, they must each pass through duct systems and be nourished and protected by secretions. External genitalia include organs responsive to stimuli and capable of sexual activity. Hormones play significant roles in the development and maintenance of the reproductive systems. The variations of hormones in ovarian and uterine (menstrual) cycles are considered in detail, along with methods of birth control. As we have done for all other systems, we discuss development of the reproductive system and normal aging changes.

As you begin your study of reproduction, carefully examine the Chapter 28 Topic Outline and Objectives. Check off each objective as you complete it. The Chapter 28 Framework provides an organizational preview of the reproductive systems. Be sure to refer to the Framework frequently and note relationships among key terms in each section.

TOPIC OUTLINE AND OBJECTIVES

A. Overview; the cell cycle in gonads

1. Describe the process of meiosis.

B. The male reproductive system

2. Describe the structure and functions of the organs of the male reproductive system.

C. The female reproductive system

3. Describe the structure and functions of the organs of the female reproductive system.

D. The female reproductive cycle

4. Compare the principal events of the ovarian and uterine cycles.

E. The human sexual response; birth control

5. Describe the similarities and differences in the sexual responses of males and females.
6. Compare the various kinds of birth control methods and their effectiveness.

F. Developmental anatomy and aging of the reproductive systems

7. Describe the development of the male and female reproductive systems.
8. Describe the effects of aging on the reproductive systems.

G. Disorders, medical terminology

WORDBYTES

Now become familiar with the language of this chapter by studying each wordbyte, its meaning, and an example of its use within a term. After you study the entire list, self-check your understanding by writing the meaning of each wordbyte on the line. As you continue through the *Learning Guide*, identify (and fill in) additional terms that contain the same wordbyte.

Wordbyte	Self-check	Meaning	Example(s)
a-	_____	without	*a*menorrhea
andro-	_____	man	*andro*gen
-arche	_____	beginning	men*arche*
cervix-	_____	neck	*cervix* of uterus
corp-	_____	body	*corp*us cavernosum
crypt-	_____	hidden	*crypt*orchidism
dys-	_____	apart	*dys*menorrhea
-genesis	_____	formation	spermato*genesis*
gyn-	_____	woman	*gyn*ecologist
hyster(o)-	_____	uterus	*hyster*ectomy
labia	_____	lips	*labia* majora
mamm-	_____	breast	*mamm*ogram
mast-	_____	breast	*mast*ectomy
meio-	_____	smaller	*meio*sis
men-	_____	month	*men*opause, *men*ses
-metrium	_____	uterus	endo*metrium*
mito	_____	thread	*mito*sis
oo-, ov-	_____	egg	*oo*genesis, *ov*um
orchi-	_____	testis	*orchi*dectomy
rete	_____	network	*rete* testis
-rrhea-	_____	a flow	dysmeno*rrhea*
salping-	_____	tube, trumpet	*salping*itis
troph-	_____	nutrition	*troph*oblast

CHECKPOINT

A. Overview; the cell cycle in gonads (pages 974–976)

A1. Define each of these groups of reproductive organs.

a. Gonad

b. Ducts

c. Accessory glands

d. Supporting structures

✓ **A2.** As you go through the chapter, note which structures of the reproductive systems are included in each of the above categories:

a. Female:

b. Male:

✓ **A3.** Do the following exercise about reproductive cell division.

a. Meiosis is a special type of nuclear division that occurs only in _____ and

is one process in the production of sex cells or _____. Fusion of gametes

during the process of _____ results in formation of a _____.

b. Cells that begin the process of meiosis are ovarian or testicular cells that have been

undergoing mitosis up to this point. Each cell initially contains _____ chromosomes, which is the *(diploid or 2n? haploid or n?)* number for humans. These

chromosomes consist of _____ homologous pairs.

c. One pair is known as the *(autosomes? sex chromosomes?)*. In females, this pair consists of *(two X chromosomes? one X and one Y chromosome?)*. The remaining 22 pairs are

known as _____.

d. By the time that cells complete meiosis, they will contain *(46? 23?)* chromosomes, the *(2n? 23?)* chromosomes, the *(2n? n?)* number.

e. Meiosis differs from mitosis in that meiosis involves *(one? two? three?)* divisions with replication of DNA only *(once? twice?)*. The first division is known as the

_____ division, whereas the second division is the

_____ division.

f. Replication of DNA occurs *(before? after?)* reduction division, resulting in doubling of

chromosome number in humans from 46 chromosomes to _____ "chromatids." (Chromatid is a name given to a chromosome while the chromosome is attached to its replicated counterpart, that is, in the doubled state.) Doubled or paired chromatids are held together by means of a *(centromere? centrosome?)*.

g. A unique event that occurs in Prophase I is the gathering of chromosomes in homo-

logous pairs, a process called _____. Note that this process *(does? does not?)* occur during mitosis. The four chromatids of the homologous pair are known as

a _____. The proximity of chromatids permits the exchange or

_____ of DNA between maternal and paternal chromatids. What is the advantage of crossing over to sexual reproduction?

h. Another factor that increases possibilities for variety among sperm or eggs you can pro-duce is the random (or chance) assortment of chromosomes as they line up on either side of the equator in Metaphase *(I? II?)*.

i. Division I reduces the chromosome number to _____ chromatids. Division II

results in the separation of doubled chromatids by splitting of _____

so that resulting cells contain only _____ chromosomes.

j. Meiosis is only one part of the process of forming sperm or eggs. Each original diploid cell in a testis will lead to production of *(1? 2? 4?)* mature sperm, while each diploid ovary cell produces *(1? 2? 4?)* mature eggs (ova) plus three small cells known as

_____.

B. The male reproductive system (pages 976–989)

✓▲ **B1.** Describe the scrotum and testes in this exercise.

a. Arrange coverings around testes from most superficial to deepest:

_____ _____ _____

D.	**Dartos: smooth muscle that wrinkles skin**
TA.	**Tunica albuginea: dense fibrous tissue**
TV.	**Tunica vaginalis: derived from peritoneum**

b. The temperature in the scrotum is about 3°C *(higher? lower?)* than the temperature in the abdominal cavity. Of what significance is this?

c. During fetal life, testes develop in the *(scrotum? posterior of the abdomen?)*.

Undescended testes is a condition known as _____. If not corrected,

this condition places the male at higher risk for _____.

d. Skeletal muscles that elevate testes closer to the warmth of the pelvic cavity are known as *(cremaster? dartos?)* muscles.

✓ **B2.** Discuss spermatogenesis in this exercise.
a. Spermatozoa are one type of gamete; name the other type: _____

Gametes are *(haploid? diploid?)*. Human gametes contain _____ chromosomes.

b. Fusion of gametes produces a cell called a _____. This cell, and all cells of the organism derived from it, contain the *(haploid? diploid?)* number of chromosomes, written as *(n? 2n?)*. These chromosomes, received from both sperm and ovum, are said to exist in _____ pairs.

c. Gametogenesis assures production of haploid gametes from otherwise diploid

humans. In males, this process, called _____-genesis, occurs in the

_____.

✓ **B3.** Refer to Figure LG 28.1 and do the following Checkpoint. Use the following list of cells for answers.

Early spermatid	**Secondary spermatocyte**
Interstitial endocrinocyte (cell of Leydig)	**Spermatogonium**
Late spermatid	**Spermatozoon (sperm)**
Primary spermatocyte	**Sustentacular (Sertoli) cell**

a. Label cells A and B and 1–6.
b. Mark cells 1–6 according to whether they are diploid (2n) or haploid (n).

c. Meiosis is occurring in cells numbered _____ and _____. Color these meiotic cells according to color ovals on the figure. Which cell undergoes reduction division

(meiosis I)? _____ Which undergoes equatorial division (meiosis II)? _____

d. From this figure, you can conclude that as cells develop and mature, they move *(toward? away from?)* the lumen of the tubule.
e. Which cells are in the process of spermiogenesis? _____ and _____

f. Which cells secrete testosterone? _____
g. Which cells produce nutrients for developing sperm, phagocytose excess spermatid

cytoplasm, form the blood-testis barrier, and produce the hormone inhibin? _____

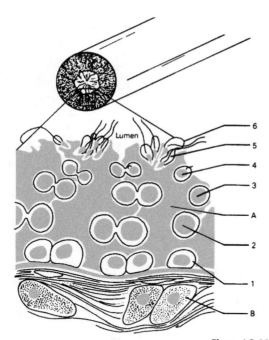

Figure LG 28.1 Diagram of a portion of a cross section of a seminiferous tubule showing the stages of spermatogenesis. Color and label as directed in Checkpoint B3.

h. Which cells lie dormant near the basement membrane until they divide mitotically beginning at puberty? They also serve as a reservoir of stem cells available for

spermatogenesis throughout life. _____

B4. Explain how cytoplasmic bridges between cells formed during spermatogenesis may facilitate survival of Y-bearing sperm and therefore permit generation of male offspring.

✓ **B5.** Describe spermatozoa (sperm) in this Checkpoint.

a. Number produced in a normal male: _____/day

b. Life expectancy of sperm once inside female reproductive tract: _____ hours

c. Portion of sperm containing nucleus: _____

d. Part of sperm containing mitochondria: _____
e. Function of acrosome:

f. Function of tail: _____

✓ **B6.** Match the hormones listed in the box with descriptions below. Answers may be used more than once.

DHT	**FSH**	**GnRH**
Inhibin	**LH**	**Testosterone**

a. Released by the hypothalamus: stimulates release of two anterior pituitary hormones:

_____

b. Inhibits release of GnRF: _____
c. Co-stimulate sustentacular cells to secrete ABP, which keeps testosterone near cells

undergoing spermatogenesis (two hormones):

_____

d. Stimulates release of testosterone:

_____

e. Principal male hormone and precursor of dihydrotestosterone (DHT):

_____

f. Testicular hormone that inhibits FSH pro-

duction: _____

g. Before birth, stimulates development of

external genitals: _____

h. At puberty, facilitate development of male sexual organs and secondary sex

characteristics; contribute to libido (two hor-

mones): _____

i. Anabolic hormones, because they stimulate protein synthesis (two hormones):

_____

B7. List five functions of androgens.

✓▲ **B8.** Refer to Figure LG 28.2. Structures in the male reproductive system are numbered in order along the pathway that sperm take from their site of origin to the point where they exit from the body. Match the structures in the figure (using numbers 1 to 9) with descriptions below.

_____ a. Ejaculatory duct

_____ b. Epididymis

_____ c. Prostatic urethra

_____ d. Spongy (cavernous) urethra

_____ e. Membranous urethra

_____ f. Ductus (vas) deferens (portion within the abdomen)

_____ g. Ductus (vas) deferens (portion within the scrotum)

_____ h. Urethral orifice

_____ i. Testis

✓ **B9.** Which structures numbered 1–9 in Figure LG 28.2 are paired? Circle their numbers:

1   2   3   4   5   6   7   8   9

Which structures are single? Write their numbers: _____

✓ **B10.** Answer these questions about the ductus (vas) deferens.

a. The ductus deferens is about _____ cm (_____ inches) long.

Figure LG 28.2 Male organs of reproduction seen in sagittal section. Structures are numbered in order along the pathway taken by sperm. Color and label as directed in Checkpoints B8, B9, B13, and B17.

KEY
O   Accessory sex glands
O   Ducts (pathway of sperm)
O   Gonad

b. It is located *(entirely? partially?)* in the abdomen. It enters the abdomen via the

    _____. Protrusion of abdominal contents through this weakened

    area is known as _____.

c. What structures besides the vas compose the seminal cord?

d. A vasectomy is performed at a point along the section of the vas that is within the

    _____. Must the peritoneal cavity be entered during this procedure?
    (Yes? No?) Will the procedure affect testosterone level and sexual desire of the male?
    (Yes? Not normally?) Why?

✓ **B11.** Which portion of the male urethra is longest? *(Prostatic? Membranous? Spongy or cavernous?)*

✓▲**B12.** Match the accessory sex glands with their descriptions.

B. Bulbourethral glands	P. Prostate gland	S. Seminal vesicles

_____ a. Paired pouches posterior to the urinary bladder

_____ b. Structures that empty secretions (along with contents of vas deferens) into ejaculatory duct

_____ c. Single doughnut-shaped gland located inferior to the bladder; surrounds and empties secretions into urethra

_____ d. Contribute about 60% of seminal fluid

_____ e. Pair of pea-sized glands located in the urogenital diaphragm

_____ f. Forms 25% of semen

_____ g. Produces PSA

✓▲ **B13.** Color parts of the male reproductive system on Figure LG 28.2 according to color code ovals. Label the *seminal vesicle, prostate gland,* and *bulbourethral gland* on the figure.

✓ **B14.** *A clinical correlation.* Explain the clinical significance of the shape and location of the prostate gland.

✓ **B15.** Complete the following exercise about semen.

a. The average amount per ejaculation is _____ mL.

b. The average range of number of sperm is _____/mL. When the count

falls below _____/mL, the male is likely to be sterile. Only one sperm fertilizes the ovum. Explain why such a high sperm count is required for fertility.

c. The pH of semen is slightly *(acid? alkaline?)*. State the advantage of this fact.

d. What is the function of *seminalplasmin?*

✓▲ **B16.** Refer to Figure 28.12 in the text and do this exercise.

a. The _____, or foreskin, is a covering over the _____

   of the penis. Surgical removal of the foreskin is known as _____.

b. The names *corpus* _____ and *corpus* _____ indicate
   that these bodies of tissue in the penis contain spaces that distend in the presence of

   excess _____. As a result, blood is prohibited from leaving the penis
   because *(arteries? veins?)* are compressed. This temporary state is known as the *(erect?
   flaccid?)* state.

c. The urethra passes through the corpus _____. The urethra functions

   in the transport of _____ and _____. During ejacula-
   tion, what prevents sperm from entering the urinary bladder and urine from entering
   the urethra?

▲ **B17.** Label these parts of the penis on Figure LG 28.2: *prepuce, glans, corpus cavernosum, corpus spongiosum,* and *bulb of the penis.*

C. The female reproductive system (pages 989–1001)

✓▲ **C1.** Refer to Figure LG 28.3 and match the structures numbered 1–5 with their
descriptions. Note that structures are numbered in order along the pathway taken by
ova from the site of formation to the point of exit from the body.

_____ a. Uterus (body)   _____ d. Uterine (Fallopian) tube

_____ b. Uterus (cervix)   _____ e. Vagina

_____ c. Ovary

✓▲ **C2.** Indicate which of the five structures in Figure LG 28.3 are paired. Circle the numbers.

   1   2   3   4   5

   Which structures are singular? Write their numbers: _____

✓ **C3.** Describe the ovaries in this Checkpoint.
a. The two ovaries most resemble:
   A. Almonds (unshelled)
   B. Peas
   C. Pears
   D. Doughnuts
b. Position of the ovaries is maintained by *(ligaments? muscles?)*. Name two or more of
   these structures.

c. Arrange in sequence according to chronological appearance, from first to develop to

   last. _____ _____ _____   CL. Corpus luteum
                                         OF. Ovarian follicle
                                         VOF. Vesicular ovarian (Graafian) follicle

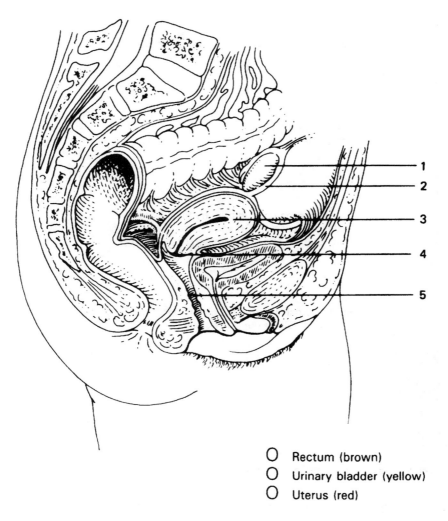

○ Rectum (brown)
○ Urinary bladder (yellow)
○ Uterus (red)

Figure LG 28.3 Female organs of reproduction. Structures are numbered in order in the pathway taken by ova. Color and label as directed in Checkpoints C1, C2, C7, C9, and C11.

✓ **C4.** Color the four cells on Figure LG 28.4 according to the color code ovals.

✓ **C5.** Describe exactly what is expelled during ovulation. Place an asterisk ([*]) next to that cell on Figure LG 28.4.

What determines whether the second meiotic (equatorial) division will occur?

✓▲ **C6.** Complete this exercise about the uterine tubes.

a. Uterine tubes are also known as _____ tubes. They are about

_____ cm (_____ inches) long.

b. Arrange these parts of the uterine tubes in order from most medial to most lateral:

_____ _____ _____

A. Ampulla   B. Infundibulum and fimbriae   C. Isthmus

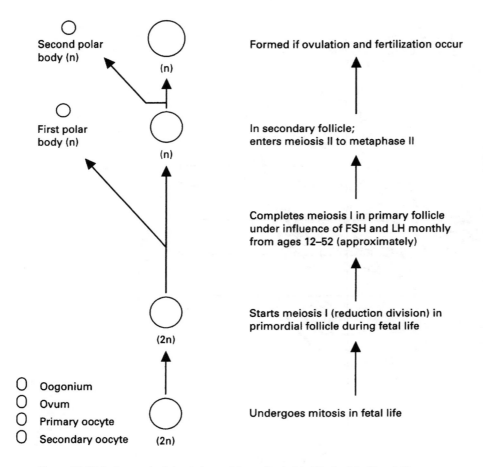

Figure LG 28.4 Oogenesis. Color and complete as directed in Checkpoints C4 and C5.

c. What structural features of the uterine tube help to draw the ovum into the tube and propel it toward the uterus?

d. List two functions of uterine tubes.

e. Define *ectopic pregnancy* and list possible causes of this condition.

▲ **C7.** Color the pelvic organs on Figure LG 28.3. Use the color code ovals on the figure.

✓▲ **C8.** Answer these questions about uterine position.
  a. The organ that lies anterior and inferior to the uterus is the _____.

   The _____ lies posterior to it.
  b. The rectouterine pouch lies *(anterior? posterior?)* to the uterus, whereas the

   _____-uterine pouch lies between the urinary bladder and the uterus.
  c. The fundus of the uterus is normally tipped *(anteriorly? posteriorly?)*. If it is malpositioned posteriorly, it would be *(anteflexed? retroflexed?)*.

  d. The _____ ligaments attach the uterus to the sacrum. The

   _____ ligaments pass through the inguinal canal and anchor into

   external genitalia (labia majora). The _____ ligaments are broad, thin sheets of peritoneum extending laterally from the uterus.
  e. Important ligaments in prevention of drooping (prolapse) of the uterus are the

   _____ ligaments, which attach the base of the uterus laterally to the pelvic wall.

✓▲ **C9.** Refer to Figure LG 28.3 and do this exercise about layers of the wall of the uterus.

  a. Most of the uterus consists of _____-metrium. This layer is *(smooth muscle? epithelium?)*. It should appear red in the figure (as directed above).
  b. Now color the peritoneal covering over the uterus green. This is a *(mucous? serous?)*

   membrane known as the _____.

  c. The innermost layer of the uterus is the _____-metrium. Which portion of it is shed during menstruation? Stratum *(basalis? functionalis?)* Which arteries supply the stratum functionalis? *(Spiral? Straight?)*

  C10. Surgical removal of the uterus is known as a _____. Contrast the radical and subtotal forms of this surgery.

✓▲ **C11.** Refer to Figure LG 28.3. In the normal position, at what angle does the uterus join

   the vagina? _____

  C12. Describe these vaginal structures.
  a. Fornix

b. Rugae

c. Hymen

✓ **C13.** The pH of vaginal mucosa is *(acid? alkaline?)*. What is the clinical significance of this fact?

✓▲ **C14.** Refer to Figure LG 28.5 and do the following exercise.
a. Label these perineal structures: *anus, clitoris, labia major, labia minor, mons pubis, urethra,* and *vagina.*
b. Draw the borders of the perineum, using landmarks labeled on the diagram. Draw a line between the right and left ischial tuberosities. Label the triangles that are formed.

    In which triangle are all of the female external genitalia located? _____
c. Draw a dotted line where an *episiotomy* incision would be performed for childbirth.

✓▲ **C15.** Describe mammary glands in this Checkpoint.

    Each breast consists of 15–20 _____ composed of lobules. Milk-

    secreting glands, known as _____, empty into a duct system that terminates at the *(areola nipple?).*

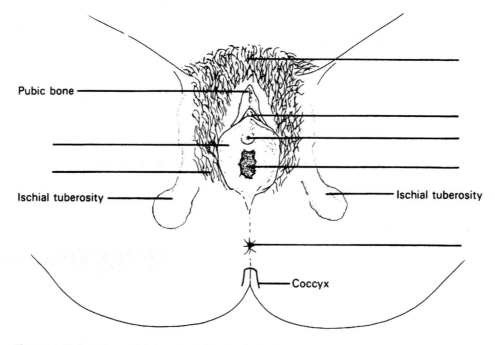

Figure LG 28.5 Perineum. Label as directed in Checkpoint C14.

D. The female reproductive cycle (pages 1001–1007)

D1. Contrast *menstrual cycle* and *ovarian cycle*.

✓ **D2.** Select hormones produced by each of the endocrine glands listed below.

E.	Estrogen	LH.	Luteinizing hormone
FSH.	Follicle-stimulating hormone	P.	Progesterone
GnRH.	Gonadotropin-releasing hormone	R.	Relaxin
I.	Inhibin		

Hypothalamus: _____   Anterior pituitary: _____

Ovary: _____

✓ **D3.** Now select the hormones listed in C2 that fit descriptions below.

_____ a. Stimulates release of FSH

_____ b. Stimulates release of LH

_____ c. Inhibits release of FSH

_____ d. Inhibits release of LH

_____ e. Stimulates development and maintenance of uterus, breasts, and secondary sex characteristics; enhances protein anabolism

_____ f. Prepares endometrium for implantation by fertilized ovum

_____ g. Dilates cervix for labor and delivery

_____ h. β-Estradiol, estrone, and estriol are forms of this hormone

_____ i. Lowers blood cholesterol level

✓ **D4.** Refer to Figure LG 28.6, and check your understanding of events in female cycles in this checkpoint.

a. The average duration of the menstrual cycle is _____ days. The menstrual phase usually lasts about _____ days. During this period about _____ mL of blood, cells, mucus, and other fluid exits from the uterus in menstrual flow. On Figure LG 28.6d, color red the endometrium in the menstrual phase, and note the changes in its thickness. Also write *menses* on the line in that phase on Figure LG 28.6d.

b. Following the menstrual phase is the _____ phase. During both of these phases _____ develop in the ovary. Usually only _____ matures each month, becoming the dominant follicle, and later the _____ (Graafian) follicle. The mature follicle contains a developing ovum that has reached metaphase of meiosis *(I? II?)* of oogenesis.

c. Follicle development occurs under the influence of the hypothalamic hormone _____, which regulates the anterior pituitary hormones _____ and later _____.

d. What are the two main functions of follicles?

Figure LG 28.6 Correlations of female cycles. (a) Anterior pituitary hormones. (b) Ovarian changes: follicles and corpus luteum. (c) Ovarian hormones. (d) Uterine (endometrial) changes. Color and label as directed in Checkpoint D4.

e. How do estrogens affect the endometrium?

For this reason, the preovulatory phase is also known as the _____ phase. Because follicles reach their peaks during this phase, it is also known as the

_____ phase. Write those labels on Figures LG 28.6d and 28.6b.

f. A moderate increase in estrogen production by the growing follicles initiates a *(positive? negative?)* feedback mechanism that *(increases? decreases?)* FSH level. It is this change in FSH level that causes atresia of the remaining partially developed follicles. Atresia means *(degeneration? regeneration?)*.

g. The principle of this negative feedback effect is utilized in oral contraceptives (See text page 1008). By starting to take "pills" (consisting of target hormones estrogen and progesterone) early in the cycle (day 5), and continuing until day 25, a woman

will maintain a very low level of the tropic hormones _____. Without these hormones, follicles and ova will not develop, and so the woman will not

_____.

h. The second anterior pituitary hormone released during the cycle is _____. The surge of this hormone results from the *(high? low?)* level of estrogens present just at the end of the preovulatory phase. This is an example of one of the very few *(positive? negative?)* feedback mechanisms in the human body.

i. The LH surge causes the release of the ovum (actually secondary oocyte), the event

known as _____. The LH surge is the basis for a home test for *(ovulation? pregnancy?)*. Color all stages of follicle development through ovulation on Figure LG 28.6b according to the color code ovals.

j. Following ovulation is the _____ phase. It lasts until day _____

of a 28-day cycle. Under the influence of tropic hormone _____, follicle cells are

changed into the corpus _____. These cells secrete two hormones:

_____ and _____. Both prepare the endometrium

for _____. What preparatory changes occur?

k. What effect do rising levels of target hormones estrogens and progesterone have upon the GnRH and LH?

l. One function of LH is to form and maintain the corpus luteum. So as LH _____-creases, the corpus luteum disintegrates, forming a white scar on the ovary known as

the corpus _____. Color these two final stages of the original follicle on Figure LG 28.6b. Also write alternative names for the postovulatory phase on lines in Figure LG 28.6b and d.

m. The corpus luteum had been secreting _____ and _____. With the demise of the corpus luteum, the levels of these hormones rapidly *(increase? decrease?)*. Because these hormones were maintaining the endometrium, endometrial tissue now deteriorates and will be shed during the next

_____ phase.

n. If fertilization should occur, estrogens and progesterone are needed to maintain the

endometrial lining. The _____ around the developing embryo

secretes a hormone named _____. It

functions much like LH in that it maintains the corpus _____, even though LH has been inhibited (step *k* above). The corpus luteum continues to secrete estrogens and progesterone for several months until the placenta itself can secrete sufficient amounts. Incidentally, hCG is present in the urine of pregnant women only

and so is routinely used to detect _____.

  o. Review hormonal changes by coloring blood levels of hormones in Figure LG 28.6a and c, according to the color code ovals. Then note uterine changes by coloring red the remainder of the endometrium in Figure LG 28.6d.

✓ **D5.** To check your understanding of the events of these cycles, list letters of the major events in chronological order beginning on day 1 of the cycle. To do this exercise it may help to list on a separate piece of paper days 1 to 28 and write each event next to the approximate day on which it occurs.

_____ _____ _____ _____ _____ _____ _____ _____ _____ _____ _____

  A. Ovulation occurs
  B. Estrogen and progesterone levels drop
  C. FSH begins to stimulate follicles
  D. LH surge is stimulated by high estrogen levels
  E. High levels of estrogens and progesterone inhibit LH
  F. Estrogens are secreted for the first time during the cycle
  G. Corpus luteum degenerates and becomes corpus albicans
  H. Endometrium begins to deteriorate, leading to next month's menses
  I. Follicle is converted into corpus hemorrhagicum and then into corpus luteum
  J. Rising level of estrogen inhibits FSH secretion
  K. Corpus luteum secretes estrogens and progesterone

✓ **D6.** Write terms that mean:
  a. Painful menstruation: _____ .

  b. Absence of monthly menstrual flow: _____ .

E. The human sexual response (pages 1007–1010)

  E1. Define *sexual intercourse* or *coitus*.

✓ **E2.** Answer these questions about the process.
  a. *(Sympathetic? Parasympathetic?)* impulses from the *(brainstem? sacral cord?)* cause dilation of arteries of the penis or the vagina, leading to the *(flaccid? erect?)* state.
  b. *(Sympathetic? Parasympathetic?)* impulses cause sperm to be propelled into the urethra of the male. Expulsion from the urethra to the exterior of the body is called

_____ .

E3. Compare and contrast the phases of sexual activity in females and males by completing this table.

Phase	Female	Male
a. Erection		
b. Lubrication		
c. Orgasm		

✓ **E4.** Describe the significance of *transudation* in female sexual response.

E5. Discuss erectile *dysfunction* in this Checkpoint.

a. Erectile dysfunction (also known as _____). List several causes of this condition.

b. Nitric oxide causes *(contraction? relaxation?)* of smooth muscles in arteries of the penis.

The result of release of nitric oxide is _____-crease of blood flow into the penis, leading to the *(erect? flaccid state?)*.

c. Many cases of erectile dysfunction are associated with nitric oxide insufficiency. Name a medication used to treat this disorder by enhancing effects of nitric oxide.

_____

✓ **E6.** Match the methods of birth control listed in the box with descriptions below.

A.	**Abstinence**		**OC.**	**Oral contraceptive**
Ch.	**Chemical methods**		**R.**	**Rhythm**
CI.	**Coitus interruptus**		**RU 486.**	**RU 486**
Con.	**Condom**		**T.**	**Tubal ligation**
D.	**Diaphragm**		**Vag.**	**Vaginal ring**
IUD.	**Intrauterine device**		**Vas.**	**Vasectomy**
N.	**Norplant**			

_____ a. Spermaticidal foams, jellies, and creams

_____ b. "The pill," a combination of progesterone and estrogens, causing decrease in GnRH, FSH, and LH so ovulation does not occur

_____ c. Removal of a portion of the ductus (vas) deferens

_____ d. Tying off of the uterine tubes

_____ e. A natural method of birth control involving abstinence during the period when ovulation is most likely

_____ f. A mechanical method of birth control in which a dome-shaped structure is placed over the cervix; accompanied by use of a jelly or cream

_____ g. Copper T 380 A placed in the uterus by a physician; prevents implantation

_____ h. A rubber sheath over the penis

_____ i. Withdrawal

_____ j. Progestin-containing capsules placed under the skin for up to 5 years

_____ k. Competitive inhibitor of progesterone; a chemical that induces miscarriage

_____ l. The 100% reliable method of avoiding pregnancy by avoidance of sexual intercourse

_____ m. Worn inside the vagina for three weeks each month, where it releases progestin and possibly estrogen

E7. Review the mechanism of action or oral contraceptives (Checkpoint D4g). Then contrast benefits and risks of use of "the pill."

F. Developmental anatomy and aging of the reproductive systems (pages 1010–1013)

✓ **F1.** Name the reproductive structures (if any) that develop from the following embryonic ducts.

a. Mesonephric duct in females: _____

b. Mesonephric duct in males: _____

c. Paramesonephric (Müller's) duct in females: _____

d. Paramesonephric (Müller's) duct in males: _____

✓ **F2.** Write *E* (endoderm) or *M* (mesoderm) next to each structure below to indicate its embryonic origin.

_____ a. Ovaries and testes

_____ b. Prostrate and bulbourethral glands

_____ c. Greater (Bartholin's) and lesser vestibular glands

✓ **F3.** Identify male homologues for each of the following.

a. Labia majora: _____

b. Clitoris: _____

c. Lesser vestibular (Skene's) glands: _____

d. Greater vestibular (Bartholin's) glands: _____

F4. Contrast these terms: *menarche/menopause*

F5. Describe normal aging changes that involve the reproductive system in:

a. Females

b. Males

G. Disorders, medical terminology (pages 1014–1016)

G1. Define *sexually transmitted disease* (STD).

G2. Describe the main characteristics of these sexually transmitted diseases by completing the table.

Disease	Causative Agent	Main Symptoms
a. Gonorrhea		
b.	*Treponema pallidum*	
c.	Type II herpes simplex virus	
d. Chlamydia		

✓ **G3.** Select disorders in the box that fit descriptions below. One answer will be used twice.

A. **Amenorrhea**	E. **Endometriosis**
Can. **Candidiasis**	G. **Gonorrhea**
Chl. **Chlamydia**	H. **Hypospadias**
CC. **Cervical cancer**	S. **Syphilis**
D. **Dysmenorrhea**	

_____ a. Painful menstruation, partly due to contractions of uterine muscle

_____ b. Possible cause of infection and possible blindness in newborns if bacteria transmitted to eyes during birth

_____ c. Spreading of uterine lining into abdominopelvic cavity via uterine tubes

_____ d. Caused by bacterium *Treponema pallidum*; may involve many systems in tertiary stage

_____ e. Absence of menstrual periods

_____ f. May be detected by a Pap smear; is linked to genital warts (human papilloma virus = HPV)

_____ g. The most prevalent STD in the U.S., leading to 20,000 cases of sterility each year

_____ h. Leading cause of PID (pelvic inflammatory disease)

_____ i. Displaced urethral opening in males

_____ j. Caused by yeast; most common form of vaginal infection

✓ **G4.** Identify two tests that screen for prostate cancer.

G5. List six symptoms of *premenstrual syndrome (PMS)*.

_____   _____

_____   _____

_____   _____

✓ **G6.** *A clinical correlation.* Mrs. Rodriguez, a 55-year-old woman, tells the nurse that she is concerned about having breast cancer. Do this exercise about the nurse's conversation with Mrs. Rodriguez.

   a. The nurse asks some questions to evaluate Mrs. Rodriguez's risk of breast cancer. Which of the following responses indicate an increased risk of breast cancer?
   A. "My older sister had cancer when she was 40 and she had the breast removed."
   B. "I've never had any kind of cancer myself."
   C. "I have two daughters; they were born when I was 36 and 38."
   D. "I don't smoke and I never did."
   E. "I do need to cut back on fatty foods; I know I eat too many of them."

   b. The nurse tells the client that the American Cancer Society recommends three methods of early detection of breast cancer. These are listed below. On the line next to each of these, write the frequency that the nurse is likely to suggest to Mrs. Rodriguez.

   1. Self breast exam (SBE) _____

   2. Mammogram _____

   3. Breast exam by physician _____

✓ **G7.** Oophorectomy is removal of a(n) _____, whereas

_____ refers to removal of a uterine (Fallopian) tube.

ANSWERS TO SELECTED CHECKPOINTS: CHAPTER 28

A2. Female: (a) Ovaries. (b) Uterine (Fallopian) tubes, uterus, and vagina. (c) Greater vestibular (Bartholin's) glands. (c) External genitalia, breasts. Male: (a) Testes. (b) Seminiferous tubules, epididymis, vas deferens, ejaculatory duct, urethra. (c) Prostate, seminal vesicles, bulbourethral glands. (d) Scrotum, penis.

A3. (a) Gonads (ovaries or testes), gametes; fertilization, zygote. (b) 46, diploid or 2n; 23. (c) Sex chromosomes; two X chromosomes; autosomes. (d) 23, n. (e) Two, once; reduction, equatorial. (f) Before, "46 × 2" (or 46 doubled); centromere. (g) Synapsis; does not; tetrad; crossing-over; permits exchange of DNA between chromatids and greatly increases possibilities for variety among sperm or eggs and thus offspring. (h) 1. (i) "23 × 2" (or 23 doubled chromatids); centromeres, 23. (j) 4, 1, polar bodies.

B1. (a) D TV TA. (b) Lower; lower temperature is required for normal sperm production. (c) Posterior of the abdomen; cryptorchidism; sterility and testicular cancer. (d) Cremaster.

B2. (a) Ova; haploid; 23. (b) Zygote; diploid, 2n; homologous. (c) Spermato, seminiferous tubules of testes.

B3. (a-b) See Figure LG 28.lA. (c)2, 3; 2; 3. (d) Toward. (e) 4, 5. (f) B. (g) A. (h) 1.

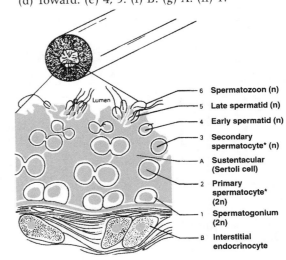

Figure LG 28.1A Diagram of a portion of a cross section of a seminiferous tubule showing the stages of spermatogenesis.

B5. (a) 300 million. (b) 48. (c) Head. (d) Midpiece. (e) A specialized lysosome that releases enzymes that help the sperm to penetrate the secondary oocyte. (f) Propels sperm.

B6. (a) GnRF. (b) Testosterone. (c) FSH and testosterone. (d) LH. (e) Testosterone. (f) Inhibin. (g) DHT. (h-i) Testosterone and DHT.

B8. (a) 5. (b) 2. (c) 6. (d) 8. (e) 7. (f) 4. (g) 3.(h) 9. (i) 1.

B9. Paired: 1–5; singular: 6–9.

B10. (a) 45, (18). (b) Partially; inguinal canal and rings; inguinal hernia. (c) Testicular artery, veins, lymphatics, nerves, and cremaster muscle. (d) Scrotum; no: not normally; hormones exit through testicular veins, which are not cut during vasectomy.

B11. Spongy or cavernous.

B12. (a) S. (b) S. (c) P. (d) S. (e) B. (f) P. (g) P.

B13. Accessory sex glands: A (seminal vesicles), B (prostate). C (bulbourethral glands); ducts, 2–9; gonad, 1.

B14. Because it surrounds the urethra (like a donut), an enlarged prostate can cause painful and difficult urination (dysuria) and bladder infection (cystitis) due to incomplete voiding.

B15. 2.5–5 ml. (b) 50–150 million; 20 million; a high sperm count increases the likelihood of fertilization, because only a small percentage of sperm ever reaches the secondary oocyte; in addition, a large number of sperm are required to release sufficient enzyme to digest the barrier around the secondary oocyte. (c) Alkaline; protects sperm against acidity of the male urethra and the female vagina. (d) Serves as an antibiotic that destroys bacteria.

B16. (a) Prepuce, glans; circumcision. (b) Cavernosum, spongiosum, blood; veins; erect. (c) Spongiosum; urine, sperm; sphincter action at base of bladder.

C1. (a) 3. (b) 4. (c) 1. (d) 2. (e) 5.

C2. 1, 2: paired; 3–5, singular.

C3. (a) A. (b) Ligaments; mesovarian, ovarian, and suspensory ligament. (c) OF VOF CL.

C4.

Figure LG 28.4A
Oogenesis.

C5. The secondary oocyte (see Figure LG 28.4A) along with the first polar body and some surrounding supporting cells are expelled. Meiosis II is completed only if the secondary oocyte is fertilized by a spermatazoon.

C6. (a) Fallopian; 10 (4). (b) C A B. (c) Currents produced by movements of fimbriae, ciliary action of columnar epithelial cells lining tubes, and peristalsis of the muscularis layer of the uterine tube wall. (d) Passageway for secondary oocyte, and sperm toward each other and site of fertilization. (e) Pregnancy following implantation at any location other than in the body of the uterus, for example, tubal pregnancy, or pregnancy on ovary, intestine, or in cervix of the uterus.

C8. (a) Bladder; rectum. (b) Posterior, vesico. (c) Anteriorly; retroflexed. (d) Uterosacral; round; broad. (e) Cardinal.

C9. (a) Myo; smooth muscle. (b) Serous perimetrium. (c) Endo; functionalis; spiral.

C11. Close to 90° angle.

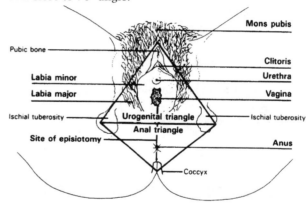

Figure LG 28.5A Perineum.

C13. Acid; retards microbial growth but may harm sperm cells.

C14. (a and c) See Figure LG 28.5A. (b) Urogenital triangle.

C15. Lobes; alveoli; nipple.

D2. Hypothalamus: GnRH; anterior pituitary: FSH, LH; Ovary: E, I, P, R.

D3. (a-b) GnRH. (c-d) I (E and P indirectly because inhibit GnRH). (e) E. (f) P and E. (g) R. (h) E. (i) E.

D4. (a) 28; 5; 50–150; see Figure LG 28.6A. (b) Preovulatory; secondary follicles; 1; mature; II. (c) GnRH, FSH, LH. (d) Secretion of estrogens and development of the ovum. (e) Estrogens stimulate replacement of the sloughed off stratum functionalis by growing (proliferating) new cells; proliferative; follicular. See Figure LG 28.6A. (f) Negative, decreases; degeneration. (g) FSH and LH; ovulate. (h) LH; high; positive. (i) Ovulation; ovulation. See Figure LG 28.6A. (j) Postovulatory; 28; LH, luteum; progesterone, estrogens; implantation; thickening by growth and secretion of glands and

Figure LG 28.6A Correlations of female cycles. (a) Anterior pituitary hormones. (b) Ovarian changes: follicles and corpus luteum. (c) Ovarian hormones. (d) Uterine (endometrial) changes. Refer to Checkpoint C4.

increased retention of fluid. (k) Inhibition of these hormones. (l) De; albicans. See Figure LG 28.6A. (m) Progesterone, estrogens; decrease; menstrual. (n) Placenta, human chorionic gonadotropin (hCG); luteum; pregnancy. (o) See Figure LG 28.6A.

D5. C F J D A I K E G B H.

D6. (a) Dysmenorrhea. (b) Amenorrhea.

E2. (a) Parasympathetic, sacral cord, erect. (b) Sympathetic; ejaculation.

E4. The vaginal wall lacks glands; most lubrication during sexual response is provided by engorgement of the vaginal wall and its capillaries which then causes liquid to seep out of blood vessels (transudation).

E6. (a) Ch. (b) OC. (c) V. (d) T. (e) R. (f) D. (g) IUD. (h) Con. (i) CI. (j) N. (k) RU 486. (l) A. (m) Vag.

F1. (a) None. (b) Ducts of testes and epididymides. ejaculatory ducts and seminal vesicles. (c) Uterus and vagina. (d) None.

F2. (a) M. (b) E. (c) E.

F3. (a) Scrotum. (b) Penis. (c) Prostate. (d) Bulbourethral glands.

G3. (a) D. (b) G (also C). (c) E. (d) S. (e) A. (f) CC. (g) Chl. (h) Chl. (i) H. (j) Can.

G4. Digital rectal exam and blood level of PSA (prostate-specific antigen).

G6. (a) A, C, and E. (b) 1: monthly; 2 and 3: annually or more often based on the decision made by physician and client.

G7. Ovary, salpingectomy.

MORE CRITICAL THINKING: CHAPTER 28

1. Explain how gonads function as both endocrine glands and exocrine glands.
2. Contrast the processes of oogenesis and spermatogenesis.
3. Briefly contrast the ovarian cycle with the menstrual cycle.
4. Contrast the roles of each of the following chemicals or structures in the ovarian and menstrual cycles: (a) GnRH and FSH; (b) estrogens and progesterone; (c) follicle and corpus luteum.
5. Tell exactly what specific hormonal change (increase or decrease of which hormone) accounts for each of the following: (a) ovulation; (b) change of the follicle to the corpus luteum; (c) demise of the corpus luteum in a nonpregnant woman; (d) maintenance of the corpus luteum for several months in a pregnant woman; (e) menstruation.
6. Identify the phase that is the "follicular, proliferative, estrogenic phase" of the menstrual cycle.
7. Tell exactly what chemical is tested for in over-the-counter tests for: (a) ovulation; (b) pregnancy. And state which specific hormones are found in oral contraceptives and in Depo-Provera injections.
8. Identify which form of birth control you think is both most effective with the least harmful side effects. State your rationale.
9. Describe the one major physiological change that normally accounts for both erection in the male and engorgement of female genitalia during sexual response.

MASTERY TEST: CHAPTER 28

Questions 1–10: Arrange the answers in correct sequence.

_____ _____ _____ 1. From most external to deepest:
   A. Endometrium
   B. Serosa
   C. Myometrium

_____ _____ _____ 2. Portions of the urethra, from proximal (closest to bladder) to distal (closest to outside of the body):
   A. Prostatic
   B. Membranous
   C. Spongy

_____ _____ _____ 3. Thickness of the endometrium, from thinnest to thickest:
   A. On day 5 of cycle
   B. On day 13 of cycle
   C. On day 23 of cycle

_____ _____ _____ 4. From anterior to posterior:
   A. Uterus and vagina
   B. Bladder and urethra
   C. Rectum and anus

_____ _____ _____ _____ 5. Order in which hormones begin to increase level, from day 1 of menstrual cycle:
   A. FSH
   B. LH
   C. Progesterone
   D. Estrogen

_____ _____ _____ _____ 6. Order of events in the female monthly cycle, beginning with day 1:
   A. Ovulation
   B. Formation of follicle
   C. Menstruation
   D. Formation of corpus luteum

_____ _____ _____ _____ 7. From anterior to posterior:
   A. Anus
   B. Vaginal orifice
   C. Clitoris
   D. Urethral orifice

_____ _____ _____ _____ 8. Pathway of milk in breasts:
   A. Alveoli
   B. Secondary tubules
   C. Mammary ducts
   D. Lactiferous ducts

_____ _____ _____ _____ _____ 9. Pathway of sperm:
   A. Ejaculatory duct
   B. Testis
   C. Urethra
   D. Ductus (vas) deferens
   E. Epididymis

_____ _____ _____ _____ _____ _____ 10. Pathway of sperm entering female reproductive system:
   A. Uterine cavity
   B. Cervical canal
   C. External os
   D. Internal os
   E. Uterine tube
   F. Vagina

Questions 11–12: Circle the letter preceding the one best answer to each question.

11. Eighty to 90% of seminal fluid (semen) is secreted by the combined secretions of:
    A. Prostate and epididymis
    B. Seminal vesicles and prostate
    C. Seminal vesicles and seminiferous tubules
    D. Seminiferous tubules and epididymis
    E. Bulbourethral glands and prostate

12. In the normal male, there are two of each of the following structures *except:*
    A. Testis   D. Ductus deferens
    B. Seminal vesicle E. Epididymis
    C. Prostate   F. Ejaculatory duct

Questions 13–15: Circle letters preceding all correct answers to each question.

13. Which of the following are produced by the testes?
    A. Spermatozoa   D. GnRH
    B. Testosterone   E. FSH
    C. Inhibin   F. LH

14. Which of the following are produced by the ovaries and then *leave* the ovaries?
    A. Follicle   D. Corpus albicans
    B. Secondary oocyte E. Estrogen
    C. Corpus luteum   F. Progesterone

15. Which of the following are functions of LH?
    A. Begin the development of the follicle
    B. Stimulate change of follicle cells into corpus luteum cells.
    C. Stimulate release of secondary oocyte (ovulation).
    D. Stimulate corpus luteum cells to secrete estrogens and progesterone.
    E. Stimulate release of GnRH.

Questions 16–20: Circle T (true) or F (false). If the statement is false, change the underlined word or phrase so that the statement is correct.

T F   16. Spermatogonia and oogonia both continue to divide throughout a person's lifetime to produce new primary spermatocytes or primary oocytes.

T F   17. The menstrual cycle refers to a series of changes in the uterus, whereas the ovarian cycle is a series of changes in the ovary.

T F   18. Meiosis is a part of both oogenesis and spermatogenesis.

T F   19. Each oogonium beginning oogenesis produces four mature ova.

T F   20. An increased level of a target hormone (such as estrogen) will inhibit release of its related releasing factor and target hormone (GnRH and FSH).

Questions 21–25: Fill-ins. Complete each sentence with the word or phrase that best fits.

_____ 21. _____ is a term that means undescended testes.

_____ 22. Oral contraceptives consist of the hormones _____ and _____.

_____ 23. The most common STD in the United States is currently _____.

_____ 24. Menstrual flow occurs as result of a sudden _____-crease in the hormones _____ and _____.

_____ 25. Three major functions of estrogens are development and maintenance of _____, protein _____, and lowering of blood _____ level.

ANSWERS TO MASTERY TEST: CHAPTER 28

Arrange

1. B C A
2. A B C
3. A B C
4. B A C
5. A D B C
6. C B A D
7. C D B A
8. A B C D
9. B E D A C
10. F C B D A E

Multiple Choice

11. B
12. C

Multiple Answers

13. A, B, C
14. B, E, F
15. B, C, D

True–False

16. F. Spermatogonia; primary spermatocytes (Oogonia do not divide in a female after her birth)
17. T
18. T
19. F. One mature ovum and three polar bodies
20. T

Fill-ins

21. Cryptorchidism
22. Progesterone and estrogens
23. Chlamydia
24. De, progesterone and estrogens
25. Female reproductive structures and secondary sex characteristics, anabolism, cholesterol

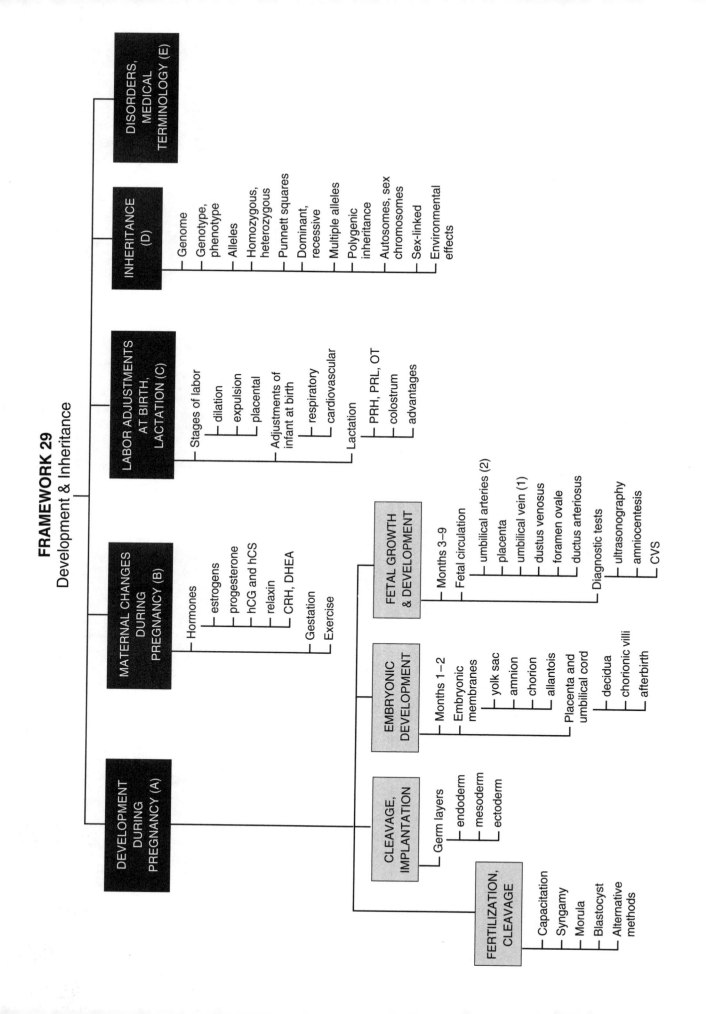

FRAMEWORK 29
Development & Inheritance

DEVELOPMENT DURING PREGNANCY (A)

- CLEAVAGE, IMPLANTATION
  - Germ layers
    - endoderm
    - mesoderm
    - ectoderm

- FERTILIZATION, CLEAVAGE
  - Capacitation
  - Syngamy
  - Morula
  - Blastocyst
  - Alternative methods

- EMBRYONIC DEVELOPMENT
  - Months 1–2
  - Embryonic membranes
    - yolk sac
    - amnion
    - chorion
    - allantois
  - Placenta and umbilical cord
    - decidua
    - chorionic villi
    - afterbirth

- FETAL GROWTH & DEVELOPMENT
  - Months 3–9
  - Fetal circulation
    - umbilical arteries (2)
    - placenta
    - umbilical vein (1)
    - ductus venosus
    - foramen ovale
    - ductus arteriosus
  - Diagnostic tests
    - ultrasonography
    - amniocentesis
    - CVS

MATERNAL CHANGES DURING PREGNANCY (B)

- Hormones
  - estrogens
  - progesterone
  - hCG and hCS
  - relaxin
  - CRH, DHEA
- Gestation
- Exercise

LABOR ADJUSTMENTS AT BIRTH, LACTATION (C)

- Stages of labor
  - dilation
  - expulsion
  - placental
- Adjustments of infant at birth
  - respiratory
  - cardiovascular
- Lactation
  - PRH, PRL, OT
  - colostrum
  - advantages

INHERITANCE (D)

- Genome
- Genotype, phenotype
- Alleles
- Homozygous, heterozygous
- Punnett squares
- Dominant, recessive
- Multiple alleles
- Polygenic inheritance
- Autosomes, sex chromosomes
- Sex-linked
- Environmental effects

DISORDERS, MEDICAL TERMINOLOGY (E)

C H A P T E R   **29**

Development and Inheritance

Twenty-eight chapters ago you began a tour of the human body, starting with the structural framework and varieties of buildings (chemicals and cells) and moving on to the global communication systems (nervous and endocrine), transportation routes (cardiovascular), and refuse removal systems (respiratory, digestive, and urinary). This last chapter synthesizes all the parts of the tour: A new individual, with all those body systems, is conceived and develops during a nine-month pregnancy. Adjustments are made at birth for survival in an independent (outside of mother) world. But parental ties are still evident in inherited features, signs of intrauterine care (mother's avoidance of alcohol or cigarettes), and provision of nourishment, as in lactation.

As you begin your study of development and inheritance, carefully examine the Chapter 29 Topic Outline and Objectives. Check off each objective as you complete it. The Chapter 29 Framework provides an organizational preview of development and inheritance. Be sure to refer to the Framework frequently and note relationships among key terms in each section. And congratulations on having completed your first tour through the wonders of the human body.

TOPIC OUTLINE AND OBJECTIVES

A. From fertilization to implantation; embryonic and fetal development

1. Explain the processes associated with fertilization, morula formation, blastocyst development, and implantation.
2. Discuss the formation of the primary germ layers and embryonic membranes as the principal events of the embryonic period.
3. List representative body structures produced by the primary germ layers.
4. Describe the formation of the placenta and umbilical cord.

B. Maternal changes during pregnancy: hormones; exercise and pregnancy

5. Describe the sources and functions of the hormones secreted during pregnancy.
6. Describe the hormonal, anatomical, and physiological changes in the mother during pregnancy.
7. Explain the effects of pregnancy on exercise, and of exercise on pregnancy.

C. Labor; adjustments at birth; physiology of lactation

8. Explain the events associated with the three stages of labor.
9. Explain the respiratory and cardiovascular adjustments that occur in an infant at birth.
10. Discuss the physiology and hormonal control of lactation.

D. Inheritance

11. Define inheritance, and explain the inheritance of dominant, recessive, polygenic, and sex-linked traits.

E. Disorders

WORDBYTES

Now become familiar with the language of this chapter by studying each wordbyte, its meaning, and an example of its use within a term. After you study the entire list, self-check your understanding by writing the meaning of each wordbyte on the line. As you continue through the *Learning Guide,* identify (and fill in) additional terms that contain the same wordbyte.

Wordbyte	Self-check	Meaning	Example(s)
-blast	_____	germ, to form	*blast*omere, *blast*ocyst
-cele	_____	hollow	blasto*cele*
-centesis	_____	puncture	amnio*centesis*
ecto-	_____	outside of	*ecto*derm, *ecto*pic
endo-	_____	inside of	*endo*derm
fertil-	_____	fruitful	*fertil*ization
lact-	_____	milk	*lact*ation
-mere	_____	part	blasto*mere*
meso-	_____	middle	*meso*derm
morula	_____	mulberry	*morula*
syn-	_____	together	*syn*gamy
-troph	_____	nourish	syncytio*troph*oblast
zygosis-	_____	a joining	*zygo*te; di*zygo*tic

CHECKPOINTS

A. From fertilization to implantation; embryonic and fetal development (pages 1022–1035)

✓ **A1.** Do this exercise about events associated with fertilization.

  a. How many sperm are ordinarily introduced into the vagina during sexual

     intercourse? _____ About how many reach the area where the

     developing ovum is located? _____

  b. The name given to the developing ovum at the time of fertilization is the

     _____. List four or more mechanisms in the female and/or male that favor the meeting of sperm with oocyte so that fertilization can occur.

     Female   **Male**

c. How long must sperm be in the female reproductive tract before they can fertilize the female gamete? Several *(minutes? hours? days?)* Describe the changes sperm undergo during this time?

These changes in the sperm are known as _____.

d. Name the usual site of fertilization. _____.
What do the corona radiata and zona pellucida have to do with fertilization?

e. What triggers the *acrosomal reaction?*

How does this reaction help the sperm to penetrate the oocyte?

f. How many sperm normally penetrate the developing egg? _____ What is the

name of this event? _____ What prevents further sperm from entering the female gamete (polyspermy)?

✓ **A2.** Describe roles of the following structures in fertilization.
  a. Upon fertilization, the secondary oocyte completes meiosis *(I? II?)*, forming the

  fertilized ovum and releasing a _____ body.
  b. A male pronucleus present in a *(secondary spermatocyte? sperm?)* combines with a female pronucleus from a *(secondary oocyte? ovum?)*. Each pronucleus contains the *(n? 2n?)* number of chromosomes which is *(23? 46?)*.
  c. The product of the fusion of the two haploid pronuclei is known as the

  _____ containing _____ chromosomes.

✓ **A3.** Contrast types of multiple births by choosing the answer that best fits each description.

DTwin. Dizygotic twins	**MTwin.** Monozygotic twins	

_____ a. Also known as fraternal twins

_____ b. Two infants derived from a single fertilized ovum that divides once

_____ c. Two infants derived from two secondary oocytes each fertilized by a different sperm

A4. Summarize the main events during the week following fertilization. Describe each of the following stages. Be sure to note how many days after fertilization each stage occurs. (For help, refer to Figures 29.2 to 29.4 of the text.)

a. Cleavage (days _____)

b. Morula (days _____)

c. Blastocyst (days _____)

A5. Draw a diagram of a blastocyst. Label: *blastocoele, inner cell mass,* and *trophoblast.* Color red the cells that will become the embryo. Color blue the cells that will become part of the placenta. (Refer to Figure 29.3b in the text, for help.)

✓ **A6.** Describe the process of implantation in this exercise.

a. Implantation is likely to occur about _____ days after ovulation, or about day

_____ of a 28-day cycle. Note that the developing baby is implanted in the uterus even before the mother "misses" the first day of her next menstrual cycle.

b. During implantation, cells of the *(inner cell mass? trophoblast?)* secrete enzymes that enable the blastocyst to burrow into the uterine lining.

c. The trophoblast is composed of two layers known as the _____-tro-

phoblast and the _____-trophoblast. The trophoblast secretes the

hormone _____, which maintains the corpus *(albicans? luteum?)*. Ultimately the
trophoblast becomes part of the *(amnion? chorion?)* portion of the placenta.

✓ **A7.** An _____ pregnancy is one that takes place in a site outside of the

uterine cavity. Where do most of those pregnancies occur? _____
List several other possible sites of such pregnancies.

✓ **A8.** The word *embryo* refers to the developing individual during the *(first two? last seven?)*
months of development in utero. Define *fetus*.

✓ **A9.** Complete this exercise about the embryonic period. Refer to Figure 29.5 in your text.
  a. Identify the structures that ultimately form from these two parts of the implanted
     blastocyst.

  Trophoblast   **Inner Cell Mass**

  Between these two portions of the blastocyst is the _____ cavity.

  b. The inner cell mass first forms the embryonic _____ consisting of
     two layers. The upper cells (closer to the implantation site toward the right on Figure

     29.5b. page 1028 of the text) are called _____-derm cells. The lower

     cells (toward the left in the figure) are called _____-derm cells.

     Eventually, a layer of _____-derm will form between these.

     Collectively the three layers are called _____ layers.

  c. The process of formation of the primary germ layers is known as _____,

     and it occurs by the end of the _____ week following fertilization.

  d. _____-derm cells line the amniotic cavity. The delicate fetal membrane
     formed from these cells that surrounds the fetus and is broken at birth is called the

     _____. Another name for this membrane is the "_____"

  (Ectoderm cells also form a number of fetal structures: see Checkpoint A11.

e. The lower cells of the embryonic disc, called _____ cells, line the

primitive _____ and for a while extend out to form the

_____ sac. This sac *(is? is not?)* as significant in humans as it is in birds.

f. Some mesoderm cells move around to line the original trophoblast cells (Figure 29.5

in the text). Together these cells form the _____ layer of the placenta.
Which membrane is more superficial? *(Amnion? chorion?)* The space between these

layers is called the _____ .

A10. Each of the primary germ layers is involved with formation of both fetal organs and
fetal membranes. Name the major structures derived from each primary germ layer:
endoderm, mesoderm, and ectoderm. Consult Table 29.1 in your text.

Endoderm	Mesoderm	Ectoderm

✓ **A11.** *For extra review.* Do this matching exercise on primary germ layer derivatives.

Ecto. Ectoderm	Endo. Endoderm	Meso. Mesoderm

_____ a. Epithelial lining of all of digestive, respiratory, and genitourinary tracts except
near openings to the exterior of the body

_____ b. Epidermis of skin, epithelial lining of
entrances to the body (such as mouth,
nose, and anus), hair, nails

_____ c. All of the skeletal system (bone, cartilage,
joint cavities)

_____ d. All skeletal and cardiac muscle; most
smooth muscle

_____ e. Blood and all blood and lymphatic vessels

_____ f. Entire nervous system, including
posterior pituitary

_____ g. Thyroid, parathyroid, thymus, and
pancreas

✓ **A12.** Write the name of each fetal membrane next to its description.
a. Originally formed from ectoderm, this membrane encloses fluid that acts as a shock

absorber for the developing baby: _____

b. Derived from mesoderm and trophoblast, this membrane becomes the principal part

of placenta: _____

c. Endoderm-lined membrane serving as exclusive nutrient supply for embryos of some

species: _____

d. A small membrane that forms umbilical blood vessels: _____

✓ **A13.** Further describe the placenta in this exercise.

a. The placenta has the shape of a _____ embedded in the wall of the

uterus. Its fetal portion is the _____ fetal membrane. Fingerlike

projections, known as _____, grow into the uterine lining. Conse-
quently maternal and fetal blood are *(allowed to mix? brought into close proximity but do
not mix?)*. Fetal blood in umbilical vessels is bathed by maternal blood in

_____ spaces.

b. The maternal aspect of the placenta is derived from the *(chorion? decidua basalis?)*.

This is a portion of the _____ -metrium of the uterus.

c. What is the ultimate fate of the placenta?

✓ **A14.** *A clinical correlation.* Following a birth, vessels of the umbilical cord are carefully

examined. A total of _____ vessels should be found, surrounded by a mucous jellylike connective tissue. The vessels should include *(1? 2?)* umbilical artery(-ies) and
(1? 2?)-umbilical vein(s). Lack of a vessel may indicate a congenital malformation.

A15. List three uses of human placentas following birth.

✓ **A16.** Study Exhibit 29.2 in your text. Write the number of the month when each of the
following events occurs.

_____ a. Heart starts to beat.

_____ b. Heartbeat can be detected.

_____ c. Backbone and vertebral canal form.

_____ d. Eyes are almost fully developed; eyelids
are still fused.

_____ e. Eyelids separate; eyelashes form.

_____ f. Limb buds develop.

_____ g. Ossification begins.

_____ h. Limb buds become distinct as arms and
legs; digits are well formed.

_____ i. Nails develop.

_____ j. Fine (lanugo) hair covers body.

_____ k. Lanugo hair is shed.

_____ l. Fetus is capable of survival.

_____ m. Subcutaneous fat is deposited.

_____ n. Testes descend into scrotum.

A17. *A clinical correlation.* Complete this Checkpoint related to pregnancy.
   a. Describe the condition known as *placenta previa*. State its frequency of occurrence, and suggest possible causes as well as consequences for the fetus and/or mother.

   b. Explain why smoking increases risk for ectopic pregnancy. List several other risk factors for ectopic pregnancy.

✓ **A18.** Complete this exercise about prenatal diagnostic tests.

   a. Amniocentesis involves withdrawal of _____ fluid. This fluid is constantly recycled through the fetus (by drinking and excretion); it *(does? does not?)* contain fetal cells and products of fetal metabolism. For what purposes is amniotic fluid collected and examined?

   For _____

   (if performed at 14–16 weeks). For _____
   (if performed at about 35 weeks).
   b. *(Amniocentesis? Ultrasound?)* is by far the most common test for determining true fetal age if the date of conception is uncertain. The technique *(is also? is not?)* used for diagnostic purposes. Fetal ultrasonography involves *(use of a transducer on? a puncture into?)* the abdomen. For this procedure the mother's urinary bladder should be *(full? empty?)*.

   c. CVS (meaning c _____ v_____ s_____)
   is a diagnostic test that can be performed as early as *(4? 8? 16? 22?)* weeks of pregnancy. It *(does? does not?)* involve penetration of the uterine cavity. Cells of the

   _____ layer of the placenta embedded in the uterus are studied.
   These *(are? are not?)* derived from the same sperm and egg that united to form the fetus. Therefore this procedure, like amniocentesis, can be used to diagnose

   _____.

B. Maternal changes during pregnancy: hormones; exercise and pregnancy (pages 1035–1039)

✓ **B1.** Complete this exercise about the hormones of pregnancy.

   a. During pregnancy, the level of progesterone and estrogens must remain *(high? low?)* in order to support the endometrial lining. During the first three or four months or so, these hormones are produced principally by the _____ located in the _____, under the influence of the tropic hormone _____. (It might be helpful to review the section on endocrine relations in Chapter 28.)

   b. Human chorionic gonadotropin (hCG) reaches its peak between the _____ and _____ month of pregnancy. Because it is present in blood, it will be filtered into _____, where it can readily be detected as an indication of pregnancy as early as _____ days after fertilization. Sometimes the urine of a woman who is not pregnant might erroneously indicate that she is pregnant. List several factors which could account for such a "false pregnancy test."

   c. In addition, the chorion produces the hormone hCS; these initials stand for _____ chorionic _____. List two functions of this hormone.

   Peak levels of hCS occur at about week *(9? 20? 32?)* of gestation.

   d. The newly found chorionic hormone CRH, if secreted in high levels early in pregnancy, is associated with *(premature? late?)* delivery. Is this hormone produced in people who are not pregnant?

   e. _____ is a hormone that relaxes pelvic joints and helps dilate the cervix near the time of birth. This hormone is produced by the _____ and _____.

✓ **B2.** Complete this exercise about maternal adaptations during pregnancy. Describe normal changes that occur.

   a. Pulse _____-creases.

   b. Blood volume _____-creases by about _____%

   c. Tidal volume _____-creases about _____%

   d. Gastrointestinal (GI) motility _____-creases, which may cause *(diarrhea? constipation?)*.

e. Pressure upon the inferior vena cava _____ -creases, which may lead to

_____.

f. Increase in weight of the uterus from _____ grams in nonpregnant state to about

_____ grams at term.

✓ **B3.** About 10–15% of all pregnant women in the United States experience PIH, which

stands for p_____-i_____ h_____.

This condition apparently results from impaired function of _____.
Write three or more signs or symptoms of PIH.

B4. Defend or dispute this statement: "All women who are pregnant or lactating should
avoid exercise because of its deleterious effects."

C. Labor; adjustments at birth; physiology of lactation (pages 1039–1044)

C1. Contrast these pairs of terms: *labor/parturition*

✓ **C2.** Do the following hormones stimulate (*S*) or inhibit (*I*) uterine contractions during
the birth process?

_____ a. Progesterone   _____ c. Oxytocin

_____ b. Estrogens

✓ **C3.** *For extra review*, discuss regulation of hormones associated with labor.
   a. During the last six months of gestation, both progesterone and estrogens are

produced in large amounts by the _____. The name progesterone
indicates that this hormone is "pro" gestation, for one reason, because it *(stimulated?
inhibits?)* uterine contractions, thereby continuing the gestation period.

   b. For birth, known as _____, to occur, levels of estrogen must
(surpass? become less than?) levels of progesterone. This change in estrogens results
from increasing levels of CRH made by the *(fetal hypothalamus? placenta?)*. This
hormone causes the fetal anterior pituitary to secrete *(ACTH? DHEA?)* which

stimulates the release of cortisol and the androgen _____ from the fetal adrenal *(cortex? medulla?)*.

c. DHEA is then converted by the placenta into *(estrogen? progesterone?)*. This hormone causes mother's uterine muscle to become more receptive to *(oxytocin? relaxin?)*.

d. Explain how oxytocin promotes labor and delivery.

Oxytocin is involved in a *(negative? positive?)* feedback loop.

e. The hormone relaxin assists in parturition by *(relaxing? tightening?)* the pubic symphysis. Relaxin and estrogen together cause *(softening? tightening?)* of the cervix.

C4. What is the "show" produced at the time of birth?

Is the "show" associated with true labor or false labor?

✓ **C5.** Identify descriptions of the three phases of labor *(first, second,* or *third)*.

_____ a. Stage of expulsion: from complete cervical dilation through delivery of the baby

_____ c. Time from onset of labor to complete dilation of the cervix; the stage of dilation

_____ b. Time after the delivery of the baby until the placenta ("afterbirth") is expelled; the placental stage

C6. *A clinical correlation.* Explain more about labor and delivery in this Checkpoint.

a. The average duration of labor is about _____ hours, whereas the duration for first

labor and delivery is about _____ hours.

b. The term _____ means painful or difficult labor. In a breech presentation, which parts of the body enter the birth canal first? _____ If

necessary, a C-section or _____ section will be required for such births. Was this procedure named such because Julius Caesar was born by C-section? *(Yes? No?)* Explain.

c. The stress of labor and delivery causes the fetus to initiate "fight-or-flight" responses in which the fetal adrenal medullae releases their hormones. Explain how these chemicals help the fetus through the birth process.

C7. Define these terms related to birth and the postnatal period.
a. *Involution*

b. *Lochia*

c. *Puerperium*

C8. What adjustments must the newborn make at birth in attempting to cope with the new environment? Describe adjustments of the following.
a. Respiratory system (What serves as a stimulus for the respiratory center of the medulla?)

The respiratory rate of a newborn is usually about *(one-third? three times?)* that of an adult.
b. Heart and blood vessels (*For extra review* go over changes in fetal circulation, Chapter 21, pages LG 454–455.)

c. Blood cell production

✓ **C9.** A premature infant ("preemie") is considered to be one who weighs less than

_____ grams (_____ pounds _____ ounces) at birth. These infants are at high

risk for RDS (or _____).

✓ **C10.** Complete this exercise about lactation.
   a. The major hormone promoting lactation is _____, which is secreted by

   the _____. During pregnancy the level of this hormone increases somewhat, but its effectiveness is limited while estrogens and progesterone are *(high? low?)*.
   b. Following birth and the loss of the placenta, a major source of estrogens and progesterone is gone, so effectiveness of prolactin *(increases? decreases?)* dramatically.
   c. The sucking action of the newborn facilitates lactation in two ways. What are they?

   d. What hormone stimulates milk ejection (let-down)? _____
   e. The time period from the start of a baby's suckling at the breast to the delivery of milk

   to the baby is about _____-seconds. Although afferent impulses (from breast to
   hypothalamus) require only *(a fraction of a second? 45 seconds?)*, passage of the oxytocin

   through the _____ to the breasts requires about _____ seconds.

C11. Contrast *colostrum* with *maternal milk* with respect to time of appearance, nutritional
   content, and presence of maternal antibodies.

C12. Defend or dispute this statement: "Breast-feeding is an effective, reliable form of birth
   control."

C13. List five or more advantages offered by breast-feeding.

D. Inheritance (pages 1044–1050)

D1. Define these terms.

a. *Inheritance*

b. *Genetic counseling*

c. *Homologues (homologous chromosomes)*

D2. Compare and contrast.

a. *Genotype/phenotype*

b. *Dominant allele/recessive allele*

c. *Homozygous/heterozygous*

d. *Angelman syndrome/Prader-Willi syndrome*

✓ **D3.** Answer these questions about genes controlling PKU. (See Figure 29.14 in the text.)

    a. The dominant gene for PKU is represented by *(P? p?)*. The dominant gene is for the *(normal? PKU?)* condition.

    b. The letters PP are an example of a genetic makeup or *(genotype? phenotype?)*. A person with such a genotype is said to be *(homozygous dominant? homozygous recessive? heterozygous?)*. The phenotype of that individual would *(be normal? be normal, but serve as a "carrier" of the PKU gene? have PKU?)*.

    c. The genotype for a heterozygous individual is _____. The phenotype is

    _____.

    d. Use a Punnett square to determine possible genotypes of offspring when both parents are Pp.

✓ **D4.** Do this exercise about genes and inheritance.

    a. Define *genome*.

    A human cell is estimated to have about _____ genes. *(Most? A small number?)* of these have been identified on the various human chromosomes.

    b. In most cases, the gene for *(having a severe disorder? being normal?)* is dominant. State

    one or more exceptions to this rule. _____

    c. What is the meaning of the term *aneuploid*?

    One example is *trisomy 21*, in which the individual has *(2? 3? 4?)* of chromosome 21.

    This alteration in the genome occurs in the condition known as _____.

    d. Sickle cell anemia (SCA) is a condition that is controlled by *(complete? incomplete?)* dominance. Match the correct genotype below with the related description. Choose from these genotypes:

A. Homozygous recessive	**C.** Heterozygous
B. Homozygous dominant	

    1. Genotype present in persons who have SCA _____

    2. Genotype present in persons who have the sickle cell trait _____

e. List the three possible alleles for inheritance of the ABO group.

_____ _____ _____ Which of

these is/are dominant? _____ List the possible genotypes of persons

who produce A antigens on red blood cells. _____ Describe the phenotype of persons with the 00 (or ii) genotype.

f. ABO blood group inheritance is known as *(multiple allele? polygenic?)* inheritance. List
two or more examples of characteristics governed by polygenic inheritance.

D5. Contrast *autosome* and *sex chromosome.*

✓ **D6.** Answer these questions about sex inheritance.
  a. All *(sperm? secondary oocytes?)* contain the X chromosome. About half of the sperm

    produced by a male contain the _____ chromosome and half contain the _____.
  b. Every cell (except those forming ova) in a normal female contains *(XX? XY?)* genes
    on sex chromosomes. All male cells (except those forming sperm) are said to be
    (XX? XY?).
  c. Who determines the sex of the child? *(Mother? Father?)* Show this by using a Punnett
    square.

  d. Name the gene on the Y chromosome that determines that a developing individual
    will be male.

✓ **D7.** Complete this exercise about X-linked inheritance.
  a. Sex chromosomes contain other genes besides those determining sex of an

    individual. Such traits are called _____ traits. Y chromosomes are
    shorter than X chromosomes and lack some genes. One of these is the gene control-

ling ability to _____ red from green colors. Thus this ability is

controlled entirely by the _____ gene.

b. Write the genotype for females of each type: color-blind, _____;

carrier, _____; normal, _____.

c. Now write possible genotypes for males: color-blind, _____; normal,

_____.

d. Determine the results of a cross between a color-blind male and a normal female.

e. Now determine the results of a cross between a normal male and a carrier female.

f. Color blindness and other X-linked, recessive traits are much more common in *(males? females?)*.

D8. Name several other examples of X-linked traits.

Discuss *fragile X syndrome*.

✓ **D9.** Complete this exercise about potential hazards to the embryo and fetus. Match the terms in the box with the related descriptions.

C. Cigarette smoking	FAS. Fetal alcohol syndrome	T. Teratogen

_____ a. An agent or influence that causes defects in the developing embryo

_____ b. May cause defects such as small head, facial irregularities, defective heart, retardation

_____ c. Linked to a variety of abnormalities such as lower infant birth weight, higher mortality rate, GI disturbances, and SIDS

E. Disorders (pages 1050–1051)

E1. Describe the following aspects of in vitro fertilization (IVF).

   a. In what year was the first live birth as a result of this procedure? *(1962? 1970? 1978? 1986?)*

   b. Briefly describe this procedure.

   c. Contrast these two forms of IVF: *embryo transfer* and *GIFT*

E2. Describe how the conditions of being underweight or overweight may affect reproductive hormones and fertility.

✓ **E3.** Fill in the term (*impotence* or *infertility*) that best fits the description. Then list several causes of each condition.

   a. Inability to fertilize an ovum: _____.

   b. Inability to attain and maintain an erection: _____ .

E4. Each of the following disorders involves a particular chromosome. Identify which chromosome is involved and then write a brief description of characteristics of individuals born with this disorder.

   a. Down syndrome involves chromosome _____ .

   b. Klinefelter's syndrome involves chromosome _____ .

   c. Turner's syndrome involves chromosome _____ .

E5. Write the technical term for morning sickness and describe who is likely to be affected by this condition.

ANSWERS TO SELECTED CHECKPOINTS: CHAPTER 29

A1. (a) Approximately 300 to 500 million; less than 1%. (b) Secondary oocyte; mechanisms in the female: peristaltic contractions and ciliary action of uterine tubes, uterine contractions promoted by prostaglandins in semen; mechanisms in the male: presence of flagella on sperm. (c) Hours; more vigorous lashing of flagella, and changes that enable plasma membranes of sperm and oocyte to better fuse; capacitation. (d) Uterine (Fallopian) tube; they are the coverings over the oocyte that sperm must penetrate. (e) ZP3, a chemical in the zona pellucida, attracts sperm and triggers this reaction; proteolytic enzymes released from the sperm's acrosome clear a path for the sperm through the zona pellucida. (f) One; syngamy; sperm penetration of the secondary oocyte triggers depolarization of the oocyte (fast block), then release of calcium ions from the oocyte that lead to inactivation of ZP3 and hardening of the zona pellucida so it is impenetrable to sperm (slow block).

A2. (a) II, polar. (b) Sperm, ovum; n, 23. (c) Zygote, 46.

A3. (a) DTwin. (b) MTwin. (c) DTwin.

A6. (a) 6, 20. (b) Trophoblast. (c) Syncytio, cyto; hCG. luteum; chorion.

A7. Ectopic; uterine tube; ovaries, abdomen, cervix portion of the uterus, and broad ligament.

A8. First two; developing individual during the last seven months of human gestation.

A9. (a) Trophoblast: part of chorion of placenta; inner cell mass: embryo and parts of placenta; amniotic. (b) Disc; ecto; endo; meso; primary germ. (c) Gastrulation, third. (d) Ecto; amnion; bag of waters. (e) Endoderm, gut, yolk; is not. (f) Chorion; chorion; extraembryonic coelom.

A11. (a) Endo. (b) Ecto. (c) Meso. (d) Meso. (e) Meso. (f) Ecto. (g) Endo.

A12. (a) Amnion. (b) Chorion. (c) Yolk sac. (d) Allantois.

A13. (a) Flattened cake (or thick pancake); chorion; chorionic villi; brought into close proximity but do not mix; intervillous. (b) Decidua basalis; endo. (c) It is discarded from mother at birth (as the "afterbirth"); separation of the decidua from the remainder of endometrium accounts for bleeding during and following birth.

A14. 3; 2; 1.

A16. (a) 1. (b) 3. (c) 1. (d) 3. (e) 6. (f) 1. (g) 2. (h) 2. (i) 3. (j) 5. (k) 9. (l) 7. (m) 8. (n) 7.

A18. (a) Amniotic; does; detecting suspected genetic abnormalities (14–16 weeks); assessing fetal maturity (35 weeks). (b) Ultrasound; is also; use of a transducer on; full. (c) Chorionic villus sampling, 8; does not; chorion; are; genetic and chromosomal defects.

B1. (a) High; corpus luteum, ovary, hCG. (b) First,

third; urine, 14; excess protein or blood in her urine, ectopic production of hCG by uterine cancer cells, or presence of certain diuretics or hormones in her urine. (c) Human (chorionic) somatomammotropin; stimulates development of breast tissue and alters metabolism during pregnancy (see page 1037 in the text); 32.) (d) Premature; yes, made by the hypothalamus, it is the releasing hormone for ACTH. (e) Relaxin; placenta, ovaries.

B2. (a) In. (b) In, 30–50. (c) In, 30–40. (d) De, constipation. (e) In, varicose veins. (f) 60–80, 900–1200.

B3. Pregnancy-induced hypertension; kidneys; hypertension, fluid retention leading to edema with excessive weight gain, headaches and blurred vision, and proteinuria.

C2. (a) I. (b) S. (c) S.

C3. (a) Chorion of the placenta; inhibits. (b) Parturition, surpass; placenta; ACTH, DHEA, cortex. (c) Estrogen; oxytocin. (d) Oxytocin stimulates uterine contractions which force the baby into the cervix. Positive; cervical stretching triggers nerve impulses to the hypothalamus where more oxytocin is released. (e) Relaxing; softening.

C5. (a) Second. (b) Third. (c) First.

C9. 2500 grams (5 pounds. 8 ounces); respiratory distress syndrome.

C10. (a) Prolactin (PRL), anterior pituitary; high. (b) Increases. (c) Maintains prolactin level by inhibiting PIH and stimulating release of PRH; stimulates release of oxytocin. (d) Oxytocin. (e) 30–60; a fraction of a second, bloodstream, 30–60.

D3. (a) P; normal. (b) Genotype; homozygous dominant; be normal. (c) Pp; normal but a carrier of PKU gene. (d) 25% PP, 50% Pp, and 25% pp.

D4. (a) The complete genetic makeup of a particular organism; 70,000–100,000; a small number. (b) Being normal; Huntington's disease (HD) or other disorders on the left side of Table 29.3 in the text. (c) Having a chromosome number other than normal (which is 46 for humans); 3; Down syndrome. (d) Incomplete; 1, A; 2,C. (e) I^A, I^B, and i; I^A and I^B; $I^A I^A$ or I^Ai; produces neither A nor B antigens on red blood cells. (f) Multiple allele; skin color, eye color, height, or body build.

D6. (a) Secondary oocytes; X, Y (b) XX; XY. (c) Father, as shown in the Punnett square (Figure 29.19 in the text), because only the male can contribute the Y chromosome. (d) SRY (sex-determining region of the Y chromosome).

D7. (a) X-linked; differentiate; X. (b) $X^c X^c$, $X^C X^c$, $X^C X^C$, (c) $X^c Y$, $X^C Y$. (d) All females are carriers and all males are normal. (e) Of females, 50% are normal and 50% are carriers; of males, 50% are normal and 50% are color-blind. (f) Males.

D9. (a) T. (b) FAS. (c) C.
E3. (a) Infertility (or sterility) (b) Impotence (erectile
       dysfunction)

MORE CRITICAL THINKING: CHAPTER 29

1. Summarize the three major events in the embryonic period following implantation.
2. Describe the layers that form the placenta and outline its protective functions.
3. Contrast the three primary germ layers and the types tissues and organs each will form.
4. Contrast use of amniocentesis and chorionic villi sampling (CVS). Describe when during gestation each procedure is likely to be performed, what is sampled, and how each test gives information about potential genetic anomalies.
5. Describe effects of the following maternal or fetal hormones upon labor and parturition: progesterone, estrogens, oxytocin, and relaxin (maternal); cortisol and epinephrine (fetal).
6. Describe the roles of the *fetal* adrenal gland in assisting the baby through labor and delivery. Be sure to include both cortical and medullary hormones.
7. Discuss effects of embryonic or fetal environment upon polygenic traits. Include effects of cigarettes, alcohol and other drugs, and radiation.

MASTERY TEST: CHAPTER 29

Questions 1–2: Arrange the answers in correct sequence.

_____ _____ _____ _____ 1. From most superficial to deepest (closest to embryo):
   A. Amnion
   B. Amniotic cavity
   C. Chorion
   D. Decidua

_____ _____ _____ _____ _____ _____ 2. Stages in development:
   A. Morula
   B. Blastocyst
   C. Zygote
   D. Fetus
   E. Embryo

Questions 3–11: Circle the letter preceding the one best answer to each question.

3. During early months of pregnancy, hCG is pro-

    duced by the _____ and it mimics the action of

    _____ .
    A. Follicle; progesterone
    B. Corpus luteum; FSH
    C. Chorion of placenta; LH
    D. Amnion of placenta: estrogens

4. These fetal characteristics—weight of 8 ounces to 1 pound, length of 10 to 12 inches, lanugo covers skin, and brown fat is forming—are associated with which month of fetal life?
    A. Third month   C. Fifth month
    B. Fourth month   D. Sixth month

5. All of the following structures are developed from mesoderm *except:*
    A. Aorta   D. Humerus
    B. Biceps muscle   E. Sciatic nerve
    C. Heart   F. Neutrophil

6. Which part of the structures surrounding the fetus is developed from maternal cells (not from fetal)?
    A. Allantois   D. Decidua
    B. Yolk sac   E. Amnion
    C. Chorion

7. All of the following events occur during the first month of embryonic development *except:*
    A. Endoderm, mesoderm, and ectoderm are formed.
    B. Amnion and chorion are formed.
    C. Limb buds develop.
    D. Ossification begins.
    E. The heart begins to beat.

8. Implantation of a developing individual (blastocyst stage) usually occurs about _____ after fertilization.
   - A. Three weeks
   - B. One week
   - C. One day
   - D. Seven hours
   - E. Seven minutes

9. The high level of estrogens present during pregnancy is responsible for all of the following *except:*
   - A. Stimulates release of prolactin
   - B. Prevents ovulation
   - C. Maintains endometrium
   - D. Prevents menstruation
   - E. Stimulates uterine contractions

10. Which of the following is a term that refers to discharge from the birth canal during the days following birth?
    - A. Autosome
    - B. Cautery
    - C. Lochia
    - D. PKU

11. Formation of the three primary germ layers is known as:
    - A. Implantation
    - B. Blastocyst formation
    - C. Gastrulation
    - D. Morula formation

Questions 12–20: Circle T (true) or F (false). If the statement is false, change the underlined word or phrase so that the statement is correct.

T F 12. Oxytocin and progesterone both stimulate uterine contractions.

T F 13. Of the primary germ layers, only the ectoderm forms part of a fetal membrane.

T F 14. Maternal blood mixes with fetal blood in the placenta.

T F 15. Amniocentesis is a procedure in which amniotic fluid is withdrawn for examination at about 8–10 weeks during the gestation period.

T F 16. Stage 2 of labor refers to the period during which the "after-birth" is expelled.

T F 17. Maternal pulse rate, blood volume, and tidal volume all normally increase during pregnancy.

T F 18. The decidua is part of the endoderm of the fetus.

T F 19. Inner cell mass cells of the blastocyst form much of the chorion of the placenta.

T F 20. A phenotype is a chemical or other agent that causes physical defects in a developing embryo.

Questions 21–25: Fill-ins. Complete each sentence with the word or phrase that best fits.

_____ 21. The newborn infant generally has a respiratory rate of _____,

pulse of _____, and a WBC count that may reach _____,

all values that are _____ than those of an adult.

_____ 22. As a prerequisite for fertilization, sperm must remain in the female reproductive tract for at least several hours so that

_____ can occur.

_____ 23. hCG is a hormone made by the _____; its level peaks at about

the end of the _____ month of pregnancy, and up to this time

it can be used to detect _____ .

_____ 24. Determine the probable genotypes of children of a couple in which

the man has hemophilia and the woman is normal. _____

_____ 25. This is the _____ mastery test question in this book.

ANSWERS TO MASTERY TEST: CHAPTER 29

Arrange

1. D C A B
2. C A B E D

Multiple Choice

3. C
4. C
5. E
6. D
7. D
8. B
9. A
10. C
11. C

True–False

12. F. Oxytocin but not progesterone stimulates.
13. F. All three layers form parts of fetal membranes: ectoderm (trophoblast) forms part of chorion; mesoderm forms parts of the chorion and the allantois; endoderm lines the yolk sac.
14. F. Does not mix but comes close to fetal blood
15. F. Amniocentesis, 14–16
16. F. Fetus
17. T
18. F. Endometrium of the uterus
19. F. Trophoblast
20. F. Teratogen

Fill-ins

21. 45, 120–160, up to 45,000, higher
22. Capacitation and dissolving of the covering over the ovum by secretion of acrosomal enzymes of sperm
23. Chorion of the placenta, second, pregnancy
24. All male children normal; all female children carriers
25. Last or final. Congratulations! Hope you learned a great deal.

A.D.A.M.® *Interactive Anatomy (AIA) Checkpoints*

Getting ready. If you are not familiar with A.D.A.M.®, read this list of directions (#1–8) which will walk you through the first AIA Checkpoint (pg LG 678). If you get lost at any point in A.D.A.M.®, return to these directions or refer to the A.D.A.M.® *User's Guide* that accompanies the software.

1. Your options are three links: *Dissectible Anatomy, Atlas Anatomy*, and *3D anatomy*. Click on *Dissectible Anatomy* (as directed by *Diss* in the *Link* column for Checkpoint A4a). When the dialog box opens, click on *Dissectible Anatomy, Female, Anterior*, and *Open*.

2. Move the mouse/cursor to the upper right hand corner of the screen towards the tiny figure (the A.D.A.M.® Navigator) and notice that a somewhat larger figure appears with a rectangle highlighting the region of the body shown on the screen. Our first Checkpoint takes place in the chest region. You can get to that region in any of three ways:

   (a) Click on/scroll the box outlining viewing area to the female's chest (thorax) area.

   (b) Move to that region by dragging the right (vertical) scroll bar up or down, or the lower (horizontal) bar right or left.

   (c) To move more slowly to the region, click on arrows on the right (vertical) scroll bar to move up/down, or on the lower (horizontal) bar to move right/left.

3. You can choose between closeups of organs or "the big picture" by using the *zoom* function. Click on the magnifying glass (at top of vertical tool bar) and then click on the portion of the chest cavity that you want to magnify. Try it a couple of times now, and use the zoom-in and zoom-out functions often in your A.D.A.M.® explorations.

4. For most exercises, you can select either gender. You can change the gender viewed by clicking on the gender icon (on the vertical tool bar just below the left lower "light bulb" icon), then clicking FEMALE or MALE.

5. Check out the three "light bulb" icons by moving the cursor/arrow to each one. In most cases you will use the *Normal Mode* where all structures stand out in living color. But *Highlight Mode* and *Extract Mode* will be useful at times.

6. The *Identify* icon (arrow at top of vertical tool bar) will allow you to test your self and get immediate feedback any time you click the arrow on a structure. You can also find a structure by typing in the term at the top of the screen or by scrolling through the structure list LM (*List Manager*) there.

7. Just to the right of the gender icon (on vertical tool bar) is your option button for *View*. Your options are ANTERIOR (front), POSTERIOR (back), LATERAL (side), MEDIAL (towards the midline), LATERAL ARM (LAT ARM),or MEDIAL ARM (MED ARM). Click on ANTERIOR now. (as indicated by *ANT* in the View column of AIA Checkpoints).

8. Now click and drag the *depth* bar (to left of screen) or click on up or down arrows there to move from surface to deeper layers of the human body. Be sure that you always scroll to the exact layer listed in this table. Due to slight variations between the Windows and Macintosh platforms, you may need to scroll one layer higher or lower if you do not see a structure on your version of A.D.A.M.® Go to layer 153 now and you are ready for the first Checkpoint.

   Enjoy your travels through A.D.A.M.®

AIA Checkpoints. Chapter 1 An Introduction to the Human Body

Check-point	Link	Gender	View	Layer	Key structure/Image/Title/Region	*Structures* (in italics) to Identify and Concepts (underlined) to Understand
A4a	Diss	F	ANT	153	Fifth rib	Identify the left *fifth* and *sixth ribs*. The space in between is the fifth intercostal space where the heart can be auscultated with a stethoscope. Now go to layer 173 and notice that the *left ventricle* of the heart lies just deep to the fifth interspace on the left side.
B4a E4	Diss	M	ANT	215	Abdomen	Go to *Highlight Options* (found to the lower right of the three "light bulb" icons) and click on Digestive system. As those organs appear in color, identify several parts of this system: *pancreas (coronal section), liver,* and *duodenum and jejunum* (parts of small intestine). Now scroll slowly from layer 215 to layer 199 and see the location of the *stomach* anterior and superior to the pancreas.
E2d	Diss	F	POST	133	Knee region.	Click on Cardiovascular system. Then find the *popliteal artery* and *vein*—two major blood vessels in the posterior of the knee.
E2i	Diss	F	LAT	145	Neck region.	Click on Skeletal system. Then identify *cervical (C1-7) vertebrae.*
E2l	Diss	F	LAT	145	Heel region.	Click on Skeletal system. Then identify the heel bone (*calcaneus*).
E3c	Diss	M	ANT	215	Abdomen	Identify the *coronal section* which divides Pancreas into anterior and posterior sections.
E4a	Diss	F	ANT	190	Upper abdomen	Scroll to layer 192 and notice that the *liver* is inferior to the *diaphragm* and that *lungs* are superior to the diaphragm.
E4g	Diss	F	LAT	0	Upper abdomen	As you move to deeper layers (from 0 to 11), notice that muscles are deep to *skin*.
F1a	Diss	F	MED	0	Head region	Click on Nervous system and see the *brain* and *spinal cord* in the cranial and vertebral regions of the dorsal cavity.
F1b	Diss	F	MED	69	Thorax/upper abdomen	Click on *Normal Mode* and identify a *lung*, the *liver* and *stomach* in the ventral cavity.
F1f	Diss	F	ANT	163	Thorax	Identify the *thymus* in the mediastinal region of the thorax. Now move to deeper layers. Remove the posterior wall of the *pericardium* at (level 175-176), and identify these mediastinal structures: *aorta, pulmonary vessels, esophagus* and *trachea.*

ID	Source	Gender	View/Type	Layer	Structure	Instruction
F1h	Diss	F	ANT	219	Pelvic cavity	Identify from anterior to posterior as you move through deeper region layers: *urinary bladder* (219), *uterus* (at 226), and *rectum* (at 227). Now go to a medial (MED) view (layer 50) and notice the position of these three organs. Change to male gender (layer 49) and notice the absence of the uterus but presence of *prostate* in the true pelvis.

Note: For Atlas Anatomy exercises, go to File, and open Atlas Anatomy; click on the correct View, Region, System, or Type, and scroll to the (underlined) structure to be identified.

ID	Source	Gender	View/Type	Layer	Structure	Instruction
F3a	Atlas		Type: Illustration		Abdominopelvic Regions	Click on Type → Illustration → Abdominopelvic regions → Open. Scroll to the upper three abdominal regions and identify organs there.
F3c	Diss	F	ANT	0	Abdomen	Identify the umbilicus in the center of the umbilical region of the abdomen. Continue to move to deeper layers, and notice that the umbilical opening passes all the way through the anterior abdominal wall.
F3e	Atlas		Type: Illustration		Abdominopelvic Quadrants	Scroll to the upper right quadrant and identify the *liver* and *gallbladder* there.

AIA Checkpoints. Chapter 4   The Tissue Level of Organization

ID	Source	Gender	View/Type	Layer	Structure	Instruction
C10	Atlas		View: Medial		Fascia in Female Pelvis (Med)	Identify the superfial layer of superficial perineal fascia containing many adipose pockets (white) combined with loose connective tissue.
C12b	Atlas		View: Posterior		*Ligaments of thoracic spine*	Identify these dense connective tissues that attach bone to bone.
C12c	Atlas		Type: Cadaver photo		Anterior Thigh	Identify the tendon of the quadriceps femoris muscle that attaches this muscle into the tibia by way of the patella.
C15a	Atlas		View: Medial		Fascia in Female Pelvis (Med)	Identify the pubic symphysis composed of dense fibrous cartilage.
D2a	Atlas		Region: Thorax		Thorax Right Mediastinum (Lateral)	Identify the pericardial sac composed largely of serous membrane and covering over the heart.
D2d	Diss	F	MED	0	Neck region	Identify the shiny mucous membrane lining the part of the throat known as the oropharynx.
D3a	Atlas		System: Digestive		Stomach 1	Identify the visceral peritoneum covering over an organ (*liver*).
D3a	Atlas		View: Medial		Fascia in Female Pelvis (Med)	Identify the parietal peritoneum lining the abdominal cavity.

Check-point	Link	Gender	View	Layer	Key structure/Image/Title/Region	*Structures* (in italics) to Identify and Concepts (underlined) to Understand

AIA Checkpoints. Chapter 5 The Integumentary System

| A5 | Diss | F | ANT | 0–157 | Umbilical region | Starting with 0, scroll to deeper layers and notice skin, *superficial fascia* (or subcutaneous tissue) which is yellow since fatty, and *deep fascia* (white) covering muscles (red). At layer 8-9, notice the extra *subcutaneous tissue* over hips of the female. Notice the *parietal peritoneum* lining the abdominal wall at layer 191: it disappears at 192. Why?. |

AIA Checkpoints. Chapter 6 Bone Tissue

B2	Diss	M	LAT	288	Hip	Identify the large *femur* bone. Notice compact bone, spongy bone, and *medullary (marrow) cavity*. Scroll to 287 to see the lateral surface of the femur. Now move to layer 286 to see *articular cartilage* over the proximal epiphysis of the femur.
B2	Atlas		Type: Radiograph		Knee (Ant/Post)	Notice the broad diaphysis and narrow epiphysis of the femur. Now identify the *tibia* just distal to the knee joint. Notice the white outer layer of compact bone. Deeper parts of the bone are composed of spongy bone which looks darker on x-ray.
C6	Diss	F	LAT ARM	64	Humerus	Identify the epiphyseal line of *humerus*, as well as compact bone, spongy bone, and the *medullary (marrow) cavity*.

AIA Checkpoints. Chapter 7 Axial Skeleton

| A3 | Diss | M | LAT | 105 | Hip region | Contrast types of bone. Identify the *femur*, a *long bone*, and then scroll distally to locate the *patella*, a *sesamoid* bone. Continue inferiorly to find the two long bones of the leg (*tibia* and *fibula*). Then find (at layer 283) the long bones (*metatarsals*) in the middle of the foot, and the short (tarsals) bones in the back of the foot, the *cuboid*, *navicular*, and *cuneiform* bones. |
| A6d | Diss | F | LAT | 0 | Head | Scroll to layer 4 and identify the marking known as the *external acoustic meatus*. Then follow the pathway of sound waves into the temporal bone as you gradually scroll to layer 119. |

A6f	Diss	M	LAT	91	Hip region	Identify the marking known as the *greater trochanter*. Notice that tendons of several muscles insert into this large bony projection. Scroll deeper (to 138 and then to 287/288) to see markings known as the *head* and *neck of the femur*. Now scroll back to 0 to see the surface view of the greater trochanter deep to skin.
B2a	Diss	F	MED	0	Head region	In this midsagittal (medial) view, notice the skull bones *protecting* the brain. If you click on skull bones and underlying brain parts, you will notice that many have identical names, such as *frontal bone* and *frontal lobe of cerebrum*.
B3 B5	3D		Skull		Rotate the skull	View the *skull* from different positions. Now go to the structure list and *temporal bone-lateral exploded*. This view will allow you to see individually most of the 28 skull bones and their markings. Go to the Structure List and click on any bone you want to identify. Examples: *lacrimal bone; sphenoid and ethmoid bones—isolated; frontal bone—posterolateral exploded*.
B4a	Diss	F	ANT	13	Head region: chin	Identify the *mandible* bone. Scroll to layer 36, and identify three other *facial bones* (*maxilla, nasal,* and *zygomatic*). Now change to the LATERAL view. As you move from layer 118 to 120, notice how thin the *mandible* is. Identify five *cranial bones* (*frontal, parietal, occipital, sphenoid,* and *temporal*).
B5 B12d	Diss	M	LAT	118	Head	Identify the following *markings* on the skull: *greater wings of sphenoid bones,* condyle (*condylar process*) of the *mandible* at the <u>temporomandibular joint</u> (TMJ), external acoustic meatus. Scroll to layer 205 where you will see the pituitary gland (*hypophysis*) in the *sella turcica* (*dorsum sellae*) of the *sphenoid* bone.
B3	3D		Skull		Rotate the skull	Go to the Structure List and click on any marking you want to see. Examples: *sella turcica; lesser wing of sphenoid; optic canal (foramen), foramen magnum; condylar process of mandible; coronal suture*.
B7	Diss	M	LAT	197	Head	Identify the *following cranial nerves* (in yellow) passing through *cranial foramina: II, III, IV,* all three branches of *V,* and *VI.* Scroll to 204 and see cranial nerves *IX, X,* and *XI* exiting from the jugular foramen.
B8	Diss	M	LAT	204	Head	Notice the *carotid canal* in the temporal bone. Scroll back to 202 to see the *internal carotid artery* coursing through this bony canal to supply blood to the brain and eye (*ophthalmic artery*). At 200, see the *internal jugular vein* exiting from the <u>jugular foramen</u>. Notice the proximity of these vessels in the neck. Can you think of an advantage to their passing through *different* foramina in the cranium?

Check-point	Link	Gender	View	Layer	Key structure/Image/Title/Region	Structures (in italics) to Identify and Concepts (underlined) to Understand
B8 B9d B7b-d	Diss	F	ANT	36	Head	Identify the *supraorbital margin* and *mental foramen*. What change occurs at layer 36-37? Proceed to layers 39–41 and note that fat and muscles protect eyes posteriorly. Locate six bones that form the *orbit*. Identify *superior* and *inferior orbital fissures*; now find *optic canals* (foramina) located in the *lesser wings of the sphenoid*.
B8g B11a	Diss	M	LAT	117	Head	Identify the *zygomatic process of the temporal bone*. Notice the arch that it forms with the *temporal process of the zygomatic bone*. Recognize a function of this bony arch as you scroll to 115.
B10a	3D		Skull		Rotate the skull	Go to the Structure List and click on: *sinuses of the skull*. Rotate to see them all. Which ones are red? Which are blue? (Check the Structure List.) Which ones are more anterior: the yellow *ethmoid cells* (sinuses) or orange *sphenoid sinuses*?
B10a B11d	Diss	M	MED	62	Head	Identify the *frontal* and *sphenoid sinuses*. Go to ANTERIOR (ANT) view 52. These two *paranasal sinuses*, as well as the *maxillary* and *ethmoid sinuses*, all drain into the nose via *meati* (spaces) formed by nasal *conchae*.
B10a,b B8d B11f	Diss	M	LAT	209	Head	Find the *palatine process of the maxilla*, just anterior to the *palatine bones*. Now scroll to layer 207 and identify the *alveolar processes of the maxilla* and the *maxillary sinus*.
B10c,e B8a	Diss	M	LAT	209	Head	Identify the three parts of the *nasal septum*: *perpendicular plate of ethmoid bone*, *vomer*, and *nasal septal cartilage*. Notice the *crista galli* just superior to the perpendicular plate.
C1	Diss	F	LAT	52	Head/neck	Identify the *hyoid bone* at chin level. Notice the many muscles anchored to it as you scroll to layer 40. The *hyoid* can also be seen at chin level on ANT view at layer 75 and 253.
C3	Diss	M	LAT	120	Neck	Scroll the 26 bones forming the "back bone," and identify regions: *cervical, thoracic,* and *lumbar vertebrae,* plus *sacrum* and *coccyx*.
C3 C5 C7b C8c	Diss	M	LAT	120	Neck	Click on the seven *cervical vertebrae*, including C1 (*atlas*) and C2 (*axis*). Identify the *spinous process* of each as well as the *nuchal ligaments*. Now go to ANT layer 270, and see the anterior of the *body* of each *cervical vertebra*, along with their *ligaments*. At 274, note that the anterior of C1 (*atlas*) that lacks a body; instead, the *dens of C2* (*axis*) is held by ligaments in that site.

D2 D3	Diss	F	ANT	75	Thorax	Identify the parts of the *sternum*: *manubrium, body,* and *xiphoid*. Find the *clavicular notch*. Identify *costal cartilages* that attach ribs to sternum
C3 D3	Diss	M	LAT	120	Thorax	Move down into the thoracic region and click on each of the 12 *thoracic vertebrae*, their *ligaments*, and related *ribs*. Identify the *costal cartilages* that link *true ribs* (*ribs 1-7*) to the sternum. Notice that <u>false ribs</u> (*ribs 8-10*) articulate with the sternum via the *costal cartilage of rib 7*. Scroll to layer 72 and to 64 to see *intercostal muscles* and *nerves* located in the <u>intercostal space</u>. Now scroll to *ribs 11* and *12* (<u>false</u> and <u>floating</u>) ribs and note their lack of anterior attachments.
C3 E1	Diss	M	POST	0	Lower back	Scroll quickly through layers and observe the many muscles and ligaments that hold the spine erect. At layer 176, click on the *spinous process, lamina* and *transverse process* of *L1 vertebra*. At layer 177 you see veins and coverings (*dura mater*) surrounding the spinal cord and spinal nerves. Now move inferiorly and scroll back out to layer 180 and 176 to observe the entire sacrum and coccyx.

AIA Checkpoints. Chapter 8 Appendicular Skeleton

A1c	Diss	F	ANT	75	Thorax	Identify the *clavicular notch* (on sternum) where the <u>appendicular skeleton</u> (*clavicle*) articulates with the <u>axial skeleton</u> (*sternum*).
A2	Diss	F	ANT	0	Neck and thorax	Notice the outline of the "collar bone" deep to the skin. Scroll deeper until you see the *clavicle* serving as a wedge and site for attachment of muscles, including the *pectoralis major, deltoid, trapezius,* and *sternocleidomastoid.*
A3 C9c, d	Diss	F	ANT	30	Neck and thorax	At layer 30 of this same view, you see a tiny part of the *scapula* which is a very posterior bone. Notice that the *coracoid process of scapula* has tendons of two muscles attached to it. Now go to LAT view 29 to see the *glenoid cavity*, and just superior to it, the *acromion process*. View these also at POST layers 15 to 73. Gradually scroll back out to 28 to see *medial* and *lateral* (axillary) *borders, spine,* and *supra-* and *infraspinous fossae*. Scroll to 21 and 18 to see muscles that lie in those fossae.
B1	Diss	F	LAT ARM	0	Shoulder/arm	Move quickly from layer 0 to 63, noting types of tissues surrounding and protecting the upper limb. Now scroll from shoulder to fingertips as you identify all bones in the upper limb (For now omit names of individual carpals.)

Check-point	Link	Gender	View	Layer	Key structure/Image/Title/Region	Structures (in italics) to Identify and Concepts (underlined) to Understand
B2	Diss	F	POST	186	Humerus (proximal end)	Identify the greater tubercle (See at layer 10 as insertion point for several muscles. Now move out to layer 9 and highlight the deltoid muscle. Notice (at layer 74) its insertion point on the lateral side of the humerus (deltoid tuberosity: not labeled).
B2 B3h	Diss	F	POST	0	Elbow	Scroll anteriorly through skin until you see (at layer 46) the olecranon. Find the ulnar nerve (yellow) on the medial (little finger) side of the elbow. This nerve becomes tingly when it is trapped and inflamed near the medial epicondyle (labeled at layer 186) of humerus ("funny bone").
B3 C9e	Diss	F	LAT ARM	63	Elbow	Identify the lateral epicondyle of humerus, and the head and radial (biceps) tuberosity of radius. From the MEDIAL ARM view (layer 71) see the trochlea of humerus in trochlear (semilunar) notch (unlabeled) of ulna.
B3b	Atlas		Region: Upper Limb		Flexed Elbow (Lat)	On this x-ray, identify the capitulum of humerus fitting into the depression in the head of radius. Find the olecranon (process) of ulna.
B2	Diss	F	POST	186	Forearm	Identify the two long bones of the forearm. Which is broader at the elbow? Which is broader at the wrist? Which bone is on the thumb side of the forearm? (Hint: the thumb can move in a radial fashion). Find the styloid process of radius.
B4a,b	Diss	F	POST	186	Wrist	Notice that only seven carpals are shown here. Where is the other one? Notice that the radius articulates almost entirely with the carpal bone known as the scaphoid. If you fall and land with much weight on your hand, the radius jams against the scaphoid. Result?
B4c	Diss	F	ANT	107	Wrist	Identify the flexor retinaculum of wrist and visualize how repetitive wrist movements can trap the median nerve and inflame these "carpal tunnel" tissues.
B5	Atlas		Type: Radiograph		Hand (Post/Ant)	Identify the bones and joints of the hand. Which bones do you actually "hold in" (because they form) "the palm of your hand"? Now count them: from first to fifth metacarpal. All but one digit has three phalanges: identify the proximal, middle, and distal phalanges. Which digit lacks a middle phalanx?
C1	Diss	F	ANT	329	Lower abdomen	Identify the bones of the pelvis: hip bones, sacrum, and coccyx. Locate the sacroiliac joint where the sacrum (part of axial skeleton) meets the ilium portion of hip bone (part of appendicular skeleton).

Code	Source	Sex	View/Type	No.	Location / Structure	Description
C2 C3	Diss	F	ANT	329	Lower abdomen	Identify the three sections of the hip bone: by selecting these on the Structure List: *ilium* (Note: not *ileum*), *ischium*, and *pubis*. Which of these is most inferior in location (and you sit on it)? Identify these markings: *pubic symphysis, iliac crest, anterior superior iliac spine* (ASIS), AIIS; *and sacral promontory.*
C2 C3 C6 C9b	Diss	F	LAT	288	Lower abdomen	Identify these markings of the hip bone: *iliac crest, ASIS, PSIS,* and *PIIS; greater sciatic notch, ischial tuberosity, ischial spine.* Notice the head of the femur located in the *acetabulum.* Notice that several of these markings are sites of muscle attachment (ASIS for *sartorius;* AIIS for the *rectus femoris* head of *quadriceps femoris,* and *ischial tuberosity* for two heads of hamstrings: see next activity).
C9b C3a	Diss	M	POST	113	Hip	Identify these strong thigh muscles (*adductors* and hamstrings: *semitendinosus, semimembranosus, and biceps femoris*) anchored to the *ischial tuberosities.* Now move to layer 121 and then 120 to see the enormous *sciatic nerve* emerge through the *greater sciatic notch.*
C8	Atlas		Region: Pelvis & Perineum		Female Bony Pelvis (Ant) / Male Bony Pelvis (Ant)	Minimize these two views and arrange them side-by-side. Contrast pelvic shapes in the two sexes. Notice the > 90° angle of the pubic arch in the female pelvis.
C8	Atlas		Region: Pelvis & Perineum		Bony Landmarks of the Male Pelvis	Identify all bony structures with pins. Does the penis emerge from superior or inferior to the *pubic symphysis* as the *urethra* passes through the pelvic floor of the male bony *outlet?*
D1 D3	Atlas		Region: Lower Limb		Bones of the Leg and Foot (Ant)	Identify each bone in the lower limb from proximal to distal. (For now omit names of tarsals.) Now identify all markings on *tibia, patella,* and *fibula* that are designated with pins. Now do the same activity on Bones of the Leg and Foot (Post).
D1	Atlas		Region: Lower Limb		Bones of the Lower Limb (Ant)	Now identify all markings on the *femur* that are designated with pins.
D1 D6	Atlas		Region: Lower Limb		Bones of the Foot (Dorsal)	Now identify all bones of the foot. Contrast shape and location of the *talus* and *calcaneus.*
D7	Atlas		Type: Radiograph		Foot (Lat)	Identify the arch of the foot, and note that it occurs in the metatarsal-tarsal regions.
D4 D8f	Diss	M	ANT	182	Hip	Identify the anterior tip of the *iliac crest* and find the two muscles (*sartorius* and *tensor fasciae latae*) attached to it. Notice *adductor* muscles attached to the *pubis.* Now move inferiorly to the knee region; scroll deeper to see the "kneecap" or *patella* at layer 293. Recognize its position just anterior to the distal epiphysis of the femur (seen at 321). Notice that the fibula does *not* articulate with femur or patella.

Check-point	Link	Gender	View	Layer	Key structure/Image/Title/Region	Structures (in italics) to Identify and Concepts (underlined) to Understand
D8d,g D3h	Diss	M	ANT	321	Knee	Scroll distally to *anterior border of tibia* (or *tibial crest*), commonly called the "shin bone". Move to the ankle region and find (at 328) the "outer ankle," or *lateral malleolus of fibula*. Now switch to LAT view and find the "heel bone" (*calcaneus*).

AIA Checkpoints. Chapter 9 Articulations

Check-point	Link	Gender	View	Layer	Key structure/Image/Title/Region	Structures (in italics) to Identify and Concepts (underlined) to Understand
A4b	Diss	M	LAT	0	Head	Scroll through coverings over the skull to layer 113. Identify the fibrous, immovable joints known as <u>sutures</u>. Identify the *coronal* and *lambdoid sutures*. At layer 116 see the *squamosal suture* and the *sphenofrontal suture*.
A4c	Diss	M	LAT	189	Head region	Observe the <u>gomphoses</u> in which teeth are anchored via their *periodontal ligaments* into spongy bone of *mandible* and *maxilla*.
A4d	Diss	M	ANT	327	Ankle	Locate the <u>syndesmosis</u> involving the *anterior tibiofibular ligament* attaching distal ends of *tibia* and *fibula*.
A5a	Diss	M	LAT	64	Humerus	Identify the *epiphyseal line of humerus*, formerly the site of <u>epiphyseal cartilage</u>—a *synchondrosis*) in a growing bone.
A5b	Atlas		Type: Radiograph		Thorax (Lat)	Amphiarthrotic <u>symphyses</u> are found at the fibrocartilaginous intervertebral discs. Which appears darker on an x-ray? *Cartilage (less dense)? Compact bone (more dense)?* State one other location of a symphysis.
A7 A8	Diss	F	LAT ARM	64	Humerus	Identify the aqua-colored *synovial capsule of shoulder joint*, the *fibrous capsule of shoulder joint*, and a cushioning, fluid-filled sac (*bursa*).
A9 C3	Diss	F	ANT	329	Knee	Now try "building" the knee joint to see its complexity. At 329-328, see the *tibia* and *fibula* with their *articular cartilages*. At 322 a protective *fat pad* is added. At 320, see the <u>intracapsular ligaments</u> such as the *anterior cruciate ligament* (ACL). In the POST view (at 171), you can see the other end of the *ACL*; at POST 170, the *posterior cruciate ligament* (PCL) is visible. Identify the *menisci* in this view also. Return to ANT (319) and see the aqua-colored *synovial* (membrane) *capsule of knee joint*. At 296, see the *infrapatellar fat pad* and the *fibrous capsule of knee joint*. <u>Extracapsular ligaments</u> lie outside of it (295-293): the *tibial* and *fibular collateral ligaments*. At 293-292 the "knee cap" (*patella*) appears to protect the joint. At 291, the strong *tendons* of the four heads of the quadriceps (*rectus femoris* and

three *vastus* heads) are seen attaching to the patella and then to the *tibial tuberosity* (185).

ID	Source	Sex / Type	View	Layer	Region	Description
A12b	Diss	F	POST	186	Wrist	Recognize why the wrist (distal end of *radius* and *scaphoid* [carpal] bone is called an ellipsoidal joint. Move to 184 and then to 64 to see support for forearm, wrist, and hand joints.
A12c A13h B2i	Diss	F	LAT ARM	329	Wrist	Identify the site of the only *saddle joint* in the body: between *first metacarpal* (base) and *trapezium* (a carpal). This one joint (in each hand) permits the movement known as *opposition*. Find this site on yourself. (*Note*: this is not the point at which your thumb (*first proximal phalanx*) meets the *metacarpal* (head).
A12d2 C4f	Atlas	Type: Radiograph			Flexed Elbow (Lat)	The elbow is the classic hinge joint. Which markings form the two sets of "hinges" in this joint? (*Hint*: see Diss → ANT elbow region → layer 329. What pivot joint is also visible here?)
A12e B1e	Atlas	Type: Radiograph			Cervical spine (Odontoid)	Select *Dens* (or odontoid process) of *C2 vertebra (axis)*. This is a site of another pivot joint (between C2 and C1) that allows you to pivot your head to say "No."
A12d3	Atlas	Type: Radiograph			Ankle (Ant/Post)	See the hinge joint between *tibia/fibula* and the *talus*. Notice that the *malleoli* help to form an inverted "U" over the *talus*.
A12d5	3D	3D: Skull			*Occipital condyle*	Notice the occipital portion of the atlanto (atlas)-occipital joint, a modified hinge joint. This joint (occipital bone-C1) allows you to flex your head to say "Yes."
A12f	Diss	M	ANT	30	Upper thorax/neck	Two planar joints are seen here, each connected by *ligaments*. Which joint is torn in "separation of the shoulder"? (*Hint*: which joint involves the bones that form the "shoulder girdle"?)
B3a	Atlas	Type: Radiograph			Flexed Cervical Spine (Lat)	Looking at the floor, this person has head and neck flexed.
B3b	Atlas	Type: Radiograph			Extended Cervical Spine (Lat)	As this person moves the head back past anatomical position (to look at the stars), the movement is *hyperextension*.
C1	Diss	M	ANT	85	Shoulder	Identify the *superior glenohumeral ligament*. At layer 125, see the *middle glenohumeral ligament* and the *coracohumeral ligament*. At 142, see a bursa. At 144, see the *synovial capsule of shoulder joint* and a *synovial tendon sheath* that reduces friction as the *biceps muscle* (seen at 81) moves. The interior structure of the joint is seen at 330.

Check-point	Link	Gender	View	Layer	Key structure/Image/Title/Region	*Structures* (in italics) to Identify and Concepts (underlined) to Understand
C1e	Diss	M	POST	9	Shoulder	Scroll to layer 11 and identify the four muscles (and their tendons) that form the rotator cuff. Three are visible here: *supra-* and *infrapinatus*, and *teres minor*. The *subscapularis* lies deep to the scapula, so is seen at POST 72-74 or at ANT 95. Because this "cuff" of muscles pulls from both anterior and posterior, they permit internal/external rotation of the humerus.
C2b	Diss	M	ANT	330	Hip joint	Identify the *head of femur* and *articular cartilage* of the acetabulum of *hip bone*. Now scroll out gradually to see blood supply of this joint (311-310), *synovial capsule*, and ligaments (at 308). Find the *iliofemoral* and *pubofemoral ligaments*. To see the *ischiofemoral ligaments*, go to POST 127. Note the many muscles that also support this joint (seen from POST 127-82).
C4a	Atlas		Type: Cadaver photo		Dissection of Medial Leg & Foot	Identify the *calcaneal (Achilles) tendon* that attaches muscle (*gastrocnemius and soleus*) to the *calcaneus* (heel) bone at the ankle joint.

AIA Checkpoints. Chapter 11 The Muscular System

Check-point	Link	Gender	View	Layer	Key structure/Image/Title/Region	*Structures* (in italics) to Identify and Concepts (underlined) to Understand
A2d,f	Atlas		Type: Radiograph		Flexed Elbow (Lat)	Identify the site of the effort (insertion) on the *tuberosity of radius*. Then find the site of the fulcrum (joint cavity between *humerus* and *radius/ulna*).
A3b	Diss	F	LAT	87	Calf/ankle region	Envision how contraction of the *gastrocnemius muscle* raises the *calcaneus* (wheelbarrow-like) in this second-class lever system.
A6	Diss	F	ANT	21	Arm	Recognize that antagonistic muscles are found on opposite sides of the body. Identify the *biceps brachii* (flexor of forearm) and then locate the *triceps brachii* (extensor of the forearm) at POST layer 10.
B2a	Diss	F	ANT	28	Umbilical region	Notice the vertical (rectus) direction of muscle fibers of the *rectus abdominis muscle*.
C1	Diss	F	ANT	31	Head	Locate these muscles of facial expression: *corrugator, orbicularis oris, levator labii superioris,* and (at layer 37) *buccinator muscles*.
C1 C3	Diss	F	LAT	126	Head	Notice the *buccinator* muscle in this view, as well as part of the *lateral rectus* (extrinsic) *muscle* of the eye. Scroll to layer 190 to see more of this muscle as well as the *inferior rectus*.

C5a,b	Diss	F	LAT	11	Head	Find the *masseter* muscle just anterior to the *parotid gland*. Scroll deeper to see more of it (at layer 51) and then scroll to the *temporalis muscle* (116). Both of these raise the jaw (close the mouth). Find the *lateral pterygoid muscle* (that forces the mouth open for prolonged dental work) at layer 123.
C7-9	Diss	F	LAT	128	Head/neck	Find the *styloglossus* muscle and then scroll in to the *hyoglossus* muscle (235). Find the *genioglossus muscles, geniohyoid,* and then the *hypoglossal nerve (CN XII)* seen at layer 40 that innervates these tongue muscles.
C10	Diss	F	LAT	21	Head/neck	Find the *sternocleidomastoid* (SCM) muscle crossing the neck diagonally from the *mastoid part of temporal bone* (layer 22). Where does the SCM insert? Hint: See ANT view (42). Do you expect to find antagonists of the SCM on the anterior or posterior of the neck? Look for the *splenius, semispinalis,* and *longissimus capitis muscles* (POST 75-76): they *extend/hyperextend* the neck.
C12g	Diss	F	ANT	20	Head/neck	Highlight the *facial nerve (CN VII)* and notice its many branches to muscles of facial expression.
D1	Diss	F	ANT	28	Umbilical region	Identify the *rectus abdominis muscle*. At layers 26 and 80, you will see the (white) connective tissue that ensheathes this muscle and forms the white line (*linea alba*) running vertically midline. At ANT 80 click on the muscle lateral to the rectus abdominis. Notice that fibers of this *transversus abdominis muscle* terminate in an *aponeurosis* that forms part of the *rectus sheath*. Scroll out to layers 26 and 24 to see the *internal oblique* (IO) and *external oblique* (EO) muscles that also have broad *aponeuroses*. Which of the three lateral layers of muscles has fibers in the direction in which you would put your hands in your jeans' pockets? (EO) Now scroll to layer 291 and identify the three lateral abdominal muscles (EO, IO, TA) from most superficial to deep.
D2	Diss	F	ANT	165	Thorax	Identify the diaphragm muscle just inferior to heart and lungs. Scroll to 190 and notice insertion of fibers of the *diaphragm* muscle into the *central tendon of diaphragm*. Scroll to layer 241, and then remove kidneys and fat at 244-245. Now you see more of the diaphragm anchored to the rib cage and lumbar vertebrae.
D2	Diss	F	ANT	143	Thorax	Identify the *external intercostal muscles;* move to 145-146 to see *internal intercostal muscles.*
D4	Atlas		Region: Pelvis & Perineum		Female Pelvic Diaphragm (Inf)	Identify *levator ani* and *coccygeus muscles.* Which is attached to the *ischial spine?* Notice other bony landmarks of the outlet of the female pelvis.

Check-point	Link	Gender	View	Layer	Key structure/Image/Title/Region	Structures (in italics) **to Identify** and Concepts (underlined) **to Understand**
D5	Atlas		Region: Pelvis & Perineum		Female Pelvic Floor (Lat)	Study the pelvic diaphragm, consisting of pelvic floor muscles and their fasciae. Also locate outlets from pelvic organs: *urethra, vagina,* and *rectum.*
E1	Diss	M	POST	0	Thorax	Scroll through skin and fascia (to 10-11) to expose muscles that move the scapulae. Locate origins and insertions (0 & I) of the *trapezius, levator scapulae,* and *rhomboids.* Since these muscles help you to keep your back "up straight" (extended) all day, an upper back massage feels good as these muscles are stretched (relaxed).
E1	Diss	M	ANT	53	Thoracic region	See the *pectoralis minor muscle* (from *ribs 3-5* to *coracoid process* of scapula. Note that contraction of the muscle helps depress scapulae. Now identify heads of the *serratus anterior muscles.* At LAT 41-45, see their attachments between *ribs #1-8* and the anterior surface of the medial border of the scapula.
E4	Diss	M	POST	9	Lower back region	Locate the origins of the *latissimus dorsi muscles,* at sacrum, iliac crest, lumbar and lower six thoracic spines. Which is more superficial at about level T6 vertebra? ("Traps" = *trapezius?* "Lats" = *latissimus dorsi?*) Now follow the "lats" to their insertion point at ANT 90, close to insertions of tendons of the *subscapularis muscle* and the "pects" = *pectoralis major muscle* (the latter at ANT 20). At ANT 19 see insertion of the "delts" = *deltoid,* the U-shaped muscle whose *origins* lie next to *insertions* of the *trapezius* (See POST 10 also.)
E5	Diss	M	ANT	17	Thoracic region	Find origins of the *pectoralis major muscle.* Now scroll from POST 30-18 to find O & I's of other muscles that move the humerus: *teres major* and *minor, infraspinatus* and *supraspinatus muscles.* Note that all attach from *scapula* to proximal *humerus.*
E6	Diss	M	LAT ARM	13	Arm	Locate the *biceps brachii, brachialis,* and *triceps brachii muscles.* Gradually scroll to layer 45 and identify O & I's of these muscles.
E7 E8a	Diss	M	ANT	106	Forearm/wrist	Identify the *flexor carpi radialis* and *flexor carpi ulnaris* muscles At layer 107, find the *flexor digitorum superficialis muscle.* Then (at layer 117) find the *flexor digitorum profundus muscle,* the deepest of all of these muscles. Do *flexors* lie on the anterior or posterior surface of the body? Notice the *flexor retinaculum* (at 110) that holds tendons of these muscles in place.
F1	Diss	M	POST	9	Middle back	Scroll inwards through "traps," "lats," and *rhomboids* to reach (at layer 17) erector spinae muscle components: *iliocostalis, longissimus,* and *spinalis* (from lateral to medial). (Try the mnemonic I L S, as in "I like spaghetti.") At layer 77, see lower

Code	Type	Sex	View	No.	Region	Description
						back levels of these muscles. Now scroll superiorward to see the various *capitis* extensions of these deep back (extensor) muscles.
G2	Diss	M	POST	11	Hip/thigh	Identify these posterior muscles that move the thigh: *gluteus maximus, gluteus medius* (at 83), and hamstrings (whose origins are seen at 83). Follow the tendons of the hamstrings (*biceps femoris, semitendinosus,* and *semimembranosus*—from lateral to medial) into the posterior of the knee (at 114). At LAT 92 and MED 14, you can see hamstring *tendons* insert on proximal *fibula* and *tibia.* To see the major antagonists (flexors) of the thigh, go to ANT 330, find the *lesser trochanter,* and then see *iliacus tendon* inserting there at 292. Next find the origins of the *iliacus* (291) and *psoas* (265) heads of the iliopsoas.
G2 G3	Diss	F	ANT	23	Hip/thigh	Find the following muscles, as well as their O & I's: *sartorius, tensor fasciae latae,* and *rectus femoris* head of the "quads." Now scroll to 188 to remove the *sartorius;* you can see clearly two large heads of the *adductors* (*longus* and *magnus*) and one small one (*brevis*). Notice two other "quads" heads: *vastus lateralis* and *vastus medialis.* Scroll to 189 to see the fourth "quads" head (*vastus intermedius*) and the "quads" *tendons* passing over the *patella* to the *tibia.* At layer 292, you will see insertions of *tendons* of adductor muscles into the *femur.*
G2	Atlas Type: Cadaver photo				Anterior Thigh	Identify these muscles: *sartorius, tensor fasciae latae,* "quads" (*rectus femoris, vastus medialis* and *vastus lateralis* seen here), *adductor longus,* and *gracilis.*
G4	Diss	M	POST	7	Hip/thigh	Find the *fascia lata* ("anatomical pantyhose") surrounding thigh muscles. Go to LAT 7-8 and see it there also. Now scroll deeper to LAT 61 and see the thickening of this fascia, known as the *iliotibial* tract, which serves as insertion site for the *tendon of gluteus maximus* muscle and the *tensor fasciae latae* muscle into the lateral surface of tibia.
G7	Atlas Type: Cadaver photo				Anterior Leg	Identify these muscles (and related tendons if shown): *tibialis anterior, gastrocnemius (medial head), peroneus longus,* and *extensor digitorum longus.*

AIA Checkpoints. Chapter 12 Nervous Tissue

Code	Type	Sex	View	No.	Region	Description
A3a,d	Diss	M	MED	34	Head	Identify the *spinal cord* and the *frontal* and *occipital* lobes of the cerebrum, all parts of the central nervous system (CNS).
A3c	Diss	M	ANT	275	Head/Neck	Identify the branches of *spinal nerves* (yellow) in the neck. Scroll inferiorly to find nerves emerging from all levels of the vertebral column. These nerves are all parts of the peripheral nervous system (PNS).

Check-point	Link	Gender	View	Layer	Key structure/Image/Title/Region	Structures (in italics) to Identify and Concepts (underlined) to Understand
A3e	Diss	M	LAT	196	Thorax	Identify the *fourth thoracic sympathetic ganglion* (yellow).

AIA Checkpoints. Chapter 13 Spinal cord and spinal nerves

Check-point	Link	Gender	View	Layer	Key structure/Image/Title/Region	Structures (in italics) to Identify and Concepts (underlined) to Understand
A3	Atlas		Type: Illustration		Spinal Cord Vessels & Meninges	Identify the three meninges (*dura mater, arachnoid,* and *pia mater*), surrounding spaces (*subarachnoid space, subdural space,* and the (yellow) fat-filled *epidural space.* Recognize the protection offered the spinal cord by the vertebra.
A3	Atlas		Type: Illustration		Sagittal Section of Lower Spine	Identify meninges as well as the *epidural space, medullary cone, filum terminale,* and *cauda equina.*
A6b	Diss	M	MED	0	Waist level	Identify the level at which a lumbar puncture is performed. This level is *(superior? inferior?)* to the end of the spinal cord.
A7	Atlas		Type: Illustration		T12 Vertebra (Sup)	Find the *gray matter* and *white matter* of the spinal cord. In which of these areas are tracts found?
C1a	Atlas		Type: Illustration		Spinal Nerves of Trunk (Ant)	Identify *cervical, thoracic,* and *lumbar spinal nerves.*
C1c	Atlas		Type: Illustration		T12 Vertebra (Sup)	Locate the intervertebral disc labeled at *nucleus pulposus* and *annulus fibrosus.* Which is more anterior (ventral) in location: *(intervertebral disc? spinal cord?).* Now locate the *ventral* and *dorsal roots of spinal nerves.*
C2b	Atlas		Type: Illustration		Isolated Cervical Plexus *(Lat)*	See the nerves of the *cervical plexus (C1-C4),* and the *phrenic nerve.*
C2a,c C3c C4d	Atlas		Type: Cadaver photo		Brachial Plexus (Lat)	Find *C5* and *C6 spinal nerves* which form the superiormost portions of this plexus. Now identify the *musculocutaneous nerve.* Follow this nerve towards the *biceps brachii* innervated by this nerve. (Note: See Diss → ANT (layers 81-82.) Notice that this nerve and the *axillary artery* both look whitish. How can they be be differentiated by appearance? Now locate the *axillary, median* and *ulnar nerves.* To see more of these, open DISS → Medial Arm (layer 7) and follow the *median* and *ulnar nerves* into the forearm and wrist.
C3b C6a	Diss	M	ANT	19	Shoulder	Locate the *deltoid muscle,* and then scroll to layer 55 to see its nerve supply *(axillary).* Scroll to 90 and identify the *radial nerve.* At POST 50, identify an extensor muscle innervated by this nerve.

Ref	Source				Type / Name	Question / Instruction
C2F C4h	Atlas				Type: Illustration Lumbar Plexus in Situ	Locate the lumbar vertebrae. Find nerves of the lumbar plexus, including the largest nerve of this plexus, the *femoral nerve*. It emerges anterior to the *iliacus* and *psoas* muscles, passes deep to the *inguinal ligament*, and then supplies heads of the *quadriceps femoris* (See at Diss → ANT (layer 66 of thigh).
C2g C3a C6g D2d	Atlas				Type: Illustration Sacral Plexus (Ant)	Identify the *sciatic nerve* from the sacral plexus. Now go to Diss → POST (layer 82–83) to see this nerve just deep to the *gluteus maximus muscle*. Scroll to layer 117 and move inferiorly into the posterior thigh. See the small branch (*common peroneal nerve*) and the large branch (*tibial nerve*). Which branch goes deep to the *gastrocnemius* and *soleus* to supply these muscles?
C6g D2	Diss	M	POST	0	Lower back	Imagine yourself as a surgical nurse assisting with a laminectomy on the *lamina* of *L4 vertebra*, related to a slipped disc causing *sciatica*. At layer 177 you see veins and coverings (*dura mater*) surrounding the spinal nerves. Go to 180. Why is the spinal cord not seen here at L4? Layer 181 is just anterior to the spinal cord and nerves. At 184 the culprit (*intervertebral disk*) is now visible anterior to the sectioned lamina.
C8	Atlas				Type: Illustration Dermatomes of Trunk (Ant)	What color are *dermatomes* supplied by *intercostal nerves (T2-T12)*? What color are *dermatomes* from the parts of the lumbar plexus (L1-L3) shown here?

AIA Checkpoints. Chapter 14 Brain and Cranial Nerves

Ref	Source				Type / Name	Question / Instruction
A1	Diss	F	MED	0	Head	Identify the major parts of the brain.
A4a	Atlas				Type: Illustration Sagittal Section of Brain	Locate the *tentorium cerebelli*. Which brain parts does it separate?
A4c	Diss	F	LAT	194	Head	Identify the part of the *dura mater* known as the *falx cerebri*. What does it separate? (*Hint:* Go to Diss → ANT, layer 48.)
A7 A8	Atlas				Type: Illustration Sagittal Section of Brain	Identify these structures in the pathway of cerebrospinal fluid (CSF): the *septum pellucidum* (between the two lateral ventricles), *third ventricle, cerebral aqueduct,* and *fourth ventricle*.
A7	Atlas				Type: Illustration Ventricles of Brain (Sup)	Identify these CSF-filled spaces within the brain. Find a *choroid plexus* shown in a *lateral ventricle*, as well as *frontal* (anterior) and *occipital* (posterior) *horns of lateral ventricles*. Notice the *septum pellucidum* between lateral ventricles.

Check-point	Link	Gender	View	Layer	Key structure/Image/Title/Region	Structures (in italics) to Identify and Concepts (underlined) to Understand
B1-5 B7 B9l	Atlas		Type: Illustration		Sagittal Section of Brain	Locate the *medulla* at the base of the skull. Now identify the *pons* and the <u>midbrain</u>. In which of these is the *cerebral aqueduct* found? Identify the *cerebellum*, with its branched tree-like (<u>arbor vitae</u>) appearance. Which ventricle lies just anterior to it? Find the slitlike *third ventricle* that lies between the two halves of the thalamus.
B7	Atlas		Type: Illustration		Ventricles of Brain (Sup)	See the (sectioned) thalamus that lies just lateral to the third ventricle. Both the thalamus and the hypothalamus (parts of the diencephalon) lie deep within the brain.
B10j	Diss	M	LAT	240	Head	Identify the *mamillary body* portions of the *hypothalamus*. This brain part also includes the posterior (white) part of the pituitary.
B11	Atlas		Type: Illustration		Sagittal Section of Brain	Find the *pineal body* (or gland) posterior to the third ventricle. It is a circumventricular organ (CVO), along with which other organs?
C1	Diss	M	ANT	70	Head	Identify cerebral cortex (gray matter), white matter, and part of a *ventricle* in the *frontal lobe* of the cerebrum.
C2 C9a, g	Atlas		Type: Illustration		Brain (Lat)	Identify the four main lobes of the cerebrum. What color is the *parietal lobe*? The *temporal lobe*? In which lobe is the <u>primary visual area</u> found? The <u>primary motor area</u>?
C3a	Diss	M	LAT	288	Head	Identify the <u>commissural fibers</u> known as the *corpus callosum*. These form the "roof" of which ventricles?
C3a C4	Atlas		Type: Illustration		Basal Ganglia (Sup)	This view shows a number of structures you have studied so far. Find the *corpus callosum, lateral ventricles, third ventricle, thalamus,* and *pineal body.* Now identify parts of the <u>basal ganglia</u>: *putamen* and *globus pallidus.*
D1 B9d D3a D4a,e	Atlas		Type: Illustration		Base of Brain (Inf)	Identify the 12 pairs of *cranial nerves (CNs)* from anterior to posterior. To which part of the brainstem are *CN V-VIII* attached? Which pair follow a pathway just inferior to the *frontal lobes*? Which ones originate from both the medulla and the brainstem? Which cranial nerves are the largest pair?
D3b	Atlas		Type: Illustration		Optic Nerve in Orbit (Lat)	Notice that the proximity of the *optic nerve* to the *pituitary gland* explains why a pituitary tumor can affect vision.
D4e	Atlas		Type: Illustration		Trigeminal Nerve-V1	Identify the three main divisions of the *trigeminal nerves (CN V).*

	Source	M	View	Layer	Location / Title	Type	Description
D3d D4f	Atlas				Facial Nerve (Lat) 1	Type: Illustration	Notice the extensive branching of the *facial nerve (CN VII)*, supplying muscles of facial expression and salivary glands.
D2i D4f	Atlas				Glossopharyngeal Nerve 1	Type: Illustration	Identify structures inervated by the *glossopharyngeal nerve (CN IX)*: the *carotid body, parotid gland,* and the posterior one third of the *tongue* (taste).
D4c	Atlas				Vagus Nerve (Lat)	Type: Illustration	Notice the extensive branching of the *vagus nerve (CN X)*, supplying head, neck, and thorax (as well as abdomen, not shown).
D4a	Diss	M	LAT	10	Thorax		Locate the *accessory nerve (CN XI)*, supplying the *trapezius* and *sternocleidomastoid*. See also at layer 12.

AIA Checkpoints. Chapter 15 Sensory, Motor, and Integrative Systems

	Source	M	View	Layer	Location / Title	Type	Description
B6b	Diss	M	ANT	202	Upper abdomen		Locate the *liver* and *gallbladder* inferior to the *diaphragm.* Now scroll to 168 and see the *phrenic nerve,* supplying the diaphragm. Move superiorward, following the pathway of this nerve. Scroll to depth 80 to see that the phrenic nerve originates in the neck (spinal nerves C3-C5), the same level at which nerves that carry pain impulses from neck and shoulder originate.
C5	Diss	M	ANT	330	Head		In this coronal (frontal) view, identify the *thalamus* where *spinothalamic tracts* terminate.
C6b D2	Atlas				Brain (Lat)	Type: Illustration	Identify the *central sulcus,* and the *postcentral gyrus* (site of the general sensory or somatosensory cortex). This region is the most anterior portion of which lobe (the pink one)? Now locate the (yellow) *precentral gyrus.* This is the primary (*motor? sensory?*) cortex.
D4	Atlas				T12 Vertebra (Sup)	Type: Illustration	Find the *gray matter* within the spinal cord. Locate the anterior horn (not labeled) of this gray matter which is the site of lower motor neurons (LMNs). Axons of LMNs then pass out through *the ventral root of the spinal nerve.*
D6	Diss	M	LAT	202	Head		Identify the *anterior cerebral artery* that lies medial to the cerebral hemisphere. Scroll to 190 to see *branches of middle cerebral arteries supplying gyri of the temporal lobe and lateral portions of frontal and parietal lobes.* Which lobe receives its blood supply from *branches of the posterior cerebral arteries?*

AIA Checkpoints. Chapter 16 Special Senses

Check-point	Link	Gender	View	Layer	Key structure/Image/Title/Region	Structures (in italics) to Identify and Concepts (underlined) to Understand
A2	Atlas		Type: Illustration		Olfactory Nerve in Nasal Cavity	Identify olfactory nerves (CN I), olfactory bulb, and olfactory tract.
C3-4 C11	Diss	M	ANT	38	Head	Identify the lacrimal gland, cornea covering pupil and iris, conjunctiva, and sclera.
C3-4	Atlas		Type: Cadaver photo		Eyeball & Extrinsic Eye Muscle	Identify the lacrimal gland, cornea, conjunctiva, and sclera.
C3b	Atlas		Type: Illustration		Extrinsic Eye Muscles (Lat)	Identify the extrinsic eye muscles.
E1 E3a	Atlas		Type: Illustration		External, Middle, & Inner Ear	Identify these ear structures: external acoustic meatus, tympanic membrane, and the tympanic cavity (middle ear).
E2	Atlas		Type: Cadaver		Ear Ossicles (Med)	Identify the malleus, incus, and stapes.
E3b	Atlas		Type: Cadaver		Dissection of Tympanic Cavity	Identify the middle ear cavity, malleus, incus, and stapedius muscle.
E6 E8 E11	Atlas		Type: Illustration		External, Middle, & Inner Ear	Identify the cochlear duct and the cochlear and vestibular divisions of CN VIII.

AIA Checkpoints. Chapter 17 Autonomic Nervous System

Check-point	Link	Gender	View	Layer	Key structure/Image/Title/Region	Structures (in italics) to Identify and Concepts (underlined) to Understand
A3-4	Atlas		Type: Illustration		Autonomic NS-Viscera 1 and Autonomic NS-Viscera 2	Contrast these two views (by minimizing and placing views side by side) of the two divisions of the autonomic nervous system (ANS). Which is shown in blue? (Sympathetic = S? Pasasympathetic = P?) Focus on the heart in each view. Notice that, like most viscera, it has dual innervation (both S and P innervation). Identify the vagus nerve (CN X). It is part of the (S? P?) division of the ANS.
B2 B4	Atlas		Type: Illustration		Autonomic NS-Viscera 1	Contrast the axons of preganglionic neurons (dotted lines: ....) with axons of postganglionic neurons (broken lines: ———). The vagus nerve consists of (pre?

ID	Source	View	Page	Region	Checkpoint
					post?)-ganglionic neurons, with terminal ganglia and postganglionic axons (broken lines) located in or near walls of viscera. The (S? P?) division has its ganglia (sites of postganglionic cell bodies) located mostly in a pair of chains or trunks parallel and close to the vertebrae (paravertebral).
B4 B8 B12c,e	Diss	M ANT	264	Neck/Thorax	Locate the sympathetic (trunk, chain, or paravertebral) ganglia (yellow) that are located segmentally (one pair per vertebra) in the (cervical? thoracic?) region. Now scroll inferiorly (to the coccyx) to see the full extent of the chain. Scroll to layer 239 to see prevertebral (or collateral) S ganglia that lie anterior to (not next to) vertebrae. Names of these individual collateral ganglia are based on names of adjacent major arteries. Find the celiac (or solar plexus), superior, and inferior mesenteric ganglia.
B4b	Atlas	Type: Cadaver photo		Dissection of Suprarenal Gland	Locate the superior mesenteric ganglion which lies just superior to the superior mesenteric artery.
B6d B9 B12j	Diss	M LAT	225	Neck/Thorax	Find the sympathetic (trunk or chain) ganglia in this view. Locate sphlanchnic nerves which connect chain ganglia to collateral ganglia.
B6c B12a	Diss	M LAT	225	Neck/Thorax	Find a white ramus and gray ramus connecting spinal cord with chain ganglia from level T1-L2 of the cord (also visible at ANT 264). Explain why no white rami are located in cervical regions or inferior to L2. (Hint: the S division is thoracolumbar outflow.)
B11a	Atlas	Type: Illustration		Ciliary Ganglion-Pathways 1	This view shows (P? S?) pathways that are all (pre? post?) ganglionic except in the eye itself. . (Hint: blue, broken lines). Identify the ciliary ganglion which is classified as terminal since it is in or close to the organ (eye) itself.
B11d B12g	Atlas	Type: Illustration		Autonomic NS-Viscera 2	Identify the vagus nerve (CN X). Follow its preganglionic (dotted) lines to see the many thoracic and abdominal viscera it innervates. Do any P nerves innervate blood vessels, sweat glands, and hair muscles of skin? (Hint: refer to Autonomic NS-Viscera 1 to see nerves to these viscera in skin.)

AIA Checkpoints. Chapter 18 The Endocrine System

Check-point	Link	Gender	View	Layer	Key structure/Image/Title/Region	Structures (in italics) to Identify and Concepts (underlined) to Understand
A3	Diss	F	ANT	157	Neck/Thorax	Contrast the locations of the *thymus gland* (layer 160) and the *thyroid gland* (layer 76). Which one is in the neck?
A3	Atlas		Type: Illustration		Glands of Head & Neck (Lat)	Locate the *thyroid and parathyroid glands*. Which are more posterior in location?
A3	Atlas		Type: Cadaver		Dissection of Suprarenal Gland	Locate the *left suprarenal (adrenal) gland* on top of the left kidney.
A3	Atlas		Type: Illustration		Pancreatic and bile ducts I	Locate the *pancreas*. Pancreatic exocrine (digestive) secretions flow into the *pancreatic duct* system (green). Find the vein (blue) that will transport pancreatic hormones such as insulin in blood that enters the liver and then passes to the heart and general circulation.
A3 C1c F8	Diss	M	LAT	222	Head	Identify the *pituitary gland*. Also locate the *pineal body*, posterior to the third ventricle.
F7	Diss	F	ANT	226	Abdominopelvis	Identify the *ovaries* where estrogens and progesterone are produced.
F10d	Diss	M	ANT	200	Abdomen	Identify the organ that produces the hormone *gastrin*. (Hint: this organ is removed in a *gastrectomy*.)

AIA Checkpoints. Chapter 19 The Cardiovascular System: The Blood

B1f, g	Diss	M	ANT	330	Abdominopelvic and Hip Regions	Identify locations of *myeloid tissue* in this view. Go to Atlas → Type → Illustration → Mediastinum (Sup) to see two more locations of hemopoiesis. Which types of blood cells are formed in these marrow areas?

AIA Checkpoints. Chapter 20 The Cardiovascular System: The Heart

						Instruction
A1a B1b	Diss	M	ANT	173	Thorax	Notice that most of the heart is made of muscle (or myocardium), which makes it a strong pump. Scroll to layer 174 to see the thickness of walls of the *left* and *right* ventricles.
A2	Diss	M	ANT	173	Thorax	Find the pointed *apex of the heart*. The apex is formed of *(right? left?)* ventricle (*Hint:* see layer 174). Scroll quickly to layer 155 to see the position of the apex deep to ribs 5 and 6. Return to 173 and notice that the *right ventricle* lies on the *diaphragm*, and that the two lungs are located laterally.
A2 B2	Atlas		Type: Illustration		Mediastinum (Sup)	Imagine looking down at your own *diaphragm* where you see the depression made by your heart (removed here). Notice the position of the heart between *sternum and vertebrae*, ideal for CPR. Notice locations of the *esophagus, aorta,* and *inferior vena cava.*
B1	Diss	M	ANT	166	Thorax	Identify the (white) fibrous layer of *pericardial sac.* Scroll to 170 to see the epicardium (or <u>visceral pericardium</u>).
B2-5	Diss	M	ANT	173	Thorax	Which three chambers of the heart are readily seen in this view? Which chamber has the thickest wall? Locate *trabeculae carnae* in that chamber. Locate the *chordae tendineae.* They are seen here attaching the *(pulmonary? tricuspid AV?)* valve into *papillary muscles.* The *pulmonary valve* prevents back flow of blood into which chamber?
B2-3	Diss	M	ANT	173	Thorax	Identify the *superior vena cava, pulmonary artery,* and *aorta.* Scroll one layer deeper (to remove the heart) and identify each of these vessels plus the *four pulmonary veins* and the *inferior vena cava.*
B8d	Atlas		Type: Illustration		Heart & Great Vessels (Ant)	The *ligamentum arteriosum* was the <u>ductus arteriosus</u> in fetal life. Which two vessels did this fetal artery (or duct) connect?
C2	Diss	M	LAT	248	Thorax	Locate the cross sections of blood vessels in the lungs. Which are blue (since they are low in O_2 and high in CO_2): cross sections of pulmonary *(arteries? veins?).* Open Atlas → Type → Radiograph → Left Lung Angiogram to see these vessels in an x-ray.
C4 C5	Atlas		Type: Illustration		Heart & Great Vessels (Ant) Heart & Great Vessels (Post)	Open and minimize both of these views of the heart; position them side by side. Identify the main *right coronary artery* and the *anterior interventricular branch* of the *left coronary artery.* Now (on POST view), find the *circumflex branch* and the *posterior interventricular branch.* Which of these latter two vessels is a branch of the *left coronary artery?* Identify the large (blue) vessel that functions like a vein to drain all coronary blood into the right atrium.

Check-point	Link	Gender	View	Layer	Key structure/Image/Title/Region	Structures (in italics) to Identify and Concepts (underlined) to Understand
C4 B3	3D				Heart	Rotate the heart to see the posterior interventricular coronary artery in the posterior coronary sulcus.
E4	Atlas				Heart Valve Auscultation Sites — Type: Illustration	Sounds of the AV (or cuspid) valves are best heard (auscultated) when the stethoscope is held over the (apex? base?) of the heart.

AIA Checkpoints. Chapter 21 The Cardiovascular System: Blood Vessels and Hemodynamics

Check-point	Link	Gender	View	Layer	Key structure/Image/Title/Region	Structures (in italics) to Identify and Concepts (underlined) to Understand
B8 B11e	Atlas				Glossopharyngeal Nerve 1 — Type: Illustration	Locate the carotid body (also the site of carotid sinuses) at the bifurcation of the common carotid artery into internal and external carotid arteries. Follow the sensory nerve pathways along CN IX to the medulla.
C2	Diss	M	LAT ARM	34	Wrist	Explain why the radial artery is most commonly used for checking pulse. Hint: Notice its proximity to the radius.
C4	Diss	M	MED ARM	0	Arm	Scroll from layer 0 to 6 and note that the brachial artery is relatively superficial in location, making it ideal for measuring blood pressure. If you scroll to layer 5, you will recognize the proximity of this artery to the brachial vein, as well as to ulnar and median nerves.
D2a,b E9	Diss	M	ANT	174	Thorax	Identify the aorta attached to the (right? left?) ventricle. The aorta lies to the (right? left?) of the superior vena cava which returns blood to the right atrium. Scroll to layer 172 and identify the ascending (ortion of the) aorta with its branches. One branch is the right coronary artery. Why is the left coronary artery not seen here? Its anterior interventricular branch is visible.
D2a,c E2d E4a	Diss	M	ANT	172	Thorax	Remove the large veins as you scroll to layer 177 and find a clear view of the arch of the aorta, as well as the descending (portion of the) aorta. Find the first artery to branch off of the aortic arch. Why is the brachiocephalic artery so named? Notice that there is only one of these arteries (It is not bilateral). Scroll back to 176 and note that the analogous veins are bilateral. At layer 68, notice the two veins that each empty blood into a brachiocephalic vein: a subclavian vein and an internal jugular vein.
D2d D3 D4	Diss	M	ANT	193	Thorax	Now see the two arteries formed from the brachiocephalic artery: right subclavian and right common carotid. Notice that vessels of these names on the left side exit directly from the aortic arch. Find the vertebral arteries branchng from each

subclavian artery. Go to LAT 228 in the neck and see these vessels carry blood to the *basilar artery* (at the base of the brain). This vessel plus *internal carotid arteries* (seen at LAT 49) supply blood to the cerebral arterial circle (of Willis).

Code	Source	M	Type/Layer	Structure	Instruction
D4	Atlas		Type: Illustration	Deep Arteries of Head (Lat)	Distinguish these arteries: *common carotid, external carotid, internal carotid, ophthalmic, vertebral, basilar, anterior cerebral, middle cerebral,* and *posterior cerebral.* Which three of these vessels are branches directly off of the Circle of Willis?
D2e,f	Diss	M	ANT 88	Thorax	Follow the pathway of blood from a *subclavian artery* to the little finger: *axillary → brachial → radial* and *ulnar).* (Notice at 103 the bifurcation of brachial into *radial* and *ulnar*). At layer 99, see the *superficial palmar arterial arch* and *proper palmar digital arteries.*
D6 / D7	Diss	M	ANT 260	Thorax	Identify parietal (*intercostal*) and visceral (*bronchial*) arteries here. Scroll to 240 and move inferior to the diaphragm. Find the left *inferior phrenic artery.* Is it categorized as parietal or visceral? Now locate major visceral branches of the *abdominal aorta: celiac trunk, superior* and *inferior mesenteric, renal* (and at layer 231), *suprarenal arteries.*
D7a	Atlas		Type: Illustration	Pancreatic & Bile Ducts 1	Distinguish the main branches of the *celiac* (artery) *trunk: hepatic artery, splenic artery,* and *left gastric artery.*
D7f	Atlas		Type: Illustration	Inferior Mesenteric Artery 1	Identify this large artery supplying blood to much of the *colon: transverse, descending,* and *sigmoid.*
D7d / D8	Atlas		Type: Radiograph	Abdominal Aortic Angiogram	Identify these arteries: *aorta;* left *renal* and left *common iliac arteries.*
D8 / D9	Diss	M	ANT 226	Hip/Thigh	Locate the bifurcation of the *aorta* at the level of the *iliac crest* (L4 vertebra). Follow the left *common iliac artery* to its own bifurcation. Notice that the *internal iliac artery* supplies pelvic organs such as *rectum* and bladder. Trace the path of blood into the left *exteral iliac artery,* and *left femoral artery.*
C2 / D11	Diss	M	ANT 226	Hip/Thigh	The femoral artery passes deep to the *vastus medialis* to form the *popliteal artery* (See at POST 156). If your femoral artery were to pass anterior (rather than posterior) to the knee joint, what would happen to the artery when you flex your leg (knee joint)? Find the *anterior* and posterior tibial arteries formed by the popliteal. Now go to ANT 296 to see *dorsalis pedis artery,* an ideal site for checking circulation (pulse) to the foot.
E1	Diss	M	ANT 175	Thorax	Identify the two major veins that return blood to the right atrium. Go to Atlas → Type → Illustration → Heart & Great Vessels (Post) to see the other structure (*coronary sinus*) returning blood to the heart.

Check-point	Link	Gender	View	Layer	Key structure/Image/Title/Region	Structures (in italics) **to Identify** and Concepts (underlined) **to Understand**
E2	Diss	F	LAT	262	Head	Identify cranial *venous sinuses*: *superior sagittal, inferior sagittal, straight,* and *transverse (cut)*. At 200 see the main vein that drains the brain, the *internal jugular*.
E2d	Atlas		Type: Cadaver photo		Dissection of Neck (Ant/Lat)	Recognize the positions of the *common carotid artery* and the *internal jugular vein* relative to the *trachea*. Which structure is most lateral?
E4a E5 E6	Diss	F	ANT	132	Left forearm	Two major factors account for the greater number of veins than arteries. Arteries may be paralleled by two veins, as exemplified by these *intraosseous vessels*, and by *ulnar* (layer 121), *radial* (99), and *brachial* (87) vessels. In addition, there are large veins such as the *basilic* (81) and *cephalic* (7) that have no arterial analogs, including extensive network of *superficial veins* (layers 5 and 1).
E8	Atlas		Type: Illustration		Posterior Thoracic Veins (Ant)	Explain why there is a gap between *superior* and *inferior vena cavae*. What protective function could the *azygos* and *hemiazygos veins* serve if a clot were blocking the superior vena cava?
E9	Atlas		Type: Illustration		Viscera of Female Pelvis (Sup)	Notice the relative positions of the *abdominal aorta* and *inferior vena cava*. How might you remember which of these vessels lies to the left and which to the right within the abdomen? Which artery appears to be at greatest risk during surgery involving the right *ureter* since the ureter passes over this vessel?
E9 Fq	Atlas		Type: Cadaver		Dissection of Suprarenal Gland	Locate the *inferior vena cava*, *left renal vein, left suprarenal (adrenal) vein,* and *left testicular vein*. Notice that no mesenteric or gastric veins carry blood directly into the vena cava. Why not? (*Hint*: scroll from ANT 213, to 215, 217, and 197. And see below.)
F1	Atlas		Type: Illustration		Mesenteric Veins Hepatic Portal Vein	Blood from GI organs drains into the *portal vein* and into the *liver*. Identify major vessels that feed into the portal vein.

AIA Checkpoints. Chapter 22 The Lymphatic System, Nonspecific Resistance to Disease, and Immunity

Check-point	Link	Gender	View	Layer	Key structure/Image/Title/Region	Structures (in italics) **to Identify** and Concepts (underlined) **to Understand**
A6	Diss	M	ANT	263	Thorax	Locate the *thoracic duct* (green) which is situated just anterior to thoracic vertebrae. Identify the two structures (seen at 260-259) that lie immediately anterior to this main lymphatic duct. Scroll to layer 82 to see the *left subclavian vein* which ultimately receives lymph from the thoracic duct. Why is it likely that lymph would empty into a vein (not an artery)?

A6e	Diss	F	LAT	22	Upper thorax	Scroll to layer 15 and 13 as you view *axillary lymph nodes* and see relationships of pectoral muscles. These nodes are biopsied in surgery for breast cancer.
A6a	Diss	F	LAT	225	Waist level	Locate the *cisternal chyli* (green) situated just anterior to lumbar vertebrae, adjacent to the sympathetic chain, and just posterior to coils of the small intestine (not visible here).
B2	Diss	M	ANT	153	Upper thorax	Scroll between 153 and 164, and notice that the *thymus* lies immediately posterior to the (*manubrium? thyroid cartilage?*). Which *vein* lies immediately behind the thymus?
B5	Atlas		Type: Illustration		Lymph Nodes of Thorax (Ant)	View the large number of *lymph nodes* along airways. Note that these can provide defense, but can also provide routes for metastasis.
B5	Atlas		Type: Illustration		Lymph Nodes of Male Pelvis	Explain why an autopsy of a man with advanced testicular cancer may reveal golf ball-sized periaortic lymph nodes.
B7	Diss	F	ANT	208	Abdomen	Locate the *spleen*. This organ lies (*anterior? posterior?*) to the stomach. (*Hint:* scroll to ANT 201.)
B8	Diss	F	LAT	223	Head/Neck	Identify the *pharyngeal tonsils* (or *adenoids*).
J3b	Atlas		Type: Illustration		Deep Lymph Flow of Neck (Lat)	Recognize why "swollen glands" (lymphadenopathy) may signal a "sore throat" or other upper respiratory infection.

AIA Checkpoints. Chapter 23 The Respiratory System

A3-5	Diss	F	LAT	208	Head/Neck	Identify the *maxillary (paranasal) sinus* and its *orifice* that drains mucus into the nose. Scroll to 210 and identify the bones and cartilage forming the *nasal septum*. At 225, view mucosa-covered *conchae* on the lateral wall of the nose. Follow the pathway of air from *vestibule of the nose* and mouth to the pharynx.
A6	Diss	F	LAT	225	Head/Neck	Gradually remove (constrictor) *pharyngeal muscles* as you scroll to 256. Identify the three parts of the pharynx: *nasopharynx, oropharynx,* and (at layer 258) *laryngopharynx*. Which part lies posterior to the mouth? *Orifices of auditory tubes* (leading to the middle ear) and also the adenoids (*pharyngeal tonsils*) are found in the (*naso? oro? laryngo?*)-pharynx.
B1	Diss	F	LAT	257	Head/Neck	Notice the *hyoid bone* at *mandible* level. Find the *thyroid cartilage* and *cricoid cartilage*. Go to 258 for a good view of the *epiglottis*.

Check-point	Link	Gender	View	Layer	Key structure/Image/Title/Region	Structures (in italics) to Identify and Concepts (underlined) to Understand
B1 B4	Atlas				Dissection of Neck (Ant/Lat) Type: Cadaver photo	Locate the *thyroid cartilage* and *cricoid cartilage* as well as the *trachea* with its cartilage rings. Note the nearby vessels: *common carotid artery* and *internal jugular vein*.
B4	Diss	F	LAT	257	Head/Neck	Which is more anterior? (*Trachea? Esophagus?*)
B6-8	Atlas				Bronchial Tree (Ant) Bronchial Tree (Post) Type: Illustration	Notice the many levels of branching of the tracheobronchial tree, and the gradual diminishment of cartilage in their walls.
B10	Diss	F	ANT	161	Thorax	Identify the *apex* and *lobes* of each lung, and the *parietal pleura*. Scroll to 248 to see parts of the *visceral pleura*. Here you can also see cross sections of *pulmonary arteries* and *veins* and *bronchi*. In a dissection, how could you distinguish these three types of structures? (Hint: Which has cartilaginous walls?) Remove the heart and *pericardial sac* at 173-175. Go to 245-248 and remove major vessels. Now identify bronchi.
B12 B10b	Atlas				Left mediastinum Type: Illustration	Identify the *left primary (main) bronchus*, and note its relationship to *pulmonary artery, pulmonary veins* and *aorta*.

AIA Checkpoints. Chapter 24 The Digestive System

Check-point	Link	Gender	View	Layer	Key structure/Image/Title/Region	Structures (in italics) to Identify and Concepts (underlined) to Understand
A2 A6 F4e	Diss	M	ANT	202	Thorax/Abdomen	Identify these parts of the GI tract: *esophagus, stomach, small intestine,* and *large intestine* (composed mainly of colon). Now locate these accessory organs: *liver* and *gallbladder*. At layer 205, see the *mucosal* lining the *rugae* of the stomach wall.
A8 A6	Diss	M	ANT	194	Thorax	Find the *parietal peritoneum* that lines the anterior abdominal wall (removed here). Go to 195 and identify the *visceral peritoneum* and its extensions: *falciform ligament, lesser* and *greater omentum*. See the *mesentery* at 208 and notice (at 209) that it serves as a pathway for mesenteric blood vessels to reach the small intestine. At 210-211, observe that the *pancreas, duodenum,* and *kidneys,* as well as the *aorta* and *inferior vena cava* (and branches) are retroperitoneal. (See Diss → MEDIAL layer 39 for a view of these retroperitoneal structures and also the *mesentery* anchored to the posterior abdominal wall.)
B1-3 B11b	Diss	M	LAT	260	Head	Find the *uvula* portion of the soft palate, and the *palatine tonsils*. Also locate the *submandibular* and *sublingual salivary glands*. At layers 16-17, see the *parotid gland*.

Code	View	Source		Type	#	Checkpoint
B2	Glands of Head & Neck (Lat)	Atlas		Type: Illustration		Does it lie superficial or deep to the *masseter muscle*? Identify the *facial nerve* (CN VII). Which salivary glands do its ANS fibers innervate?
B4a B11a	Head	Diss	M	LAT	233	Locate the *parotid gland*. Notice that its *duct* that empties into the mouth above the *upper second molar*. Scroll from 233 to 245 to see *extrinsic muscles of the tongue* (*hyoglossus* and *genioglossus*) and their nerve supply: *hypoglossal* (CN XII)/(layer 39).
B4b	Surface of Tongue (Dorsal)	Atlas		Type: Illustration		Identify three types of *papillae*: *circumvallate, filiform, and fungiform*. Which papillae contain most receptors for bitter taste?
B7	Head	Diss	M	ANT	46	Find the following types of teeth: *incisors, canines, premolar, and molars*. How many premolars are in an entire set of adult teeth? AT LAT 196, identify the *crown, roots, and pulp cavity of the upper first molar*. Which *paranasal sinus* appears most likely to become infected if this molar is abcessed?
B10	Thorax	Diss	M	LAT	254	The *esophagus* lies (*anterior? posterior?*) to the *trachea*. At 204, notice the (yellow) nerves which are branches of the cranial nerve that stimulates most GI organs. Identify that nerve here and at 173 where you see the esophagus attached to the *stomach*.
C1a, b	Abdomen	Diss	M	ANT	203	Identify these parts of the *stomach*: *lesser and greater curvatures, rugae, cardiac part, and pyloric part*. Scroll to 205 to see *rugae*. Note that the *pyloric sphinter* leads directly into the (*duodenum? jejunum?*).
C1b	Stomach 1	Atlas		Type: Illustration		Identify the *greater omentum*. To which organs is it attached? *Hint*: one is the *transverse colon*.
D2a, b	Pancreatic & Bile Ducts 1	Atlas		Type: Illustration		Identify the *pancreas* and its *ducts* emptying into the *hepatopancreatic ampulla* of the *duodenum*. Which part of the pancreas nudges up next to the *spleen*? (*Head? Tail?*)
D5 D6	Liver (Post)	Atlas		Type: Illustration		Identify the portal triad inferior to the liver: *hepatic artery, portal vein, and bile ducts*.
D5c	Upper abdomen	Diss	M	LAT	157	Notice that the *inferior vena cava* (IVC) "tunnels" through the liver, but does not receive blood from the liver (and other GI organs via the *portal vein*) until the IVC is exiting the liver at its superior surface. There the hepatic veins empty into the liver. An embolus in the leg is more likely to travel to and "get stuck" in the (*liver? pulmonary artery and its branches?*)

Check-point	Link	Gender	View	Layer	Key structure/Image/Title/Region	Structures (in italics) to Identify and Concepts (underlined) to Understand
D6	Diss	M	ANT	198	Upper abdomen	Locate the *gallbladder* and its *cystic duct* (green) here and at layer 202. Scroll to 219 to follow the pathway of *bile* from *cystic* and *common hepatic ducts* → *common bile duct* → duodenum.
E1 F1-2	Atlas			Type: Cadaver photo	Abdominal Viscera	Follow the pathway of food and wastes as you identify these digestive organs: *stomach, jejunum, ileum, cecum, ascending colon, transverse colon,* and *sigmoid colon.* Also locate the *liver, gallbladder,* and part of the *greater omentum.* Why is the duodenum not seen here? (Hint: see AIA activity above for A8 and A6.) See parts of intestines in Diss → ANT 207 also.
F3 F4g	Atlas Atlas			Type: Illustration	Large Intestine in Situ	Identify parts of the large intestine, including the *rectum.* Note attachment of the *appendix* to the *cecum.*
F4a, c	Diss	M	ANT	201	Abdomen	Identify these structures on the colon: *taeniae* and *epiploic appendages.*

AIA Checkpoints. Chapter 26 The Urinary System

Check-point	Link	Gender	View	Layer	Key structure/Image/Title/Region	Structures (in italics) to Identify and Concepts (underlined) to Understand
A2-3 E1-2	Diss	F	ANT	225	Abdomen	Identify kidneys (covered with *renal fascia*), ureters, and urinary bladder at 218. Scroll through *perirenal fat* (at 229-230) to see sectioned kidneys at 235. Differentiate the *renal cortex, medulla,* and *pelvis.*
A3 A5	Atlas			Type: Illustration	Renal Arteries	Locate the *renal capsule,* a *renal papilla* and *renal calyx* (*minor calix*), *renal columns* and *renal pyramids.* Follow the pathway of blood through kidneys. Identify the *renal artery* → *interlobar* → *arcuate* → *interlobular arteries.*
A5-7	Atlas			Type: Illustration	Diagram of Nephron	Identify these blood vessels in the pathway of blood through nephron: *afferent arteriole* → *glomerulus* → *efferent arteriole* → *peritubular capillaries.* Find the parts of a renal tubule (in beige), as well as a *collecting tubule.*
A10	Atlas			Type: Illustration	Diagram of Renal Glomerulus	Identify parts of the *juxtaglomerular apparatus* (*JGA*). Find *macula densa cells* (beige) where the ascending limb of loop of Henle enters the *distal convoluted tubule.* Now locate the juxtaglomerular cells (lavender) in the wall of the (*afferent? efferent?*) arteriole.
B2	Atlas			Type: Illustration	Diagram of Renal Glomerulus	Identify the *podocytes* adjacent to glomerular capillaries. Podocytes are parts of the (*parietal? visceral?*) layer of Bowman's capsule. Now locate the parietal layer

Code	Source				Structure	Instructions
E3	Diss	F	ANT	233	Abdomen	of this capsule where filtrate is collected before it is passed into the *proximal convoluted tubule.* Recognize the location of *kidneys* at waist level; they lie very *(anterior? posterior?)* Follow *ureters* inferiorly and notice their proximity to blood and lymph vessels as you scroll to 217 to see the *urinary bladder.*
E5b	Atlas	Type: Illustration			Female Pelvic Organs (Med) 1	Identify the *urethra* anterior to the *vagina.* Note that the short length of the urethra permits greater access of microbes to the bladder of females compared to males.
E5 G3	Atlas	Type: Illustration			Male Pelvic Organs (Med) 1	Identify the *neck of the bladder,* site of the voluntary sphincter that controls micturition. Locate the three parts of the *urethra* within the *prostate, urogenital diaphragm,* and *corpus spongiosum* of the *penis.* Explain why enlargement (hypertrophy) of the *prostate* is likely to interfere with urinary output.

AIA Checkpoints. Chapter 28 The Reproductive Systems

Code	Source				Structure	Instructions
B1 B8	Atlas	Type: Illustration			Male Pelvic Organs (Med) 1	Locate these layers of the scrotum: *dartos tunic* and *tunica vaginalis.* Identify the *testis, epididymis,* and *cremaster muscles.*
B12 B13	Atlas	Type: Illustration			Male Pelvic Organs (Lat)	Identify the *ductus* (vas) *deferens* passing around the *urinary bladder, seminal vesicles,* and *prostate.*
B16	Diss	M	MED	0	Penis	Identify these parts of the penis: *glans, prepuce, corpus cavernosum, corpus spongiosum,* and *spongy part of urethra.*
B17	Atlas	Type: Illustration			Male Superficial Perineal Space I	Identify the *tuberosities of the ischium* (ischial tuberosities). Now locate the *bulbospongiosus* and *isciocavernosus muscles* in the *bulb of the penis.*
C1-2 C6	Atlas	Type: Illustration			Female Pelvic Organs (Ant)	Identify female pelvic organs: the *uterus, uterine tubes* (including *isthmus* and *infundibulum*). Also locate the *ureters* and *urinary bladder.*
C1 C7 C8	Diss	F	MED	0	Pelvis (female)	Note relationships among female pelvic organs: *urinary bladder* and *urethra, uterus* and *vagina,* and *rectum.* Identify the *fundus, body,* and *cervix* of the uterus.
C8	Atlas	Type: Illustration			Viscera of Female Pelvis (Sup)	Locate the *urinary bladder, uterus, uterine tubes, ovaries,* and *rectum.* Identify the *round, broad,* and *ovarian ligaments, vesicouterine pouch,* and *rectouterinepouch.*

Check-point	Link	Gender	View	Layer	Key structure/Image/Title/Region	Structures (in italics) to Identify and Concepts (underlined) to Understand
C8	Atlas		Type: Cadaver photo		Contents of Female Pelvis (Sup)	Contrast organs and ligaments in the cadaver with the illustration you just viewed.
C9 C11	Atlas		Type: Illustration		Female Pelvic Organs (Med) 1	Identify these layers of the uterus: *myometrium* and *endometrium*. Which one makes up most of the uterus? Note the angle formed between the *vagina* and the *uterus*.
C14a	Atlas		Type: Illustration		Female Superficial Perineal Space I	Identify the *body of the clitoris, urethral and vaginal orifices, anus, vestibule,* and *hymen*.
C14b	Atlas		Type: Illustration		Triangles of Perineum (Female)	Locate the *tuberosity of the ischium, urogenital and anal triangles*. Identify the *labium majus (major)* and *labium minus (minor)* and other external genitalia visible here.
C15	Diss	F	Ant	0	Thoracic region	Identify these parts of the mammary glands: *nipple,* and *areola*. Scroll to layer 5 to see *lactiferous ducts*. Go to Diss → LAT to see these ducts, as well as *adipose issue,* and *lobes of mammary glands*.